D0205820

Graduate Texts in Mathematics 93

Graduate Texts in Mathematics

A Selection

60 Arnold. Mathematical Methods in Classical Mechanics.
61 Whitehead. Elements of Homotopy Theory.
62 Kargapolov/Merzljakov. Fundamentals of the Theory of Groups.
63 Bollabas. Graph Theory.
64 Edwards. Fourier Series. Vol. I. 2nd ed.
65 Wells. Differential Analysis on Complex Manifolds. 2nd ed.
66 Waterhouse. Introduction to Affine Group Schemes.
67 Serre. Local Fields.
68 Weidmann. Linear Operators in Hilbert Spaces.
69 Lang. Cyclotomic Fields II.
70 Massey. Singular Homology Theory.
71 Farkas/Kra. Riemann Surfaces.
72 Stillwell. Classical Topology and Combinatorial Group Theory.
73 Hungerford. Algebra.
74 Davenport. Multiplicative Number Theory. 2nd ed.
75 Hochschild. Basic Theory of Algebraic Groups and Lie Algebras.
76 Iitake. Algebraic Geometry.
77 Hecke. Lectures on the Theory of Algebraic Numbers.
78 Burris/Sankappanavar. A Course in Universal Algebra.
79 Walters. An Introduction to Ergodic Theory.
80 Robinson. A Course in the Theory of Groups.
81 Forster. Lectures on Riemann Surfaces.
82 Bott/Tu. Differential Forms in Algebraic Topology.
83 Washington. Introduction to Cyclotomic Fields.
84 Ireland/Rosen. A Classical Introduction Modern Number Theory.
85 Edwards. Fourier Series: Vol. II. 2nd ed.
86 van Lint. Introduction to Coding Theory.
87 Brown. Cohomology of Groups.
88 Pierce. Associative Algebras.
89 Lang. Introduction to Algebraic and Abelian Functions. 2nd ed.
90 Brondsted. An Introduction to Convey Polytopes.
91 Beardon. On the Geometry of Decrete Groups.
92 Diestel. Sequences and Series in Banach Spaces.
93 Dubrovin/Fomenko/Novikov. Modern Geometry—Methods and Applications Vol. I.
94 Warner. Foundations of Differentiable Manifolds and Lie Groups.
95 Shiryayev. Probability, Statistics, and Random Processes.
96 Zeidler. Nonlinear Functional Analysis and Its Applications I: Fixed Points Theorem.

B. A. Dubrovin
A. T. Fomenko
S. P. Novikov

Modern Geometry—Methods and Applications

Part I. The Geometry of Surfaces, Transformation Groups, and Fields

Translated by Robert G. Burns

With 45 Illustrations

Springer-Verlag
New York Berlin Heidelberg Tokyo

B. A. Dubrovin
c/o VAAP—Copyright Agency of the U.S.S.R.
B. Bronnaya 6a
Moscow 103104
U.S.S.R.

A. T. Fomenko
3 Ya Karacharavskaya
d.b. Korp. I. Ku. 35
109202 Moscow
U.S.S.R.

S. P. Novikov
L. D. Landau Institute
for Theoretical Physics
Academy of Sciences of the U.S.S.R.
Vorobevskoe Shosse, 2
117334 Moscow
U.S.S.R.

R. G. Burns (*Translator*)
Department of Mathematics
Faculty of Arts
York University
4700 Keele Street
Downsview, ON, M3J 1P3
Canada

AMS Subject Classifications: 49-01, 51-01, 53-01

Library of Congress Cataloging in Publication Data
Dubrovin, B. A.
Modern geometry—methods and applications.
(Graduate texts in mathematics; 93–)
"Original Russian edition published by Nauka in 1979."
Contents: pt. 1. The geometry of surfaces, transformation groups, and fields. —
Bibliography: p.
Vol. 1 includes index.
1. Geometry. I. Fomenko, A. T. II. Novikov, Sergeĭ
Petrovich. III. Title. IV. Series: Graduate texts in
mathematics; 93, etc.
QA445.D82 1984 516 83-16851

This book is part of the Springer Series in Soviet Mathematics.

Original Russian edition: *Sovremennaja Geometria: Metody i Priloženia.* Moskva: Nauka, 1979.

Typeset by Composition House Ltd., Salisbury, England.
Printed and bound by R. R. Donnelley & Sons, Harrisonburg, Virginia.
Printed in the United States of America.

9 8 7 6 5 4 3 2 1

ISBN 0-387-90872-2 Springer-Verlag New York Berlin Heidelberg Tokyo
ISBN 3-540-90872-2 Springer-Verlag Berlin Heidelberg New York Tokyo

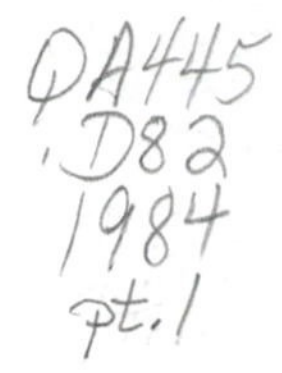

Preface*

Up until recently, Riemannian geometry and basic topology were not included, even by departments or faculties of mathematics, as compulsory subjects in a university-level mathematical education. The standard courses in the classical differential geometry of curves and surfaces which were given instead (and still are given in some places) gradually came to be viewed as anachronisms. However, there has been hitherto no unanimous agreement as to exactly how such courses should be brought up to date, that is to say, which parts of modern geometry should be regarded as absolutely essential to a modern mathematical education, and what might be the appropriate level of abstractness of their exposition.

The task of designing a modernized course in geometry was begun in 1971 in the mechanics division of the Faculty of Mechanics and Mathematics of Moscow State University. The subject-matter and level of abstractness of its exposition were dictated by the view that, in addition to the geometry of curves and surfaces, the following topics are certainly useful in the various areas of application of mathematics (especially in elasticity and relativity, to name but two), and are therefore essential: the theory of tensors (including covariant differentiation of them); Riemannian curvature; geodesics and the calculus of variations (including the conservation laws and Hamiltonian formalism); the particular case of skew-symmetric tensors (i.e. "forms") together with the operations on them; and the various formulae akin to Stokes' (including the all-embracing and invariant "general Stokes formula" in n dimensions). Many leading theoretical physicists shared the mathematicians' view that it would also be useful to include some facts about

* Parts II and III are scheduled to appear in the Graduate Texts in Mathematics series at a later date.

manifolds, transformation groups, and Lie algebras, as well as the basic concepts of visual topology. It was also agreed that the course should be given in as simple and concrete a language as possible, and that wherever practicable the terminology should be that used by physicists. Thus it was along these lines that the archetypal course was taught. It was given more permanent form as duplicated lecture notes published under the auspices of Moscow State University as:

Differential Geometry, Parts I and II, by S. P. Novikov, Division of Mechanics, Moscow State University, 1972.

Subsequently various parts of the course were altered, and new topics added. This supplementary material was published (also in duplicated form) as

Differential Geometry, Part III, by S. P. Novikov and A. T. Fomenko, Division of Mechanics, Moscow State University, 1974.

The present book is the outcome of a reworking, re-ordering, and extensive elaboration of the above-mentioned lecture notes. It is the authors' view that it will serve as a basic text from which the essentials for a course in modern geometry may be easily extracted.

To S. P. Novikov are due the original conception and the overall plan of the book. The work of organizing the material contained in the duplicated lecture notes in accordance with this plan was carried out by B. A. Dubrovin. This accounts for more than half of Part I; the remainder of the book is essentially new. The efforts of the editor, D. B. Fuks, in bringing the book to completion, were invaluable.

The content of this book significantly exceeds the material that might be considered as essential to the mathematical education of second- and third-year university students. This was intentional: it was part of our plan that even in Part I there should be included several sections serving to acquaint (through further independent study) both undergraduate and graduate students with the more complex but essentially geometric concepts and methods of the theory of transformation groups and their Lie algebras, field theory, and the calculus of variations, and with, in particular, the basic ingredients of the mathematical formalism of physics. At the same time we strove to minimize the degree of abstraction of the exposition and terminology, often sacrificing thereby some of the so-called "generality" of statements and proofs: frequently an important result may be obtained in the context of crucial examples containing the whole essence of the matter, using only elementary classical analysis and geometry and without invoking any modern "hyperinvariant" concepts and notations, while the result's most general formulation and especially the concomitant proof will necessitate a dramatic increase in the complexity and abstractness of the exposition. Thus in such cases we have first expounded the result in question in the setting of the relevant significant examples, in the simplest possible language appropriate, and have postponed the proof of the general form of the result, or omitted it altogether. For our treatment of those geometrical questions more closely bound up with modern physics, we analysed the physics literature: books on quantum field theory (see e.g. [35], [37]) devote considerable portions of their beginning sections to describing, in physicists' terms, useful facts about the most important concepts associated with the higher-dimensional calculus of variations and the simplest representations of Lie groups; the books [41], [43] are devoted to field theory in its geometric aspects; thus, for instance, the book [41] contains an extensive treatment of Riemannian geometry from the physical point of view, including much useful concrete material. It is interesting to look at books on the mechanics of continuous media and the theory of rigid bodies ([42], [44], [45]) for further examples of applications of tensors, group theory, etc.

In writing this book it was not our aim to produce a "self-contained" text: in a standard mathematical education, geometry is just one component of the curriculum; the questions of concern in analysis, differential equations, algebra, elementary general topology and measure theory, are examined in other courses. We have refrained from detailed discussion of questions drawn from other disciplines, restricting ourselves to their formulation only, since they receive sufficient attention in the standard programme.

In the treatment of its subject-matter, namely the geometry and topology of manifolds, Part II goes much further beyond the material appropriate to the aforementioned basic geometry course, than does Part I. Many books have been written on the topology and geometry of manifolds: however, most of them are concerned with narrowly defined portions of that subject, are written in a language (as a rule very abstract) specially contrived for the particular circumscribed area of interest, and include all rigorous foundational detail often resulting only in unnecessary complexity. In Part II also we have been faithful, as far as possible, to our guiding principle of minimal abstractness of exposition, giving preference as before to the significant examples over the general theorems, and we have also kept the interdependence of the chapters to a minimum, so that they can each be read in isolation insofar as the nature of the subject-matter allows. One must however bear in mind the fact that although several topological concepts (for instance, knots and links, the fundamental group, homotopy groups, fibre spaces) can be defined easily enough, on the other hand any attempt to make nontrivial use of them in even the simplest examples inevitably requires the development of certain tools having no forbears in classical mathematics. Consequently the reader not hitherto acquainted with elementary topology will find (especially if he is past his first youth) that the level of difficulty of Part II is essentially higher than that of Part I; and for this there is no possible remedy. Starting in the 1950s, the development of this apparatus and its incorporation into various branches of mathematics has proceeded with great rapidity. In recent years there has appeared a rash, as it were, of nontrivial applications of topological methods (sometimes in combination with complex algebraic geometry) to various problems of modern theoretical physics: to the quantum theory of specific fields of a geometrical nature (for example, Yang–Mills and chiral fields), the theory of fluid crystals and superfluidity, the general theory of relativity, to certain physically important nonlinear wave equations (for instance, the Korteweg–de Vries and sine–Gordon equations); and there have been attempts to apply the theory of knots and links in the statistical mechanics of certain substances possessing "long molecules". Unfortunately we were unable to include these applications in the framework of the present book, since in each case an adequate treatment would have required a lengthy preliminary excursion into physics, and so would have taken us too far afield. However, in our choice of material we have taken into account which topological concepts and methods are exploited in these applications, being aware of the need for a topology text which might be read (given strong enough motivation) by a young theoretical physicist of the modern school, perhaps with a particular object in view.

The development of topological and geometric ideas over the last 20 years has brought in its train an essential increase in the complexity of the algebraic apparatus used in combination with higher-dimensional geometrical intuition, as also in the utilization, at a profound level, of functional analysis, the theory of partial differential equations, and complex analysis; not all of this has gone into the present book, which pretends to being elementary (and in fact most of it is not yet contained in any single textbook, and has therefore to be gleaned from monographs and the professional journals).

Three-dimensional geometry in the large, in particular the theory of convex figures and its applications, is an intuitive and generally useful branch of the classical geometry of surfaces in 3-space; much interest attaches in particular to the global problems of the theory of surfaces of negative curvature. Not being specialists in this field we were unable to extract its essence in sufficiently simple and illustrative form for inclusion in an elementary text. The reader may acquaint himself with this branch of geometry from the books [1], [4] and [16].

Of all the books on the topology and geometry of manifolds, the classical works *A Textbook of Topology* and *The Calculus of Variations in the Large*, of Seifert and Threlfall, and also the excellent more modern books [10], [11] and [12], turned out to be closest to our conception in approach and choice of topics. In the process of creating the present text we actively mulled over and exploited the material covered in these books, and their methodology. In fact our overall aim in writing Part II was to produce something like a modern analogue of Seifert and Threlfall's *Textbook of Topology*, which would however be much wider-ranging, remodelled as far as possible using modern techniques of the theory of smooth manifolds (though with simplicity of language preserved), and enriched with new material as dictated by the contemporary view of the significance of topological methods, and of the kind of reader who, encountering topology for the first time, desires to learn a reasonable amount in the shortest possible time. It seemed to us sensible to try to benefit (more particularly in Part I, and as far as this is possible in a book on mathematics) from the accumulated methodological experience of the physicists, that is, to strive to make pieces of nontrivial mathematics more comprehensible through the use of the most elementary and generally familiar means available for their exposition (preserving however, the format characteristic of the mathematical literature, wherein the statements of the main conclusions are separated out from the body of the text by designating them "theorems", "lemmas", etc.). We hold the opinion that, in general, understanding should precede formalization and rigorization. There are many facts the details of whose proofs have (aside from their validity) absolutely no role to play in their utilization in applications. On occasion, where it seemed justified (more often in the more difficult sections of Part II) we have omitted the proofs of needed facts. In any case, once thoroughly familiar with their applications, the reader may (if he so wishes), with the help of other sources, easily sort out the proofs of such facts for himself. (For this purpose we recommend the book [21].) We have, moreover, attempted to break down many of these omitted proofs into soluble pieces which we have placed among the exercises at the end of the relevant sections.

In the final two chapters of Part II we have brought together several items from the recent literature on dynamical systems and foliations, the general theory of relativity, and the theory of Yang–Mills and chiral fields. The ideas expounded there are due to various contemporary researchers; however in a book of a purely textbook character it may be accounted permissible not to give a long list of references. The reader who graduates to a deeper study of these questions using the research journals will find the relevant references there.

Homology theory forms the central theme of Part III.

In conclusion we should like to express our deep gratitude to our colleagues in the Faculty of Mechanics and Mathematics of M.S.U., whose valuable support made possible the design and operation of the new geometry courses; among the leading mathematicians in the faculty this applies most of all to the creator of the Soviet school of topology, P. S. Aleksandrov, and to the eminent geometers P. K. Raševskiĭ and N. V. Efimov.

We thank the editor D. B. Fuks for his great efforts in giving the manuscript its final shape, and A. D. Aleksandrov, A. V. Pogorelov, Ju. F. Borisov, V. A. Toponogov and V. I. Kuz'minov who in the course of reviewing the book contributed many useful comments. We also thank Ja. B. Zel'dovič for several observations leading to improvements in the exposition at several points, in connexion with the preparation of the English and French editions of this book.

We give our special thanks also to the scholars who facilitated the task of incorporating the less standard material into the book. For instance the proof of Liouville's theorem on conformal transformations, which is not to be found in the standard literature, was communicated to us by V. A. Zorič. The editor D. B. Fuks simplified the proofs of several theorems. We are grateful also to O. T. Bogojavlenskiĭ, M. I. Monastyrskiĭ, S. G. Gindikin, D. V. Alekseevskiĭ, I. V. Gribkov, P. G. Grinevič, and E. B. Vinberg.

Translator's acknowledgments. Thanks are due to Abe Shenitzer for much kind advice and encouragement, and to Eadie Henry for her excellent typing and great patience.

Contents

CHAPTER 2

The Theory of Surfaces

CHAPTER 3

Tensors: The Algebraic Theory

CHAPTER 6
The Calculus of Variations in Several Dimensions. Fields and Their Geometric Invariants

CHAPTER 1

Geometry in Regions of a Space. Basic Concepts

§1. Co-ordinate Systems

We begin by discussing some of the concepts fundamental to geometry. In school geometry—the so-called "elementary Euclidean" geometry of the ancient Greeks—the main objects of study are various metrical properties of the simplest geometrical figures. The basic goal of that geometry is to find relationships between lengths and angles in triangles and other polygons. Knowledge of such relationships then provides a basis for the calculation of the surface areas and volumes of certain solids. The central concepts underlying school geometry are the following: the length of a straight line segment (or of a circular arc); and the angle between two intersecting straight lines (or circular arcs).

The chief aim of analytic (or co-ordinate) geometry is to describe geometrical figures by means of algebraic formulae referred to a Cartesian system of co-ordinates of the plane or 3-dimensional space. The objects studied are the same as in elementary Euclidean geometry: the sole difference lies in the methodology. Again, differential geometry is the same old subject, except that here the subtler techniques of the differential calculus and linear algebra are brought into full play. Being applicable to general "smooth" geometrical objects, these techniques provide access to a wider class of such objects.

1.1. Cartesian Co-ordinates in a Space

Our most basic conception of geometry is set out in the following two paragraphs:

(i) We do our geometry in a certain space consisting of points $P, Q, \ldots$.
(ii) As in analytic geometry, we introduce a system of co-ordinates for the space. This is done by simply associating with each point of the space an ordered n-tuple $(x^1, \ldots, x^n)$ of real numbers—the *co-ordinates* of the point—in such a way as to satisfy the following two conditions:
   (a) Distinct points are assigned distinct n-tuples. In other words, points P and Q with co-ordinates $(x^1, \ldots, x^n)$ and $(y^1, \ldots, y^n)$ are one and the same point if and only if $x^i = y^i$, $i = 1, \ldots, n$.
   (b) Every possible n-tuple $(x^1, \ldots, x^n)$ is used, i.e. is assigned to some point of the space.

1.1.1. Definition. A space furnished with a system of Cartesian co-ordinates satisfying conditions (a) and (b) is called an *n-dimensional Cartesian space*,† and is denoted by $\mathbb{R}^n$. The integer n is called the *dimension* of the space.

We shall often refer somewhat loosely to the n-tuples $(x^1, \ldots, x^n)$ themselves as the points of the space. The simplest example of a Cartesian space is the real number line. Here each point has just one co-ordinate x^1, so that $n = 1$, i.e. it is a 1-dimensional Cartesian space. Other examples, familiar from analytic geometry, are provided by Cartesian co-ordinatizations of the plane (which is then a 2-dimensional Cartesian space), and of ordinary (i.e. 3-dimensional) space (Figure 1). These Cartesian spaces are completely adequate for solving the problems of school geometry.

We shall now consider a less familiar but extremely important example of a Cartesian space. Modern physics teaches us that time and space are not separate, non-overlapping concepts, but are merged in a 4-dimensional "space-time continuum." The following mathematical formulation of the natural ordering of phenomena turns out to be extraordinarily convenient.

The points of our space–time continuum are taken to be events. We assign to each event an ordered quadruple (t, x^1, x^2, x^3) of real numbers, where t is the "instant in time" when the event occurs, and x^1, x^2, x^3 are the co-ordinates of the "spatial location" of the event. With this co-ordinatization, the space-time continuum becomes a 4-dimensional Cartesian space, and we then set aside our interpretation of the co-ordinates (t, x^1, x^2, x^3) as times and locations of the events. The 3-dimensional space of classical geometry is then simply the hyperspace defined by an equation $t = \text{const}$. The course, or path, in space-time, of an object which can be regarded abstractly at every instant of time as a point (a so-called "point-particle"),

† This terminology is perhaps unconventional. We hope that the reader will not find it too disconcerting.

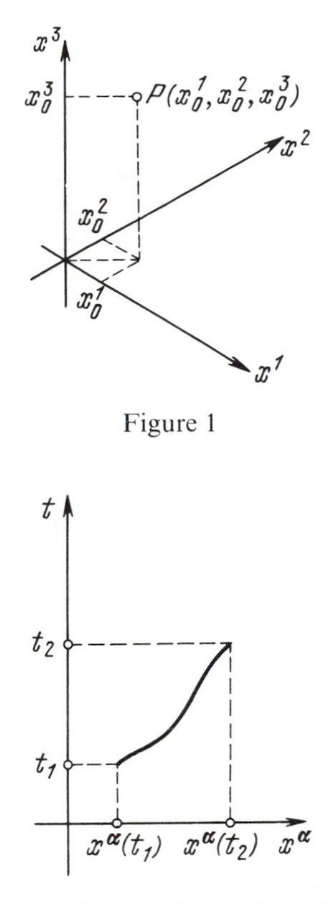

Figure 1

Figure 2. The world-line of an object.

is then identified with a curve segment (or arc) $x^\alpha(t)$, $\alpha = 1, 2, 3$, $t_1 \leq t \leq t_2$, in 4-dimensional space. We call this curve the world-line of the point-particle (Figure 2). We shall be considering also 3-dimensional and even 2-dimensional space–time continua, co-ordinatized by triples (t, x^1, x^2) and pairs (t, x^1) respectively, since for these spaces it is easier to draw intelligible pictures.

1.2. Co-ordinate Changes

Suppose that in an n-dimensional Cartesian space we are given a real-valued function $f(P)$, i.e. a function assigning a real number to each point P of the space. Since each point of the space comes with its n co-ordinates we can think of f as a function of n real variables: if $P = (x^1, \ldots, x^n)$, then $f(P) = f(x^1, \ldots, x^n)$. We shall be concerned only with continuous (usually even continuously differentiable) functions $f(x^1, \ldots, x^n)$. At times the functions we consider will not be defined for every point of the space $\mathbb{R}^n$, but only on portions, or, more precisely, "regions" of it.

1.2.1. Definition. A *region*, or *region without boundary* ("open set" in other terminology), is a set D of points in $\mathbb{R}^n$ such that together with each of its points P_0, D also contains all points sufficiently close to P_0; more precisely, for each point $P_0 = (x_0^1, \ldots, x_n^1)$ in the region D, there is a number $\varepsilon > 0$ such that all points $P = (x^1, \ldots, x^n)$ satisfying the inequalities

$$|x^i - x_0^i| < \varepsilon, \qquad i = 1, \ldots, n,$$

also lie in D.

1.2.2. Definition. A *region with boundary* is obtained from a region D (without boundary) by simply adjoining all boundary points (i.e. points not in D, yet having points of D arbitrarily close to them). The *boundary* of the region is just the set of boundary points.

The simplest example of a region without boundary is the whole space $\mathbb{R}^n$. Another simple example is afforded by the set of points (x_1, x_2) of the plane for which $x_1^2 + x_2^2 < \rho^2$ (the open disc of radius $\rho > 0$). The corresponding region with boundary consists of those points (x_1, x_2) satisfying $x_1^2 + x_2^2 \leq \rho^2$. The manner of definition of this region is typical in the specific sense indicated by the following theorem.

1.2.3. Theorem. *Let $f_1(P), \ldots, f_m(P)$ $(P = (x^1, \ldots, x^n))$ be continuous functions defined on the space $\mathbb{R}^n$. Then the set D of all points P satisfying the inequalities*

$$f_1(P) < 0, f_2(P) < 0, \ldots, f_m(P) < 0$$

is a region without boundary.

PROOF. Suppose $P_0 = (x_0^1, \ldots, x_0^n)$ lies in D, i.e. $f_1(P_0) < 0, \ldots, f_m(P_0) < 0$. By the property of "preservation of sign" of continuous functions we have that for each j there is a number $\varepsilon_j > 0$ such that $f_j(P) < 0$ for all $P = (x^1, \ldots, x^n)$ satisfying $|x^i - x_0^i| < \varepsilon_j$, $i = 1, \ldots, n$. Putting $\varepsilon = \min(\varepsilon_1, \ldots, \varepsilon_m)$, we then see that D certainly contains all points $(x^1, \ldots, x^n)$ satisfying $|x^i - x_0^i| < \varepsilon$. Hence D is a region without boundary. □

Remark. If a segment of a continuous curve is such that all of its interior points are in the region D: $f_j(P) < 0$, $j = 1, \ldots, m$, then in view of the concontinuity of the f_j, its end-points must satisfy $f_j(P) \leq 0$; i.e. travel along such a segment will only get us to points P satisfying $f_j(P) \leq 0$, $j = 1, \ldots, m$. If the functions $f_1, \ldots, f_m$ satisfy certain simple analytic conditions (which we shall specify in Part II), then it follows that every point P satisfying $f_j(P) \leq 0$, $j = 1, \ldots, m$, can be reached in this way. Thus under these conditions the solutions of the inequalities $f_j(P) \leq 0$, $j = 1, \ldots, m$, form a region with boundary.

We mention here also the frequently encountered and very important idea of a bounded region of a space, i.e. a region all of whose points are less than a certain fixed distance from the origin of co-ordinates.

Cartesian co-ordinates $(x^1, \ldots, x^n)$ assigned to $\mathbb{R}^n$ obviously furnish, in particular, a co-ordinatization of each region D, except that if D is not the whole space $\mathbb{R}^n$, then of course the n-tuples corresponding to points of D will not take on all possible values; it still makes sense of course to talk about the continuity and differentiability of functions defined only on the region D.

Suppose another system of co-ordinates $(z^1, \ldots, z^n)$ is given for the same region. We can write

$$\begin{aligned} x^i &= x^i(z^1, \ldots, z^n), \qquad i = 1, 2, \ldots, n, \\ z^j &= z^j(x^1, \ldots, x^n), \qquad j = 1, 2, \ldots, n. \end{aligned} \tag{1}$$

These equations mean simply that each point of the region has associated with it both its "old" co-ordinates $(x^1, \ldots, x^n)$, and its "new" co-ordinates $(z^1, \ldots, z^n)$, so that the new co-ordinates can be thought of as functions of the old ones, and conversely.

We first of all investigate *linear changes of co-ordinates* of the space:

$$x^i = \sum_{j=1}^{n} a^i_j z^j, \qquad i = 1, \ldots, n \tag{2}$$

(or more briefly $x^i = a^i_j z^j$, where here (as in the sequel) it is understood that repeated indices (here j only) are summed over). From linear algebra we know that the z_i are expressible in terms of the x_i if and only if the matrix $A = (a^i_j)$ has an inverse $B = A^{-1} = (b^i_j)$. This inverse matrix is defined by the equations $b^i_j a^j_k = \delta^i_k$, where again summation over the repeated index j is understood, and the *Kronecker symbol* δ^i_k is defined by

$$\delta^i_k = \begin{cases} 1 & \text{for } i = k, \\ 0 & \text{for } i \neq k. \end{cases}$$

In (2) the Cartesian co-ordinates $x^1, \ldots, x^n$ of the point P are expressed in terms of the new co-ordinates $z^1, \ldots, z^n$ by means of the matrix $A = (a^i_j)$; the equations (2) can be rewritten more compactly as

$$X = AZ, \qquad X = (x^1, \ldots, x^n), \qquad Z = (z^1, \ldots, z^n)$$

(where in the first equation, X and Z are written as column vectors). The equations (2) tell us that if $x^1, \ldots, x^n$ are the co-ordinates assigned to P in the original co-ordinate system, then in the new co-ordinate system, P is assigned co-ordinates $z^1, \ldots, z^n$ satisfying those equations. We have seen that A must be invertible (or in other words be nonsingular, or, in yet other words, have nonzero determinant), so that the new co-ordinates can be expressed in terms of the old:

$$Z = BX, \qquad z^j = b^j_k x^k \tag{3}$$

(where summation over k is implicit).

We return to the general situation where $x^i = x^i(z^1, \ldots, z^n)$, $i = 1, \ldots, n$, except that now we shall assume that the functions $x^i(z^1, \ldots, z^n)$ are continuously differentiable (i.e. have continuous first-order partial derivatives, or, more briefly, are "smooth").

We assume that every point of the region under scrutiny gets assigned new co-ordinates, or, in other words, that to each n-tuple $(x^1_0, \ldots, x^n_0)$ of the region there corresponds at least one n-tuple $(z^1_0, \ldots, z^n_0)$ such that

$$x^i_0 = x^i(z^1_0, \ldots, z^n_0), \; i = 1, \ldots, n.$$

1.2.4. Definition. A point $P = (x^1_0, \ldots, x^n_0)$ is called an *ordinary* or *nonsingular point of the co-ordinate system* $(z^1, \ldots, z^n)$ if the matrix

$$A = (a^i_j) = \left(\frac{\partial x^i}{\partial z^j}\right)_{z^1 = z^1_0, \ldots, z^n = z^n_0} \tag{4}$$

(where $z^1_0, \ldots, z^n_0$ satisfy $x^i(z^1_0, \ldots, z^n_0) = x^i_0$, $i = 1, \ldots, n$) has nonzero determinant (i.e. is nonsingular).

The matrix A is called the *Jacobian matrix* of the given transformation of co-ordinates, and is denoted by $\hat{J} = (\partial x/\partial z)$. The determinant of the Jacobian matrix is called simply the *Jacobian*, and is denoted by J:

$$J = \det\left(\frac{\partial x}{\partial z}\right) = \det \hat{J}.$$

The following theorem, known as the "Inverse Function Theorem" (a particular case of the general "Implicit Function Theorem"), should be familiar from courses in mathematical analysis.

1.2.5. Theorem. *Suppose we have a change of co-ordinate systems where, as above, the old co-ordinates are expressed in terms of the new by* $x^i = x^i(z)$, $i = 1, \ldots, n$, *and let* $x^i_0 = x^i(z^1_0, \ldots, z^n_0)$, $i = 1, \ldots, n$, *be the co-ordinates of some point with the property that* $J = \det(\partial x/\partial z) \neq 0$ *at* $z^1 = z^1_0, \ldots, z^n = z^n_0$. *Then for some sufficiently small neighbourhood of* (*i.e. region about*) *the point* $(x^1_0, \ldots, x^n_0)$ *we shall have that: the co-ordinates* $z^1, \ldots, z^n$ *of points of that neighbourhood are expressible in terms of* $x^1, \ldots, x^n$, *say* $z^i = z^i(x)$, *where, in particular,* $z^i_0 = z^i(x^1_0, \ldots, x^n_0)$, $i = 1, \ldots, n$; *and at each point of the neighbourhood the matrix* $(b^i_j) = (\partial z^i/\partial x^j)$ (*the Jacobian matrix of the inverse transformation*) *is the inverse of the matrix* $(a^k_l) = (\partial x^k/\partial z^l)$; *i.e.*

$$\frac{\partial z^i}{\partial x^j}\frac{\partial x^j}{\partial z^k} = \delta^i_k \tag{5}$$

(*with, as usual, summation over the repeated index understood*).

For the case $n = 1$ this becomes the following simple statement: If $x = x(z)$, and if $x_0 = x(z_0)$ is such that $dx/dz \neq 0$ at $z = z_0$, then on some sufficiently small interval about x_0, z can be expressed in terms of x, say $z = z(x)$, with in particular $z_0 = z(x_0)$, and, throughout the interval, $(dz/dx)(dx/dz) = 1$.

What does Theorem 1.2.5 convey in the special case, already considered, where the transformation of co-ordinates is linear? Here the transformation is given by $X = AZ$, i.e. $x^i = a^i_j z^j$, so that since $\partial x^i/\partial z^k = a^i_k$, the Jacobian matrix $\partial x/\partial z$ is just the constant matrix A. Thus in this case Theorem 1.2.5 reduces to the previously mentioned well-known fact that if $\det A \neq 0$, then the transformation is invertible on the whole space, and $Z = BX$ where B is the inverse of A.

The three further examples which follow are all taken from analytic geometry of two and three dimensions.

1.2.6. Examples. (a) It is often convenient to use *polar co-ordinates* r, φ, of the plane. Rectangular Cartesian co-ordinates are expressed in terms of these by

$$x^1 = r\cos\varphi, \qquad x^2 = r\sin\varphi. \tag{6}$$

(Here we allow only $r \geq 0$.) Thus for all integers k the pairs (r, φ) and $(r, \varphi + 2k\pi)$ represent the same point $P = (x^1, x^2)$. Thus in order that there be a unique φ for each P we impose the requirement that $0 \leq \varphi < 2\pi$. Note also that the pairs $(0, \varphi)$ all represent a single point, namely the (common) origin; thus at the origin we might expect the transformation (6) to behave badly. Let us verify that the origin is indeed a singular (i.e. non-ordinary) point of the system of polar co-ordinates. The Jacobian matrix is

$$A = \begin{pmatrix} \dfrac{\partial x^1}{\partial r} & \dfrac{\partial x^1}{\partial \varphi} \\[2ex] \dfrac{\partial x^2}{\partial r} & \dfrac{\partial x^2}{\partial \varphi} \end{pmatrix} = \begin{pmatrix} \cos\varphi & -r\sin\varphi \\ \sin\varphi & r\cos\varphi \end{pmatrix}. \tag{7}$$

Hence the Jacobian is

$$J = \det A = r \geq 0,$$

so it is zero only at the origin. Expressing r as a function of x^1 and x^2, we get $r = \sqrt{(x^1)^2 + (x^2)^2}$, which is not differentiable at $x^1 = 0$, $x^2 = 0$. On the other hand, in the region $\{(r, \varphi) \mid r > 0, 0 < \varphi < 2\pi\}$, there are no singular points, and the new co-ordinates correspond one-to-one to the points.

(b) The rectangular Cartesian co-ordinates x^1, x^2, x^3 of 3-dimensional "Euclidean" space are expressed in terms of the *cylindrical co-ordinates* $z^1 = r$, $z^2 = \varphi$, $z^3 = z$ by

$$x^1 = r\cos\varphi, \qquad x^2 = r\sin\varphi, \qquad x^3 = z. \tag{8}$$

Here the equation $r = 0$ defines the z-axis, and it is along this straight line that the co-ordinate system "misbehaves," in the sense that the Jacobian matrix

$$A = \begin{pmatrix} \cos\varphi & -r\sin\varphi & 0 \\ \sin\varphi & r\cos\varphi & 0 \\ 0 & 0 & 1 \end{pmatrix} \tag{9}$$

has zero determinant there (and only there). In the region $r > 0$ this co-ordinate system has no singular points. As in 1.2.6(a) the co-ordinate φ is single-valued provided we impose the restriction $0 \le \varphi < 2\pi$.

(c) Finally we consider *spherical co-ordinates* $z^1 = r$, $z^2 = \theta$, $z^3 = \varphi$ in Euclidean 3-space (Figure 3). In this case

$$\begin{aligned} x^1 &= r\cos\varphi\sin\theta, \qquad x^2 = r\sin\varphi\sin\theta, \\ x^3 &= r\cos\theta; \qquad r \ge 0, \quad 0 \le \theta \le \pi, \quad 0 \le \varphi < 2\pi. \end{aligned} \tag{10}$$

Hence the Jacobian matrix is

$$A = \begin{pmatrix} \cos\varphi\sin\theta & r\cos\varphi\cos\theta & -r\sin\varphi\sin\theta \\ \sin\varphi\sin\theta & r\sin\varphi\cos\theta & r\cos\varphi\sin\theta \\ \cos\theta & -r\sin\theta & 0 \end{pmatrix}, \tag{11}$$

and the Jacobian $J = \det A$ is $J = r^2 \sin\theta$. Thus the Jacobian is zero only when $r = 0$, or $\theta = 0, \pi$. We conclude that in the region $r > 0$, $0 < \theta < \pi$, $0 < \varphi < 2\pi$, the spherical co-ordinates are single-valued and there are no singular points of the system. The points defined by $r = 0$ (θ, φ arbitrary), and by $\theta = 0, \pi$ (r, φ arbitrary) are singular points of the spherical co-ordinate system.

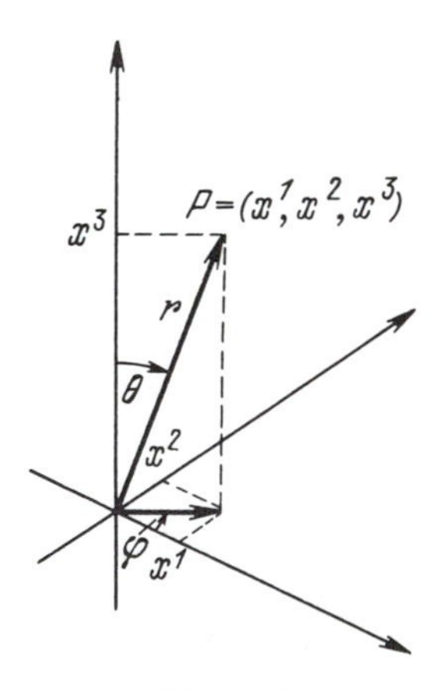

Figure 3

§2. Euclidean Space

We now supplement the rudimentary idea of geometry considered in the preceding section with the two concepts basic to geometry, namely the length of a curve segment in space, and the angle between two curves at a point where they intersect. Our intuitive ideas of length and angle are determined by the fact that we live in a space which is (to a certain approximation) 3-dimensional Euclidean, i.e. which can be co-ordinatized by Cartesian co-ordinates with special properties. We shall now describe these special properties.

2.1. Curves in Euclidean Space

Suppose we have a 3-dimensional Cartesian space where the square of the length l of a straight line segment joining any point $P = (x^1, x^2, x^3)$ to any point $Q = (y^1, y^2, y^3)$ is given by

$$l^2 = (x^1 - y^1)^2 + (x^2 - y^2)^2 + (x^3 - y^3)^2.$$

We call such a Cartesian space *Euclidean* (of 3 dimensions) and call these Cartesian co-ordinates *Euclidean co-ordinates.*

The reader will recall from courses in linear algebra that it is often convenient to associate vectors with the points of Euclidean space. With each point P we associate the vector (or "arrow") with its tail at O (the origin of co-ordinates), and its tip at P. This vector is called the *radius vector* of the point P, and the co-ordinates (x^1, x^2, x^3) of P are the *co-ordinates* or *components* of the vector. Vectors $\xi = (x^1, x^2, x^3)$, $\eta = (y^1, y^2, y^3)$ can be added co-ordinate-wise to yield the vector $\xi + \eta$ with co-ordinates $(x^1 + y^1, x^2 + y^2, x^3 + y^3)$. A vector can also be multiplied (co-ordinate-wise) by any real number (called a "scalar" in this context). The *unit vectors* e_1, e_2, e_3 with co-ordinates $(1, 0, 0)$, $(0, 1, 0)$, $(0, 0, 1)$ respectively, clearly have length 1; we shall see later on that they are also mutually perpendicular. Any vector $\xi = (x^1, x^2, x^3)$ can be expressed as a unique linear combination of these unit vectors: $\xi = x^1e_1 + x^2e_2 + x^3e_3$.

We define n-dimensional Euclidean space analogously. Thus an n-dimensional Euclidean space may be regarded as a linear space (i.e. vector space) for which the square of the distance l between any two points (or tips of radius vectors) $\xi = (x^1, \ldots, x^n)$ and $\eta = (y^1, \ldots, y^n)$ is given by

$$l^2 = \sum_{i=1}^{n} (x^i - y^i)^2.$$

As we have seen, the case $n = 3$ corresponds to "ordinary" Euclidean space. The case $n = 2$ corresponds to the Euclidean plane, while the Euclidean spaces of dimension $n > 3$ are simply generalizations to higher dimensions.

Of fundamental importance is the scalar product of a pair of vectors in Euclidean n-space.

2.1.1. Definition. The *Euclidean scalar product* of two (real) vectors $\xi = (x^1, \ldots, x^n)$ and $\eta = (y^1, \ldots, y^n)$ is the number

$$\langle \xi, \eta \rangle = \sum_{i=1}^{n} x^i y^i. \tag{1}$$

The quintessential properties of the scalar product are the following:

(i) $\langle \xi, \eta \rangle = \langle \eta, \xi \rangle$;
(ii) $\langle \lambda_1 \xi_1 + \lambda_2 \xi, \eta \rangle = \lambda_1 \langle \xi_1, \eta \rangle + \lambda_2 \langle \xi_2, \eta \rangle$ for any real numbers λ_1, λ_2;
(iii) $\langle \xi, \xi \rangle > 0$ if $\xi \neq 0$.

As noted above, Cartesian co-ordinates $x^1, \ldots, x^n$ in terms of which the scalar product has the form (1), are called *Euclidean co-ordinates.*

We shall now see that lengths and angles in Euclidean n-space are expressible in terms of the scalar product. Thus the length (or norm) of a vector ξ, which we denote by $|\xi|$, is given by $|\xi| = \sqrt{\langle \xi, \xi \rangle}$. Similarly the square of the distance between the points P and Q with radius vectors $\xi = (x^1, \ldots, x^n)$ and $\eta = (y^1, \ldots, y^n)$ respectively is just the scalar product of the vector $\xi - \eta$ with itself. Hence property (iii) can be interpreted as saying that any non-zero vector has positive length.

The reader is no doubt familiar from analytic geometry with the formula for the angle between two vectors $\xi = (x^1, \ldots, x^n)$ and $\eta = (y^1, \ldots, y^n)$, namely

$$\cos \varphi = \frac{\langle \xi, \eta \rangle}{\sqrt{\langle \xi, \xi \rangle \langle \eta, \eta \rangle}} = \frac{\langle \xi, \eta \rangle}{|\xi||\eta|}, \qquad 0 \leq \varphi \leq \pi. \tag{2}$$

We conclude from this that the two basic geometrical concepts, namely length and angle, can be expressed in terms of a single concept, namely the scalar product. Subsequently, when we come to deal with general spaces, we shall take some scalar product satisfying (i), (ii) and (iii) as the basic concept, in terms of which the geometrical structure is defined.

Suppose now that we have a segment (i.e. an arc) of a curve in Euclidean n-space given in parametric form:

$$x^1 = f^1(t), \ldots, x^n = f^n(t), \tag{3}$$

where the parameter t varies from a to b, and the $f^i(t)$ are smooth functions of t. The *tangent* or *velocity vector* of the curve at the instant t is the vector

$$v(t) = \left(\frac{df^1}{dt}, \ldots, \frac{df^n}{dt} \right). \tag{4}$$

2.1.2. Definition. The *length* of this curve segment (or arc) is

$$l = \int_a^b \sqrt{\langle v(t), v(t) \rangle}\, dt = \int_a^b |v(t)|\, dt. \tag{5}$$

In words: *the length of the arc is defined as the integral from a to b of the norm of its velocity vector.*†

We next define the angle between two curves at a point where they intersect. Suppose the curves are given as $x^i = f^i(t)$, $i = 1, \ldots, n$, and $x^i = g^i(t)$, $i = 1, \ldots, n$, and that they intersect when $t = t_0$ (i.e. $f^i(t_0) = g^i(t_0)$, $i = 1, \ldots, n$). Denote the respective tangent vectors to the curves by

$$v = \left(\frac{df^1}{dt}, \ldots, \frac{df^n}{dt}\right)\bigg|_{t=t_0},$$
$$w = \left(\frac{dg^1}{dt}, \ldots, \frac{dg^n}{dt}\right)\bigg|_{t=t_0}. \tag{6}$$

2.1.3. Definition. The *angle* between the two curves at the point of their intersection (at $t = t_0$) is that angle φ satisfying $0 \leq \varphi \leq \pi$ and

$$\cos\varphi = \frac{\langle v, w\rangle}{|v||w|}. \tag{7}$$

In courses in mathematical analysis Definitions 2.1.2 and 2.1.3 are usually treated as important results which have to be inferred from simpler facts. On the other hand one can, as in our treatment here, simply regard them as definitions; of course we must then check that these definitions fall into line with our previous ideas of arc length and of angle between two curves, in Euclidean space. Such verification will incidentally also provide further support for the modern point of view of geometry, namely that the appropriate initial concept on which to base all geometry is that of the scalar product operation on pairs of tangent vectors.

What do our earlier ideas of length and angle amount to? Careful recollection reveals that for us the elemental concept of length was that of a straight line. Proceeding from there we took for the length of a polygonal arc (i.e. a "broken straight line segment") the sum of the lengths of the straight line segments composing it. Thence, imitating the definition (encountered, perhaps, in high school) of the circumference of a circle, we arrived at the definition of arc length for more general curves: we represent the arc we are studying as the limit of a sequence of polygonal arcs, and define its length as the limit (when it exists) of the lengths of those approximating polygonal arcs. From analytic geometry we know that the length of a straight line segment joining the origin to the point $(y^1, \ldots, y^n)$ (i.e. the norm of this as a vector) is $\sqrt{(y^1)^2 + \cdots + (y^n)^2}$ (this is in essence Pythagoras' theorem). At school we were taught that the circumference of a circle of radius R is $2\pi R$. Direct use of Definition 2.1.2 yields the same answers, as we shall now see.

† We make no attempt at an axiomatic treatment of concept of length. Rather than deduce the uniqueness of this definition from a set of axioms for length, we simply regard the definition itself as an axiom.

2.1.4. Examples. (a) *The straight line segment.* For simplicity we suppose (as above) that one end of our segment is at the origin. Its (simplest) equations are then $x^i = y^i t$, $i = 1, \ldots, n$, $0 \leq t \leq 1$. When $t = 0$ the co-ordinates x^i are all zero, while when $t = 1$ we have $x^i = y^i$ for all i, i.e. we are at the tip of the vector. According to our definition (2.1.2), the length l of our straight line segment is given by

$$l = \int_0^1 \sqrt{\left(\frac{dx^1}{dt}\right)^2 + \cdots + \left(\frac{dx^n}{dt}\right)^2}\, dt = \sqrt{(y^1)^2 + \cdots + (y^n)^2},$$

which is the usual formula for the length of a straight line segment.

(b) *The circle.* The usual parametric equations of the circle (in the plane) of radius R and with centre the origin, are: $x^1 = R\cos t$, $x^2 = R\sin t$, where $0 \leq t \leq 2\pi$. Here the tangent vector $v(t) = (-R\sin t, R\cos t)$, and so by Definition 2.1.2, the circumference is

$$l = \int_0^{2\pi} \sqrt{R^2\sin^2 t + R^2\cos^2 t}\, dt = 2\pi R. \tag{8}$$

Thus for the circle also, our definition of length gives the answer it should.

It is clear that our definition also satisfies the requirement that the length of an arc made up of several non-overlapping segments, be the sum of the lengths of those segments.

The formula (5) for arc length has one apparent flaw: it seems to depend on the parametrization $x^i = f^i(t)$, $i = 1, \ldots, n$, $a \leq t \leq b$, of the curve segment. To put it kinematically, if $(f^1(t), \ldots, f^n(t))$ represents our position on the curve at time t, then our speed at time t is $|v(t)|$, which enters into the formula (5). What will happen if we trace out the same curve segment (from the point $P = (f^1(a), \ldots, f^n(a))$ to the point $Q = (f^1(b), \ldots, f^n(b))$) at different speeds? Will our arc length formula (5) give us the same number?

To be more precise, suppose we have a new parameter τ varying from a' to b' ($a' \leq \tau \leq b'$), and that our old parameter t is expressed as a function $t = t(\tau)$ of τ, where $t(a') = a$, $t(b') = b$, and $dt/d\tau > 0$. (The last condition is the natural one that, whichever of the two parametrizations we use, we should move along the curve in the same direction.) Then our curve has the new parametrization:

$$x^i = f^i(t) = f^i(t(\tau)) = g^i(\tau), \qquad i = 1, \ldots, n. \tag{9}$$

The rate at which we trace out the curve relative to the parameter τ is:

$$w(\tau) = \left(\frac{dg^1}{d\tau}, \ldots, \frac{dg^n}{d\tau}\right), \qquad a' \leq \tau \leq b'. \tag{10}$$

With the new parametrization our arc length formula (5) gives

$$l' = \int_{a'}^{b'} |w(\tau)|\, d\tau. \tag{11}$$

We shall show that

$$l' = \int_{a'}^{b'} |w(\tau)|\, d\tau = l = \int_a^b |v(t)|\, dt.$$

The norm of the vector $w(\tau)$ is:

$$|w(\tau)| = \sqrt{\sum_{i=1}^{n}\left(\frac{dg^i}{d\tau}\right)^2} = \sqrt{\sum_{i=1}^{n}\left(\frac{df^i}{dt}\frac{dt}{d\tau}\right)^2} = \left|\frac{dt}{d\tau}\right|\sqrt{\sum_{i=1}^{n}\left(\frac{df^i}{dt}\right)^2} = \frac{dt}{d\tau}|v(t)|,$$

since $dt/d\tau > 0$. Hence

$$l' = \int_{a'}^{b'} |w(\tau)|\, d\tau = \int_{a'}^{b'} |v(t(\tau))| \frac{dt}{d\tau}\, d\tau = \int_a^b |v(t)|\, dt,$$

as required.

We conclude that: *The length of an arc of a curve is independent of the speed at which the arc is traced out.*

Thus our definition of length (2.1.2) satisfies all the requirements imposed by our intuitive ideas of that concept.

2.1.5. Example. Suppose a curve in the plane happens to be the graph of a function $x^2 = f(x^1)$. Then x^1 will serve as parameter: $x^1 = t$, $x^2 = f(t)$. The tangent vector is then $v = (1, df/dx^1)$, and the length of the segment of the graph above the interval $a \le x^1 \le b$ is

$$l = \int_a^b |v|\, dt = \int_a^b \sqrt{1 + \left(\frac{df}{dx^1}\right)^2}\, dx^1. \tag{12}$$

A curve given by $x^1 = x^1(t), \ldots, x^n = x^n(t)$ is called *smooth* if $x^1(t), \ldots, x^n(t)$ have continuous derivatives and $v(t) \ne 0$ for all t in the specified interval of values of t. For any smooth arc there is a natural parameter, namely the length l traced out from some point. Since for any pair of numbers a, b in the range of values of l we have $\int_a^b |v(l)|\, dl = b - a$, it follows that $|v(l)| = 1$.

Suppose that in Euclidean n-space with Euclidean co-ordinates $(x^1, \ldots, x^n)$, we are given a new system of (not necessarily Euclidean) co-ordinates $(z^1, \ldots, z^n)$, and that $x^i = x^i(z^1, \ldots, z^n)$, $i = 1, \ldots, n$. Suppose also that we are given a curve whose parametric equations in terms of the new co-ordinates are: $z^i = z^i(t)$, $i = 1, \ldots, n$. Then we can get from these a parametrization of the curve in terms of the original, Euclidean, co-ordinates, namely

$$x^j = x^j(z(t)) = h^j(t), \qquad j = 1, \ldots, n.$$

We define the velocity or tangent vector of the curve relative to the co-ordinates $(z^1, \ldots, z^n)$ to be the vector $v_z(t) = (v_z^1, \ldots, v_z^n)$, where

$$v_z^j = \frac{dz^j}{dt}, \qquad j = 1, \ldots, n. \tag{13}$$

Relative to the original co-ordinates $(x^1, \ldots, x^n)$, the tangent vector is $v = v_x = (dh^1/dt, \ldots, dh^n/dt)$. The vectors $v_z(t)$ and $v_x(t)$ are actually the same vector since each of them is the velocity (relative to t) at which the curve is being traced out, evaluated at the point P, represented by $(z^1(t), \ldots, z^n(t))$ in the new co-ordinate system, and by $(h^1(t), \ldots, h^n(t))$ in the old.

The n-tuples $(dz^1/dt, \ldots, dz^n/dt)$ and $(dh^1/dt, \ldots, dh^n/dt)$ are simply the representations of this vector relative to the two different co-ordinate systems (z) and (x). Let us now see how the co-ordinates of the velocity vector transform under the given co-ordinate change. We have

$$v_x^i = \frac{dh^i}{dt} = \frac{\partial x^i}{\partial z^j}\frac{dz^j}{dt} = \frac{\partial x^i}{\partial z^j} v_z^j, \tag{14}$$

where, as always, summation over the repeated index j is understood. Hence the square of the length of the tangent vector is

$$|v_x|^2 = \sum_{i=1}^n \left(\frac{dh^i}{dt}\right)^2 = \sum_{i=1}^n \left(\frac{\partial x^i}{\partial z^j} v_z^j\right)^2 = g_{jk} v_z^j v_z^k, \tag{15}$$

where the symbol g_{jk} is defined by

$$g_{jk} = \sum_{i=1}^n \frac{\partial x^i}{\partial z^j}\frac{\partial x^i}{\partial z^k}. \tag{16}$$

To summarize: *For any system of co-ordinates* $(z^1, \ldots, z^n)$, *where* $x = x(z)$ (x *being Euclidean co-ordinates*), *the scalar square of the tangent vector* $v_z = (dz^1/dt, \ldots, dz^n/dt)$ *to the curve* $z^i = z^i(t)$, $i = 1, \ldots, n$, *is given by*

$$|v_z|^2 = |v_x|^2 = g_{jk}\frac{dz^j}{dt}\frac{dz^k}{dt}, \tag{17}$$

where $g_{jk} = \sum_{i=1}^n (\partial x^i/\partial z^j)(\partial x^i/\partial z^k)$.

The definition of arc length (2.1.2) and the equations (17), together yield a general formula for arc length in any system of co-ordinates $(z^1, \ldots, z^n)$.

2.2. Quadratic Forms and Vectors

We have just seen (in (14)) that the components of the tangent vector to a curve transform under a co-ordinate change $x = x(z)$ according to the rule

$$v_x^i = \frac{\partial x^i}{\partial z^j} v_z^j, \tag{18}$$

or, more briefly,

$$v_x = A v_z,$$

where $A = \partial x/\partial z$ is the Jacobian matrix of the co-ordinate change, defined in §1. Note that in arriving at the rule (18) no use was made of the assumption that the co-ordinates $(x^1, \ldots, x^n)$ were Euclidean. Using (18) as a model, we now give a definition, superseding any previous ones, of what we shall mean by a vector.

2.2.1. Definition. A *vector* at the point $P = (x_0^1, \ldots, x_0^n)$ is, relative to an arbitrary co-ordinate system $(x^1, \ldots, x^n)$, an n-tuple $(\xi^1, \ldots, \xi^n)$ of numbers, which transforms under a co-ordinate change $x = x(z)$ from the co-ordinates $(x^1, \ldots, x^n)$ to co-ordinates $(z^1, \ldots, z^n)$, according to the rule

$$\xi^i = \left.\frac{\partial x^i}{\partial z^j}\right|_{z^k = z_0^k} \zeta^j, \tag{19}$$

where $x^i(z_0^1, \ldots, z_0^n) = x_0^i$, $i = 1, \ldots, n$, and $(\zeta^1, \ldots, \zeta^n)$ is the representation of the vector relative to the co-ordinates $(z^1, \ldots, z^n)$.

It should be emphasized that the "meat" of this definition is in the form of the rule of transformation (19).

By way of contrast, we now consider another frequently encountered geometrical object, namely the gradient of a function. According to the usual definition the gradient of a real-valued function $f(x^1, \ldots, x^n)$ is the "vector" represented by

$$\operatorname{grad} f = \left(\frac{\partial f}{\partial x^1}, \ldots, \frac{\partial f}{\partial x^n}\right) \tag{20}$$

in the Cartesian co-ordinates $x^1, \ldots, x^n$. What does the gradient look like in terms of different co-ordinates $z^1, \ldots, z^n$, where $x = x(z)$? We have

$$\operatorname{grad} f(x^1(z), \ldots, x^n(z)) = \left(\frac{\partial f}{\partial z^1}, \ldots, \frac{\partial f}{\partial z^n}\right),$$

$$\frac{\partial f}{\partial z^i} = \frac{\partial f}{\partial x^j}\frac{\partial x^j}{\partial z^i}, \qquad i = 1, \ldots, n.$$

Writing briefly $\xi_j = \partial f/\partial x^j$, $\eta_i = \partial f/\partial z^i$, $i = 1, \ldots, n$, we get

$$\eta_i = \frac{\partial x^j}{\partial z^i}\xi_j. \tag{21}$$

We thus see that under a co-ordinate change the gradient of a function transforms differently from a vector; we call such an entity a *covector*.

We now turn to quadratic forms. Suppose the system of co-ordinates $(x^1, \ldots, x^n)$ is Euclidean, and that $\xi_1 = (\xi_1^1, \ldots, \xi_1^n)$ and $\xi_2 = (\xi_2^1, \ldots, \xi_2^n)$ are two vectors both originating from the point $P = (x_0^1, \ldots, x_0^n)$. If in a new system of co-ordinates $(z^1, \ldots, z^n)$ with $x = x(z)$, $x(z_0) = x_0$, these vectors have components $(\eta_1^1, \ldots, \eta_1^n)$ and $(\eta_2^1, \ldots, \eta_2^n)$ respectively, then the relationship between the old and new components is, by the very definition of vector (2.2.1), given by the equations

$$\xi_1^i = a_j^i \eta_1^j, \qquad \xi_2^i = a_j^i \eta_2^j,$$

where (a_j^i) is the Jacobian matrix evaluated at $z^k = z_0^k$, $k = 1, \ldots, n$. The scalar product of the vectors ξ_1 and ξ_2 in the original (Euclidean) co-ordinate system is

$$\langle \xi_1, \xi_2 \rangle = \sum_{i=1}^n \xi_1^i \xi_2^i = \delta_{ij} \xi_1^i \xi_2^j. \tag{22}$$

In terms of the components relative to the new co-ordinate system this is

$$\langle \xi_1, \xi_2 \rangle = \sum_{i=1}^n (a_j^i \eta_1^j)(a_j^i \eta_2^j) = g_{jk} \eta_1^j \eta_2^k, \tag{23}$$

where

$$g_{jk} = \sum_{i=1}^n a_j^i a_k^i = \delta_{sq} a_j^s a_k^q. \tag{24}$$

The coefficients g_{jk} occurring here are the same as those which we encountered in the preceding subsection (see (15)), in solving the problem of calculating arc length in any system of co-ordinates. (This is not surprising as there the matrix $G = (g_{ij})$ occurred in the expression for the scalar square of a certain vector in terms of new co-ordinates.) In the language of matrices formula (24) can be rewritten as

$$G = A^{\mathrm{T}} A, \tag{25}$$

where T denotes the operation of transposition of matrices. Let us now see how the components or entries g_{ij} of the matrix G transform under a further change of co-ordinates. Thus let $y^1, \ldots, y^n$ be yet other co-ordinates for the same region, and let $z^j = z^j(y^1, \ldots, y^n)$, $j = 1, \ldots, n$. Write $C = (c_j^i) = (\partial z^i / \partial y^j)$. Then by Definition 2.2.1 the components $(\zeta_1^1, \ldots, \zeta_1^n)$, $(\zeta_2^1, \ldots, \zeta_2^n)$ of the vectors ξ_1, ξ_2 relative to the co-ordinates $y^1, \ldots, y^n$, satisfy

$$\eta_1^i = c_j^i \zeta_1^j, \qquad \eta_2^i = c_j^i \zeta_2^j. \tag{26}$$

Denote by (h_{ij}) the matrix which arises in expressing the scalar product $\langle \xi_1, \xi_2 \rangle$ in terms of these co-ordinates; then

$$\langle \xi_1, \xi_2 \rangle = h_{kl} \zeta_1^k \zeta_2^l = g_{ij} \eta_1^i \eta_2^j. \tag{27}$$

Substituting for the η_1^i, η_2^i from (26), this gives us

$$h_{kl} \zeta_1^k \zeta_2^l = (c_k^i g_{ij} c_l^j)(\zeta_1^k \zeta_2^l), \tag{28}$$

whence

$$h_{kl} = c_k^i g_{ij} c_l^j, \tag{29}$$

since (28) holds for any vectors ξ_1, ξ_2 (originating at P). In matrix language (29) becomes $H = C^{\mathrm{T}}GC$.

2.2.2. Definition. Let $z^1, \ldots, z^n$ be co-ordinates for a region of a space. A *quadratic form* (on vectors) at a point $P = (z_0^1, \ldots, z_0^n)$ of the region, relative to these co-ordinates, is a family of numbers g_{ij}, $i, j = 1, \ldots, n$, satisfying $g_{ij} = g_{ji}$, and transforming under a change to co-ordinates $y^1, \ldots, y^n$ where $z = z(y)$ and $z_0^i = z^i(y_0^1, \ldots, y_0^n)$, according to the rule

$$h_{kl} = \left.\frac{\partial z^i}{\partial y^k}\right|_{y^s = y_0^s} g_{ij} \left.\frac{\partial z^j}{\partial y^l}\right|_{y^s = y_0^s}, \tag{30}$$

where the h_{kl}, $k, l = 1, \ldots, n$, $h_{kl} = h_{lk}$, are the "components" (or "coefficients") of the quadratic form relative to the new co-ordinates $y^1, \ldots, y^n$. (As already noted, (30) can be rewritten in matrix notation as $H = C^{\mathrm{T}}GC$.)

If we are given at a point P a quadratic form g_{ij} which transforms in accordance with (30), then we can define a bilinear form $\{\xi, \eta\}$ on pairs of vectors originating at P, by setting

$$\{\xi, \eta\} = g_{ij}\xi^i\eta^j.$$

(We obtain from this a quadratic form on individual vectors, given by $\{\xi, \xi\} = g_{ij}\xi^i\xi^j$.) It follows from the transformation rule (30) that this bilinear form (and so also its quadratic restriction) does not depend on the choice of co-ordinate system, but only on the point P and the pair ξ, η of vectors (cf. (27)).

§3. Riemannian and Pseudo-Riemannian Spaces

3.1. Riemannian Metrics

We have already discussed the concept of length, or, as they say, "metric," in a space or region of a space: The length of an arc of a smooth curve $x^i = x^i(t)$ in n-dimensional space with co-ordinates $(x^1, \ldots, x^n)$ is defined (by analogy with 2.1.2.) to be

$$l = \int_a^b |\dot{x}(t)|\, dt, \qquad \dot{x} = v = \frac{dx}{dt}. \tag{1}$$

This definition assumes beforehand that we know what is meant by the length of the tangent vector $\dot{x}(t)$ to the curve at each point of the curve. For the metric to be "Riemannian" we shall require first of all that the formula for the square of the length of a vector ξ originating from a point P take the form

$$|\xi|^2 = g_{ij}\xi^i\xi^j, \tag{2}$$

where $\xi^1, \ldots, \xi^n$ are the components of ξ relative to the given co-ordinate system, and the numbers g_{ij} depend on P (and on the co-ordinate system). Thus $|\xi|^2$ is a quadratic function of the vector ξ in the sense of the last paragraph of the preceding section. In order that the length of a vector should not depend on the choice of co-ordinate system, the g_{ij} must perforce transform under a co-ordinate change like the coefficients of a quadratic form, i.e. according to the rule (30) of the preceding section. We are thus led to the following definition.

3.1.1. Definition. A *Riemannian metric* in a region of the space $\mathbb{R}^n$ is a positive definite quadratic form defined on vectors originating at each point P of the region and depending smoothly on P.

If, using 2.2.2, we spell out what is meant by "positive definite quadratic form," then this definition takes on the following more explicit form:

3.1.1′. Definition. A *Riemannian metric* in a region of a space, relative to arbitrary co-ordinates $(z^1, \ldots, z^n)$, is a family of smooth functions $g_{ij} = g_{ji}(z^1, \ldots, z^n)$, $i, j = 1, \ldots, n$, with the following two properties: (i) the matrix (g_{ij}) is positive definite; (ii) if $(y^1, \ldots, y^n)$ are new co-ordinates for the region, and $z^i = z^i(y^1, \ldots, y^n)$, $i = 1, \ldots, n$, then relative to these new co-ordinates the Riemannian metric is represented by the family of functions $g'_{ij} = g'_{ji}(y^1, \ldots, y^n)$, $i, j = 1, \ldots, n$, given by

$$g'_{ij} = \frac{\partial z^k}{\partial y^i} g_{kl} \frac{\partial z^l}{\partial y^j}. \tag{3}$$

("Positive definiteness" of the matrix (g_{ij}) means simply that $g_{ij}\xi^i\xi^j > 0$ for non-zero vectors ξ, i.e. that the quadratic form is positive definite.)

Given a Riemannian metric as in 3.1.1′, we define *arc length* of a curve $z^i = z^i(t)$ by

$$l = \int_a^b \sqrt{g_{ij}(z(t)) \frac{dz^i}{dt} \frac{dz^j}{dt}}\, dt. \tag{4}$$

Before proceeding to the definition of angle, we define the "scalar product" of a pair of vectors originating at a single point.

3.1.2. Definition. Let $\xi = (\xi^1, \ldots, \xi^n)$ and $\eta = (\eta^1, \ldots, \eta^n)$ be two vectors at the point $P = (z_0^1, \ldots, z_0^n)$. Their *scalar product* $\langle \xi, \eta \rangle$ is defined by

$$\langle \xi, \eta \rangle = g_{ij}(z_0^1, \ldots, z_0^n)\xi^i\eta^j. \tag{5}$$

Note that the transformation rules (19) and (30) of §2 ensure that the scalar product of two vectors attached to a point is independent of the choice of co-ordinate system. Our definition of angle now takes on the familiar form: If we have two curves $z^i = f^i(t)$ and $z^i = h^i(t)$ which intersect when $t = t_0$, then the *angle between the curves* (at $t = t_0$) is the unique φ satisfying $0 \leqq \varphi \leqq \pi$, and

$$\cos \varphi = \frac{\langle \xi, \eta \rangle}{|\xi||\eta|},$$

where ξ, η are the respective tangent vectors to the curves at $t = t_0$, and $|\xi| = \sqrt{\langle \xi, \xi \rangle}$. Note that φ is real since $|\langle \xi, \eta \rangle| \leq |\xi||\eta|$ (a form of the "Cauchy–Bunjakovskiĭ inequality"). To see this observe first that for any real number a,

$$\langle a\xi + \eta, a\xi + \eta \rangle = a^2\langle \xi, \xi \rangle + 2a\langle \xi, \eta \rangle + \langle \eta, \eta \rangle,$$

so that since our quadratic form is positive we must have the discriminant $\langle \xi, \eta \rangle^2 - \langle \xi \cdot \xi \rangle\langle \eta, \eta \rangle \leq 0$.

3.1.3. Example. What form does the metric take in the various co-ordinate systems for a Euclidean space?

(i) $n = 2$. Relative to Euclidean co-ordinates $x^1 = x$, $x^2 = y$, we have:

$$g_{ij} = \delta_{ij} = \begin{cases} 1 & \text{for } i = j \\ 0 & \text{for } i \neq j \end{cases}; \qquad (g_{ij}) = \begin{pmatrix} 1 & 0 \\ 0 & 1 \end{pmatrix}.$$

Relative to polar co-ordinates $z^1 = r$, $z^2 = \varphi$, where, as usual, $x^1 = r \cos \varphi$, $x^2 = r \sin \varphi$ (see 1.2.6(a)), we have, after a little computation,

$$g'_{ij} = \begin{pmatrix} 1 & 0 \\ 0 & r^2 \end{pmatrix}.$$

Thus the length l of an arc given by $r = r(t)$, $\varphi = \varphi(t)$, $a \leq t \leq b$, is

$$l = \int_a^b \sqrt{\left(\frac{dr}{dt}\right)^2 + r^2\left(\frac{d\varphi}{dt}\right)^2}\, dt$$

(ii) $n = 3$. Relative to Euclidean co-ordinates x^1, x^2, x^3, we have as before $g_{ij} = \delta_{ij}$.

Relative to cylindrical co-ordinates $y^1 = r$, $y^2 = \varphi$, $y^3 = z$ (see 1.2.6(b)) we get

$$(g'_{ij}) = \begin{pmatrix} 1 & 0 & 0 \\ 0 & r^2 & 0 \\ 0 & 0 & 1 \end{pmatrix},$$

whence

$$l = \int_a^b \sqrt{\left(\frac{dr}{dt}\right)^2 + r^2\left(\frac{d\varphi}{dt}\right)^2 + \left(\frac{dz}{dt}\right)^2}\, dt.$$

In spherical co-ordinates $y^1 = r$, $y^2 = \theta$, $y^3 = \varphi$ (see 1.2.6(c)), we get

$$g'_{ij} = \begin{pmatrix} 1 & 0 & 0 \\ 0 & r^2 & 0 \\ 0 & 0 & r^2 \sin^2 \theta \end{pmatrix},$$

$$l = \int_a^b \sqrt{\left(\frac{dr}{dt}\right)^2 + r^2\left(\frac{d\theta}{dt}\right)^2 + r^2 \sin^2 \theta \left(\frac{d\varphi}{dt}\right)^2}\, dt.$$

Often the metric is given via the formula for the square of the differential dl, serving as a suggestive mnemonic:

$$dl^2 = g_{ij}\, dz^i\, dz^j. \tag{6}$$

(Strictly speaking, dz^i is defined by $dz^i = \dot{z}^i\, dt$ (and similarly for dl), where $z^i = z^i(t)$ is the curve under study.)

Returning to our examples of co-ordinate systems in the Euclidean plane and space, we have:

$$\left.\begin{array}{l} \text{In rectangular Cartesian co-ordinates, } dl^2 = \sum_{i=1}^{n} (dx^i)^2; \\ \text{In polar co-ordinates, } dl^2 = (dr)^2 + r^2(d\varphi)^2; \\ \text{In cylindrical co-ordinates, } dl^2 = (dr)^2 + r^2(d\varphi)^2 + (dz)^2; \\ \text{In spherical co-ordinates, } dl^2 = (dr)^2 + r^2[(d\theta)^2 + \sin^2 \theta (d\varphi)^2]. \end{array}\right\} \tag{7}$$

3.1.4. Definition. A metric $g_{ij} = g_{ji}(z)$ is said to be *Euclidean* if there exist co-ordinates $x^1, \ldots, x^n$, $x^i = x^i(z)$, such that

$$\det\left(\frac{\partial x^i}{\partial z^i}\right) \neq 0, \qquad g_{ij} = \sum_{k=1}^{n} \frac{\partial x^k}{\partial z^i} \frac{\partial x^k}{\partial z^j}.$$

It follows from (3) that relative to the co-ordinates $x^1, \ldots, x^n$, we shall have at all points of our region $g'_{ij} = \delta_{ij}$. These co-ordinates are termed *Euclidean co-ordinates.* In Example 3.1.3 we were merely representing the Euclidean metric relative to various co-ordinate systems. In Chapter 2 we shall see examples of Riemannian metrics which are not Euclidean.

3.2. The Minkowski Metric

If in the definition (3.1.1′) we replace the requirement that the matrix (g_{ij}) be positive definite by the conditions that at all points the quadratic form $g_{ij}\xi^i\xi^j$ be indefinite (i.e. take both positive and negative values), but have fixed index of inertia (see below) and still have rank n (i.e. $\det(g_{ij}) \neq 0$), we then arrive at the definition of a *pseudo-Riemannian metric.*

Let g_{ij}^0 be the values of g_{ij} at a particular point $P = (z_0^1, \dots, z_0^n)$, i.e. $g_{ij}^0 = g_{ij}(z_0^1, \dots, z_0^n)$. It is a well-known fact of linear algebra that there is a linear "change of variables" $\xi^i = \lambda_k^i \eta^k$, under which the quadratic form $g_{ij}^0 \xi^i \xi^j$ takes on the canonical form

$$\eta_1^2 + \cdots + \eta_p^2 - \eta_{p+1}^2 - \cdots - \eta_n^2,$$

where p depends only on the quadratic form $g_{ij}^0 \xi^i \xi^j$. (This is Sylvester's "Law of Inertia." As an exercise, prove it! It can be found in most elementary text-books on linear algebra.) By assumption (or considerations of continuity) the integer p, the so-called "index of inertia" of the quadratic form, is the same for all points P of our region, so that we can define unambiguously the *type* of the (variable) quadratic form $g_{ij}\xi^i\xi^j$ to be the pair (p, q) where $p + q = n$. Note that in general it will not be possible to change co-ordinates so that in terms of these co-ordinates, $g_{ij}\xi^i\xi^j$ is in canonical form at all points of a neighbourhood of P.

3.2.1 Definition. Let $g_{ij} = g_{ji}(z)$ be a pseudo-Riemannian metric. We shall say that this metric is *pseudo-Euclidean* if there exist co-ordinates $x^1, \dots, x^n$, $x^i = x^i(z)$, $\det(\partial x^i/\partial z^j) \neq 0$, such that

$$g_{ij} = \frac{\partial x^1}{\partial z^i}\frac{\partial x^1}{\partial z^j} + \cdots + \frac{\partial x^p}{\partial z^i}\frac{\partial x^p}{\partial z^j} - \frac{\partial x^{p+1}}{\partial z^i}\frac{\partial x^{p+1}}{\partial z^j} - \cdots - \frac{\partial x^n}{\partial z^i}\frac{\partial x^n}{\partial z^j}.$$

It follows from (3) that, relative to these co-ordinates,

$$g'_{ij} = 0 \quad \text{for } i \neq j,$$

$$g'_{ii} = 1 \quad \text{for } i \leq p, \qquad g'_{ii} = -1 \quad \text{for } i \geq p+1.$$

The co-ordinates $x^1, \dots, x^n$ are called *pseudo-Euclidean co-ordinates of type* (p, q), where $q = n - p$. We can define a pseudo-Euclidean metric of type (p, q) on the space $\mathbb{R}^n$ of ordered n-tuples $(x^1, \dots, x^n)$ of reals by setting

$$\langle \xi, \eta \rangle_{p,q} = \xi^1\eta^1 + \cdots + \xi^p\eta^p - \xi^{p+1}\eta^{p+1} - \cdots - \xi^n\eta^n, \tag{8}$$

for vectors $\xi = (\xi^1, \dots, \xi^n)$ and $\eta = (\eta^1, \dots, \eta^n)$; the "natural" co-ordinates $x^1, \dots, x^n$ will then be pseudo-Euclidean. The space $\mathbb{R}^n$ equipped with this metric is also referred to as pseudo-Euclidean, and is denoted by $\mathbb{R}_{p,q}^n$.

Note that we may suppose $p \leq [n/2]$ since for our purposes the quadratic form $-g_{ij}$ does not differ in any essential way from g_{ij}.

The space $\mathbb{R}_{1,3}^4$ has special significance. This is the "Minkowski space" of the special theory of relativity. In that theory it is postulated that the space–time continuum, which we considered at the beginning of §1, be the Minkowski space $\mathbb{R}_{1,3}^4$. Recall that in §1 we assigned to each point of the space–time continuum Cartesian co-ordinates t, x^1, x^2, x^3, where the co-ordinate t has the dimension of time, and the x^i the dimension of length. The pseudo-Euclidean co-ordinates are then taken to be $x^0 = ct, x^1, x^2, x^3$, where c is a constant (the speed of light *in vacuo*) with the dimensions of velocity, namely length/time.

Thus the square d_l^2 of an element of length is given by

$$d_l^2 = (dx^0)^2 - (dx^1)^2 - (dx^2)^2 - (dx^3)^2. \tag{9}$$

Given two points (or "events") $P_1 = (x_1^0, x_1^1, x_1^2, x_1^3)$ and $P_2 = (x_2^0, x_2^1, x_2^2, x_2^3)$, we define their separation in space–time, or the *space–time interval* between them, to be the quantity

$$|P_1 - P_2|^2 = (x_1^0 - x_2^0)^2 - (x_1^1 - x_2^1)^2 - (x_1^2 - x_2^2)^2 - (x_1^3 - x_2^3)^2. \tag{10}$$

The space–time separation of two distinct events P_1, P_2 can be positive, negative, or even zero (see §6).

We conclude this section by considering a useful co-ordinatization of a region of the space $\mathbb{R}_{1,2}^3$, namely by pseudo-spherical co-ordinates, which are defined as follows. Let x^0, x^1, x^2 denote pseudo-Euclidean co-ordinates in $\mathbb{R}_{1,2}^3$. Then *pseudo-spherical co-ordinates* ρ, χ, φ are defined by

$$\left.\begin{aligned} x^0 &= \rho \cosh \chi, \\ x^1 &= \rho \sinh \chi \cos \varphi, \\ x^2 &= \rho \sinh \chi \sin \varphi, \end{aligned}\right\} \qquad \begin{aligned} -\infty < \rho &< \infty, \\ 0 \le \chi &< \infty, \\ 0 \le \varphi &< 2\pi. \end{aligned} \tag{11}$$

Hence

$$(x^0)^2 - (x^1)^2 - (x^2)^2 = \rho^2 \ge 0.$$

Consequently co-ordinates ρ, χ, φ are assigned only to points in the region defined by $(x^0)^2 - (x^1)^2 - (x^2)^2 > 0$, i.e. in the interior of the cone $(x^0)^2 = (x^1)^2 + (x^2)^2$ (Figure 4). All points of this region except those on the x^0-axis are ordinary points of the pseudo-spherical co-ordinate system. In that region (with the x^0-axis removed) the square of an element of length is given by

$$dl^2 = d\rho^2 - \rho^2[(d\chi)^2 + \sinh^2 \chi(d\varphi)^2]. \tag{12}$$

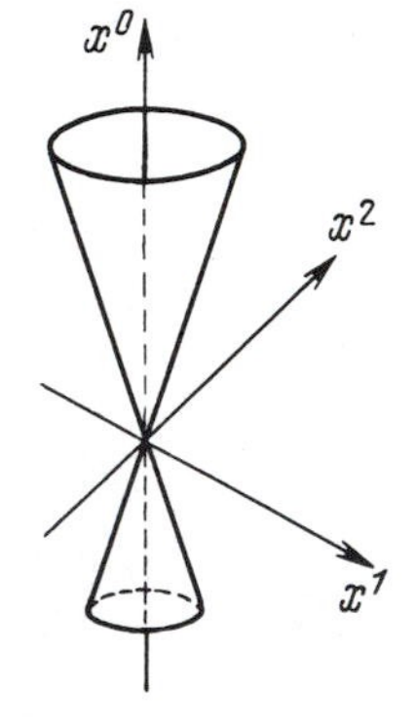

Figure 4

One can also assign pseudo-spherical co-ordinates to points outside the cone by means of the formulae

$$\left.\begin{aligned} x^0 &= \rho \sinh \chi, \\ x^1 &= \rho \cosh \chi \cos \varphi, \\ x^2 &= \rho \cosh \chi \sin \varphi. \end{aligned}\right\} \qquad \rho > 0 \tag{13}$$

This is, however, less important for applications.

§4. The Simplest Groups of Transformations of Euclidean Space

4.1. Groups of Transformations of a Region

Suppose that Ω_x and Ω_z are two regions of an n-dimensional Cartesian space, and that we co-ordinatize these regions anew with co-ordinates $x^1, \ldots, x^n$ and $z^1, \ldots, z^n$ respectively. Suppose further that with each point $(z^1, \ldots, z^n)$ of Ω_z we associate a point $(x_1, \ldots, x_n)$ in a one-to-one and onto manner, so that both $x^i = x^i(z^1, \ldots, z^n)$ and $z^i = z^i(x^1, \ldots, x^n)$, $i = 1, \ldots, n$. We call such a one-to-one map from the region Ω_x onto the region Ω_z a *transformation* of Ω_x onto Ω_z, if the functions $x^i(z^1, \ldots, z^n)$, $z^i(x^1, \ldots, x^n)$ satisfy the usual requirement of smoothness. The bijectivity (one-to-oneness) then entails that the matrix $(\partial x/\partial z)$ (with inverse $(\partial z/\partial x)$) is non-singular at all points of Ω_z (cf. Theorem 1.2.5).

If the regions Ω_x and Ω_z are one and the same, say $\Omega_x = \Omega_z = \Omega$, then we speak simply of a *transformation of the region* Ω. One may in this case think of the transformation as merely a change of co-ordinates for the region Ω, with the property that the old co-ordinates $x^1, \ldots, x^n$ are (smooth) functions of the new co-ordinates $z^1, \ldots, z^n$, and conversely.

We now recall for the reader the concept of a "group." Consider a set G together with two operations: one binary, associating with each ordered pair g, h of elements of G an element of G (called their *product*) denoted by $g \circ h$; and one unary, associating with each element g of G an element of G denoted by g^{-1} (the *inverse* of g). This is called a *group* if the following conditions hold:

(i) $(f \circ g) \circ h = f \circ (g \circ h)$;
(ii) there exists an element $1 \in G$ (the identity element of G) such that $1 \circ g = g \circ 1 = g$ for all $g \in G$;
(iii) $g \circ (g^{-1}) = 1$ for all $g \in G$.

The importance of this concept for us lies in the fact that the totality of all transformations of a given region Ω forms a group under the operation of composition of functions. Thus if φ is the transformation

$$x = x(z), \tag{1}$$

and ψ is the transformation

$$z = z(y), \tag{2}$$

then $\varphi \circ \psi$ is defined to be the composite transformation

$$x = x(z(y)). \tag{3}$$

The transformation φ^{-1} is defined by

$$z = z(x), \tag{4}$$

i.e. the solution of (1) for the z^j in terms of the x^i. The role of 1 is played by the identity transformation

$$x^i = z^i, \qquad i = 1, \ldots, n. \tag{5}$$

It is then easy to verify that conditions (i), (ii), (iii) are satisfied.

In what follows our attention will be centered not on the group of all transformations of Ω, but on certain subgroups which preserve geometrical quantities. (A *subgroup* of a group is a subset forming a group under the same (i.e. restricted) operation.) Suppose that our region Ω is endowed with some Riemannian or pseudo-Riemannian metric, given relative to the co-ordinates $x^1, \ldots, x^n$ as a symmetric non-singular matrix (g_{ij}) where $g_{ij} = g_{ji}(x^1, \ldots, x^n)$. In terms of new co-ordinates $z^1, \ldots, z^n$, with $x^i = x^i(z^1, \ldots, z^n)$, the same metric is given by the functions $g'_{ij} = g'_{ij}(z^1, \ldots, z^n)$, where

$$g'_{ij} = \frac{\partial x^k}{\partial z^i} g_{kl} \frac{\partial x^l}{\partial z^j}. \tag{6}$$

4.1.1. Definition. The transformation $x^i = x^i(z^1, \ldots, z^n)$ is called an *isometry* (or a *motion of the given metric*) if

$$g'_{ij}(z^1, \ldots, z^n) = g_{ij}(x^1(z), \ldots, x^n(z)). \tag{7}$$

Thus an isometry preserves the form of the scalar product (Definition 3.1.2). The following simple fact is almost immediate.

4.1.2. Proposition. *The set of all motions of a given metric is a group.*

(Indeed, if two transformations φ and ψ preserve the metric, then so does their composite, and so do their inverses. That the identity transformation preserves the metric is obvious.)

This group is called the *group of motions*, or *isometries*, of the given metric. This concept ranks in importance with that of the scalar product. It is in a sense the group of symmetries of the metrized region.

4.2. Transformations of the Plane

(a) Let x^1, x^2 be Cartesian co-ordinates of a space (i.e. of the plane). The simplest example of a transformation of the plane is that of a translation of the plane as a whole along some vector $\xi = (\xi^1, \xi^2)$. Thus in terms of co-ordinates this transformation is given by

$$x^1 = z^1 + \xi^1, \qquad x^2 = z^2 + \xi^2. \tag{8}$$

The product of two translations, one along the vector ξ, and the other through η, has the form

$$x^1 = z^1 + (\xi^1 + \eta^1), \qquad x^2 = z^2 + (\xi^2 + \eta^2),$$

which is again a translation (along the vector $\xi + \eta$). The inverse of the transformation (8) is

$$z^1 = x^1 - \xi^1, \qquad z^2 = x^2 - \xi^2, \tag{9}$$

which is just the translation along the vector $-\xi$. The identity transformation is also a translation (along the zero vector). Thus the translations of the plane form a group. We have just seen that there is a one-to-one correspondence between translations and vectors, with the property that to the product of two translations corresponds the sum of the corresponding vectors, and to the inverse of a translation corresponds the negative of the corresponding vector. In the language of groups, such an operation-preserving correspondence is called an *isomorphism*; we have, therefore, that the group of all translations of the plane is isomorphic to the (additive) group of vectors in the plane (i.e. as far as their group-theoretical properties are concerned, the two groups are identical). This group is abelian (i.e. commutative) since $\xi + \eta = \eta + \xi$.

(b) We next describe *dilations* (or *homotheties*) of the plane; these are, except for the identity transformation, not isometries. In terms of co-ordinates a typical dilation has the form

$$x^1 = \lambda z^1, \qquad x^2 = \lambda z^2, \tag{10}$$

where λ is any non-zero real number. The product of two dilations with factors λ and μ respectively, has the form

$$x^1 = \lambda\mu y^1, \qquad x^2 = \lambda\mu y^2. \tag{11}$$

The inverse of the transformation (10) is given by

$$z^1 = \frac{x^1}{\lambda}, \qquad z^2 = \frac{x^2}{\lambda}, \tag{12}$$

which is again a dilation (by a factor $1/\lambda$). It follows that the set of all dilations of the plane is once again a group, and that this group is isomorphic to the (abelian) group of non-zero reals under multiplication.

(c) If we follow a translation by a dilation, or the other way around, then we obtain a transformation of the form

$$x^1 = \lambda z^1 + \xi^1, \qquad x^2 = \lambda z^2 + \xi^2, \qquad \lambda \neq 0. \tag{13}$$

If we follow the transformation (13) by one of the same form, say $z^1 = \mu y^1 + \eta^1$, $z^2 = \mu y^2 + \eta^2$, then we obtain as the product of these transformations

$$\begin{aligned} x^1 &= (\lambda\mu)y^1 + (\xi^1 + \lambda\eta^1), \\ x^2 &= (\lambda\mu)y^2 + (\xi^2 + \lambda\eta^2), \end{aligned} \tag{14}$$

a transformation of the same form as those composing it. If we denote the transformation (13) by the ordered pair (λ, ξ) determining it, and the second transformation, similarly, by (μ, η), then the rule for composing transformations, given by (14), translates into the following rule for multiplying pairs:

$$(\lambda, \xi) \circ (\mu, \eta) = (\lambda\mu, \xi + \lambda\eta). \tag{15}$$

(Note that the first member of each pair is restricted to being a non-zero real, while the second member is a vector in the plane.) It is easy to see that the inverse of a transformation (λ, ξ) is a transformation of the same form, in fact the one represented by the pair $(1/\lambda, -(1/\lambda)\xi)$. Consequently those transformations which are dilations followed by translations form a group. This group is not abelian since if we calculate the product $(\mu, \eta) \circ (\lambda, \xi)$ using (15), we get $(\lambda\mu, \eta + \mu\xi)$, which for suitable choice of λ, μ, η, ξ differs from the right hand side of (15).

Remark. The group G say, which we have been considering, contains as subgroups both the group of translations and the group of dilations of the plane. The former subgroup is normal in G, with factor group isomorphic to the group of dilations. The multiplication rule (15) shows that G is a semi-direct product of the group of translations of the plane by the multiplicative group of non-zero reals, i.e. by the group of dilations.

(d) We next turn to linear transformations of the plane. These have the form

$$\begin{aligned} x^1 &= az^1 + bz^2, \\ x^2 &= cz^1 + dz^2, \end{aligned} \quad \text{or} \quad \begin{pmatrix} x^1 \\ x^2 \end{pmatrix} = \begin{pmatrix} a & b \\ c & d \end{pmatrix} \begin{pmatrix} z^1 \\ z^2 \end{pmatrix}. \tag{16}$$

Each such transformation is determined by its matrix $\begin{pmatrix} a & b \\ c & d \end{pmatrix}$. The transformation is invertible (i.e. z^1, z^2 can be expressed in terms of x^1, x^2) precisely if this matrix is non-singular (or, in other words, if its determinant $ad - bc$

is non-zero). If we have another non-singular matrix $\begin{pmatrix} a' & b' \\ c' & d' \end{pmatrix}$ corresponding to the linear transformation

$$\begin{aligned} z^1 &= a'y^1 + b'y^2, \\ z^2 &= c'y^1 + d'y^2, \end{aligned} \tag{17}$$

then as the result of composing the transformations (16) and (17), we get the transformation

$$\begin{aligned} x^1 &= (aa' + bc')y^1 + (ab' + bd')y^2, \\ x^2 &= (ca' + dc')y^1 + (cb' + dd')y^2, \end{aligned} \tag{18}$$

which is again linear, with matrix the product $\begin{pmatrix} a & b \\ c & d \end{pmatrix}\begin{pmatrix} a' & b' \\ c' & d' \end{pmatrix}$. We conclude that the group of linear transformations of the plane is isomorphic to the group of all 2×2 non-singular matrices with real entries (the *general linear group of degree* 2 *over* $\mathbb{R}$, denoted by $GL(2, \mathbb{R})$). This also is a non-abelian group.

(e) If we combine the linear transformations and the translations (i.e. form the subgroup generated by these) we obtain the *affine group*. Thus the general *affine transformation* has the form

$$\begin{aligned} x^1 &= az^1 + bz^2 + \xi^1, \\ x^2 &= cz^1 + dz^2 + \xi^2, \end{aligned} \quad \text{or} \quad x = Az + \xi, \tag{19}$$

where $\det A = ad - bc \neq 0$, and $\xi = (\xi^1, \xi^2)$ is a vector. The importance of the affine transformations resides in the fact that they are precisely those transformations which preserve collinearity. (We leave it as an exercise for the reader to prove that such a transformation necessarily has the form (19).) Clearly the affine transformation (19) is determined by the pair (A, ξ) consisting of the matrix A and the planar vector ξ. With this representation of affine transformations, the rule for composing them becomes

$$(A, \xi) \cdot (B, \eta) = (AB, \xi + A\eta). \tag{20}$$

It follows from this that the planar affine group is a semi-direct product of the additive group of planar vectors by the group $GL(2, \mathbb{R})$.

(f) Suppose now that there is defined on the plane a Euclidean metric with respect to which x^1, x^2 are Euclidean co-ordinates; thus relative to these co-ordinates the metric is given by $g_{ij} = \delta_{ij}$. It is not difficult to show that an isometry of the Euclidean plane, being distance-preserving, is necessarily a collineation, i.e. affine. The converse question of which of the affine transformations are isometries is thus of great interest, and we shall now answer it.

With respect to the new co-ordinates z^1, z^2 (as in (19)) the metric is given by the matrix (g'_{ij}), where

$$g'_{ij} = \frac{\partial x^k}{\partial z^i}\delta_{kl}\frac{\partial x^l}{\partial z^j} = \sum_{k=1}^{2}\frac{\partial x^k}{\partial z^i}\frac{\partial x^k}{\partial z^j}. \tag{21}$$

The Jacobian matrix of the affine transformation (19) is just $A = \begin{pmatrix} a & b \\ c & d \end{pmatrix}$; hence

$$(g'_{ij}) = \begin{pmatrix} a^2 + c^2 & ab + cd \\ ab + cd & b^2 + d^2 \end{pmatrix}.$$

The condition for (19) to be an isometry is that $g'_{ij} = \delta_{ij}$, i.e. that

$$a^2 + c^2 = 1, \qquad ab + cd = 0, \qquad b^2 + d^2 = 1. \tag{22}$$

In matrix notation these equations can be expressed compactly as $A^{\mathrm{T}}A = 1$, which is just the defining condition for *orthogonality* of A. We conclude, therefore, that an affine transformation is an isometry precisely if its matrix A is orthogonal.

To find a more explicit form for A we now solve the equations (22). Since $a^2 + c^2 = 1$ we can find an angle φ, $0 \le \varphi < 2\pi$, such that $a = \cos\varphi$, $c = \sin\varphi$. The remaining two equations then yield the two possibilities

$$d = \cos\varphi, \qquad b = -\sin\varphi, \quad \text{or} \quad d = -\cos\varphi, \qquad b = \sin\varphi.$$

Thus to each value of φ there correspond two orthogonal matrices:

$$A = \begin{pmatrix} \cos\varphi & \sin\varphi \\ -\sin\varphi & \cos\varphi \end{pmatrix}, \tag{23}$$

$$A = \begin{pmatrix} \cos\varphi & \sin\varphi \\ -\sin\varphi & \cos\varphi \end{pmatrix}\begin{pmatrix} 1 & 0 \\ 0 & -1 \end{pmatrix}. \tag{24}$$

The first of these matrices represents a rotation about the origin of the plane as a whole through the angle φ; it has determinant 1. The second matrix, which has determinant -1, corresponds to a rotation through the angle φ, followed by a reflection in one of the co-ordinate axes:

$$z^1 = y^1,$$
$$z^2 = -y^2.$$

The transformations represented by matrices of the first type (23) form a subgroup of the group of all isometries. We shall call them *proper* or *direct* isometries. (The term "orientation-preserving" is also used of them.)

4.2.1. Lemma. (i) *Every proper isometry of the Euclidean plane is either a rotation about some point, or a translation.*

(ii) *Every isometry of the form* $z \mapsto Az + \xi$, *with* $\det A = -1$, *can be realized as a reflection in some straight line followed by a translation along the axis of reflection (i.e. a "glide-reflection").*

PROOF. (i) A proper isometry has the form

$$x = Az + \xi, \qquad A = \begin{pmatrix} \cos\varphi & \sin\varphi \\ -\sin\varphi & \cos\varphi \end{pmatrix}, \tag{25}$$

If $\varphi = 0$, then $A = 1$, and we get a translation. Suppose that $A \neq 1$. Since we wish to show that in this case we have a rotation, with the centre of rotation in mind we look for a fixed point, i.e. for a point z_0 such that $z_0 = Az_0 + \xi$, or, equivalently,

$$(1 - A)z_0 = \xi. \tag{26}$$

From the form of A given in (25) and from the fact that $\varphi \neq 0$, it follows that the matrix $(1 - A)$ is non-singular, so that the equation (26) has a unique solution z_0. By transferring the origin of co-ordinates to z_0 (by means of a translation) we then see that the transformation is a rotation about z_0, establishing (i).

(ii) Consider now an isometry (19) with $\det A = -1$. By rotating the axes suitably we may assume (see (24)) that

$$A = \begin{pmatrix} 1 & 0 \\ 0 & -1 \end{pmatrix}, \tag{27}$$

i.e. that A is the matrix of the reflection in the first co-ordinate axis (axis of abscissas). Thus with respect to these rotated co-ordinate axes the original transformation becomes

$$\begin{pmatrix} z^1 \\ z^2 \end{pmatrix} \longmapsto \begin{pmatrix} z^1 + \xi^1 \\ -z^2 + \xi^2 \end{pmatrix}. \tag{28}$$

We now make a further change in the co-ordinate system, introducing new co-ordinates y^1, y^2 given in terms of z^1, z^2 by

$$z^1 = y^1, \qquad z^2 = y^2 + \tfrac{1}{2}\xi^2. \tag{29}$$

i.e. we translate our system of axes a distance $\frac{1}{2}\xi^2$ along the axis of ordinates. With respect to the translated axes the isometry is given by

$$\begin{pmatrix} y^1 \\ y^2 \end{pmatrix} \longmapsto \begin{pmatrix} y^1 + \xi^1 \\ -y^2 \end{pmatrix},$$

where the fact that the isometry is a glide-reflection (in the y^1-axis) is now evident. (Note that in particular when $\xi^1 = 0$ we get a reflection.) This completes the proof of the lemma. □

To summarize: *there are three types of isometries of the Euclidean plane, namely translations, rotations, and glide-reflections (including in particular reflections).*

(g) To conclude we consider the group generated by the isometries together with the dilations. To be more specific, we shall examine the effect on the Euclidean metric of transformations of the form

$$x = \lambda Bz + \xi, \quad \text{where } BB^{\mathrm{T}} = 1. \tag{30}$$

In this case the Jacobian matrix has the form $A = \lambda B$ where B is orthogonal. It is easy to calculate that, relative to the co-ordinates z^1, z^2, the Euclidean metric beomes

$$g'_{ij} = \lambda^2 g_{ij} = \lambda^2 \delta_{ij}, \tag{31}$$

i.e. the components g_{ij} of the metric are all multiplied by the same number. Such affine transformations are called *conformal.* It is not difficult to see that every conformal transformation has the form (30). For suppose that the affine transformation (19) is conformal; then arguing as in (f) we deduce that the matrix $(1/\lambda)A$ is orthogonal. Setting $B = (1/\lambda)A$ we arrive at the form (30), as desired.

Suppose that in (30) $\det B = 1$. Then as before there is a φ, $0 \le \varphi < 2\pi$, such that

$$B = \begin{pmatrix} \cos\varphi & \sin\varphi \\ -\sin\varphi & \cos\varphi \end{pmatrix}.$$

In terms of the complex variables $v = z^1 + iz^2$ and $w = x^1 + ix^2$, equation (30) can be rewritten as

$$w = (x^1 + ix^2) = \lambda(\cos\varphi - i\sin\varphi)(z^1 + iz^2) + (\xi^1 + i\xi^2),$$

or, equivalently,

$$w = \alpha v + \beta, \tag{32}$$

where $\alpha = \lambda e^{-i\varphi}$, $\beta = \xi^1 + i\xi^2$. From this equation we see that the proper conformal transformations of the Euclidean plane (i.e. those with $\det B = 1$) can be regarded as the complex affine transformations

$$v = z^1 + iz^2 \mapsto w = x^1 + ix^2$$

of the complex "line" $\mathbb{C} = \mathbb{C}^1$.

Next we consider the case $\det B = -1$. We saw in (f) above that such a matrix corresponds to a transformation realizable as a rotation of the plane about the origin followed by a reflection in the axis of abscissas:

$$x^1 = z^1,$$
$$x^2 = -z^2.$$

In terms of the complex variables v and w this reflection can be expressed as

$$w = \bar{v} = z^1 - iz^2.$$

Putting this together with the previous case, we conclude that *the general affine conformal transformation of the Euclidean plane has (in terms of the complex variables v, w) either the form*

$$w = \alpha v + \beta \quad \text{or} \quad w = \alpha\bar{v} + \beta, \tag{33}$$

where α and β are arbitrary complex numbers with $\alpha \neq 0$. Moreover the transformations (33) *with* $|\alpha| = 1$ *are just the isometries of the Euclidean plane, while those with α real are the "translations followed by dilations" analysed in* (c) *above.*

4.3. The Isometries of 3-Dimensional Euclidean Space

As in the planar case it follows without difficulty from their distance-preserving property that the isometries of Euclidean 3-space also are affine. Thus, in particular, those fixing the origin are linear; we shall therefore first distinguish from among the linear transformations of Euclidean 3-space those that are isometries. Let $A = (a^i_j)$ be the matrix (of degree 3) of a linear transformation which is also an isometry:

$$x = Az; \qquad x^i = a^i_j z^j. \tag{34}$$

(We are assuming the co-ordinates x^1, x^2, x^3 to be Euclidean; i.e. that relative to these co-ordinates the metric is given by $g_{ij} = \delta_{ij}$.) As before, the Jacobian matrix $(\partial x^i/\partial z^j)$ coincides with A. Hence in terms of the co-ordinates z^1, z^2, z^3 the metric is (g'_{ij}) where

$$g'_{ij} = a^k_i \delta_{kl} a^l_j = a^1_i a^1_j + a^2_i a^2_j + a^3_i a^3_j.$$

Thus the assumption that the transformation (34) is an isometry, is equivalent to the condition

$$a^1_i a^1_j + a^2_i a^2_j + a^3_i a^3_j = \delta_{ij}. \tag{35}$$

Equation (35) signifies simply that, given that the unit vectors $e_1 = (1, 0, 0)$, $e_2 = (0, 1, 0)$, $e_3 = (0, 0, 1)$ are orthonormal (i.e. $\langle e_i, e_j \rangle = \delta_{ij}$, or, in yet other words, that the co-ordinates x^1, x^2, x^3 are Euclidean), then so must the vectors $Ae_i = a^k_i e_k$ be orthonormal.

In matrix notation (35) can be rewritten as $A^{\mathrm{T}}A = 1$. Thus, analogously to the planar case, we have that the transformation (34) is an isometry precisely if A is orthogonal. Since $\det A^{\mathrm{T}} = \det A$, it follows that $\det A = \pm 1$. It is easy to verify that the orthogonal matrices are exactly the matrices leaving invariant the quadratic form $(x^1)^2 + (x^2)^2 + (x^3)^2$; i.e. the matrices A for which $\langle x, x \rangle = \langle Ax, Ax \rangle$ for all x. Using this we can prove the following

4.3.1. Lemma. *A linear transformation with orthogonal matrix A leaves at least one straight line invariant. Such an invariant straight line is either fixed pointwise, or reflected (in the origin) by the linear transformation.*

PROOF. If v is a direction-vector for an invariant straight line, then we must have

$$Av = \lambda v \tag{36}$$

for some (non-zero) real λ. In the language of linear algebra, (36) means that v is an eigenvector of A corresponding to the eigenvalue λ. The latter is a root of the characteristic polynomial of the matrix A; i.e. the number λ satisfies

$$\det(A - \lambda.1) = 0. \tag{37}$$

The left-hand side of this equation is a cubic polynomial in λ with real coefficients, and therefore has at least one real root λ_0 say. (Note that $\lambda_0 \neq 0$ since otherwise from (37) we would have $\det A = 0$.) Let v_0 be any eigenvector corresponding to λ_0. Then

$$\langle v_0, v_0 \rangle = \langle Av_0, Av_0 \rangle = \lambda_0^2 \langle v_0, v_0 \rangle,$$

whence $\lambda_0 = \pm 1$. This completes the proof. □

Let w be any vector orthogonal to the eigenvector v_0, i.e. $\langle w, v_0 \rangle = 0$. Then the vector Aw will also be orthogonal to v_0 since

$$\langle v_0, Aw \rangle = \pm \langle Av_0, Aw \rangle = \pm \langle v_0, w \rangle = 0.$$

We conclude that: *The plane through the origin normal to a straight line invariant under the transformation corresponding to the orthogonal matrix A, is also invariant under that transformation.*

We now choose Euclidean co-ordinates x^1, x^2, x^3 so that the x^3-axis is invariant, i.e. is parallel to the vector v_0. We see from 4.3.1 that relative to these co-ordinates the transformation has matrix

$$\hat{A} = \begin{pmatrix} a & b & 0 \\ c & d & 0 \\ 0 & 0 & \pm 1 \end{pmatrix},$$

where of course, for the same reasons as before, $\hat{A}$ is orthogonal. It follows that $\begin{pmatrix} a & b \\ c & d \end{pmatrix}$ is orthogonal, and therefore (see (23), (24) above) has the form

$$\begin{pmatrix} a & b \\ c & d \end{pmatrix} = \begin{pmatrix} \cos\varphi & \sin\varphi \\ -\sin\varphi & \cos\varphi \end{pmatrix} \quad \text{or} \quad \begin{pmatrix} \cos\varphi & -\sin\varphi \\ -\sin\varphi & -\cos\varphi \end{pmatrix}.$$

Thus depending on the signs of λ_0 and of $ad - bc$ we have the following two types of isometries fixing the origin.

(i) *Rotations about some axis.* With the x^3-axis as the axis of rotation, the matrices of such isometries have the form

$$\hat{A} = \begin{pmatrix} \cos\varphi & \sin\varphi & 0 \\ -\sin\varphi & \cos\varphi & 0 \\ 0 & 0 & 1 \end{pmatrix}, \qquad \det\hat{A} = \lambda_0(ad - bc) = 1. \tag{38}$$

This in essence includes the case $\lambda_0 = -1, ad - bc = -1$, since by choosing the co-ordinates x^1, x^2 suitably we may assume $\begin{pmatrix} a & b \\ c & d \end{pmatrix} = \begin{pmatrix} 1 & 0 \\ 0 & -1 \end{pmatrix}$, and then relative to the co-ordinates x^1, x^2, x^3, the matrix of the transformation will be

$$\begin{pmatrix} 1 & 0 & 0 \\ 0 & -1 & 0 \\ 0 & 0 & -1 \end{pmatrix},$$

which represents the rotation about the x^1-axis through the angle π.

(ii) *Rotatory reflections.* An isometry of this type is the composite of a rotation about some axis (passing through the origin) followed by the reflection in the plane through the origin normal to that axis. By choosing co-ordinates suitably it can always be arranged that the matrix of such a transformation have the form

$$A = \begin{pmatrix} \cos\varphi & \sin\varphi & 0 \\ -\sin\varphi & \cos\varphi & 0 \\ 0 & 0 & -1 \end{pmatrix}, \qquad \det A = -1. \tag{39}$$

We halt for a moment our discussion of the isometries of Euclidean 3-space, to introduce some standard notation. The group (under matrix multiplication) of all 3×3 orthogonal matrices is normally denoted by $O(3)$. Those orthogonal matrices of determinant $+1$ form a subgroup of $O(3)$, the *special orthogonal group*, denoted by $SO(3)$. Note incidentally that every matrix in $SO(3)$, since it fixes at least one non-zero vector, has 1 as an eigenvalue. By (i) above the matrices in $SO(3)$ correspond to all possible rotations of the space about all possible axes through the origin.

We now resume our development, and consider general (i.e. not necessarily origin-preserving) isometries of Euclidean 3-space. As remarked above, such an isometry is necessarily affine, i.e. of the form

$$z \mapsto Az + \xi. \tag{40}$$

That the matrix A is orthogonal follows as before. We now translate our system of co-ordinates along a vector y_0, thereby obtaining a new system of co-ordinates y^1, y^2, y^3, linked to the old by $z = y + y_0$. In terms of the new co-ordinates the transformation (40) takes the form

$$y \mapsto Ay + (A - 1)y_0 + \xi. \tag{41}$$

We first consider the case that A is as in (39), i.e. corresponds to a rotatory reflection. In this case the matrix $(A - 1)$ is non-singular, so that there is a vector y_0 such that

$$(1 - A)y_0 = \xi, \qquad y_0 = (1 - A)^{-1}\xi. \tag{42}$$

With this choice of y_0, the transformation (41) is simply a rotatory reflection:

$$y \mapsto Ay, \qquad A = \begin{pmatrix} \cos\varphi & \sin\varphi & 0 \\ -\sin\varphi & \cos\varphi & 0 \\ 0 & 0 & -1 \end{pmatrix}.$$

Next suppose A has the form (38), i.e. corresponds to a rotation about an axis. Suppose also that $A \neq 1$, i.e. $\varphi \neq 0$ (the contrary case corresponds to a simple translation). In this case the matrix $(A - 1)$ is singular. The equation (42) for the unknown vector y_0, when written out in terms of components, becomes

$$\begin{cases} (1 - \cos\varphi)y_0^1 - \sin\varphi\, y_0^2 = \xi^1, \\ \sin\varphi\, y_0^1 + (1 - \cos\varphi)y_0^2 = \xi^2, \\ \qquad\qquad\qquad\qquad 0 = \xi^3. \end{cases}$$

If $\xi^3 \neq 0$ this system of equations has no solutions. However since by assumption $\varphi \neq 0$, the first two equations of the system do have a unique solution for y_0^1 and y_0^2. If we then choose the third co-ordinate y_0^3 of y_0 arbitrarily, the transformation (41) takes the form

$$y \mapsto Ay + (0, 0, \xi^3), \qquad A = \begin{pmatrix} \cos\varphi & \sin\varphi & 0 \\ -\sin\varphi & \cos 0 & 0 \\ 0 & 0 & 1 \end{pmatrix},$$

where $\eta = (0, 0, \xi^3)$ is directed along the axis of the rotation represented by A, so that $A\eta = \eta$. It follows that in this case the motion can be likened to that of a screw. We conclude that: *The proper isometries* (*i.e. those with* $\det A = +1$) *of Euclidean* 3-*space are the screw-displacements* (*which include, in particular, as "extreme" cases, the rotations and translations*).

Thus the screw-displacements together with the rotatory reflections account for all isometries of Euclidean 3-space.

4.4. Further Examples of Transformation Groups

(a) Having dealt with dimensions 2 and 3, we now proceed to dimension n, and consider the group of isometries of Euclidean n-space fixing the origin. As before we find that these isometries are just those linear transformations whose matrices (which now have degree n) are orthogonal. The group of these matrices (which is of course isomorphic to the group of isometries they represent) is denoted by $O(n)$. Thus an isometry fixing the origin has the form

$$z \mapsto Az, \qquad A^{\mathrm{T}}A = 1, \qquad \det A = \pm 1. \tag{43}$$

As before, the matrices A with determinant $+1$ (or the corresponding isometries) form a subgroup, the *special orthogonal group of degree n*, denoted by $SO(n)$. By analogy with the 2-dimensional case the group $SO(n)$ is also called the *n-dimensional rotation group*, and its elements *rotations* (though they are not all rotations in any common-sense meaning of the word).

We shall limit ourselves here to showing that the group $SO(n)$ is *connected*, i.e. given any two rotations A_0, $A_1 \in SO(n)$, there is in the space $SO(n)$ a curve segment $A(t)$, $0 \leq t \leq 1$ (i.e. a continuous family of orthogonal matrices of determinant 1, or, in other words, with entries continuous real-valued functions of t, and $\det A(t) = 1$ for $0 \leq t \leq 1$), such that $A(0) = A_0$ and $A(1) = A_1$.

It clearly suffices to prove this in the case $A_0 = 1$, the identity matrix. It is a well-known fact of linear algebra that after a suitable (linear) change of co-ordinate systems, the matrix A in (43) will have the following form:

$$\begin{pmatrix} \Box & & & & & 0 \\ & \ddots & & & & \\ & & \Box & & & \\ & & & \pm 1 & & \\ & 0 & & & \ddots & \\ & & & & & \pm 1 \end{pmatrix}, \tag{44}$$

where the ith block has the form

$$\Box = \begin{pmatrix} \cos \varphi_i & \sin \varphi_i \\ -\sin \varphi_i & \cos \varphi_i \end{pmatrix}, \qquad 0 \leq \varphi_i < 2\pi.$$

(Note that we established this canonical form for real orthogonal matrices of degrees 2 and 3 in §§4.2, 4.3.)

We may therefore assume that the matrix $A(1)$ has the form (44), with an even number of the diagonal entries -1 since $\det A(1) = 1$. We then define $A(t)$ to be the matrix obtained from $A(1)$ by replacing φ_i in each block in (44) by $t\varphi_i$. By re-indexing the co-ordinates appropriately, the matrix $A(0)$ can be brought into the form

$$A(0) = \begin{pmatrix} \Box & & & & & 0 \\ & \ddots & & & & \\ & & \Box & & & \\ & & & 1 & & \\ & 0 & & & \ddots & \\ & & & & & 1 \end{pmatrix}, \tag{45}$$

where each block has the form

$$\square = \begin{pmatrix} -1 & 0 \\ 0 & -1 \end{pmatrix}.$$

It only remains to show that the matrix (45) can be connected to the identity matrix. This is not difficult: if we replace each block in (45) by

$$\begin{pmatrix} \cos t & \sin t \\ -\sin t & \cos t \end{pmatrix},$$

then when $t = \pi$ we obtain the matrix (45), while putting $t = 0$ gives us the identity matrix. This essentially completes the proof of connectedness of $SO(n)$.

It is not difficult to see that the full orthogonal group $O(n)$ is not connected, since if A_0 and A_1 have determinants 1 and -1 respectively then they cannot be joined by a continuous arc in $O(n)$: if they were so joined, by say $A(t)$, $0 \leqq t \leqq 1$, then $\det A(t)$ would be a continuous real-valued function of t taking only the values 1 and -1, and with $\det A(0) = 1$, $\det A(1) = -1$.

(b) *The Galilean group.* Galileo's famous principle of relativity, fundamental to classical mechanics, asserts the following: Under replacement of a fixed (rectangular) system of co-ordinates x^1, x^2, x^3 (or "frame of reference") by a frame of reference x'^1, x'^2, x'^3 moving in a straight line and uniformly (i.e. with constant velocity) relative to the fixed frame, all the laws of classical mechanics preserve their form. In other words (and more precisely) the laws of classical mechanics are invariant under Galilean transformations:

$$\begin{aligned} x'^1 &= x^1 - vt, \\ x'^2 &= x^2, \\ x'^3 &= x^3, \qquad t' = t. \end{aligned} \tag{46}$$

Here the moving frame is assumed to be moving with speed v in the direction of the positive x^1-axis (Figure 5). A frame of reference related to a fixed frame

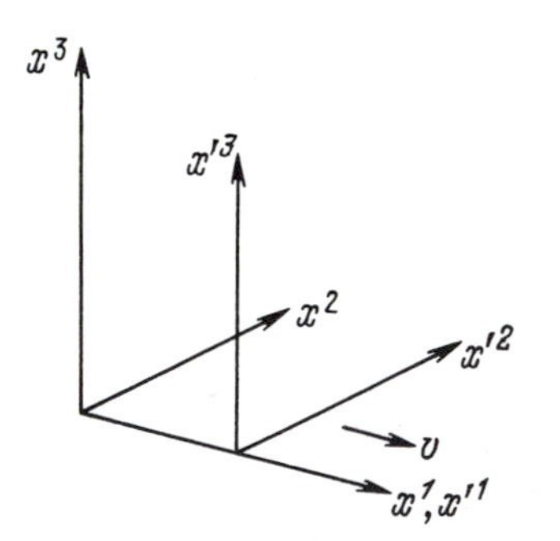

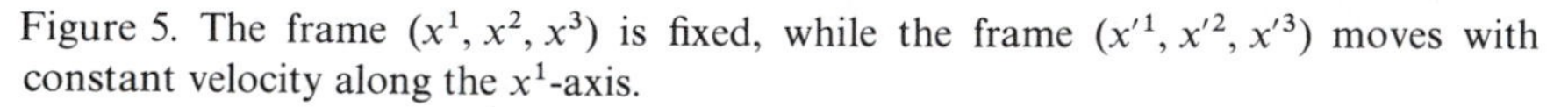
Figure 5. The frame (x^1, x^2, x^3) is fixed, while the frame (x'^1, x'^2, x'^3) moves with constant velocity along the x^1-axis.

via a transformation like (46), i.e. moving with constant velocity relative to the fixed frame, is called an *inertial frame of reference*. We now make an important remark concerning the physical interpretation of the last equation $t' = t$ of (46): In classical, as opposed to relativistic, mechanics, the time t (as also the pervading space $\mathbb{R}^3$) has attributed to it a quality of absoluteness, i.e. the time interval Δt between two events is assumed to be the same when measured by any observer equipped with a clock, no matter how that observer moves through space. It is as if there were a clock at each point in space in synchrony with all the other clocks, from which the time can be read simultaneously. However from a purely geometrical point of view the physical interpretation is irrelevant: a Galilean transformation is just a set of equations like (46) for changing from one co-ordinate system (t, x^1, x^2, x^3) to another system (t', x'^1, x'^2, x'^3). The general Galilean transformation, preserving the form of the laws of classical mechanics, is given by

$$\begin{aligned} x' &= Ax + x_0 - vt, \\ t' &= t, \end{aligned} \tag{47}$$

where A is an orthogonal matrix, and v is a velocity vector. The *Galilean group* is then the group of all such transformations. (It is easy to see that they do indeed form a group.)

(c) A less obvious example of a transformation group, arising again from mechanics, is afforded by:

$$\begin{aligned} x &\to \alpha x, \\ t &\to \beta t, \end{aligned} \qquad \alpha^3 = \beta^2. \tag{48}$$

The group consisting of these transformations arises in connexion with Kepler's third law, which states that the squares of the periods of revolution of planets about the sun are directly proportional to the cubes of their shortest distances from the sun (i.e. distances at perihelion). This law can be regarded as a consequence of the fact that the behaviour of such a mechanical system, consisting of a particle moving in a Newtonian attractive field of potential $\varphi = a/r$, is invariant under the transformation (48), in the sense that under such a transformation any possible trajectory or orbit goes into another. (Here $x = (x^1, x^2, x^3)$ is the position of a planet at time t.)

4.5. Exercises

1. Let $Q(x) = b_{ij}x^ix^j$, where $b_{ij} = b_{ji}$, be a quadratic form, and $B(x, y) = b_{ij}x^iy^j$ the corresponding bilinear form. Show that a linear transformation A preserves the bilinear form (in the sense that $B(Ax, Ay) = B(x, y)$ for all vectors x, y) if and only if it preserves the quadratic form (in the sense that $Q(Ax) = Q(x)$ for all x).
2. An isometry of Euclidean n-space is necessarily affine.

3. The n-dimensional affine group is isomorphic to the group of all matrices of degree $n+1$ of the form

$$\begin{pmatrix} A & \xi \\ 0 & 1 \end{pmatrix},$$

where A is an $n \times n$ non-singular matrix, and ξ is a column n-vector.

4. Find a matrix representation of the Galilean group.

§5. The Serret–Frenet Formulae

5.1. Curvature of curves in the Euclidean Plane

Let x, y be Euclidean co-ordinates for the Euclidean plane, and let e_1, e_2 denote, as usual, the unit basis vectors $(1, 0)$, $(0, 1)$. With each point $P = (x, y)$ we associate its radius-vector $r = xe_1 + ye_2$ with tail at the origin and tip at P. The length of the vector r is given by the Euclidean formula

$$|r| = \sqrt{\langle r, r \rangle} = \sqrt{x^2 + y^2}\,.$$

Suppose we are given a smooth curve

$$r = r(t); \qquad x = x(t), \qquad y = y(t),$$

where what is intended by this notation is that the point of the curve corresponding to the value t of the parameter, has radius-vector $r(t) = x(t)e_1 + y(t)e_2$. The length of the curve segment between $t = a$ and $t = b$ is, according to our definition (2.1.2),

$$l = \int_a^b \sqrt{(\dot{x})^2 + (\dot{y})^2}\, dt = \int_a^b dl, \quad \text{where } \dot{x} = \frac{dx}{dt}, \quad \dot{y} = \frac{dy}{dt},$$

where the differential of length $dl = |v|dt$. (Recall that $|v| = \sqrt{\langle v, v \rangle}$, where $v = \dot{x}e_1 + \dot{y}e_2$ is the tangent or velocity vector.) We shall now change our notation slightly, and for $v = dr/dt$ write v_t, indicating explicitly thereby the parameter with respect to which the tangent vector is calculated. We shall often find it convenient to consider curves parametrized by the natural length parameter l:

$$x = x(l), \qquad y = y(l).$$

In this case $v = v_l = (dx/dl)e_1 + (dy/dl)e_2$, and $|v_l| = 1$ (see §2.1). In what follows, in addition to the velocity $v_t = dr/dt$, the *acceleration* $w_t = d^2r/dt^2 = \dot{v}_t$ will play an important role. The following simple lemma will find much use later on.

5.1.1. Lemma. *In Euclidean space, a variable vector $v = v(t)$ of constant length $|v| =$ const., has the property that v and $\dot{v}$ are orthogonal. (And conversely if v and $\dot{v}$ are orthogonal for all t, then $|v(t)|$ is constant.)*

PROOF. Writing $v = v^1 e_1 + v^2 e_2$, we have that $\langle v, v\rangle = |v|^2 = (v^1)^2 + (v^2)^2$, whence

$$0 = \frac{d}{dt}|v|^2 = \frac{d}{dt}\langle v, v\rangle = \langle \dot{v}, v\rangle + \langle v, \dot{v}\rangle = 2\langle v, \dot{v}\rangle,$$

so that $\langle v, \dot{v}\rangle = 0$, as claimed. We leave the proof of the converse to the reader. □

Remark. It is worth noting separately the fact revealed here that in Euclidean space the formula

$$\frac{d}{dt}\langle v, w\rangle = \langle \dot{v}, w\rangle + \langle v, \dot{w}\rangle \tag{1}$$

holds for any two vectors $v(t)$, $w(t)$.

Applying Lemma 5.1.1. to a curve parametrized by the natural parameter $t = l$, and taking $v = dr(l)/dl$, so that $|v| = 1$, we get

5.1.2. Corollary. *In Euclidean space the velocity vector $v(t)$ and the acceleration vector $w(t) = dv/dt$ are orthogonal if $t = l$, the natural parameter.*

5.1.3. Definition. The *curvature $k(t)$ of a curve $r(t)$* is the magnitude of the acceleration vector, i.e. $k(t) = |w(t)|$, under the condition that the curve is parametrized by the natural parameter, i.e. $t = l$.

It is immediate that

$$\frac{dv}{dl} = kn = \frac{d^2r}{dl^2}, \tag{2}$$

where $n = n(t)$ is the unit vector normal to the curve (i.e. normal to v), in the direction of $w(t)$ if $w(t) \neq 0$, i.e.

$$n = \frac{w}{|w|} = \frac{1}{\sqrt{(d^2x/dl^2)^2 + (d^2y/dl^2)^2}}\left(\frac{d^2x}{dl^2}e_1 + \frac{d^2y}{dl^2}e_2\right). \tag{3}$$

We define the *radius of curvature* R of the curve at t to be $1/k(t)$.

Does this precise definition of curvature chime with our intuitive ideas of that concept? The following two simple examples support an affirmative answer.

5.1.4. Examples. (a) *The curvature of a straight line is zero.* To see this note first that the parametric equations of a straight line in terms of the natural parameter have the form $x = x_0 + al$, $y = y_0 + bl$ (where $1 = |v|^2 = a^2 + b^2$). Thus in this case

$$v = \frac{dx}{dl} e_1 + \frac{dy}{dl} e_2 = ae_1 + be_2$$

is a constant vector, so that $w = dv/dl = 0$, i.e. $k = 0$, $R = \infty$.

(b) *The curvature of a circle of radius R is the constant* $1/R$. For the natural parametric equations of a circle of radius R and centre (x_0, y_0) are:

$$x = x_0 + R\cos\left(\frac{l}{R}\right), \qquad y = y_0 + R\sin\left(\frac{l}{R}\right),$$

Since R is constant, it follows that

$$\frac{d^2x}{dl^2} = -\frac{\cos(l/R)}{R}, \qquad \frac{d^2y}{dl^2} = -\frac{\sin(l/R)}{R},$$

whence $|w| = 1/R = k$.

Formula (2) is one of the two "Serret–Frenet formulae" for plane curves:

5.1.5. Theorem (Serret–Frenet Formulae). *Given the parametric equation $r = r(l)$ of a curve, in terms of the natural parameter l, the following formulae hold:*

$$\begin{aligned} \frac{dv}{dl} &= w = kn, \\ \frac{dn}{dl} &= -kv, \end{aligned} \tag{4}$$

where $n(=w/|w|$ if $w \neq 0)$ is the unit normal vector.

PROOF. We have only to prove the second of the formulae (4). Since n, being a unit vector, has constant length 1, Lemma 5.1.1 tells us that n and dn/dl are orthogonal. Since by definition of n the vectors v and n are also orthogonal (and since our underlying space is the Euclidean plane), it follows that dn/dl and v have the same or opposite directions (or else $dn/dl = 0$), so that in any case

$$\frac{dn}{dl} = \alpha v,$$

where, since $|v| = |v_l| = 1$, we must have $|\alpha| = |dn/dl|$. What is α? By (1) we have

$$0 = \frac{d}{dl}\langle v, n\rangle = \left\langle \frac{dv}{dl}, n \right\rangle + \left\langle v, \frac{dn}{dl} \right\rangle = k + \alpha\langle v, v\rangle = k + \alpha.$$

Since $\langle v, n\rangle = 0$, we deduce that $\alpha = -k$, completing the proof. ☐

What is the geometrical import of the Serret–Frenet formulae? From those formulae and Taylor's theorem we have

$$\begin{aligned} v + \Delta v &= v + (\Delta l)\frac{dv}{dl} + O(\Delta l^2) = v + k(\Delta l)n + O(\Delta l^2), \\ n + \Delta n &= n + (\Delta l)\frac{dn}{dl} + O(\Delta l^2) = n - k(\Delta l)v + O(\Delta l^2). \end{aligned} \tag{5}$$

Writing $\Delta\varphi$ for $k\,\Delta l$, and using

$$\cos(\Delta\varphi) = 1 + O(\Delta\varphi^2),$$
$$\sin(\Delta\varphi) = \Delta\varphi + O(\Delta\varphi^2),$$

we get from (5) that

$$\begin{aligned} v + \Delta v &\approx \cos(\Delta\varphi)v + \sin(\Delta\varphi)n, \\ n + \Delta n &\approx -\sin(\Delta\varphi)v + \cos(\Delta\varphi)n, \end{aligned} \tag{6}$$

for small Δl, i.e. in going from the orthonormal frame (v, n) to the frame $(v + \Delta v, n + \Delta n)$, the first frame is rotated (approximately) through the small angle $\Delta\varphi = k\,\Delta l$. In other words the Serret–Frenet formulae (4) can be thought of as embodying the fact that the orthonormal frame (v, n) undergoes a rotation as we move from the point on the curve corresponding to l to the nearby point corresponding to $l + \Delta l$, with accuracy of the order of the second power of small quantities (Δl^2). This is sometimes also expressed by means of the formula

$$k = \left|\frac{d\varphi}{dl}\right|, \tag{7}$$

where φ denotes the angle through which the vector v (and also n) is rotated in moving along the curve.

In the definition of curvature, as also in the Serret–Frenet formulae and their geometrical interpretation above, the parameter was for good reason always taken to be the natural one. It is however natural to ask how one goes about calculating the curvature of a plane curve parametrized as $r(t) = (x(t), y(t))$ where t is not necessarily natural. (It is easy to choose the parameter t so that $|v_t|$ is not constant, so that in view of Lemma 5.1.1 we cannot expect that for all parameters t the velocity and acceleration vectors v_t and $\dot{v}_t = w_t$ will be perpendicular.) We want first an expression for $v_l = dr(t(l))/dl$ in terms of derivatives with respect to t. From earlier (2.1.2) we know that $dl = |v_t|\,dt$. Hence $dt/dl = 1/|v_t|$ (assuming, as in §2.1, that $v_t \neq 0$), and so

$$v_l = \frac{dr}{dl} = \frac{dr}{dt}\cdot\frac{dt}{dl} = \frac{v_t}{|v_t|}, \tag{8}$$

a not unexpected conclusion, since v_l is the unit tangent vector. Now by definition the curvature $k = |w_l|$, where $w_l = d^2r/dl^2 = dv_l/d_l$. From (8) we have

$$w_l = \frac{d}{dl}\left(\frac{v_t}{|v_t|}\right) = \frac{d}{dt}\left(\frac{v_t}{|v_t|}\right) \cdot \frac{dt}{dl} = \frac{1}{|v_t|}\frac{d}{dt}\frac{v_t}{|v_t|}$$

$$= \frac{1}{|v_t|^2}\left(\frac{dv_t}{dt} - \frac{v_t}{|v_t|}\frac{d|v_t|}{dt}\right) = \frac{1}{|\dot{r}|^2}\left(\ddot{r} - \frac{\dot{r}}{2|\dot{r}|^2}\frac{d|\dot{r}|^2}{dt}\right),$$

using $\dot{r} = v_t$. Hence by (1)

$$w_l = \frac{1}{|\dot{r}|^2}\left(\ddot{r} - \frac{\langle \dot{r}, \ddot{r} \rangle}{|\dot{r}|^2}\dot{r}\right). \tag{9}$$

Hence

$$k = \frac{1}{|\dot{r}|^2}\left|\ddot{r} - \frac{\langle \dot{r}, \ddot{r} \rangle}{|\dot{r}|^2}\dot{r}\right|. \tag{10}$$

In terms of the components $x(t)$, $y(t)$ of $r(t)$, (9) becomes

$$w_l = \frac{1}{\dot{x}^2 + \dot{y}^2}\left(\ddot{x} - \dot{x}\frac{\dot{x}\ddot{x} + \dot{y}\ddot{y}}{\dot{x}^2 + \dot{y}^2}\right)e_1 + \frac{1}{\dot{x}^2 + \dot{y}^2}\left(\ddot{y} - \dot{y}\frac{\dot{x}\ddot{x} + \dot{y}\ddot{y}}{\dot{x}^2 + \dot{y}^2}\right)e_2. \tag{11}$$

From this (or from (10)) we get, after a little manipulation,

$$|w_l|^2 = k^2 = \frac{(\ddot{x}\dot{y} - \ddot{y}\dot{x})^2}{(\dot{x}^2 + \dot{y}^2)^3}. \tag{12}$$

Extraction of the non-negative square root then gives us the desired formula for k. We formulate this important result as

5.1.6. Theorem. *Let $x = x(t)$, $y = y(t)$ be the equations of a curve in terms of any parameter t for which $\dot{x}^2 + \dot{y}^2 \neq 0$. Then (under obvious conditions on the functions $x(t)$, $y(t)$) the curvature is given by*

$$k = \frac{|\ddot{x}\dot{y} - \ddot{y}\dot{x}|}{(\dot{x}^2 + \dot{y}^2)^{3/2}}. \tag{13}$$

Notice that in the numerator of (13) stands the absolute value of the determinant of the matrix $\begin{pmatrix} \dot{x} & \dot{y} \\ \ddot{x} & \ddot{y} \end{pmatrix}$.

5.2. Curves in Euclidean 3-Space. Curvature and Torsion

From the definition of length (2.1.2), we have that for any curve in Euclidean space given in terms of Euclidean co-ordinates by parametric equations $x = x(t)$, $y = y(t)$, $z = z(t)$ (i.e. $r = r(t)$),

$$dl = |\dot{r}|\,dt = |v_t|\,dt = \sqrt{\dot{x}^2 + \dot{y}^2 + \dot{z}^2}\,dt. \tag{14}$$

We shall assume (as initially in the planar case) that $t = l$, the natural parameter, since it is in terms of l that the basic concepts of curvature and torsion are most conveniently defined. Thus our curve is given by $r = r(l)$, or $x = x(l)$, $y = y(l)$, $z = z(l)$, and we shall use the dot to indicate, for the time being, derivatives with respect to l; thus

$$v_l = \dot{r} = \dot{x}e_1 + \dot{y}e_2 + \dot{z}e_3, \qquad w_l = \dot{v}_l = \ddot{r} = \ddot{x}e_1 + \ddot{y}e_2 + \ddot{z}e_3.$$

We define curvature as in the planar case:

5.2.1. Definition. The *curvature* k of a space curve is the absolute value of the acceleration with respect to the natural parameter l; i.e. $k = |w_l| = |\ddot{r}|$. The *radius of curvature* is the reciprocal of the curvature, i.e. $R = 1/k$.

We know from Lemma 5.1.1 that since $|v_l| = 1$, the velocity vector v_l and acceleration vector w_l are orthogonal. Thus for each value of l for which $w_l \neq 0$, the vectors v_l and w_l, when supplemented by a third vector perpendicular to both of them, will provide a natural orthogonal reference frame for the whole of Euclidean 3-space. It is in terms of this third vector that we shall shortly define the concept of "torsion" of a space curve. It is clear that some additional concept is needed, since in general the curvature is by itself not sufficient for characterizing a space curve. Consider for example the curve which winds itself round a cylinder, the familiar circular helix (Figure 6):

$$x = R\cos t, \qquad y = R\sin t, \qquad z = t.$$

In addition to its curvature this curve also undergoes a twisting (or "torsion") about its (continually changing) tangent line.

Before proceeding, we recall for the reader the operation, familiar from the linear algebra of Euclidean 3-space, of the *vector*, or *cross*, *product* of an ordered pair of vectors. If ξ, η are space vectors:

$$\xi = \xi^i e_i, \qquad \eta = \eta^i e_i,$$

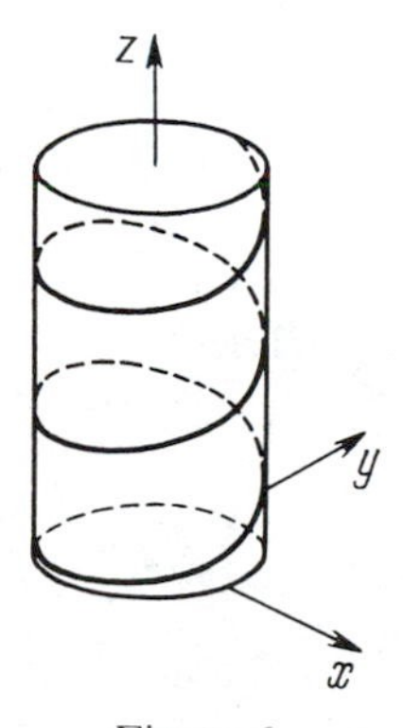

Figure 6

where e_1, e_2, e_3 form an orthonormal basis ($e_i \perp e_j$, $|e_i| = 1$), then we define the vector product $\gamma = [\xi, \eta]$, $\gamma = \gamma^i e_i$, by setting

$$\gamma^1 = \xi^2\eta^3 - \xi^3\eta^2, \gamma^2 = \xi^3\eta^1 - \xi^1\eta^3, \gamma^3 = \xi^1\eta^2 - \xi^2\eta^1.$$

This definition is easiest recalled by expanding the matrix

$$\gamma = \begin{pmatrix} e_1 & e_2 & e_3 \\ \xi^1 & \xi^2 & \xi^3 \\ \eta^1 & \eta^2 & \eta^3 \end{pmatrix}$$

(whose first-row entries are vectors, while those of the other two rows are numbers) about the first row, as for conventional matrices:

$$\gamma = \begin{pmatrix} \xi^2 & \xi^3 \\ \eta^2 & \eta^3 \end{pmatrix} e_1 - \begin{pmatrix} \xi^1 & \xi^3 \\ \eta^1 & \eta^3 \end{pmatrix} e_2 + \begin{pmatrix} \xi^1 & \xi^2 \\ \eta^1 & \eta^2 \end{pmatrix} e_3. \tag{15}$$

It is easy to verify that

$$[\xi, \eta] = -[\eta, \xi], [\xi_1 + \xi_2, \eta] = [\xi_1, \eta] + [\xi_2, \eta], \qquad [\lambda\xi, \eta] = \lambda[\xi, \eta], \tag{16}$$

and also "Jacobi's identity"

$$[[\xi, \eta], \zeta] + [[\zeta, \xi], \eta] + [[\eta, \zeta], \xi] = 0. \tag{17}$$

Recall also the following two facts from analytic geometry: The vector $[\xi, \eta]$ is perpendicular to the plane spanned by the vectors ξ, η, i.e. to all vectors of the form $\lambda\xi + \mu\eta$; and, secondly, its length is given by

$$|[\xi, \eta]| = |\xi||\eta| \sin \varphi, \tag{18}$$

where φ is the angle between ξ and η if these are non-zero; this angle can of course be calculated from

$$\cos \varphi = \frac{\langle \xi, \eta \rangle}{|\xi||\eta|}, \qquad 0 \leq \varphi \leq \pi. \tag{19}$$

Thus, in particular, if ξ, η lie in the (x, y)-plane, then their vector product is perpendicular to that plane, i.e. is directed along the positive or negative z-axis, and then $[\xi, \eta] = (\xi^1\eta^2 - \xi^2\eta^1)e_3$. Thus in this case we have from (18) that $|[\xi, \eta]| = |\xi^1\eta^2 - \eta^1\xi^2| = |\xi||\eta| \sin \varphi$, so that in terms of the vector product, we can rewrite the formula (13) for the curvature of the plane curve $x = x(t)$, $y = y(t)$, where t is any parameter, as

$$k = \frac{|\ddot{x}\dot{y} - \ddot{y}\dot{x}|}{(\dot{x}^2 + \dot{y}^2)^{3/2}} = \frac{|[\dot{r}, \ddot{r}]|}{|\dot{r}|^3}. \tag{20}$$

Thus the general formula (13) for the curvature of a plane curve relates the curvature k to the length of the vector product $[\dot{r}, \ddot{r}]$.

As we saw in the preceding subsection (§5.1) the curvature of a curve $x = x(l)$, $y = y(l)$ lying in the (x, y)-plane is, by virtue of the Serret–Frenet formulae, naturally related to the angular velocity of the frame (v_l, n_l) in a direction (i.e. with axis of rotation) perpendicular to the (x, y)-plane (see (7)). With this in mind we return to our space curve $r = r(l)$, $r = (x, y, z)$, $x = x(l)$, $y = y(l)$, $z = z(l)$. Since l is natural we have $|v_l| = 1$ and $v_l \perp w_l$. We shall suppose also that $w_l \neq 0$ (and call points where $w_l = 0$ *singular points* of the curve). Write briefly $w = w_l$, $v = v_l$, and, as before, write $n = w/|w|$. We now introduce the promised third vector b orthogonal to both v_l and n: define $b = [v, n]$. We call n the *principal normal* to the curve, and b the *binormal.* Note that $|b| = |v|\,|n| \sin \varphi = 1$, and, as already remarked, $b \perp v$, $b \perp n$. We thus have an orthonormal basis $(v(l), n(l), b(l))$ at each point of the curve where $w_l \neq 0$, i.e. at each non-singular point of the curve.

We shall find useful the following simple lemma, the analogue for vector products of the formula (1). It follows almost immediately from the usual formula of Leibniz for differentiating a product of two real-valued functions: $(fg)' = f'g + fg'$.

5.2.2. Lemma. *For any two vectors $\xi(t), \eta(t)$ in Euclidean 3-space, the following analogue of the Leibniz product rule holds:*

$$\frac{d}{dt}[\xi, \eta] = \left[\frac{d\xi}{dt}, \eta\right] + \left[\xi, \frac{d\eta}{dt}\right]. \tag{21}$$

(Since the vector product is a non-commutative operation, attention must be paid to the order of the factors in the three vector products in (21).)

5.2.3. Theorem (Serret–Frenet Formulae for Space Curves). *For any curve $r = r(l)$ in Euclidean 3-space, where l is the natural parameter, the following formulae hold:*

$$\begin{aligned} dv/dl &= kn, \\ dn/dl &= -\varkappa b - kv, \\ db/dl &= \varkappa n. \end{aligned} \tag{22}$$

Before giving the proof we make a few remarks. To begin with, note that the first of these formulae is just the definition of k. Secondly, the number $\varkappa(l)$ (which may be positive, negative or 0) is called the *torsion* of the space curve (at the value l of the natural parameter). From the third of the equations (22) we see that its magnitude $|\varkappa| = |db/dl|$. If the curve is planar (lies in some plane), then b is a constant vector, so that $db/dl = 0$, and therefore $\varkappa = 0$.

Proof of Theorem 5.2.3. We begin by proving the third formula $db/dl = \varkappa n$. Since $b = [v, n]$, we have by Lemma 5.2.2 that

$$\dot{b} = [\dot{v}, n] + [v, \dot{n}]. \tag{23}$$

Since $|n| = 1$, Lemma 5.1.1 (which holds in Euclidean space of any dimension) tells us that $n \perp \dot{n}$. Hence $\dot{n} = \alpha v + \beta b$ for some numbers α, β. Hence

$$[v, \dot{n}] = [v, \alpha v + \beta b] = \alpha[v, v] + \beta[v, b] = \beta[v, b] = -\beta n.$$

(To see that $[v, b] = -n$, i.e. $[[v, n], v] = n$, show that $[[e_1, e_2], e_1] = e_2$.) What is $[\dot{v}, n]$? From the first of the formulae, namely $\dot{v} = kn$ (which needs no proof since it is just the definition of k), we deduce that $[\dot{v}, n] = k[n, n] = 0$. Thus substituting in (23) we get

$$\frac{db}{dl} = -\beta n = \varkappa n,$$

i.e. $\varkappa$ is defined as $-\beta$.

To establish the second formula in (22) we calculate $\dot{n}$. Since $n = [b, v]$, it follows that

$$\begin{aligned}\dot{n} = \frac{d}{dl}[b, v] &= [\dot{b}, v] + [b, \dot{v}] = [\varkappa n, v] + [b, kn] \\ &= \varkappa[n, v] + k[b, n] = -\varkappa b - kv,\end{aligned}$$

and the theorem is proved. □

The operation of differentiation with respect to l, of the vectors v, n, b, described by the Serret–Frenet formulae (22), can be expressed in matrix form relative to the basis v, n, b. If we take e_1, e_2, e_3 to be v, n, b respectively, then we have

$$\frac{de_i}{dl} = b_i^j e_j, \qquad i, j = 1, 2, 3,$$

where the matrix $B = (b_i^j)$ is given by

$$B = \begin{pmatrix} 0 & k & 0 \\ -k & 0 & -\varkappa \\ 0 & \varkappa & 0 \end{pmatrix}. \tag{24}$$

Note that this matrix is skewsymmetric, as is its analogue in the planar case. (The significance of this formulation of the Serret–Frenet formulae will be revealed in the next subsection (§5.3).) We conclude by bringing to the reader's attention the fact that the curvature and torsion of a space curve form a complete set of invariants for the curve. More precisely, we have:

(i) A curve in the Euclidean plane is determined up to an isometry of the plane by the equation $k = k(l)$ expressing curvature as a function of length. This equation is called the *natural* or *intrinsic equation of the planar curve.*

(ii) A curve in Euclidean 3-space is determined up to an isometry of 3-space by the equations $k = k(l)$, $\varkappa = \varkappa(l)$, the *natural* or *intrinsic equations of the space curve.*

Proofs of these theorems may be found in the textbook of P. K. Raševskiĭ [19]. (Alternatively, see [28], p. 29 *et seqq.*)

5.3. Orthogonal Transformations Depending on a Parameter

Let $A = (a_{ij})$, $i, j = 1, \ldots, n$, be an orthogonal matrix, i.e. $A^T A = 1$, or, equivalently,

$$\sum_{i=1}^{n} a_{ij} a_{ik} = \delta_{jk}, \tag{25}$$

where in addition the a_{ij} are all functions of a parameter t (so that what we really have is a family of orthogonal matrices, one for each value of t). (We saw the importance of orthogonal matrices in §4—they represent the isometries of Euclidean spaces fixing the origin.) We shall further assume that $a_{ij}(0) = \delta_{ij}$, i.e. $A(0) = 1$.

5.3.1. Lemma. *If $A(0) = 1$, then the matrix $B = dA/dt|_{t=0}$ is skewsymmetric.*

PROOF. On differentiating both sides of equation (25) with respect to t, we obtain

$$0 = \frac{d}{dt}\delta_{jk} = \sum_{i=1}^{n} (\dot{a}_{ij} a_{ik} + a_{ij} \dot{a}_{ik}).$$

Putting $t = 0$ and using the hypothesis $a_{ij}(0) = \delta_{ij}$, we get

$$0 = \sum_{i=1}^{n} (\dot{a}_{ij} \delta_{ik} + \delta_{ij} \dot{a}_{ik}) = b_{kj} + b_{jk},$$

completing the proof. □

Using this lemma and the result of §4 that an isometry of Euclidean space fixing the origin is an orthogonal linear transformation, we can give a quick proof of the Serret–Frenet formulae (Theorem 5.2.3). Let $r = r(l)$ be a curve in Euclidean 3-space, where l is the natural parameter, and take $e_1(l) = v(l)$, $e_2(l) = n(l)$, $e_3(l) = b(l)$, where v, n, b are as in the preceding section. If we think of the frames $e_1(l)$, $e_2(l)$, $e_3(l)$ and $e_1(l + \Delta l)$, $e_2(l + \Delta l)$, $e_3(l + \Delta l)$ as being attached to the origin, then since they are both orthonormal there is an orthogonal transformation taking the first into the second. Hence there is a matrix (a_{ij}) (whose entries depend on both l and Δl) such that

$$e_i(l + \Delta l) = \sum_{j=1}^{3} a_{ij}(l, \Delta l) e_j(l), \qquad \sum_{i=1}^{3} a_{ij}(l, \Delta l) a_{ik}(l, \Delta l) = \delta_{jk}. \tag{26}$$

On differentiating the first of these equations with respect to Δl, and then setting $\Delta l = 0$, we obtain

$$\dot{e}_i(l) = \sum_{j=1}^{3} b_{ij} e_j(l),$$

where by Lemma 5.3.1 (with Δl playing the role of t), the matrix $B = (b_{ij})$ is skewsymmetric, i.e. B has the form

$$B = \begin{pmatrix} 0 & b_{12} & b_{13} \\ -b_{12} & 0 & b_{23} \\ -b_{13} & -b_{23} & 0 \end{pmatrix}.$$

By the definition of the curvature k, and since $e_1 = v(l)$, $e_2 = n(l)$, we have that

$$\dot{e}_1(l) = \frac{de_1}{dl} = \frac{dv}{dl} = kn = ke_2.$$

Hence $b_{12} = k$ and $b_{13} = 0$, so that

$$B = \begin{pmatrix} 0 & k & 0 \\ -k & 0 & -b_{32} \\ 0 & b_{32} & 0 \end{pmatrix}.$$

Writing $\varkappa$ for b_{32}, we obtain the Serret–Frenet formulae (22) (in the matrix form (24)).

The Serret–Frenet formulae for the planar case can be proven similarly.

5.4. Exercises

1. Find the "Serret–Frenet basis" v, n, b, and the curvature and torsion of the circular helix $r = (a \cos t, a \sin t, ct)$, $a > 0$, $c \neq 0$.

2. Find the curvature and torsion of the curves:
   (i) $r = e^t(\sin t, \cos t, 1)$;
   (ii) $r = a(\cosh t, \sinh t, t)$.

3. Find the curvature of the ellipse $x^2/a^2 + y^2/b^2 = 1$, at the points $(a, 0)$, $(0, b)$.

4. Find the curvature and torsion of the curves:
   (i) $r = (t^2 \sqrt{\frac{3}{2}}, 2\text{-}t, t^3)$;
   (ii) $r = (3t\text{-}t^3, 3t^2, 3t + t^3)$.

5. Prove that a curve with identically zero curvature is a straight line.

6. Prove that a curve with identically zero torsion lies in some plane. Find the equation of the plane.

7. Describe the class of curves with constant curvature and torsion: $k(l) \equiv \text{const}$, $\varkappa(l) \equiv \text{const}$.

8. Describe the class of curves with constant torsion: $\varkappa(l) \equiv \text{const}$.

9. Prove that a curve $r = r(t)$ is planar if and only if $(\dot{r}, \ddot{r}, \dddot{r}) = 0$. (Here (ξ, η, ζ) denotes the "scalar triple product" $\langle \xi, [\eta, \zeta] \rangle$, which can be shown to be zero if and only if the vectors ξ, η, ζ are coplanar.)

10. Let S be the area (if finite) of the region bounded by a plane curve and a straight line parallel to the tangent line to the curve at a point on the curve, and at a distance h from it. Express $\lim_{h\to 0}(S^2/h^3)$ in terms of the curvature of the curve.

11. Show that for a smooth closed curve $C: r = r(l)$

$$\int_C (r\dot{k} - \varkappa b)\, dl = 0.$$

12. Show that the Serret–Frenet formulae can be put into the form

$$\dot{v} = [\zeta, v], \qquad \dot{n} = [\zeta, n], \qquad \dot{b} = [\zeta, b],$$

where ζ is a certain vector (called "Darboux' vector"). Find this vector.

13. Solve the equation $dr/dt = [\omega, r]$ where $r = r(t)$ and ω is a constant vector.

14. Prove that the curvature and torsion of a curve $r = r(l)$ are proportional (i.e. $k = c\varkappa$ for some constant c) if and only if there is a constant vector u such that $\langle u, v\rangle = \text{const}$.

15. Suppose that $r = r(l)$ is a curve with the property that every plane orthogonal to the curve (i.e. spanned by $n(l)$ and $b(l)$) passes through a fixed point x_0. Show that the curve lies on (the surface of) a sphere with centre at x_0.

16. Prove that a curve $r = r(l)$ lies on a sphere of radius R if and only if $\varkappa$ and k satisfy

$$R^2 = \frac{1}{k^2}\left(1 + \frac{(dk/dl)^2}{(\varkappa k)^2}\right).$$

17. Show that

$$\varkappa = -\frac{(\dot{r}, \ddot{r}, \dddot{r})}{|[\dot{r}, \ddot{r}]|}. \qquad \text{(Cf. Exercise 9.)}$$

18. With any smooth curve $r = r(l)$ we can associate the curve $r = n(l)$ (where, as usual, $n(l)$ is the principal normal to the curve at the point on it corresponding to the value l of the parameter.) If l^* denotes the natural parameter for the latter curve, show that

$$\frac{dl^*}{dl} = \sqrt{k^2 + \varkappa^2}.$$

19. Let $r = r(l)$ be a space curve. Write

$$A = A(l) = \begin{pmatrix} 0 & k(l) & 0 \\ -k(l) & 0 & \varkappa(l) \\ 0 & -\varkappa(l) & 0 \end{pmatrix} = (a^i_j(l)).$$

Let the vectors $r_j = r_j(l)$ be the unique solutions of the system of equations

$$\frac{dr_j}{dl} = a^i_j r_i, \qquad j = 1, 2, 3,$$

satisfying the initial condition that $r_1(0), r_2(0), r_3(0)$ coincide with a fixed orthonormal basis.

(i) Show that the basis $r_1(l), r_2(l), r_3(l)$ is orthonormal for all l.

(ii) Define $r^*(l) = r_0 + \int_0^l r_1(l)\,dl$. Show that then $r_1(l) = v^*(l)$, $r_2(l) = n^*(l)$, $r_3(l) = b^*(l)$, where v^*, n^*, b^* are the tangent, normal and binormal to the curve $r = r^*(l)$, and show that the curvature and torsion of this curve are the same as those of the original, namely $k(l)$, $\varkappa(l)$.

20. Show that if a curve lies on a sphere and has constant curvature then it is a circle.

§6. Pseudo-Euclidean Spaces

6.1. The Simplest Concepts of the Special Theory of Relativity

Recall that the pseudo-Euclidean space $\mathbb{R}^n_{p,q}$, $p + q = n$, is by definition (3.2.1) the space equipped with co-ordinates $x^1, \dots, x^n$ in terms of which the square of the norm of a vector $\xi = (\xi^1, \dots, \xi^n)$ is given by the formula

$$|\xi|^2 = \langle \xi, \xi \rangle = \sum_{i=1}^{p} (\xi^i)^2 - \sum_{i=1}^{q} (\xi^{p+i})^2. \tag{1}$$

As already noted (in §3.2) for $n = 4$, $p = 1$, this space is termed *space–time of the special theory of relativity*, or *Minkowski space*. We denote this space by $\mathbb{R}^4_1$ rather than $\mathbb{R}^4_{1,3}$. We shall extend this term to cover also the spaces $\mathbb{R}^n_{1,n-1} = \mathbb{R}^n_1$, i.e. we shall call these spaces also Minkowski spaces (one for each dimension n).

By (1) the square of the length of a vector $\xi = (\xi^0, \xi^1, \xi^2, \xi^3)$ in $\mathbb{R}^4_1$ is given by

$$|\xi|^2 = \langle \xi, \xi \rangle = (\xi^0)^2 - (\xi^1)^2 - (\xi^2)^2 - (\xi^3)^2, \tag{2}$$

which quantity, as already noted in §3.2, may be positive, negative or zero. Those vectors ξ for which $|\xi| = 0$, form in $\mathbb{R}^4_1$ a cone called the *isotropic* or *light cone* (in Figure 7 the analogous cone in $\mathbb{R}^3_1$ is shown). Vectors inside this cone are just those whose squared length is positive, $|\xi|^2 > 0$; they are called *time-like vectors*. Those outside the cone are the ones with negative squared length, $|\xi|^2 < 0$, and are said to be *space-like*. In Figure 7 the time-like vectors are denoted by ξ_+, and the space-like vectors by ξ_-; vectors which, like ξ_0, lie on the light cone, have zero length and are called *isotropic* or *light vectors*.

We now consider the world-line of an arbitrary material point-particle (see §1.1). Such a particle will have a world-line in $\mathbb{R}^4_1$ of the form

$$x_0 = ct, \qquad x^1 = x^1(t), \qquad x^2 = x^2(t), \qquad x^3 = x^3(t). \tag{3}$$

Here the curve $x^1 = x^1(t)$, $x^2 = x^2(t)$, $x^3 = x^3(t)$ is just the usual trajectory of the point-particle in Euclidean 3-space $\mathbb{R}^3$. The tangent vector to the world-line (3) is given by

$$\xi = (c, \dot{x}^1, \dot{x}^2, \dot{x}^3). \tag{4}$$

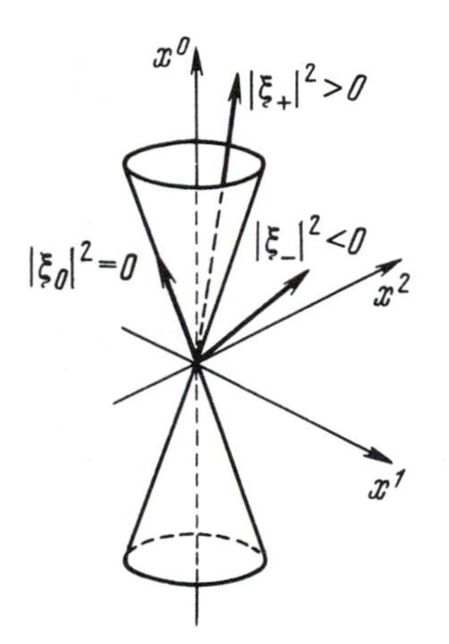

Figure 7. The isotropic cone $(x^0)^2 - (x^1)^2 - (x^2)^2 = 0$, in the space $\mathbb{R}^3_1$.

(Note again that $\dot{x}^1$, $\dot{x}^2$, $\dot{x}^3$ are just the components of the velocity of the particle through ordinary space.) It is one of the basic postulates or assumptions of the special theory of relativity that it is impossible for a material object to have speed greater than the speed of light, i.e. $|v| \le c$. Hence

$$c^2 - (\dot{x}^1)^2 - (\dot{x}^2)^2 - (\dot{x}^3)^2 \ge 0, \tag{5}$$

or, in other words, the vector ξ has either to be time-like or isotropic. For the world-line of a photon of light the vector ξ is isotropic since $|v| = c$. This is the reason for using the name "light" cone synonymously with "isotropic" cone. It turns out that the only particles which can have world-lines with isotropic tangent vectors, are those with zero mass, like photons, neutrinos, and others among the elementary particles. The world-lines of particles with mass always have time-like tangent vectors. It follows that the world-line of a particle with mass must lie wholly inside the light cone (see Figure 8;

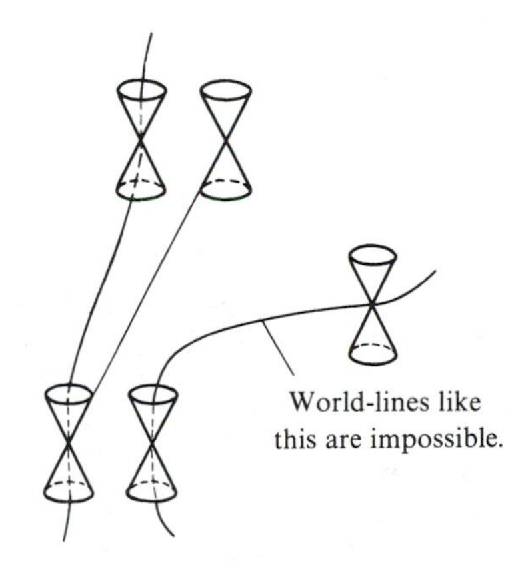

Figure 8. World-lines of particles with and without mass.

each point of the space is to be thought of as being the apex of a light cone). For time-like curves (i.e. for those curves whose tangent vector is always time-like) the concept of length can be defined analogously to that in Euclidean geometry. Thus if a curve in $\mathbb{R}_1^4$ is given by the equations

$$x^0 = x^0(\tau), \qquad x^1 = x^1(\tau), \qquad x^2 = x^2(\tau), \qquad x^3 = x^3(\tau),$$

where the condition corresponding to being time-like is satisfied, i.e.

$$|\xi|^2 = (\dot{x}^0)^2 - (\dot{x}^1)^2 - (\dot{x}^2)^2 - (\dot{x}^3)^2 > 0,$$

then the length l of the curve between $\tau = a$ and $\tau = b$ is given by

$$l = \int_a^b |\xi|\, d\tau = \int_a^b \sqrt{(\dot{x}^0)^2 - \sum_{\alpha=1}^{3} (\dot{x}^\alpha)^2}\, d\tau. \tag{6}$$

In the special theory of relativity the quantity l/c, with l defined as in (6), is called the *proper time* elapsed for the particle. As in the Euclidean case we have $|v(l)| = 1$, and call the parameter l the *natural parameter* for the time-like world-line.

If a point-particle moves through Euclidean 3-space with constant velocity (v^1, v^2, v^3), i.e.

$$x^0 = ct, \qquad x^1 = v^1 t, \qquad x^2 = v^2 t, \qquad x^3 = v^3 t, \tag{7}$$

then, writing briefly v for $|v|$, we have from (6) that

$$dl = \sqrt{c^2 - v^2}\, dt = \sqrt{1 - \frac{v^2}{c^2}}\, dx^0,$$

whence

$$l = x^0 \sqrt{1 - \frac{v^2}{c^2}}. \tag{8}$$

Thus in particular if $v = 0$, i.e. if the particle does not move relative to the frame x^1, x^2, x^3, then its proper time is $l/c = x^0/c = t$.

6.2. Lorentz Transformations

In §4.4(b) we noted that in Newtonian (i.e. classical) mechanics, time is assumed to have an absolute character, so that in particular the time interval between two events is the same when measured with respect to any inertial reference frame; in other words, for any two observers (equipped with clocks) moving uniformly and in a straight line relative to a fixed frame, the time interval between the events will turn out to be the same. We used the term "Galilean transformation" for transformations associated with changes from one inertial frame of reference to another which, firstly, have the property just described of preserving time intervals, and, secondly, preserve the Euclidean metric $dl^2 = (dx^1)^2 + (dx^2)^2 + (dx^3)^2$ (see equations (46), or the more general (47), in §4).

In the special theory of relativity, which is our present interest, the "Lorentz transformations" assume the role which in classical mechanics is played by the Galilean transformations. Thus a *Lorentz transformation* tells us how to change from one reference frame ct, x^1, x^2, x^3 for the Minkowski space $\mathbb{R}_1^4$ to another ct', x'^1, x'^2, x'^3, where the system (x') moves relative to (x) with constant velocity. The form of the Lorentz transformation is determined by the fact that it is required to preserve the Minkowski metric, i.e. to preserve the quadratic form $dl^2 = c^2(dt)^2 - (dx^1)^2 - (dx^2)^2 - (dx^3)^2$, and also (but this is not so important) to fix the origin (i.e. when $t = x^1 = x^2 = x^3 = 0$, then also $t' = x'^1 = x'^2 = x'^3 = 0$).

We wish to know what the Lorentz transformations look like. To this end we first investigate the isometry group of $\mathbb{R}_1^4$ (the *Poincaré group*). We begin by considering the space $\mathbb{R}_1^2$; thus initially we want to discover the general form of an isometry of $\mathbb{R}_1^2$ which fixes the origin of a given pseudo-Euclidean co-ordinate system x^0, x^1 (i.e. a co-ordinate system with respect to which the metric $g_{\alpha\beta}$ is given at all points by $g_{00} = 1, g_{11} = -1, g_{12} = g_{21} = 0$). Once again we leave it as an exercise to show that such isometries are necessarily linear transformations, i.e. if $x^0 \to x'^0$, $x^1 \to x'^1$ is an origin-fixing transformation of $\mathbb{R}_1^2$ for which, always, $(x^0)^2 - (x^1)^2 = (x'^0)^2 - (x'^1)^2$, then it is linear. This assumed, we can write our isometry as

$$\begin{aligned} x^0 &= ax'^0 + bx'^1, \\ x^1 &= cx'^0 + dx'^1. \end{aligned} \tag{9}$$

Since the transformation (9) is supposed to be an isometry, we have as before (§4.1)

$$(g_{\alpha\beta}) = G = A^T G A, \tag{10}$$

where $A = \begin{pmatrix} a & b \\ c & d \end{pmatrix}$. Since $\det A^T = \det A$, and the determinant of a product is the product of the determinants of the factors, we deduce from (10) that $(\det A)^2 = 1$, whence $\det A = \pm 1$. Since we know the $g_{\alpha\beta}$, we can rewrite (10) as a system of three equations:

$$a^2 - c^2 = 1, \qquad ab - cd = 0, \qquad b^2 - d^2 = -1. \tag{11}$$

It is clear from these that $a \neq 0$. If we set $\beta = c/a$, then direct solution of (11) yields

$$a = \pm \frac{1}{\sqrt{1 - \beta^2}}, \qquad c = a\beta, \qquad d = \pm \frac{1}{\sqrt{1 - \beta^2}}, \qquad b = d\beta. \tag{12}$$

Hence the matrices A corresponding to Lorentz transformations of $\mathbb{R}_1^2$ are just those of the form

$$A = \pm\begin{pmatrix} \dfrac{1}{\sqrt{1-\beta^2}}, & \pm\dfrac{\beta}{\sqrt{1-\beta^2}} \\ \dfrac{\beta}{\sqrt{1-\beta^2}}, & \pm\dfrac{1}{\sqrt{1-\beta^2}} \end{pmatrix}. \tag{13}$$

In terms of a new parameter ψ defined by $\beta = \tanh \psi$, (13) takes the form

$$A = \pm\begin{pmatrix} \cosh\psi & \pm\sinh\psi \\ \sinh\psi & \pm\cosh\psi \end{pmatrix}, \tag{14}$$

whence it is clear that the group of isometries of $\mathbb{R}_1^2$ fixing the origin, is just the group of hyperbolic rotations through angles ψ. Every hyperbolic rotation fixes (i.e. maps onto itself) the set of points satisfying $|\xi|^2 = \text{const.}$, i.e. $(x^0)^2 - (x^1)^2 = \text{const.}$ In particular therefore the hyperbolic rotations fix the isotropic cone $|\xi|^2 = 0$, and the "unit pseudo-sphere" $(x^0)^2 - (x^1)^2 = 1$, which is, of course, a hyperbola (see Figure 9).

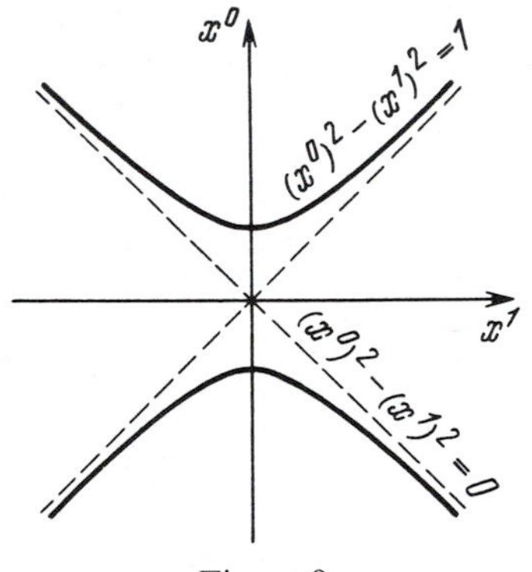

Figure 9

Recall that the group of orthogonal transformations of Euclidean space is not connected, but has two connected components, one consisting of the proper (or direct) transformations, and the other of the improper (or opposite) transformations (see §4.4(a)). The group of isometries of the pseudo-Euclidean plane $\mathbb{R}_1^2$ fixing the origin, is more complicated in the sense that it falls into four maximal connected pieces, consisting respectively of the matrices of the following four types:

$$\begin{pmatrix} \cosh\psi & \sinh\psi \\ \sinh\psi & \cosh\psi \end{pmatrix}, \quad \begin{pmatrix} \cosh\psi & -\sinh\psi \\ \sinh\psi & -\cosh\psi \end{pmatrix},$$
$$\begin{pmatrix} -\cosh\psi & \sinh\psi \\ -\sinh\psi & \cosh\psi \end{pmatrix}, \quad \begin{pmatrix} -\cosh\psi & -\sinh\psi \\ -\sinh\psi & -\cosh\psi \end{pmatrix}. \tag{15}$$

Thus in particular each of these matrices can be connected by a (continuous) arc to that one of the four matrices

$$1 = \begin{pmatrix} 1 & 0 \\ 0 & 1 \end{pmatrix}, \quad P = \begin{pmatrix} 1 & 0 \\ 0 & -1 \end{pmatrix},$$

$$T = \begin{pmatrix} -1 & 0 \\ 0 & 1 \end{pmatrix}, \quad PT = \begin{pmatrix} -1 & 0 \\ 0 & -1 \end{pmatrix},$$

in its component (but there is no arc connecting any two of the matrices $1, P, T, PT$).

The transformations in the first two connected components represented in (15), in contrast with the last two, do not change the direction of flow of time t (interpreting x^0 as ct); for this reason these transformations are said to be *orthochronous*. Thus the connected component containing the identity transformation consists of just those isometries (fixing the origin) which are both orthochronous and proper (i.e. whose matrices have determinant $+1$).

We shall denote the group we have just been considering, i.e. the group of isometries of $\mathbb{R}_1^2$ which fix the origin, by $O(1, 1)$. More generally we shall use the notation $O(p, q)$, $p + q = n$, for the group of all pseudo-orthogonal transformations of the space $\mathbb{R}_{p,q}^n$; thus $O(p, q)$ is the group of linear transformations A (or matrices relative to an appropriate basis) preserving the pseudo-Euclidean scalar product:

$$\langle A\xi, A\eta \rangle = \langle \xi, \eta \rangle = \xi^1\eta^1 + \cdots + \xi^p\eta^p - \cdots - \xi^n\eta^n.$$

Of particular importance are the groups $O(1, n-1)$ of isometries of n-dimensional Minkowski space, fixing the origin. It turns out that the above-mentioned decomposition of $O(1, 1)$ as the union of exactly four pieces (i.e. maximal connected components) is typical: all of the groups $O(1, n-1)$ have this property. We shall not prove this here, but rather limit ourselves to the following weaker result.

6.2.1. Lemma. *There is an epimorphism φ from the group $O(1, n-1)$ onto the group $C_2 \times C_2$, the direct product of the cyclic group $\{+1, -1\}$ (under multiplication), with itself.*

PROOF. Let e_0 be the unit vector in the direction of the positive x^0-axis, and for each A in $O(1, n-1)$ set

$$\varphi(A) = (\det A, \operatorname{sgn}\langle e_0, Ae_0 \rangle). \tag{16}$$

For this definition of φ to make sense, we need to show that $\langle e_0, Ae_0 \rangle \neq 0$. Since $\langle Ae_0, Ae_0 \rangle = \langle e_0, e_0 \rangle = 1$, the x^0-component of Ae_0 (call it α) cannot be zero. Hence $\langle e_0, Ae_0 \rangle = \alpha \neq 0$, as required. That the map φ is onto and a homomorphism (i.e. $\varphi(AB) = \varphi(A)\varphi(B)$)† is easily verified; we leave it to the reader. □

† It follows from this defining property that if φ is a homomorphism from a group G_1 to a group G_2, then $\varphi(1) = 1$, and $\varphi(g^{-1}) = \varphi(g)^{-1}$ for all g in G_1. A homomorphism $\varphi: G_1 \to G_2$ is often called a "representation" of the group G_1 in the group G_2.

If, by analogy with the planar case, we define a transformation A in $O(1, n-1)$ to be *proper* if $\varphi(A) = (1, 1)$, where φ is as in (16), then we can say that the kernel of φ consists of just the proper isometries. It turns out that this kernel is precisely the maximal connected component of $O(1, n-1)$ containing the identity transformation, but we shall not prove this here. What can be inferred from the lemma is that the group $O(1, n-1)$ has at least four connected components.

As in the planar case the transformations A in $O(1, n-1)$ for which $\operatorname{sgn} \langle e_0, Ae_0 \rangle = 1$, are called *orthochronous* (since they do not change the direction of flow of time $t = x^0/c$).

We obtain the full group of isometries of the pseudo-Euclidean space $\mathbb{R}_1^n$ by combining the isometries in $O(1, n-1)$ with the translations.

As promised, we now apply our knowledge of the isometries of the pseudo-Euclidean spaces (or more particularly of $O(1, 1)$) to the special theory of relativity. Our particular aim is to find the form of a Lorentz transformation, i.e. of a transformation arising from a change from a reference frame ct, x^1, x^2, x^3 to another ct', x'^1, x'^2, x'^3 (where the system (x') moves with constant velocity relative to the system (x)), which fixes the origin, and (more significantly) preserves the quadratic form $(ct)^2 - (x^1)^2 - (x^2)^2 - (x^3)^2$. For simplicity (so as not to obscure the essential nature of the general Lorentz transformation) we suppose that the primed system x'^1, x'^2, x'^3, moves in the positive x^1-direction with (constant) speed v. We then have $x^2 = x'^2$, $x^3 = x'^3$; hence the transformation has the form

$$\begin{gathered} x^0 = ct = \alpha(ct') + \beta(x'^1), \\ x^1 = \gamma(ct') + \delta x'^1, \qquad x^2 = x'^2, \qquad x^3 = x'^3, \end{gathered} \tag{17}$$

where $A = \begin{pmatrix} \alpha & \beta \\ \gamma & \delta \end{pmatrix}$ belongs to $O(1, 1)$.

If the speed v is reduced to zero, then the transformation (17) should become the identity, so that the matrix A must lie in the connected component of the identity of $O(1, 1)$, i.e. must have the form

$$A = \begin{pmatrix} \cosh \psi & \sinh \psi \\ \sinh \psi & \cosh \psi \end{pmatrix}.$$

Hence

$$\begin{gathered} ct = ct' \cosh \psi + x'^1 \sinh \psi, \\ x^1 = ct' \sinh \psi + x'^1 \cosh \psi. \end{gathered} \tag{18}$$

We wish to investigate the motion of the origin of the (x') system relative to the system ct, x^1, x^2, x^3. Thus let O' denote the origin of the frame x'^1, x'^2, x'^3

at various times t'. (Note that O' is not a point in $\mathbb{R}^4_1$, but the whole ct'-axis.) Then for O', since $x'^1 = 0$, the equations (18) simplify to

$$\begin{aligned} ct &= ct' \cosh \psi, \\ x^1 &= ct' \sinh \psi, \end{aligned} \tag{19}$$

i.e. the point with co-ordinates $(ct', 0, 0, 0)$ in the primed system has co-ordinates $(ct' \cosh \psi, ct' \sinh \psi, 0, 0)$ in the unprimed system. On dividing the second equation in (19) by the first, we get $x^1/ct = \tanh \psi$. But x^1/t is just v, the speed of O' along the x^1-axis, so that

$$\tanh \psi = v/c,$$

whence

$$\sinh \psi = \frac{v/c}{\sqrt{1 - v^2/c^2}}, \qquad \cosh \psi = \frac{1}{\sqrt{1 - v^2/c^2}}.$$

Substituting in (18) we get finally

$$\begin{aligned} t &= \left(t' + \left(\frac{v}{c^2}\right) x'^1\right) \frac{1}{\sqrt{1 - v^2/c^2}}, \\ x^1 &= (x'^1 + vt') \frac{1}{\sqrt{1 - v^2/c^2}}. \end{aligned} \tag{20}$$

Transformations of the form (20) (supplemented by $x^2 = x'^2$, $x^3 = x'^3$) are what are generally known as Lorentz transformations.

Let us now consider the physical consequences of the form (20) taken by the Lorentz transformations. Suppose that the speed v is small in comparison with the speed of light c, i.e. $v/c \ll 1$. From (20) it follows that as $v/c \to 0$, the Lorentz transformation goes into the Galilean transformation (46) of §4: $t = t'$, $x^1 = x'^1 + vt$, $x^2 = x'^2$, $x^3 = x'^3$. In other words, for frames moving with small velocities relative to one another, the relativistic formulae are approximately classical. However for relative speeds comparable with the speed of light, the two theories will have radically different consequences. We shall now illustrate these large differences in the two theories by contrasting (in the manner of most popularizations of Einstein's special theory of relativity) some of the predictions of the relativistic theory with the expectations of common sense (which is rooted in the classical theory). The effects in question are the shortening of the dimensions of uniformly moving objects in the direction of their motion, the retardation of moving clocks† (and the related effect that events which are simultaneous relative to one reference frame, can occur at different times as observed from other frames).

† The reader curious about these "popular" effects might wish to ponder their exact physical meaning, and the extent to which they are observable.

Denote the two frames by $K(ct, x^1, x^2, x^3)$ and $K'(ct', x'^1, x'^2, x'^3)$. Suppose that we have a rod fixed relative to the frame K and lying along the x^1-axis, and which has ordinary (i.e. Euclidean) length l (as measured by an observer also fixed relative to K). Thus if x_1^1 and x_2^1 are the x^1-co-ordinates of the ends of the rod (at any time t), then $l = x_2^1 - x_1^1$. It seems reasonable to take as the "ordinary" length l' of the rod in K' (i.e. its length as it appears to an observer fixed relative to K') to be the difference between the x'^1-co-ordinates, $x_1'^1$ and $x_2'^1$ say, of the ends of the rod at the same time t' (as measured in K'). By (20), at time t' we have

$$x_1^1 = \frac{x_1'^1 + vt'}{\sqrt{1 - v^2/c^2}}, \qquad x_2^1 = \frac{x_2'^1 + vt'}{\sqrt{1 - v^2/c^2}}, \tag{21}$$

where we are using the fact that x_1^1 and x_2^1 are the x^1-co-ordinates of the ends of the rod at all times t, i.e. that the rod is fixed relative to K. Since $l' = x_2'^1 - x_1'^1$, we get from (21) that

$$l = \frac{l'}{\sqrt{1 - v^2/c^2}}. \tag{22}$$

(Note that l' is not an invariant 4-dimensional pseudo-Euclidean length, except in the sense that it is the distance between the projections of the two events $(ct', x_1'^1, 0, 0)$, $(ct', x_2'^1, 0, 0)$ onto the hyperplane (i.e. 3-dimensional subspace) co-ordinatized by x'^1, x'^2, x'^3.) From (22) we conclude that the rod has greatest length relative to a frame of reference in which it does not move. To an observer in a frame relative to which the rod moves with speed v (uniformly and in a straight line), its length will appear to be reduced by the factor $\sqrt{1 - v^2/c^2}$ (the "Lorentz contraction").

We next deduce from (20) that a clock moving relative to an observer will appear to go more slowly than a fixed one. Suppose we have a clock fixed at the origin O' of the co-ordinate system x'^1, x'^2, x'^3. Let t_1' and t_2' be two readings taken from this clock in K'; thus relative to an observer fixed in K', the time interval elapsed between the two readings will be $\Delta t' = t_2' - t_1'$. Considering the two readings as events (with K'-co-ordinates $(t_1', 0, 0, 0)$ and $(t_2', 0, 0, 0)$, we can calculate from (20) their co-ordinates relative to K; in particular their time co-ordinates relative to K are, since $x'^1 = 0$ for both events,

$$t_1 = \frac{t_1'}{\sqrt{1 - v^2/c^2}}, \quad t_2 = \frac{t_2'}{\sqrt{1 - v^2/c^2}},$$

whence

$$t_2 - t_1 = \Delta t = \frac{\Delta t'}{\sqrt{1 - v^2/c^2}}. \tag{23}$$

Thus $\Delta t > \Delta t'$, i.e. to an observer in the frame K the clock moving with the frame K' appears to be retarded.

Thirdly we deduce from (20) that it is possible for two events to be simultaneous in the reference frame K', yet not simultaneous relative to K. Thus let A_1 and A_2 be two events which are simultaneous relative to K' (both occurring at time t'), and have different x'^1-co-ordinates $x_1'^1$, $x_2'^1$, $x_1'^1 \neq x_2'^1$; for simplicity we may suppose that the co-ordinates relative to K' of these events are $(t', x_1'^1, 0, 0)$ and $(t', x_2'^1, 0, 0)$ respectively. Then the co-ordinates relative to K of these events are $(t_1, x_1^1, 0, 0)$ and $(t_2, x_2^1, 0, 0)$ where, by the first equation in (20),

$$t_1 = \frac{t_1' + (v/c^2)x_1'^1}{\sqrt{1 - v^2/c^2}}, \qquad t_2 = \frac{t_2' + (v/c^2)x_2'^1}{\sqrt{1 - v^2/c^2}}. \tag{24}$$

Since $t_1' = t_2' = t'$, it follows that

$$t_1 - t_2 = \frac{v/c^2}{\sqrt{1 - v^2/c^2}}(x_1'^1 - x_2'^1) \neq 0,$$

i.e. to an observer in the frame K, the two events will not be simultaneous.

Note that if $t_1' > t_2'$ then for sufficiently large $x_2'^1 - x_1'^1$, we shall get from (24) that $t_1 < t_2$, i.e. the order of occurrence of the two events is reversed in the two reference frames; superficially this might suggest that the order of cause and effect may be reversed relative to a suitably chosen reference frame. We invite the reader to attempt to resolve this apparent paradox.

Finally we deduce the rule for "addition" of parallel velocities. If a point P is moving along the x'^1-axis of the frame K' with speed w' relative to K', then what is the speed w of P relative to the frame K? Since $w' = dx'^1/dt'$, $w = dx^1/dt$, and, by (20),

$$dt = \frac{dt' + (v/c^2)\,dx'^1}{\sqrt{1 - v^2/c^2}}, \qquad dx^1 = \frac{v\,dt' + dx'^1}{\sqrt{1 - v^2/c^2}},$$

we have that

$$w = \frac{dx^1}{dt} = \frac{v + w'}{1 + vw'/c^2}. \tag{25}$$

If $v, w' \ll c$, then $w \approx w' + v$, i.e. for small speeds we get (approximately) the usual formula of classical mechanics for addition of relative (parallel) speeds. Note also that if $w' = c$, then for any $v < c$, the formula (25) gives the speed w of the point P relative to the frame K as

$$\frac{v + c}{1 + vc/c^2} = c,$$

and we have come full circle to one of the original starting points of Einstein's special theory of relativity, namely that the speed of light is the same relative to all frames of reference moving with constant velocities relative to one another.

6.3. Exercises

1. Define a "vector product" in the space $\mathbb{R}_1^3$ by
$$\xi \times \eta = (\xi^1\eta^2 - \xi^2\eta^1, \xi^0\eta^2 - \xi^2\eta^0, \xi^1\eta^0 - \xi^0\eta^1),$$
where $\xi = (\xi^0, \xi^1, \xi^2)$, $\eta = (\eta^0, \eta^1, \eta^2)$.

   (i) Verify that for the basis vectors e_0, e_1, e_2 (where e_0 is time-like) we have
$$e_0 \times e_1 = -e_2, \qquad e_0 \times e_2 = e_1, \qquad e_1 \times e_2 = e_0.$$

   (ii) Show that $\times$ is a bilinear antisymmetric operation and that Jacobi's identity holds:
$$\xi_1 \times (\xi_2 \times \xi_3) + \xi_3 \times (\xi_1 \times \xi_2) + \xi_2 \times (\xi_3 \times \xi_1) = 0.$$

   (iii) Show that this vector product is preserved by the proper transformations in $O(1, 2)$.

2. Let $r = r(l)$ be a time-like curve in $\mathbb{R}_1^3$ such that $(\dot{r}(l))^2 = (\dot{r}^0)^2 - (\dot{r}^1)^2 - (\dot{r}^2)^2 \equiv 1$, and $\dot{r}^0 > 0$. Define vectors v, n, b by $v = \dot{r}$, $\dot{v} = kn$, $b = n \times v$. Prove the pseudo-Euclidean analogues of the Serret–Frenet formulae:
$$\dot{v} = kn,$$
$$\dot{v} = kv - \varkappa b,$$
$$\dot{b} = \varkappa n.$$

3. Establish a result analogous to that of Lemma 5.3.1, for the derivative of a transformation in $O(1, 2)$ depending on a parameter.

4. Solve in $\mathbb{R}_1^3$ the equation $\dot{r} = \omega \times r$, where ω is a constant vector.

5. Show that an orthogonal complement in $\mathbb{R}_1^{n+1}$ of a time-like vector is a space-like hyperplane (a hyperplane is an n-dimensional subspace). What are the possible orthogonal complements of a space-like vector? Of a light vector?

CHAPTER 2
The Theory of Surfaces

§7. Geometry on a Surface in Space

7.1. Co-ordinates on a Surface

A surface in a 3-dimensional space is the simplest object having its own, internal or "intrinsic," geometry. What exactly do we mean by this?

In our investigation (in Chapter 1) of curves in Euclidean space, we were led to the metrical invariants curvature and torsion of a curve; it is clear however that these are invariants rather of the way the curve is situated in space, than internal to the curve, i.e. they are extrinsic invariants. A curve $r = r(t)$ has no internal metrical invariants, since essentially the only candidate for this status is arc length l, the natural parameter, defined by

$$l = \int_{t_0}^{t_1} |v_t|\, dt, \qquad v_t = \dot{r} = (\dot{x}, \dot{y}, \dot{z}),$$

and clearly this is by itself inadequate for distinguishing the curve from, for instance, a straight line, i.e. we can co-ordinatize a straight line with the same parameter l in such a way that distances along both curve and straight line are measured in the same way.

For surfaces the situation is different: it is impossible to co-ordinatize the sphere (or even a piece of the sphere) so that the formula for distance on the sphere in terms of these co-ordinates, is the same as the usual distance formula in the Euclidean plane.

What mathematical guises do surfaces assume? Surfaces in a 3-dimensional Cartesian space are given in three different ways:

(i) (the simplest) as the graph of a function of two variables

$$z = f(x, y);$$

(ii) more generally as the graph of the equation

$$F(x, y, z) = 0;$$

(iii) by parametric equations (analogous to those for a curve)

$$r = r(u, v);$$

or, in more detail,

$$x = x(u, v), \qquad y = y(u, v), \qquad z = z(u, v),$$

where the pair (u, v) ranges over some region in the plane, the domain of values of the parameters u, v. We shall often refer suggestively to the parameters u, v as "co-ordinates" for the surface.

As in Chapter 1, unless expressly stated otherwise, all functions will be assumed to be continuously differentiable.

7.1.1. Definition. We say that the surface given by the equation $F(x, y, z) = 0$ is *non-singular at the point* $P = (x_0, y_0, z_0)$ on it (i.e. satisfying $F(x_0, y_0, z_0) = 0$), if the gradient of F at P is non-zero:

$$\frac{\partial F}{\partial x} e_1 + \frac{\partial F}{dy} e_2 + \frac{\partial F}{\partial z} e_3 \neq 0 \quad \text{at } x = x_0, \quad y = y_0, \quad z = z_0.$$

By the Implicit Function Theorem, if say $\partial F/\partial z|_{x_0, y_0, z_0} \neq 0$, then near the point $P(x_0, y_0, z_0)$ on the surface, the equation $F(x, y, z) = 0$ can be solved for z as a (once again continuously differentiable) function of x and y; i.e. there exists a (continuously differentiable) function $z = f(x, y)$, such that $z_0 = f(x_0, y_0)$, and, in a neighbourhood of (i.e. region containing) the point (x_0, y_0), $F(x, y, f(x, y)) = 0$. Formulae for the partial derivatives of f are obtained as follows. On differentiating the identity $F(x, y, f(x, y)) \equiv 0$ implicitly with respect to x, we obtain (using the "chain rule," and bearing in mind that $z = f(x, y)$):

$$\frac{\partial F}{\partial z}\frac{\partial f}{\partial x} + \frac{\partial F}{\partial x}\frac{\partial x}{\partial x} = 0,$$

whence

$$\frac{\partial f}{\partial x} = -\frac{\partial F/\partial x}{\partial F/\partial z}. \tag{1}$$

Similarly

$$\frac{\partial f}{\partial y} = -\frac{\partial F/\partial y}{\partial F/\partial z}. \tag{2}$$

(Alternatively one may derive (1) and (2) as follows: It can be shown that, under our prevailing assumptions,

$$\begin{aligned}\Delta F &= F(x + dx, y + dy, z + dz) - F(x, y, z)\\ &= F_x\, dx + F_y\, dy + F_z\, dz + o\left(\sqrt{dx^2 + dy^2 + dz^2}\right),\end{aligned}$$

where F_x, F_y, F_z are evaluated at (x, y, z). (This is what is usually taken to be the definition of "differentiability" of the function F at the point (x, y, z), the geometric idea being that the function F is "locally" almost affine.) Then if (x, y, z) and $(x + dx, y + dy, z + dz)$ are both on the surface $F(x, y, z) = 0$, we get

$$F_x\, dx + F_y\, dy + F_z\, dz = o(\sqrt{dx^2 + dy^2 + dz^2}), \tag{3}$$

which tells us that the surface is "locally" approximable by the plane ("tangent" plane—see below) through (x, y, z) with normal (F_x, F_y, F_z). Setting $dy = 0$ in (3), using the fact that $F(x, y, z) = 0$ defines a function $z = f(x, y)$ in a neighbourhood of (x, y, z), dividing both sides of (3) by dx, and, finally, letting $dx \to 0$, we get the formula (1).)

From Definition 7.1.1 and the Implicit Function Theorem it follows that around each non-singular point of a surface given by $F(x, y, z) = 0$, there is a neighbourhood such that the piece of the surface enclosed by that neighbourhood is the graph of a function (where if not z, then either x or y may serve as the dependent variable). Hence locally, i.e. in a neighbourhood of a non-singular point, the surface is given by the parametric equations $z = f(u, v)$, $x = u$, $y = v$ (around the point $x_0 = u_0$, $y_0 = v_0$). This is often expressed as follows: *In a neighbourhood of a non-singular point of a surface there exist local co-ordinates* u, v.

Suppose now that a surface is given parametrically:

$$r = r(u, v); \qquad x = x(u, v), \qquad y = y(u, v), \qquad z = z(u, v). \tag{4}$$

7.1.1′. Definition. A point $P = (x_0, y_0, z_0) = (x(u_0, v_0), y(u_0, v_0), z(u_0, v_0))$ on the surface given by (4), is said to be *non-singular* if the matrix

$$A = \begin{pmatrix} \dfrac{\partial x}{\partial u} & \dfrac{\partial y}{\partial u} & \dfrac{\partial z}{\partial u} \\ \dfrac{\partial x}{\partial v} & \dfrac{\partial y}{\partial v} & \dfrac{\partial z}{\partial v} \end{pmatrix}_{u_0, v_0}$$

has rank 2.

The next theorem tells us that Definitions 7.1.1 and 7.1.1′ are "locally" equivalent, but, more significantly, we can conclude from it that locally, i.e. in a neighbourhood of a non-singular point of a surface, the above three ways of presenting a surface equationally are equivalent.

7.1.2. Theorem. *If a surface is given parametrically (as in (4)) and the point $P(x_0, y_0, z_0)$ corresponding to (u_0, v_0) is non-singular, then there is an equation $F(x, y, z) = 0$ which defines the surface in a neighbourhood of P, and has the property that $(\text{grad } F)_{x_0, y_0, z_0} \neq 0$.*

PROOF. Since P is non-singular, by 7.1.1′ the matrix A has rank 2; hence, by renaming x, y, z if necessary, we may suppose that the determinant

$$\frac{\partial x}{\partial u}\frac{\partial y}{\partial v} - \frac{\partial x}{\partial v}\frac{\partial y}{\partial u} \neq 0.$$

However this determinant is the Jacobian of the transformation $(x, y) \mapsto (u, v)$, evaluated at the point (u_0, v_0). Since it is non-zero we deduce from the Inverse Function Theorem (1.2.5) that there is a neighbourhood of the point (x_0, y_0) (where $x_0 = x(u_0, v_0)$, $y_0 = y(u_0, v_0)$), on which the restriction of the transformation $(x, y) \mapsto (u, v)$, has an inverse and on which, in addition, the matrix $\begin{pmatrix} u_x & v_x \\ u_y & v_y \end{pmatrix}$ is the inverse of $\begin{pmatrix} x_u & y_u \\ x_v & y_v \end{pmatrix}$. If we define $f(x, y)$ by

$$f(x, y) = z(u(x, y), v(x, y)),$$

then the equation $z = f(x, y)$ (or rather $z - f(x, y) = 0$) will fulfil the claims of the theorem. □

7.1.3. Examples. (a) The ellipsoid $x^2/a^2 + y^2/b^2 + z^2/c^2 = 1$ has no singular points, but is not globally the graph of a function (though of course locally it is). It has no global parametrization with respect to which all its points are non-singular.

(b) The hyperboloid of one sheet $x^2/a^2 + y^2/b^2 - z^2/c^2 = 1$ is not globally the graph of a function, but does have global parameters $u = z$, $v = \varphi$ (where φ is the usual polar angle defined in terms of x and y) yielding a parametrization with no singular points.

(c) The two-sheeted hyperboloid $-x^2/a^2 - y^2/b^2 + z^2/c^2 = 1$. Each sheet is the graph of a function of the form $z = f(x, y)$ (and therefore has a parametrization with no singular points).

(d) The cone $x^2/a^2 + y^2/b^2 - z^2/c^2 = 0$ has $(0, 0, 0)$ as a singular point.

We now turn to the general n-dimensional Cartesian space where we shall understand a surface to be given by a system of equations

$$f_1(x^1, \ldots, x^n) = 0, \ldots, f_{n-k}(x^1, \ldots, x^n) = 0 \qquad (5)$$

defined in a region of the space.

7.1.4. Definition. A point $P(x_0^1, \ldots, x_0^n)$ of the surface (5) is said to be *non-singular* if the $(n-k) \times n$ matrix

$$\left(\frac{\partial f_i}{\partial x^j}\right)_{x^i = x_0^i}$$

has rank $(n-k)$.

7.1.5. Lemma. *Local co-ordinates may be defined in a neighbourhood of a non-singular point of the surface* (5).

PROOF. By re-indexing $x^1, \ldots, x^n$, if necessary, we may assume that the $(n-k) \times (n-k)$ minor

$$\left(\frac{\partial f_i}{\partial x^q}\right)_{x^q = x_0^q}, \qquad q = k+1, \ldots, n,$$

has non-zero determinant. Hence by the general form of the Implicit Function Theorem, as applied to the transformation $f_i(x^1, \ldots, x^n)$ from $\mathbb{R}^n$ to $\mathbb{R}^{n-k}$, there exist $(n-k)$ functions

$$g^{k+1}(x^1, \ldots, x^k), \ldots, g^n(x^1, \ldots, x^k)$$

defined on a neighbourhood of $(x_0^1, \ldots, x_0^k)$, with

$$x_0^{k+1} = g^{k+1}(x_0^1, \ldots, x_0^k), \ldots, x_0^n = g^n(x_0^1, \ldots, x_0^k),$$

and having the property that

$$f_i(x^1, \ldots, x^k, g^{k+1}, \ldots, g^n) = 0, \qquad i = 1, \ldots, n-k,$$

throughout the neighbourhood. Thus we may take as parameters $u^1 = x^1, \ldots, u^k = x^k$, in terms of which the desired parametric equations for a neighbourhood of the surface about the point P are

$$x^1 = u^1, \ldots, x^k = u^k, \qquad x^{k+1} = g^{k+1}(u^1, \ldots, u^k), \ldots, x^n = g^n(u^1, \ldots, u^k).$$

□

An important special case is that of a *hypersurface* in an n-dimensional space, defined as the set of points whose co-ordinates satisfy a single equation

$$f(x^1, \ldots, x^n) = 0. \tag{6}$$

Analogously to the 3-dimensional case, the defining condition for non-singularity of a point $(x_0^1, \ldots, x_0^n)$ on the hypersurface (6) (i.e. satisfying $f(x_0^1, \ldots, x_0^n) = 0$) is taken to be

$$\operatorname{grad} f = \left(\frac{\partial f}{\partial x^1}, \ldots, \frac{\partial f}{\partial x^n}\right)_{x^i = x_0^i} \neq 0.$$

7.2. Tangent Planes

Suppose a surface in $\mathbb{R}^3$ is given in parametric form $r = r(u, v)$, $r = (x, y, z)$, where u, v are the parameters, or, in other terminology, co-ordinates on the surface. A curve $u = u(t)$, $v = v(t)$ in the (u, v)-region defines a curve $r(t) = r(u(t), v(t))$ lying on our surface in $\mathbb{R}^3$. The tangent vector $\dot{r}(t)$ to the curve, has the form

$$\dot{r}(t) = r_u \dot{u} + r_v \dot{v}, \qquad r_u = \frac{\partial r}{\partial u}, \quad r_v = \frac{\partial r}{\partial v}. \tag{7}$$

By Definition 7.1.1′ a non-singular point of the surface is one at which the vectors $r_u = (x_u, y_u, z_u)$ and $r_v = (x_v, y_v, z_v)$ are linearly independent. Putting this together with (7), which tells us that every vector tangent to our surface is a linear combination of the vectors r_u and r_v, we deduce that the totality of vectors tangent to our surface at a given non-singular point P forms a 2-dimensional subspace having the pair r_u, r_v as a basis. We call this subspace the *tangent plane* to the surface at the given point. (Note that in terms of the basis r_u, r_v the components of the tangent vector $\dot{r}$ are the quantities $\dot{u}$, $\dot{v}$.)

Suppose now that a surface in $\mathbb{R}^3$ is given to us instead via the equation $F(x, y, z) = 0$. If $r = r(t) = (x(t), y(t), z(t))$ is a curve on this surface, i.e. if $F(x(t), y(t), z(t)) = 0$ for the specified interval of values of t, then $(d/dt)F(x(t), y(t), z(t)) = 0$ for those t, whence, applying the chain rule yet again, (or using (3)), we get

$$F_x \dot{x} + F_y \dot{y} + F_z \dot{z} = 0. \tag{8}$$

At a non-singular point $P(x_0, y_0, z_0)$ of the surface we have (by definition), grad $F = (F_x, F_y, F_z) \neq 0$. Hence writing A, B, C for F_x, F_y, F_z respectively, evaluated at P, we see from (8) that every vector $(\dot{x}, \dot{y}, \dot{z})$ tangent to the surface at P is in the plane $Ax + By + Cz = 0$. Hence there is a tangent plane to the surface at the non-singular point P: it is the 2-dimensional subspace consisting of all vectors (x, y, z) satisfying $Ax + By + Cz = 0$. We now repeat these two arguments (showing that a unique tangent plane exists at each non-singular point of a surface) in n dimensions.

Thus firstly suppose that in n-dimensional Cartesian space with co-ordinates $x^1, \ldots, x^n$, we are given a k-dimensional surface in parametric form:

$$\begin{aligned} x^1 &= x^1(z^1, \ldots, z^k), \\ &\cdots\cdots\cdots\cdots \\ x^n &= x^n(z^1, \ldots, z^k). \end{aligned} \tag{9}$$

(We call this surface "k-dimensional" since we think of the k parameters $z^1, \ldots, z^k$ as co-ordinates for it.) As in the case $n = 3$, a family of functions $z^j = z^j(t)$, $j = 1, \ldots, k$ (defining a curve in a space co-ordinatized by

$z^1, \ldots, z^k$) yields, on substitution into (9), a curve lying in the surface (9) with tangent vector

$$v = (\dot{x}^1, \ldots, \dot{x}^n) = \dot{z}^1 b_1 + \cdots + \dot{z}^k b_k, \tag{10}$$

where the vectors b_j are given by

$$b_j = \left(\frac{\partial x^1}{\partial z^j}, \ldots, \frac{\partial x^n}{\partial z^j}\right), \qquad j = 1, \ldots, k. \tag{11}$$

By analogy with Definition 7.1.1′ (which applies when $n = 3, k = 2$), we shall say that a point lying on the surface (9) is a *non-singular point of the surface* if the k vectors in (11) are linearly independent. Continuing this analogy with the 3-dimensional case, we then define the *tangent plane to the surface* (9) at a non-singular point P, to be the k-dimensional subspace of $\mathbb{R}^n$ spanned by the k independent vectors in (11); from (10) we then see that all vectors tangent to the surface at the non-singular point P, lie in the tangent plane to the surface at P, so that this definition of the tangent plane is the appropriate one.

If on the other hand a k-dimensional surface is given to us in terms of a system of $(n - k)$ equations

$$\begin{gathered} f_1(x^1, \ldots, x^n) = 0, \\ \cdots\cdots\cdots\cdots\cdots\cdots \\ f_{n-k}(x^1, \ldots, x^n) = 0, \end{gathered} \tag{12}$$

then the components of a vector $(\dot{x}^1, \ldots, \dot{x}^n)$ tangent to the surface (i.e. tangent to a curve $x^1 = x^1(t), \ldots, x^n = x^n(t)$ lying in the surface) will satisfy the system of $(n - k)$ linear equations

$$\sum_{l=1}^{n} \frac{\partial f_j}{\partial x^l} \dot{x}^l = 0, \qquad j = 1, \ldots, n - k. \tag{13}$$

If we now define (by analogy with 7.1.1 and as in 7.1.4) a *non-singular point of the surface* to be one at which the coefficient matrix $(\partial f_j / \partial x^l)$ of the system (13) has rank $(n - k)$, then at such a point P the subspace of $\mathbb{R}^n$ made up of all solutions of (13), will have dimension k and will contain all vectors $(\dot{x}^1, \ldots, \dot{x}^n)$ tangent to the surface at P; hence we take that subspace to be the *tangent plane to the surface* at the non-singular point P.

The reader will be prepared to believe that Theorem 7.1.2 has an analogue in n dimensions, so that "locally" the above two definitions of a non-singular point of a surface are the same, and of course the two ways of arriving at the tangent plane at such a point yield the same plane!

To summarize: the geometric significance of the property of being non-singular lies in the fact that the tangent plane to a surface at a non-singular point has dimension the same as that of the surface. This is the number of parameters in a parametric representation of the surface, or, if the surface is given by a system of equations (12), then it is the difference between the dimension of the underlying space and the number of these equations.

7.3. The Metric on a Surface in Euclidean Space

Up to this point our study of surfaces has been completely metric-free. We shall now assume that the spaces in which our surfaces lie are Euclidean.

To begin with we shall work in Euclidean 3-space, with Euclidean co-ordinates x, y, z (see 3.1.4). Thus suppose we wish to study a surface in this space which (or at least a relevant piece of which) is given parametrically by $r = r(u, v)$, $r = (x, y, z)$, where u, v are co-ordinates on the surface. In what follows any points of the surface which come up for consideration will be tacitly assumed to be non-singular (see 7.1.1′).

If we think of the surface as a space co-ordinatized by u, v, then it is natural to ask how one goes about measuring length in this space, i.e. in terms of the co-ordinates u, v. To be more precise, suppose we have a curve $r = r(t) = (x(t), y(t), z(t))$ on our surface. Since this curve lies in Euclidean space, by its length we shall naturally mean its length in that space, given by

$$l = \int_a^b \sqrt{\dot{x}^2 + \dot{y}^2 + \dot{z}^2}\, dt. \tag{14}$$

Since the curve lies on the surface, we can find a pair of functions $u = u(t)$, $v = v(t)$, such that

$$\begin{aligned} x &= x(u(t), v(t)) = x(t), \\ y &= y(u(t), v(t)) = y(t), \\ z &= z(u(t), v(t)) = z(t); \end{aligned} \tag{15}$$

i.e. the equations $u = u(t)$, $v = v(t)$ determine the curve. If we regard the surface as a space co-ordinatized by u, v, then these equations ($u = u(t)$, $v = v(t)$) take on the role of parametric equations of the curve (15) in that space, in terms of the co-ordinates u, v. From this point of view our question becomes: In the space co-ordinatized by u, v (i.e. in the surface) what metric $g_{ij} = g_{ij}(u, v)$ will give us the same result for the length of the curve $u = u(t)$, $v = v(t)$, as the formula (14)? In other words we are asking for the metric $g_{ij} = g_{ij}(u, v)$ on the surface, for which, for all a, b in the interval of values of t,

$$\int_a^b |v|\, dt = \int_a^b \sqrt{\dot{x}^2 + \dot{y}^2 + \dot{z}^2}\, dt, \tag{16}$$

where $|v|^2 = g_{ij}\dot{x}^i\dot{x}^j$, $x^1 = u$, $x^2 = v$, and $\dot{x}, \dot{y}, \dot{z}$ are as in (14). (Note that for (16) to make sense, the metric g_{ij} must be Riemannian (see 3.1.1)). It follows from (16) that for all t

$$\dot{x}^2 + \dot{y}^2 + \dot{z}^2 = g_{ij}\dot{x}^i\dot{x}^j = E\dot{u}^2 + 2F\dot{u}\dot{v} + G\dot{v}^2, \tag{17}$$

where we have put $g_{11} = E$, $g_{22} = G$, $g_{21} = g_{12} = F$. Since from (15) we have $\dot{x} = x_u\dot{u} + x_v\dot{v}$ etc., on substituting in (17) we obtain

$$\begin{aligned} g_{11} &= E = x_u x_u + y_u y_u + z_u z_u, \\ g_{12} &= F = x_u x_v + y_u y_v + z_u z_v, \\ g_{22} &= G = x_v x_v + y_v y_v + z_v z_v. \end{aligned} \tag{18}$$

Using

$$r_u = x_u e_1 + y_u e_2 + z_u e_3, \qquad r_v = x_v e_1 + y_v e_2 + z_v e_3,$$

we can rewrite (18) briefly as

$$g_{ij} = \langle r_{x^i}, r_{x^j} \rangle, \qquad x^1 = u, \quad x^2 = v. \tag{19}$$

We have thus arrived at the desired metric $g_{ij}(u, v) = (E, F, G)$ (where E, F, G are given in (18)) in the co-ordinates u, v of the surface. This Riemannian metric on the surface, which in terms of differentials has the form

$$g_{ij}\, dx^i\, dx^j = E(du)^2 + 2F(du\, dv) + G(dv)^2,$$

is often also called the *first fundamental form on the surface*. We shall also say that this metric is *induced* on the surface.

It follows that if the surface is given to us in the form $z = f(x, y)$, then by (18) the metric induced on the surface, in terms of the co-ordinates $u = x$, $v = y$, is given by

$$g_{11} = 1 + f_x^2, \qquad g_{12} = g_{21} = f_x f_y, \qquad g_{22} = 1 + f_y^2. \tag{20}$$

Thirdly, and finally, how do we calculate the induced metric, and with respect to what co-ordinates do we calculate it, on a surface given by $F(x, y, z) = 0$? In this case we confine ourselves to a neighbourhood of a non-singular point P of the surface (i.e. grad $F \neq 0$ at P). We may suppose without loss of generality that $\partial F/\partial z \neq 0$ at P. Then as we showed at the beginning of §7.1, the equation $F(x, y, z) = 0$ defines a function $z = f(x, y)$ implicitly in a neighbourhood of P, and in that neighbourhood, $f_x = -F_x/F_z$, $f_y = -F_y/F_z$ (see (1), (2)). Substituting in (20) we find that, in this neighbourhood of P, the induced metric on the surface in terms of the co-ordinates x, y is given by

$$g_{11} = 1 + \frac{F_x^2}{F_z^2}, \qquad g_{12} = g_{21} = \frac{F_x F_y}{F_z^2}, \qquad g_{22} = 1 + \frac{F_y^2}{F_z^2}. \tag{21}$$

We conclude that, however a surface is defined, we can calculate the metric induced on the surface with respect to appropriate co-ordinates defined on it, and then, if so desired, calculate in terms of that metric, lengths of segments of curves lying on the surface. We emphasize that these lengths are simply the ordinary Euclidean lengths as measured in Euclidean 3-space, the space in which the surface is situated.

Note that if the surface is given in the form $r = r(u, v)$, i.e. $x = x(u, v)$, $y = y(u, v)$, $z = z(u, v)$, then one often expresses (17) in the brief form

$$dl^2 = dx^2 + dy^2 + dz^2 = g_{ij}\, dx^i\, dx^j, \qquad x^1 = u, \quad x^2 = v, \tag{22}$$

where $g_{11} = E$, $g_{12} = g_{21} = F$, $g_{22} = G$ are given by (18), and where dl is an element of arc length of a curve on the surface.

It is possible for the induced metric $g_{ij}(x^1, x^2)$ on the surface to be itself Euclidean (of dimension 2, naturally); i.e. that there exist a system of co-ordinates $\bar{u} = \bar{u}(x^1, x^2)$, $\bar{v} = \bar{v}(x^1, x^2)$ on the surface, such that

$$(d\bar{u})^2 + (d\bar{v})^2 = g_{ij}\, dx^i\, dx^j.$$

7.3.1. Example. The metric induced on a cylinder is Euclidean. For, suppose the cylinder given by the equation $f(x, y) = 0$ (in which z does not appear); then the surface co-ordinates $\bar{u} = z$, $\bar{v} = l$, where l is arc length measured along the curve $f(x, y) = 0$ in the plane $z = 0$, are Euclidean co-ordinates for the cylinder, i.e.

$$(dx^2 + dy^2 + dz^2)|_{f(x, y)=0} = dz^2 + dl^2.$$

(This is easy to see intuitively by imagining the cylinder unrolled flat onto a plane.)

We digress briefly to introduce the concept of the *unit normal* to a surface at a point. By (8), if a surface is given in the form $F(x, y, z) = 0$, then at any point the vector grad $F = (F_x, F_y, F_z)$ is perpendicular to every tangent vector to the surface at that point. Thus at a non-singular point of the surface, grad F is a non-zero vector normal to the surface at that point. If on the other hand the surface is given to us in parametric form $r = r(u, v)$, $r = (x, y, z)$, then the vectors

$$\xi = r_u = r_{x^1} = x_u e_1 + y_u e_2 + z_u e_3,$$

$$\eta = r_v = r_{x^2} = x_v e_1 + y_v e_2 + z_v e_3$$

are both tangent to the surface (see (7)), so that if they are linearly independent (as they will be at a non-singular point of the surface—see 7.1.1′) then their vector product $[\xi, \eta]$ furnishes us with a non-zero vector perpendicular to the surface (at that non-singular point).

In equipping our surface with the Riemannian metric g_{ij} in the manner described above, we have conferred on it, as it were, the status of a self-contained metric space; thus we can (as already mentioned) use this metric to calculate arc length of a curve $u = u(t)$, $v = v(t)$ on the surface, and we can also use it to measure the angle between two such curves at a point where they intersect. Thus if $(u_1(t), v_1(t))$ and $(u_2(t), v_2(t))$ are two curves on the surface, and if $\eta_1 = (\dot{u}_1, \dot{v}_1)$, $\eta_2 = (\dot{u}_2, \dot{v}_2)$ are the tangent vectors to these curves at a point where they intersect, then the angle $\varphi (0 \le \varphi \le \pi)$ between the curves at that point is given by

$$\cos \varphi = \frac{\langle \eta_1, \eta_2 \rangle}{|\eta_1| |\eta_2|} \qquad (|\eta_1|^2 = \langle \eta_1, \eta_1 \rangle, |\eta_2|^2 = \langle \eta_2, \eta_2 \rangle),$$

where the scalar product $\langle \ , \ \rangle$ is defined (as usual for a metric space) by

$$\langle \eta_1, \eta_2 \rangle = g_{ij} \eta_1^i \eta_2^j \qquad (\eta_k^1 = \dot{u}_k, \eta_k^2 = \dot{v}_k, k = 1, 2).$$

It is clear from (17) that this angle φ will be the same as the Euclidean angle between the curves $r = r(u_1(t), v_1(t))$ and $r = r(u_2(t), v_2(t))$ (assuming our surface given in the form $r = r(u, v)$) in Euclidean 3-space.

We now derive the formula for the metric on a k-dimensional surface lying in n-dimensional Euclidean space. Since, at least in a neighbourhood of a non-singular point of such a surface, co-ordinates (i.e. parameters) may be defined on it (this was shown at the end of §7.1), we may assume that our surface is given in parametric form $x^i = x^i(z^1, \dots, z^k)$, $i = 1, \dots, n$. As before a curve $z^j = z^j(t)$, $j = 1, \dots, k$, in the k-dimensional parameter-space, defines a curve

$$x^i(t) = x^i(z^1(t), \dots, z^k(t)), \qquad i = 1, \dots, n, \tag{23}$$

in Euclidean n-space, lying on the surface. Arc length along the curve (23) is given by the formula

$$\begin{aligned} l = \int_a^b |\dot{x}| \, dt &= \int_a^b \sqrt{\sum_{i=1}^n (\dot{x}^i)^2} \, dt \\ &= \int_a^b \sqrt{\sum_{h=1}^n \frac{\partial x^h}{\partial z^i} \frac{\partial x^h}{\partial z^j} \dot{z}^i \dot{z}^j} \, dt. \end{aligned}$$

Thus as the metric defined on the surface we take

$$g_{ij}(z^1, \dots, z^k) = \sum_{h=1}^n \frac{\partial x^h}{\partial z^i} \frac{\partial x^h}{\partial z^j}. \tag{24}$$

As in the 3-dimensional case, we shall say that the metric defined by (24) is *induced* on the surface by the Euclidean metric of the ambient Euclidean space. The induced metric is in general not Euclidean.

Finally, note that in a region of a hypersurface $F(x^1, \dots, x^n) = 0$, where, say, $\partial F / \partial x^n \neq 0$, the induced metric (24) takes the form

$$g_{ij} = \delta_{ij} + \frac{(\partial F / \partial x^i)(\partial F / \partial x^j)}{(\partial F / \partial x^n)^2}.$$

We leave the verification of this to the reader.

7.4. Surface Area

From a second course in the calculus the reader will recall that the area of a region U of the Euclidean plane with Euclidean co-ordinates x, y, is given (or rather defined) by the double integral

$$\sigma(U) = \iint_U dx\,dy \tag{25}$$

(if this integral exists, i.e. if the boundary of the region is not too "wild"), and also that given a one-to-one change of co-ordinates

$$x = x(u, v), \qquad y = y(u, v), \tag{26}$$

the formula for $\sigma(U)$ in terms of the new co-ordinates u, v is

$$\sigma(U) = \iint_V |x_u y_v - x_v y_u|\,du\,dv, \tag{27}$$

where V is the region of the (u, v)-plane corresponding to the region U of the (x, y)-plane. Of course the integrand in (27) is just the absolute value $|J|$ of the Jacobian of the transformation (25), so that (27) can be rewritten as

$$\sigma(U) = \iint_V |J|\,du\,dv.$$

The question we wish to consider is the following one: How does one go about calculating the area of a region not necessarily in a plane, but on an arbitrary surface in Euclidean 3-space? To be more precise, given a surface $r = r(u, v)$, $r = (x, y, z)$ in Euclidean 3-space, and its induced Riemannian metric

$$dl^2 = g_{ij}\,dx^i\,dx^j, \qquad x^1 = u, \quad x^2 = v, \tag{28}$$

we want a formula analogous to (25) for the area of a region on the surface. The determinant g of the matrix (g_{ij}) will play a leading role in our answer. We know that $g > 0$ since the metric (28) is Riemannian. As for arc length (see §3.1(4)), so also for surface area we simply present the definition as a *fait accompli*, and then after the fact show that it is the appropriate one. Let U be a region (whose boundary is not too "wild") of a surface $r = r(u, v)$, $r = (x, y, z)$ in Euclidean 3-space. Since each region of the surface is determined by a region in the (u, v)-plane it will occasionally be convenient (as it was when we were considering curves on a surface) to think of U rather as the region in the (u, v)-plane which determines it. With this flexible interpretation of U our definition of surface area is as follows.

7.4.1. Definition. The *area* of the region U on the surface $r = r(u, v)$, $r = (x, y, z)$ is given by

$$\sigma(U) = \iint_U \sqrt{g}\, du\, dv, \tag{29}$$

where the integral is taken over the region in the (u, v)-plane which determines U.

By analogy with (28) we introduce the useful abbreviated expression

$$d\sigma = \sqrt{g}\, du\, dv,$$

where $d\sigma$ is called the "element" or "differential" of area on the surface with metric g_{ij}.

We now turn to the promised justification of Definition 7.4.1.

To begin with, let ξ, η be vectors in the Euclidean plane, and consider the parallelogram consisting of all (tips of) vectors $\lambda\xi + \mu\eta$, $0 \le \lambda, \mu \le 1$. It is easy to verify that the area σ of this parallelogram is given by

$$\sigma = |\det A|, \quad \text{where } A = \begin{pmatrix} \xi^1 & \xi^2 \\ \eta^1 & \eta^2 \end{pmatrix}, \tag{30}$$

and where ξ^1, ξ^2 and η^1, η^2 are the components of ξ and η relative to an orthonormal basis e_1, e_2, i.e. $\xi = \xi^1 e_1 + \xi^2 e_2$, $\eta = \eta^1 e_1 + \eta^2 e_2$.

Now suppose we are working in any 2-dimensional inner product space (over the reals), i.e. in any 2-dimensional vector space equipped with an inner (i.e. scalar) product. Let e_1, e_2 be an orthonormal basis for this space. (Such bases always exist.) It is then natural to define the area of the parallelogram determined by two vectors ξ, η by analogy with the Euclidean situation; thus we shall take (30) as defining the area of this parallelogram in our present more general context of an inner product space. Write $\langle \xi, \eta \rangle = g_{12}$, $\langle \xi, \xi \rangle = g_{11}$, $\langle \eta, \eta \rangle = g_{22}$. For reasons which will appear later, we wish now to express the area σ of the parallelogram determined by ξ, η, in terms of these g_{ij}. To this end observe first that

$$g_{11} = (\xi^1)^2 + (\xi^2)^2,$$
$$g_{12} = \xi^1\eta^1 + \xi^2\eta^2,$$
$$g_{22} = (\eta^1)^2 + (\eta^2)^2,$$

which can be expressed compactly in matrix notation as

$$(g_{ij}) = AA^{\mathrm{T}},$$

where A is as in (30). Taking determinants of both sides, we get $\det(g_{ij}) = (\det A)^2$. From this and (30) we obtain our desired result, which we state as

7.4.2. Lemma. *The area of the parallelogram determined by the vectors ξ, η of the inner product space, is $\sqrt{g}$ where $g = \det(g_{ij}) = g_{11}g_{22} - g_{12}^2$.*

Now the integrand in the definition (7.4.1) of surface area is also denoted by $\sqrt{g}$. What is the connexion with Lemma 7.4.2? In order to answer this (and complete our justification of 7.4.1) we first recall, without proofs, the essential ideas underlying the concept of the double integral, and the related concept of area of a planar region.

Thus let U be a region in a plane which we shall suppose for the moment to be Euclidean, with Euclidean co-ordinates $x^1 = u$, $x^2 = v$. Suppose also that the boundary of U is a piecewise smooth curve. Partition the whole plane into small non-overlapping rectangles by means of lines parallel to the x^1-axis and x^2-axis, and denote by Δu the largest width, and by Δv the largest height, of rectangles in the partition. Clearly the area of U is then greater than (or equal to) the sum of the areas of those rectangles whose interiors are in U; call them "interior" rectangles. Repeat for all possible partitions.

7.4.3. Definition. The *area* of the region U is the limit of the sum of the areas of all interior rectangles as $\Delta u + \Delta v \to 0$, if this limit exists.

Suppose next that we are given in addition a function $f(u, v)$ which is continuous on U. We shall now sketch the definition of the double integral of f over U. Partition the plane into rectangles as before, and for each interior rectangle S_α, take the value $f(u_\alpha, v_\alpha)$ of f at the centre of S_α, and with these values form the "Riemann sum"

$$S_p(f, U) = \sum_\alpha f(u_\alpha, v_\alpha)\sigma(S_\alpha),$$

over all rectangles S_α interior to U. (Here p denotes the particular partition.) Repeat for all possible partitions p.

7.4.4. Definition. The limit of $S_p(f, U)$ as $\Delta u + \Delta v \to 0$ (if this limit exists), is called the *double integral* of f over U, and is denoted by $\iint_U f(u, v)\, du\, dv$.

(In particular if $f(u, v) \equiv 1$, then $\iint_U du\, dv$ is clearly just the area of U as defined in 7.4.3, in terms of the Euclidean co-ordinates u, v.)

Now suppose that in the above the (u, v)-plane is not necessarily Euclidean, but rather has defined on it a more general Riemannian metric, which, relative to the co-ordinates $u = x^1$, $v = x^2$, has the form $dl^2 = g_{11}\, du^2 + 2g_{12}\, du\, dv + g_{22}\, dv^2$. In this more general situation we partition the (u, v)-plane into parallelograms the directions of whose sides are given by the vectors $\xi = (1, 0)$, $\eta = (0, 1)$. If a parallelogram S_α has sides Δu_α, Δv_α, then by Lemma 7.4.2 its area is approximately (for small Δu, Δv) $\sqrt{g}\, \Delta u_\alpha \Delta v_\alpha$, where g is evaluated at the centre (u_α, v_α) of S_α. Hence

$$\sum_\alpha S_\alpha \approx \sum_\alpha \sqrt{g(u_\alpha, v_\alpha)}\, \Delta u_\alpha \Delta v_\alpha$$

(where the summations are over parallelograms interior to U), and this makes it very plausible (and indeed it is true) that as $\Delta u + \Delta v \to 0$,

$$\sum_{\alpha} S_\alpha \to \iint_U \sqrt{g}\, du\, dv,$$

provided the limit exists. This completes the justification of Definition 7.4.1.

We conclude the section by re-expressing the formula (29) for surface area in three ways, corresponding to the three ways of presenting a surface.

7.4.5. Theorem. (i) *If a surface is given in the form $z = f(x, y)$, and if U is a region on this surface with projection onto the (x, y)-plane V, then the formula for surface area can be put into the form*

$$\sigma(U) = \iint_V \sqrt{1 + f_x^2 + f_y^2}\, dx\, dy. \tag{31}$$

(ii) *If a surface is given by the equation $F(x, y, z) = 0$, and if U is a region on the surface which projects in a one-to-one fashion onto a region V of the (x, y)-plane, then the formula for the area of U becomes*

$$\sigma(U) = \iint_V \frac{|\operatorname{grad} F|}{|F_z|} dx\, dy, \tag{32}$$

provided $F_z = \partial F/\partial z \neq 0$ for all (x, y, z) in U.

(iii) *If a surface is given in parametric form $r = r(u, v)$, then*

$$\sigma(U) = \iint_V |[r_u, r_v]|\, du\, dv, \tag{33}$$

where V is the region in the (u, v)-plane giving rise to U, and $[r_u, r_v]$ is the vector product of r_u and r_v.

PROOF. (i) By (20) the metric g_{ij} induced on a surface $z = f(x, y)$ is given by

$$g_{11} = 1 + f_x^2, \qquad g_{12} = f_x f_y, \qquad g_{22} = 1 + f_y^2, \qquad u = x, \quad v = y.$$

Hence $\sqrt{g} = \sqrt{g_{11}g_{22} - g_{12}^2} = \sqrt{1 + f_x^2 + f_y^2}$, and (31) is then immediate from 7.4.1.

(ii) For a region of a surface $F(x, y, z) = 0$ throughout which $F_z \neq 0$, we have, by (21),

$$g_{11} = 1 + \frac{F_x^2}{F_z^2}, \qquad g_{12} = \frac{F_x F_y}{F_z^2}, \qquad g_{22} = 1 + \frac{F_y^2}{F_z^2}, \qquad u = x, \quad v = y.$$

Hence

$$\sqrt{g} = \sqrt{1 + \frac{F_x^2}{F_z^2} + \frac{F_y^2}{F_z^2}} = \frac{|\operatorname{grad} F|}{|F_z|}.$$

(iii) By (19), the induced metric g_{ij} on a surface $r = r(u, v)$ is given by

$$g_{12} = \langle r_u, r_v \rangle, \qquad g_{11} = \langle r_u, r_u \rangle, \qquad g_{22} = \langle r_v, r_v \rangle.$$

It is then easy to verify that $g = g_{11}g_{22} - g_{12}^2 = |[r_u, r_v]|^2$, whence the result. □

Using Definition 7.4.1 to give us a definition of area in an arbitrary 2-dimensional space with a Riemannian metric, we see that the concept of area, like arc length and angle, is defined in terms of the scalar product of vectors attached to each point, or, what amounts to the same thing, the metric g_{ij}.

7.5. Exercises

1. The torus T^2 in Euclidean 3-space can be realized as the surface of revolution obtained by revolving a circle about a straight line which it does not intersect, and which lies in the same plane. Find parametric equations for this torus, and calculate the induced metric on it.

2. Find the first fundamental form of the ellipsoid of revolution

$$\frac{x^2}{a^2} + \frac{y^2 + z^2}{b^2} = 1.$$

3. Find the metric induced on the surface of revolution

$$r(u, \varphi) = (\rho(u) \cos \varphi, \rho(u) \sin \varphi, z(u)).$$

Verify that its meridians ($\varphi =$ const.) and circles of latitude ($u =$ const.) form an orthogonal net. Find those curves which bisect the angles between the meridians and circles of latitude.

4. Find those curves on a sphere which intersect the meridians at a fixed angle α ("loxodromes"). Find the length of a loxodrome.

5. Let $F(x, y, z)$ be a smooth homogeneous function of degree n (i.e. $F(cx, cy, cz) = c^n F(x, y, z)$). Prove that away from the origin the induced metric on the conical surface $F(x, y, z) = 0$ is Euclidean.

§8. The Second Fundamental Form

8.1. Curvature of Curves on a Surface in Euclidean Space

Suppose we are given a surface in Euclidean 3-space and a non-singular point (x_0, y_0, z_0) on it. We shall assume initially that the z-axis is perpendicular to the tangent plane to the surface at the point (x_0, y_0, z_0), in which case the x-axis and y-axis will be parallel to it. The surface may then be given locally about (i.e. in a neighbourhood of) the point (x_0, y_0, z_0) by an equation of the form $z = f(x, y)$, where $z_0 = f(x_0, y_0)$ and

$$\left.\frac{\partial f}{\partial x}\right|_{\substack{x=x_0\\y=y_0}} = \left.\frac{\partial f}{\partial y}\right|_{\substack{x=x_0\\y=y_0}} = 0, \quad \text{i.e. grad } f\,|_{\substack{x=x_0\\y=y_0}} = 0. \tag{1}$$

We now consider the second differential of the function $z = f(x, y)$ (which we now assume has continuous second derivatives), i.e. $d^2 l = f_{xx}\,dx^2 + 2f_{xy}\,dx\,dy + f_{yy}\,dy^2$. The matrix of this quadratic form, i.e. the matrix (a_{ij}) where $a_{ij} = f_{x^i x^j}$, $x^1 = x$, $x^2 = y$, is known as the *Hessian* of the function f.

8.1.1. Definition. Given a surface $z = f(x, y)$, and a point (x_0, y_0, z_0) on it at which grad $f = 0$, we define the *principal curvatures* of the surface at the point to be the eigenvalues of the matrix (a_{ij}) evaluated at the point. (These eigenvalues are real since (a_{ij}) is symmetric.) In addition we call $\det(a_{ij})$ at the point the *Gaussian curvature* K, and the trace of (a_{ij}), $\operatorname{tr}(a_{ij})$, at the point, the *mean curvature*, of the surface at (x_0, y_0, z_0).

Thus the eigenvalues k_1, k_2 say, of (a_{ij}) evaluated at (x_0, y_0, z_0) are the principal curvatures, while the Gaussian curvature is $K = k_1 k_2 = a_{11}a_{22} - a_{12}a_{21}$, and the mean curvature is $a_{11} + a_{22} = k_1 + k_2$. As we shall see later (in Chapter 4), the Gaussian curvature of the surface is an "intrinsic" invariant of the surface, i.e. depends only on the internal metrical properties of the surface; this is Gauss' famous "Theorema Egregium."

The disadvantage of Definition 8.1.1 is that it tells us what the curvatures at a point of the surface are in terms of a rather special co-ordinate system, influenced by the point under scrutiny in that the z-axis was chosen perpendicular to the surface at the point. To define these quantities in terms of arbitrary co-ordinates, we turn to the theory of the curvature of curves lying in the surface.

Thus suppose the surface given in parametric form

$$r = r(u, v). \tag{2}$$

We saw in §7.3 that the vector $[r_u, r_v]$ is normal to the surface at each non-singular point; hence if m denotes a unit vector in the same direction, then $[r_u, r_v] = |[r_u, r_v]|m$. If now $r = r(u(t), v(t))$ is a curve on the surface, then we have $\dot{r} = r_u \dot{u} + r_v \dot{v}$, and

$$\ddot{r} = (r_{uu}\dot{u}^2 + 2r_{uv}\dot{u}\dot{v} + r_{vv}\dot{v}^2) + (r_u \ddot{u} + r_v \ddot{v}).$$

Using $r_u \perp m$ and $r_v \perp m$, we deduce that

$$\langle \ddot{r}, m\rangle = \langle r_{uu}, m\rangle \dot{u}^2 + 2\langle r_{uv}, m\rangle \dot{u}\dot{v} + \langle r_{vv}, m\rangle \dot{v}^2 = b_{11}\dot{u}^2 + 2b_{12}\dot{u}\dot{v} + b_{22}\dot{v}^2. \tag{3}$$

From (3) we conclude that: *In terms of local co-ordinates $u = x^1$, $v = x^2$ the projection $\langle \ddot{r}, m\rangle$ of the acceleration vector $\ddot{r}$ on the normal to the surface, is a quadratic form in the components $\dot{u}$, $\dot{v}$ of the tangent vector.*

If we put $b_{11} = L$, $b_{12} = M$, $b_{22} = N$, then in terms of differentials (3) can be rewritten as

$$\langle \ddot{r}, m\rangle\, dt^2 = b_{ij}\, dx^i\, dx^j = L(du)^2 + 2M\, du\, dv + N(dv)^2. \tag{4}$$

We call this expression for $\langle \ddot{r}, m\rangle\, dt^2$, the *second fundamental form of the surface* (2). (Note that its coefficients L, M, N depend only on the point of the surface under scrutiny, and not on the curve $r = r(u(t), v(t))$.)

Suppose now that the curve $r = r(u(t), v(t))$ is parametrized by the natural parameter, i.e. $t = l$, where l is arc length. (Thus with this parametrization the curve is a "unit-speed curve.") By the first of the Serret–Frenet formulae (see 5.2.3), we have

$$\ddot{r} = \frac{d^2 r}{dl^2} = kn,$$

where n is the principal normal to the curve, and k is its curvature. Hence $\langle \ddot{r}, m\rangle = k\langle n, m\rangle = k \cos \theta$, where θ is the angle between the unit vectors m and n. From this and (4) we obtain

$$k \cos \theta (dl)^2 = \langle \ddot{r}, m)\, dl^2 = L(du)^2 + 2M\, du\, dv + N(dv)^2 = b_{ij}\, dx^i\, dx^j.$$

Comparing this with the formula for the induced metric on the surface:

$$dl^2 = g_{ij}\, dx^i\, dx^j = E(du)^2 + 2F\, du\, dv + G(dv)^2,$$

where $u = x^1$, $v = x^2$, we conclude that

$$k \cos \theta = \frac{b_{ij}\dot{x}^i \dot{x}^j}{g_{ij}\dot{x}^i \dot{x}^j},$$

where here the dots denote differentiation with respect to any parameter t. In the convenient differential notation we may rewrite this as

$$k \cos \theta = \frac{b_{ij}\, dx^i\, dx^j}{g_{ij}\, dx^i\, dx^j}, \qquad x^1 = u, \quad x^2 = v, \tag{5}$$

(although, as usual with this notation, we need to bear in mind that strictly speaking here dx^i is not an independent differential, but just $\dot{x}^i\, dt$).

We restate (5) in words, as a

8.1.2. Theorem. *The curvature at a point of a curve on a surface in Euclidean 3-space, when multiplied by the cosine of the angle between the normal to the surface and the principal normal of the curve at the point, is the same as the ratio of the second and first fundamental forms in the components of the tangent vector to the curve at the point.*

8.1.3. Corollary. *If the curve is obtained by sectioning the surface with a plane normal to the surface at the point, then (since* $\cos\theta = \pm 1$*) the corresponding "normal curvature" is given by*

$$k = \frac{b_{ij}\dot{x}^i\dot{x}^j}{g_{ij}\dot{x}^i\dot{x}^j} \tag{6}$$

(where here we allow k to be negative if m and n have opposite directions).

8.2. Invariants of a Pair of Quadratic Forms

Our main achievement so far in this chapter, is the association with each point of a surface in Euclidean 3-space, of the pair of quadratic ("fundamental") forms

$$dl^2 = g_{ij}\,dx^i\,dx^j, \tag{7}$$

$$\langle \ddot{r}, m\rangle\,dt^2 = b_{ij}\,dx^i\,dx^j, \tag{8}$$

the first of which is positive definite. Our aim is to deduce from these quadratic forms invariants of the surface. What are the known algebraic invariants of a pair of quadratic forms?

To answer this, consider any pair of quadratic forms in the plane with co-ordinates x, y, having matrices relative to these co-ordinates

$$G = \begin{pmatrix} g_{11} & g_{12} \\ g_{21} & g_{22} \end{pmatrix}, \qquad Q = \begin{pmatrix} b_{11} & b_{12} \\ b_{21} & b_{22} \end{pmatrix}, \tag{9}$$

(using the same symbols as in (7) and (8)), where G is positive definite and $g_{12} = g_{21}$, $b_{12} = b_{21}$. (Note that in contrast with the general context of (7) and (8), the g_{ij} and b_{ij} are assumed here to be fixed numbers.) Next consider the equation

$$\det(Q - \lambda G) = 0, \tag{10}$$

which, when written out in detail, becomes

$$(b_{11} - \lambda g_{11})(b_{22} - \lambda g_{22}) - (b_{12} - \lambda g_{12})^2 = 0. \tag{11}$$

We call the roots λ_1, λ_2 of this equation *the eigenvalues of the pair of quadratic forms.*

For each $i = 1, 2$ we look for solutions of the system of equations

$$\left.\begin{aligned} (b_{11} - \lambda_i g_{11})\xi_i^1 + (b_{12} - \lambda_i g_{12})\xi_i^2 &= 0, \\ (b_{12} - \lambda_i g_{12})\xi_i^1 + (b_{22} - \lambda_i g_{22})\xi_i^2 &= 0, \end{aligned}\right\} \tag{12}$$

in the unknowns ξ_i^1, ξ_i^2. Since λ_1, λ_2 satisfy (11), the coefficient matrix of each of the systems (12) is singular, so that there exist non-trivial solutions, say

$$f_1 = \xi_1^1 e_1 + \xi_1^2 e_2, \qquad f_2 = \xi_2^1 e_1 + \xi_2^2 e_2,$$

and all solutions will be scalar multiples of the appropriate one of f_1, f_2 (i.e. the solution space of (12) has dimension 1). The directions of the vectors f_1 and f_2 are called the *principal directions* in the plane of the pair of quadratic forms, that of f_1 corresponding to λ_1, and that of f_2 to λ_2.

We now equip our plane with a scalar product defined in terms of the positive definite quadratic form with matrix G, by setting $\langle e_i, e_j\rangle = g_{ij}$, where $e_1 = (1, 0)$, $e_2 = (0, 1)$.

8.2.1. Lemma. *If the eigenvalues of the pair of quadratic forms* (9) *are distinct, then the principal directions are orthogonal.*

PROOF. We wish to show that the two vectors $f_1 = \xi_1^1 e_1 + \xi_1^2 e_2$ and $f_2 = \xi_2^1 e_1 + \xi_2^2 e_2$ are orthogonal, i.e. that

$$\langle f_1, f_2\rangle = g_{ij}\xi_1^i\xi_2^j = 0.$$

Multiplying the first of the equations (12), with $i = 1$, by ξ_2^1, and the second by ξ_2^2, we obtain

$$(b_{11} - \lambda_1 g_{11})\xi_1^1\xi_2^1 + (b_{12} - \lambda_1 g_{12})\xi_1^2\xi_2^1 = 0,$$
$$(b_{11} - \lambda_1 g_{12})\xi_1^1\xi_2^2 + (b_{22} - \lambda_1 g_{22})\xi_1^2\xi_2^2 = 0.$$

On adding these equations we get

$$b_{ij}\xi_1^i\xi_2^j - \lambda_1 g_{ij}\xi_1^i\xi_2^j = 0. \tag{13}$$

If now we take $i = 2$ in the equations (12), and multiply the first of them by ξ_1^1 and the second by ξ_1^2, and again add the resulting equations, we obtain

$$b_{ij}\xi_1^i\xi_2^j - \lambda_2 g_{ij}\xi_1^i\xi_2^j = 0. \tag{14}$$

Finally, subtracting (13) from (14), and using the hypothesis that $\lambda_1 \neq \lambda_2$, we obtain the desired conclusion. □

This lemma may be regarded as a variant of the theorem from linear algebra which states that a real quadratic form assumes diagonal form relative to a suitable orthogonal basis; hence a real quadratic form can be brought to diagonal form by a rotation of the given orthogonal co-ordinate system.

8.3. Properties of the Second Fundamental Form

We now return to the first and second fundamental forms of a surface $r = r(u, v)$ in Euclidean 3-space:

$$dl^2 = g_{ij}dx^i\,dx^j, \tag{15}$$

$$\langle \ddot{r}, m\rangle\,dt^2 = b_{ij}dx^i\,dx^j, \qquad x^1 = u, \quad x^2 = v. \tag{16}$$

Recall that the ratio of these quadratic forms in the components of the tangent vector to the section of the surface by a plane orthogonal to the surface at a point (i.e. a "normal section"), gives the curvature of the section (i.e. "normal curvature") at that point (Corollary 8.1.3).

Having at our disposal the concepts introduced in the preceding subsection, we can now formulate the general definition of the principal, Gaussian, and mean curvatures of a surface at a point.

8.3.1. Definition. The eigenvalues of the pair of quadratic forms (15) and (16) (see §8.2) are called the *principal curvatures* of the surface at the point under investigation. The product of the principal curvatures is called the *Gaussian curvature* K of the surface at the point, and their sum the *mean curvature*.†

8.3.2. Example. As at the beginning of §8.1 (see especially 8.1.1), we suppose we are given our surface in the form $z = f(x, y)$, and further that at the point (x_0, y_0, z_0) of it under study, $f_x = f_y = 0$ (i.e. that the z-axis is perpendicular to the tangent plane to the surface at (x_0, y_0, z_0)). Putting $x = u$, $y = v$, the first and second fundamental forms at the point (x_0, y_0, z_0) are given by

$$g_{11} = 1, \qquad g_{22} = 1, \qquad g_{12} = g_{21} = 0 \qquad (g_{ij} = \delta_{ij}), \tag{17}$$

$$\begin{aligned} L = b_{11} &= \langle r_{uu}, m\rangle = f_{xx}(x_0, y_0), \\ M = b_{12} &= \langle r_{uv}, m\rangle = f_{xy}(x_0, y_0), \\ N = b_{22} &= \langle r_{vv}, m\rangle = f_{yy}(x_0, y_0). \end{aligned} \tag{18}$$

(Here of course the unit normal m to the surface at (x_0, y_0, z_0) is in the direction of the positive z-axis.) Thus at the point under scrutiny, the second fundamental form of the surface is

$$b_{11}\,dx^2 + 2b_{12}\,dx\,dy + b_{22}\,dy^2 = f_{xx}\,dx^2 + 2f_{xy}\,dx\,dy + f_{yy}\,dy^2 = d^2 f,$$

so that since $g_{ij} = \delta_{ij}$, the principal curvatures are the eigenvalues λ_1, λ_2 of the matrix

$$\begin{pmatrix} f_{xx} & f_{xy} \\ f_{xy} & f_{yy} \end{pmatrix}, \tag{19}$$

i.e. the roots of the equation $(f_{xx} - \lambda)(f_{yy} - \lambda) - (f_{xy})^2 = 0$, and the Gaussian curvature is $f_{xx} f_{yy} - (f_{xy})^2$, the determinant of the Hessian (19) of the function f. The mean curvature is $f_{xx} + f_{yy} = \lambda_1 + \lambda_2$. Thus our general Definition 8.3.1 does agree with the earlier one (8.1.1) where the various curvatures were defined in terms of a co-ordinate system dependent (as in the present example) very much on the particular point under consideration.

† It is perhaps more usual to define the mean curvature as the arithmetic mean of the principal curvatures.

Continuing with our example, let us now consider the principal directions at our point, of the pair of quadratic forms given by (17) and (18). As defined in the preceding subsection (see (12)), these are (the directions of) the non-zero vectors in the (x, y)-plane $f_1 = (\xi_1^1, \xi_1^2)$ and $f_2 = (\xi_2^1, \xi_2^2)$, where ξ_i^1, ξ_i^2 are non-trivial solutions of the following system of linear equations ($i = 1, 2$):

$$(f_{xx} - \lambda_i \delta_{11})\xi_i^1 + (f_{xy} - \lambda_i \delta_{12})\xi_i^2 = 0,$$
$$(f_{xy} - \lambda_i \delta_{12})\xi_i^1 + (f_{yy} - \lambda_i \delta_{22})\xi_i^2 = 0.$$

If we assume that the eigenvalues λ_1, λ_2 are distinct, then by Lemma 8.2.1, we have $f_1 \perp f_2$ (in the usual Euclidean sense since $g_{ij} = \delta_{ij}$). Hence we can replace the co-ordinate system x, y by a new system x', y', with axes in the principal directions, obtained from the old system by means of a rotation of the (x, y)-plane through an angle φ. Then in terms of the new co-ordinates x', y', z, we have

$$z = f(x(x', y'), y(x', y')),$$

where $x = x' \cos \varphi + y' \sin \varphi$, $y = -x' \sin \varphi + y' \cos \varphi$. Relative to these new co-ordinates, the second fundamental form becomes (at the point (x_0, y_0, z_0))

$$\lambda_1(dx')^2 + \lambda_2(dy')^2.$$

By Corollary 8.1.3, the curvature at our point (x_0, y_0, z_0) of a normal section of the surface through that point (i.e. the "normal curvature" at the point) is therefore

$$k = \frac{\lambda_1(\dot{x}')^2 + \lambda_2(\dot{y}')^2}{(\dot{x}')^2 + (\dot{y}')^2}, \tag{20}$$

where we are taking the normal section to have parametric equations $x' = x'(t)$, $y' = y'(t)$, $z = z(t)$.

Clearly the tangent vector to this normal section at the point (x_0, y_0, z_0) is $e = (\dot{x}', \dot{y}', 0)$. Hence the angle α between this tangent vector and the x'-axis satisfies

$$\cos^2 \alpha = \frac{(\dot{x}')^2}{(\dot{x}')^2 + (\dot{y}')^2}, \qquad \sin^2 \alpha = \frac{(\dot{y}')^2}{(\dot{x}')^2 + (\dot{y}')^2}. \tag{21}$$

From (20) and (21) we deduce that at the point (x_0, y_0, z_0), at least for our special choice of co-ordinates,

$$k = \lambda_1 \cos^2 \alpha + \lambda_2 \sin^2 \alpha.$$

This formula, called "Euler's formula," holds in any system of co-ordinates. We shall now prove this. (Thus our example ends here, although the argument we have used actually comprises the bulk of the proof.)

8.3.3. Theorem. *The curvature of a normal section* (i.e. *the* "*normal curvature*") *through a point on a surface, is given by the formula*

$$k = \lambda_1 \cos^2 \alpha + \lambda_2 \sin^2 \alpha, \tag{22}$$

where λ_1, λ_2 are the principal curvatures of the surface at the point, and α is the angle between the tangent vector to the normal section at the point, and the principal direction corresponding to λ_1.

PROOF. In Example 8.3.2 above we deduced Euler's formula in the case where the surface is given in the form $z = f(x, y)$, and the z-axis is perpendicular to the surface at the point (x_0, y_0, z_0) being studied. However since nothing of this special co-ordinate system is involved in the formula, for any particular (non-singular) point of the surface we simply choose a co-ordinate system with these properties (as we can, locally at least, by §7.1), i.e. so that the z-axis is normal to the surface at the point, and the x- and y-axes mutually orthogonal, and parallel to the tangent plane through the point (we may even choose them parallel to the principal directions). Then in a neighbourhood of the point the surface can be given by a function $z = f(x, y)$, with, of course, $f_x = f_y = 0$ at the point. Moreover if the x-axis and y-axis are chosen in the principal directions then $f_{xy} = f_{yx} = 0$, and $\lambda_1 = f_{xx}$, $\lambda_2 = f_{yy}$ at the point. We may then argue as in the example, provided $\lambda_1 \neq \lambda_2$ (which is assumed implicitly in the theorem since otherwise the principal directions remain undefined). This completes the proof. □

Note that if we reverse the direction of the z-axis in the above, then the signs of the principal curvatures are also reversed.

8.3.4. Corollary. *If λ_1 and λ_2 are the principal curvatures of a surface at a point and $\lambda_1 > \lambda_2$, then λ_1 and λ_2 are respectively the greatest and smallest values of the normal curvature (i.e. curvature of a normal section) at the point. (Here the signs are taken into account—see Corollary* 8.1.3.)

We next derive useful formulae for the second fundamental form and the Gaussian curvature in the case that the surface is given in the form $z = f(x, y)$ (but where we are no longer assuming a particular direction for the z-axis at any point). In that case, with $x = u$, $y = v$, and $r(u, v) = (u, v, f(u, v))$ we have

$$r_u = (1, 0, f_x), \qquad r_v = (0, 1, f_y),$$
$$[r_u, r_v] = (-f_x, -f_y, 1),$$

$$r_{uu} = (0, 0, f_{xx}), \qquad r_{uv} = (0, 0, f_{xy}), \qquad r_{vv} = (0, 0, f_{yy}),$$

$$m = \frac{[r_u, r_v]}{|[r_u, r_v]|} = \frac{(-f_x, -f_y, 1)}{\sqrt{1 + f_x^2 + f_y^2}}, \tag{23}$$

$$L = b_{11} = \frac{f_{xx}}{\sqrt{1 + f_x^2 + f_y^2}}, \qquad M = b_{12} = b_{21} = \frac{f_{xy}}{\sqrt{1 + f_x^2 + f_y^2}},$$

$$N = b_{22} = \frac{f_{yy}}{\sqrt{1 + f_x^2 + f_y^2}}.$$

Hence

$$b_{ij}\,dx^i\,dx^j = \frac{1}{\sqrt{1 + f_{x^1}^2 + f_{x^2}^2}}\, f_{x^i x^j}\,dx^i\,dx^j), \tag{24}$$

$$x^1 = u = x, \qquad x^2 = v = y.$$

Recall also the formulae (20) of §7 for the g_{ij}:

$$\begin{gathered} g_{11} = 1 + f_x^2, \qquad g_{12} = g_{21} = f_x f_y, \qquad g_{22} = 1 + f_y^2, \\ g = g_{11}g_{22} - g_{12}^2 = 1 + f_x^2 + f_y^2. \end{gathered} \tag{25}$$

We shall use formulae (24) and (25) in the proof of the following

8.3.5. Theorem. *The Gaussian curvature of a surface is equal to the ratio of the determinants of the second and the first fundamental forms:*

$$K = \frac{b_{11}b_{22} - b_{12}^2}{g_{11}g_{22} - g_{12}^2}. \tag{26}$$

In particular, if the surface is given in the form $z = f(x, y)$, then we have

$$K = \frac{f_{xx}f_{yy} - f_{xy}^2}{(1 + f_x^2 + f_y^2)^2}. \tag{27}$$

PROOF. By definition, the principal curvatures λ_1, λ_2 are the solutions of equation (10) above:

$$\det(Q - \lambda G) = 0,$$

where $Q = (b_{ij})$ is the matrix of the second fundamental form, and $G = (g_{ij})$. Since the first fundamental form is positive definite, its matrix G is non-singular. Hence

$$\det(Q - \lambda G) = \det G \det(G^{-1}Q - \lambda \cdot 1),$$

from which it follows that λ_1 and λ_2 are just the eigenvalues of the matrix $G^{-1}Q$. Since the product of the eigenvalues of a matrix is the same as the determinant (a fact we have used before!), we deduce that

$$K = \lambda_1\lambda_2 = \det(G^{-1}Q) = \frac{\det Q}{\det G},$$

which proves (26).

To get formula (27), simply substitute in (26) the expressions for the b_{ij} and g_{ij} given in (24) and (25). □

8.3.6. Corollary. *If a surface is given in the form $z = f(x, y)$, then the sign of the Gaussian curvature K is the same as the sign of the determinant $f_{xx}f_{yy} - f_{xy}^2$ of the Hessian of $f(x, y)$.*

8.3.7. Example. Suppose that we are given a surface in the form $z = f(x, y)$, where the function $f(x, y)$ is a solution of Laplace's equation $f_{xx} + f_{yy} = 0$. Then for all points of the surface we shall have $f_{xx}f_{yy} - f_{xy}^2 \leqq 0$, since $f_{xx} = -f_{yy}$. Hence by Corollary 8.3.6, at all points of the surface where at least one of the partial derivatives f_{xx}, f_{xy} is non-zero, we shall have that the Gaussian curvature $K < 0$.

Finally we explain the geometric significance of the sign of the Gaussian curvature at a point. As usual we choose a co-ordinate system x, y, z, with mutually orthogonal co-ordinate axes, such that the z-axis is perpendicular to the surface at the point (x_0, y_0, z_0). Then locally (about the point) the surface is given by a function $z = f(x, y)$, where $f_x = f_y = 0$. We consider three cases:

(i) $K > 0; \lambda_1 > 0, \lambda_2 > 0$. Since the smallest value of the normal curvature is positive, all values of it are positive, i.e. at the point the surface bends upwards in all directions, so the function $f(x, y)$ has a local minimum at the point (x_0, y_0) (see Figure 10(a)).
(ii) $K > 0;\ \lambda_1 < 0,\ \lambda_2 < 0$. The function $f(x, y)$ has a local maximum at (x_0, y_0) (see Figure 10(b)).
(iii) $K < 0$. In this case we must have $\lambda_1 > 0, \lambda_2 < 0$, or $\lambda_1 < 0, \lambda_2 > 0$. The function has a "saddle" or "pass" at the point (Figure 10(c)).

We conclude that: *If $K > 0$ there is a neighbourhood of the point throughout which the surface lies on one side of the tangent plane at the point. If $K < 0$ then the surface intersects the tangent plane at the point arbitrarily close to the point.*

If the Gaussian curvature is positive at every point of a surface, then we say that the surface is *strictly convex*.

Note that in the above, by using Corollary 8.3.6, and the fact that λ_1, λ_2 are the eigenvalues of the Hessian of $f(x, y)$ (since $g_{ij} = \delta_{ij}$ at the point), it is easy to see that cases (i) and (ii) are respectively equivalent to: $K > 0$; $f_{xx}, f_{yy} > 0$, and $K > 0; f_{xx}, f_{yy} < 0$. In this form, (i), (ii) and (iii) comprise the familiar "second-partials test" for determining the nature of a surface in a neighbourhood of a point at which the tangent plane is "horizontal."

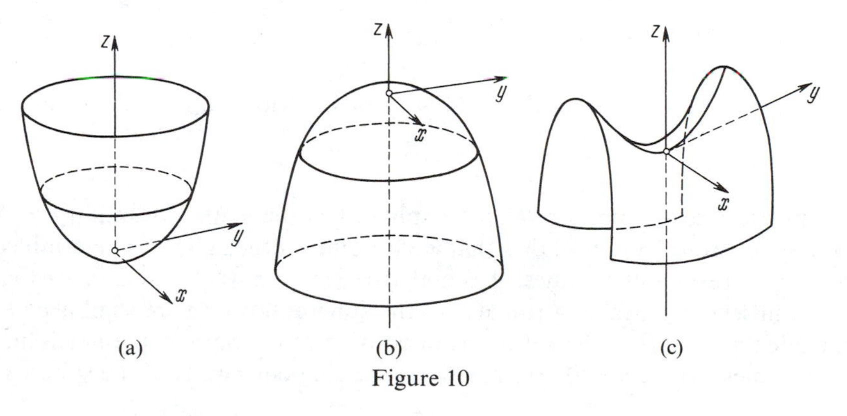

Figure 10

8.4. Exercises

1. Find the surface all of whose normals intersect at a single point.

2. Calculate the second fundamental form for the surface of revolution

$$r(u, \varphi) = (x(u), \rho(u) \cos \varphi, \rho(u) \sin \varphi), \qquad \rho(u) > 0.$$

3. Calculate the Gaussian and mean curvatures of a surface given by an equation of the form

$$z = f(x) + g(y).$$

4. Prove that if the Gaussian and mean curvatures of a surface (as usual in Euclidean 3-space), are identically zero, then the surface is a plane.

5. Show that for the surface $z = f(x, y)$, the mean curvature is given by

$$H = \operatorname{div}\left(\frac{\operatorname{grad} f}{\sqrt{1 + |\operatorname{grad} f|^2}}\right).$$

6. Let S denote the surface swept out (i.e. "generated") by the tangent vector to a given curve with curvature $k(l)$. Prove that if the curve is twisted, but in such a way as to preserve $k(l)$, then the metric on the surface S is also preserved.

7. If the metric on a surface has the form

$$dl^2 = A^2\, du^2 + B^2\, dv^2, \qquad A = A(u, v), \qquad B = B(u, v),$$

then its Gaussian curvature is given by

$$K = -\frac{1}{AB}\left[\left(\frac{A_v}{B}\right)_v + \left(\frac{B_u}{A}\right)_u\right].$$

8. Show that the only surfaces of revolution with zero mean curvature are the plane and the catenoid (which is the surface obtained by revolving the curve $x = [\cosh(ay + b)]/a$ about the y-axis).

§9. The Metric on the Sphere

The equation of the sphere $S^2 \subset \mathbb{R}^3$ of radius R with centre at the origin is

$$x^2 + y^2 + z^2 = R^2. \tag{1}$$

In spherical co-ordinates r, θ, φ the sphere has the simple equation $r = R$ (with θ, φ arbitrary). It follows that each point on the sphere is determined by the corresponding values of θ and $\varphi (0 \leq \theta \leq \pi, 0 \leq \varphi < 2\pi)$, so that θ, φ will serve as local co-ordinates of the sphere; however we shall need to exclude two points of the sphere from consideration, namely the north and south poles (where $\theta = 0, \pi$) since these are singular points of the spherical

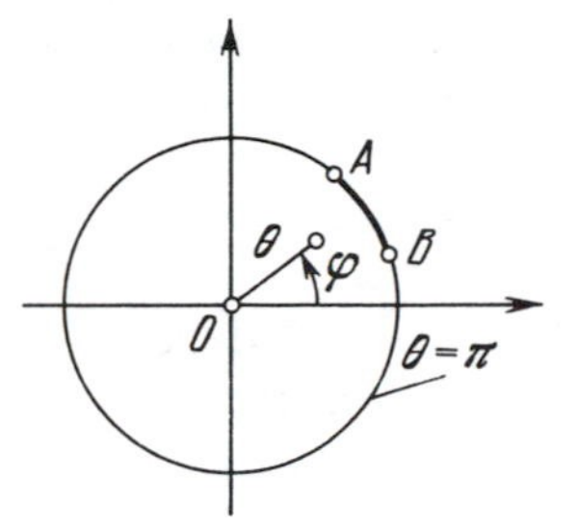

Figure 11. The (θ, φ)-plane. The arc AB of the circle of radius $\pi(\theta = \pi)$ has zero length as measured using the metric (3); i.e. the boundary $\theta = \pi$ of the disc, is pinched to a point, yielding the 2-sphere.

co-ordinate system (see 1.2.6(c)). We know from §3.1(7) that in terms of spherical co-ordinates the Euclidean metric takes the form

$$dl^2 = dx^2 + dy^2 + dz^2 = dr^2 + r^2(d\theta^2 + \sin^2 \theta \, d\varphi^2). \tag{2}$$

On the surface $r = R$, the differential dr is zero, so that the metric induced on the sphere is given by

$$dl^2 = R^2(d\theta^2 + \sin^2 \theta \, d\varphi^2) \tag{3}$$

(i.e. this is the square of an element of arc length of a curve $\rho = \rho(t)$ on the sphere). Here $0 \le \theta \le \pi$, $0 \le \varphi < 2\pi$ (see Figure 11).

Note that, since $\sin \theta \approx \theta$ for small θ, around the north pole the metric (3) is given approximately by

$$\frac{dl^2}{R^2} = d\theta^2 + \theta^2(d\varphi)^2,$$

i.e. near the north pole the metric on the sphere is approximately Euclidean, with polar co-ordinates θ, φ (as in Figure 11).

We now consider the *stereographic projection* of the sphere onto the plane (see Figure 12 which shows a section of the sphere by a plane through its centre). The stereographic projection sends a point (θ, φ) on the sphere to the point with polar co-ordinates (r, φ) in the plane, as shown in Figure 12, from which it is clear that $\varphi = \varphi$, $r = R \cot(\theta/2)$. Rewriting the metric (3)

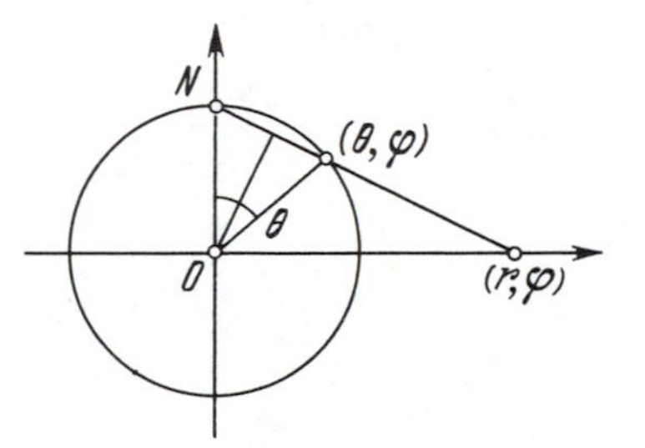

Figure 12. The stereographic projection: $(\theta, \varphi) \mapsto (r, \varphi)$.

in terms of these new co-ordinates r, φ, and then in terms of the usual Euclidean co-ordinates x, y of the plane, we obtain

$$dl^2 = \frac{4R^4}{(R^2 + r^2)^2}(dr^2 + r^2\,d\varphi^2) = \frac{4R^4}{(R^2 + x^2 + y^2)^2}(dx^2 + dy^2), \quad (4)$$

i.e. the metric on the sphere is obtained from the Euclidean metric on the plane by multiplying the latter by the function $4R^4/(R^2 + x^2 + y^2)^2$:

$$dl^2_{\text{sphere}} = \frac{4R^4}{(R^2 + x^2 + y^2)^2}\,dl^2_{\text{plane}}. \quad (5)$$

Example. We shall now calculate (in terms of the metric on the sphere) the circumference of a circle, and the area of a disc on the sphere, both of radius ρ. We may suppose that centre of our circle-on-the-sphere is at the north pole N, i.e. at the point $\theta = 0$ (see Figure 13). Then the radius ρ of the circle is equal to $R\theta_0$, so that the equation of the circle is $\theta = \rho/R$, $0 \le \varphi < 2\pi$. The disc of radius ρ is the set of points θ, φ satisfying $\theta < \rho/R$, $0 \le \varphi < 2\pi$.

We first calculate the circumference of the circle. On the circle, $\theta = \rho/R$ ($=$ const.) so we have

$$dl^2 = R^2(d\theta^2 + \sin^2\theta\,d\varphi^2) = R^2\sin^2\frac{\rho}{R}(d\varphi)^2,$$

and then its circumference l_ρ is

$$l_\rho = \int_0^{2\pi} R\sin\left(\frac{\rho}{R}\right)d\varphi = 2\pi R\sin\left(\frac{\rho}{R}\right). \quad (6)$$

As expected, we see that the equator (where $\rho = R(\pi/2)$) has the greatest length, and that when $\rho = R\pi$, the circle has become a point (the south pole), and has length zero. From the formula (6) it follows that the ratio of the circumference of the circle to its radius ρ is always less than 2π:

$$\frac{l_\rho}{\rho} = 2\pi\,\frac{\sin(\rho/R)}{\rho/R} < 2\pi \qquad (\text{since } \rho > 0).$$

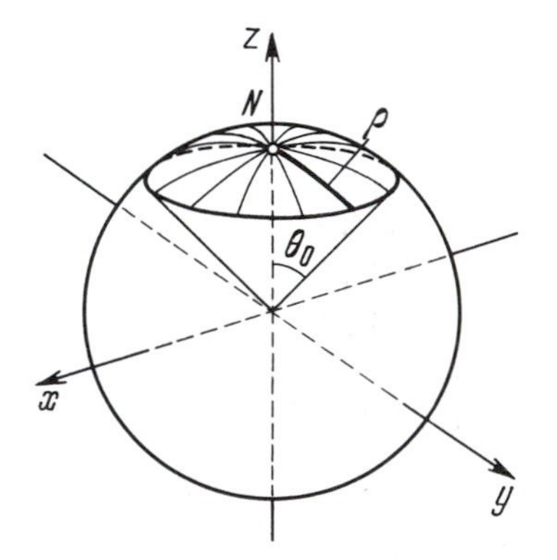

Figure 13. The disc of radius ρ on the sphere.

We next calculate the area of the disc of radius ρ. From (3) we see that $g = g_{11}g_{22} - g_{12}^2 = R^4 \sin^2 \theta$, so that by Definition 7.4.1 the area σ_ρ of the disc is

$$\sigma_\rho = \iint\limits_{0 \le \theta \le \rho/R} R^2 \, |\sin \theta| \, d\theta \, d\varphi = R^2 \int_0^{\rho/R} \sin \theta \, d\theta \int_0^{2\pi} d\varphi$$

$$= 2\pi R^2 \left(1 - \cos \frac{\rho}{R}\right), \tag{7}$$

where we have used the fact that $\rho \le \pi R$, and also the standard result that a double integral is equal to the corresponding repeated integral, provided only that both integrals exist.

Note that in particular when $\rho = \pi R$, the disc is the whole sphere, whence we obtain the familiar formula $4\pi R^2$ for the surface area of the sphere.

Note also that for small ρ we have $\sin(\rho/R) \approx \rho/R$, and $1 - \cos(\rho/R) \approx \rho^2/2R^2$, whence

$$l_\rho = 2\pi R \sin \frac{\rho}{R} \approx 2\pi \rho,$$

$$\sigma_\rho = 2\pi R^2 \left(1 - \cos \frac{\rho}{R}\right) \approx \pi \rho^2,$$

i.e. for small circles on the sphere the formulae for circumference and area are approximately the same as those for circles in the Euclidean plane (which is no surprise to earth-dwellers!). With this we conclude our example.

We next calculate the Gaussian and mean curvatures for our sphere of radius R. Observe first that the normal sections are the circles of radius R on the sphere (the so-called "great circles" on the sphere). It follows that the normal curvature is everywhere and in all directions just the constant R^{-1} (see 5.1.3). Hence the two principal curvatures are both equal to R^{-1} everywhere, so that the Gaussian curvature is $1/R^2$, and the mean curvature is $2/R$.

Finally we investigate the group of motions of the metric on the sphere (i.e. the group of isometries of the sphere). Any orthogonal transformation of Euclidean 3-space maps our sphere onto itself. It is clear that an isometry of Euclidean 3-space preserving the sphere must induce an isometry of the sphere (since arc length on the sphere is a special case of arc length in the ambient 3-space). Hence the group of isometries of the sphere S^2 contains $O(3)$, the full group of orthogonal transformations of Euclidean 3-space (see §4.3). It turns out to be the case that $O(3)$ is the full isometry group of S^2, i.e. every isometry of S^2 is induced by an orthogonal transformation of $\mathbb{R}^3$. (A strict proof of this requires, however, the concept of a geodesic, which we shall introduce in Chapter 4.) We note finally that it can be shown that the values of three parameters need to be given for an orthogonal transformation to be fully determined; in other words the orthogonal group $O(3)$ is a 3-dimensional space (see §14.1).

§10. Space-like Surfaces in Pseudo-Euclidean Space

10.1. The Pseudo-Sphere

Consider the 3-dimensional pseudo-Euclidean space $\mathbb{R}_1^3$, with co-ordinates (t, x, y) in terms of which the pseudo-Euclidean metric has the form

$$dl^2 = dt^2 - dx^2 - dy^2. \tag{1}$$

By the *pseudo-sphere of radius* R in the space $\mathbb{R}_1^3$, we mean the set of points satisfying the equation

$$t^2 - x^2 - y^2 = R^2. \tag{2}$$

Figure 14 shows this surface (the "two-sheeted hyperboloid") as it is usually represented in ordinary space. The pseudo-sphere lies in the interior of the light cone $t^2 - x^2 - y^2 = 0$, and in terms of the pseudo-spherical co-ordinates ρ, χ, φ (see §3.2(11)), has equations

$$\rho = R \qquad \text{(upper half)},$$

$$\rho = -R \qquad \text{(lower half)}.$$

For most of what follows we shall restrict our attention to the upper half only of the hyperboloid, for which $\rho = R$. Recall (from §3.2(12)) that in pseudo-spherical co-ordinates the metric (1) takes the form

$$dl^2 = -\rho^2(d\chi^2 + \sinh^2 \chi\, d\varphi^2) + d\rho^2. \tag{3}$$

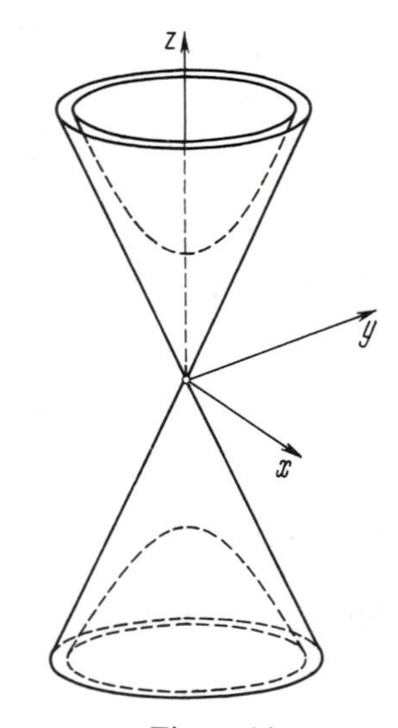

Figure 14

Since ρ is constant $(=R)$ on the (upper half of the) hyperboloid, the metric induced on it by that of the ambient pseudo-Euclidean space is given by

$$-dl^2 = R^2(d\chi^2 + \sinh^2 \chi \, d\varphi^2). \tag{4}$$

Thus the quadratic form defining the metric induced on the pseudo-sphere is negative definite, so that the pseudo-sphere $t^2 - x^2 - y^2 = R^2$ in $\mathbb{R}^3_1$ is a *space-like hypersurface*, by which we mean that the tangent vectors to this surface are all space-like (see §6.1).

Analogously to the stereographic projection of the sphere onto the plane, described in the preceding section, one can define a *stereographic projection of the pseudo-sphere* onto the plane. (We define the *centre* of the pseudo-sphere to be the origin, and the *north* and *south poles* to be the points with co-ordinates $(R, 0, 0)$ and $(-R, 0, 0)$ respectively.) The stereographic projection f, defined in Figure 15, maps the upper half of the pseudo-sphere onto the open disc $x^2 + y^2 < R^2$. If the point P on the pseudo-sphere has co-ordinates (t, x, y) (where $t > 0$), and the point $f(P)$ on the plane has co-ordinates (u, v), then from Figure 15 it is easy to see that

$$\frac{x}{u} = \frac{t + R}{R}, \qquad \frac{y}{v} = \frac{t + R}{R},$$

whence

$$x = u\left(1 + \frac{t}{R}\right), \qquad y = v\left(1 + \frac{t}{R}\right).$$

On substituting for x and y in the equation of the pseudo-sphere $t^2 - x^2 - y^2 = R^2$, and solving the resulting quadratic equation for $t > 0$, we get

$$t = -R\left(1 + \frac{2R^2}{u^2 + v^2 - R^2}\right), \tag{5}$$

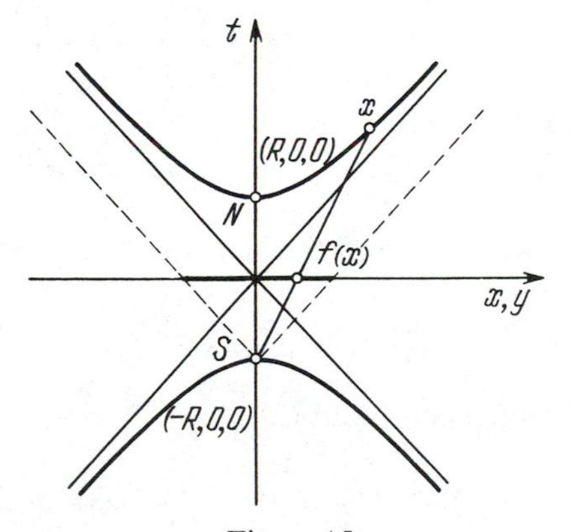

Figure 15

whence

$$x = \frac{2R^2u}{R^2 - u^2 - v^2}, \qquad y = \frac{2R^2v}{R^2 - u^2 - v^2}. \tag{6}$$

Formulae (5) and (6) for the stereographic projection, give us the co-ordinates (t, x, y) of a general point on the surface, in terms of the parameters u, v. Analogously to the case of the sphere, we now express the induced metric in terms of the co-ordinates (u, v). To this end, first recall (from §3.2(11)) that

$$t = R\cosh\chi, \qquad x = R\sinh\chi\cos\varphi, \qquad y = R\sinh\chi\sin\varphi. \tag{7}$$

From the first equation of (7) together with (5), we get, putting $r^2 = u^2 + v^2$,

$$\cosh\chi = -\left(1 + \frac{2R^2}{(r^2 - R^2)}\right),$$

whence

$$\sinh\chi\, d\chi = \frac{4R^2r}{(r^2 - R^2)^2}\, dr, \tag{8}$$

while from (6) and (7) we have

$$\sinh^2\chi = \frac{x^2 + y^2}{R^2} = \frac{4R^2r^2}{(R^2 - r^2)^2}. \tag{9}$$

Using (8) and (9) it is then straightforward to get from (4) to

$$-dl^2 = \frac{4R^4}{(R^2 - r^2)^2}(dr^2 + r^2\, d\varphi^2),$$

that is,

$$-dl^2 = \frac{4R^4}{(R^2 - u^2 - v^2)^2}(du^2 + dv^2). \tag{10}$$

We see that, as in the case of the sphere (see (5) of the preceding section), in terms of the co-ordinates u, v, the metric on the pseudo-sphere is obtained from the metric on the Euclidean plane by multiplying the latter by a function (of u and v), i.e. in the obvious extended sense of the word, these metrics are "proportional."

If we drop the minus sign from the right-hand side of (4) we obtain the metric

$$dl^2 = R^2(d\chi^2 + \sinh^2\chi\, d\varphi^2) \tag{11}$$

on the upper half of the hyperboloid (2).

10.1.1. Definition. The metric (11) on the upper half of the hyperboloid (2) is called the *Lobachevsky metric*.

In terms of the co-ordinates u, v, the metric (11) takes the form

$$dl^2 = \frac{4R^4}{(R^2 - u^2 - v^2)^2}(du^2 + dv^2), \tag{12}$$

where $u^2 + v^2 < R^2$, since the upper half of the hyperboloid projects stereographically onto the open disc defined by $u^2 + v^2 < R^2$. That open disc equipped with the metric (12), is called the *Poincaré model of Lobachevsky's geometry*.

For the sake of comparison, we gather together the different forms of the metrics on the sphere S^2 and the Lobachevskian plane L^2 in the following table (where $R = 1$):

S^2	L^2
$d\theta^2 + \sin^2\theta(d\varphi)^2$	$d\chi^2 + \sinh^2\chi(d\varphi)^2$
$4\dfrac{dx^2 + dy^2}{(1 + x^2 + y^2)^2}$	$4\dfrac{dx^2 + dy^2}{(1 - x^2 - y^2)^2}, \quad x^2 + y^2 < 1$

We shall now express the Lobachevsky metric in terms of yet other co-ordinates (or, to put it differently, we shall give yet another model of the Lobachevskian plane). It is a well-known fact of the theory of functions of a complex variable, that there is a linear-fractional transformation of the complex plane which sends the upper half-plane onto the unit disc; for instance the transformation $z = (1 + iw)/(1 - iw)$ will do the trick (see Figure 16). If we write $z = u + iv$, $w = x + iy$, then this linear-fractional transformation can be regarded as introducing new co-ordinates x, y ($y > 0$) on the open unit disc. Direct computation shows that in terms of these new co-ordinates the Lobachevsky metric takes the form

$$dl^2 = \frac{dx^2 + dy^2}{y^2}, \qquad y > 0. \tag{13}$$

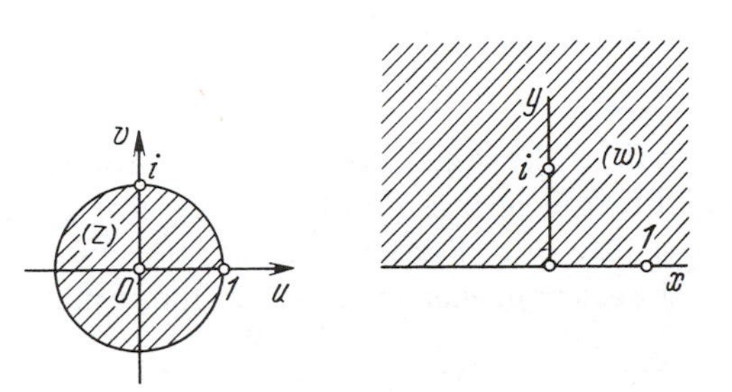

Figure 16

Regarded as a metric on the open upper half-plane, (13) is known as the metric of *Klein's model of Lobachevskian geometry*.

To conclude this subsection, we consider the group of isometries of the Lobachevskian plane. Every pseudo-orthogonal transformation of the space $\mathbb{R}_1^3$ (i.e. element of $O(1, 2)$—see §6.2) preserves the quadratic form $t^2 - x^2 - y^2$, and therefore maps the pseudo-sphere onto itself. However since the Lobachevskian plane consists of the upper half of the pseudo-sphere only, we must exclude those transformations in $O(1, 2)$ which interchange the two halves. We conclude that the group of motions of the Lobachevskian plane contains the group of orthochronous (see §6.2) transformations in $O(1, 2)$. We shall show in Chapter 4 of Part II that in fact these exhaust the isometries of L^2. It follows that the Lobachevskian plane, like the sphere, has a 3-parameter group of motions.

10.2. Curvature of Space-like Curves in $\mathbb{R}_1^3$

As mentioned above (in §10.1), when we say that a surface in $\mathbb{R}_1^3$ is "space-like" we mean that every vector tangent to it is space-like (in the sense of §6.1), or, what comes to the same thing, that the metric induced on the surface by the Minkowski metric $dl^2 = dt^2 - dx^2 - dx^2$, is negative definite.

We shall assume that our space-like surface is given to us in the form $t = f(x, y)$. Then since $dt = f_x\, dx + f_y\, dy$, the metric induced on the surface is given by

$$-dl^2 = -(dt^2 - dx^2 - dy^2) = (1 - f_x^2)\, dx^2 - 2f_x f_y\, dx\, dy + (1 - f_y^2)\, dy^2$$
$$= g_{ij} dx^i\, dx^j, \qquad x^1 = x, \quad x^2 = y,$$

whence $\det(g_{ij}) = 1 - f_x^2 - f_y^2$. The eigenvalues of the matrix (g_{ij}) are easily calculated to be 1 and $1 - f_x^2 - f_y^2$, so that for the quadratic form $dl^2 = g_{ij}\, dx^i\, dx^j$ to be negative definite, we require

$$1 - f_x^2 - f_y^2 (= \det(g_{ij})) > 0.$$

The unit normal to the surface is then

$$m = \frac{(1, f_x, f_y)}{\sqrt{1 - f_x^2 - f_y^2}}$$

(Note that this is a time-like vector. We leave it to the reader to verify that it is indeed normal to the surface.) As for the case of a surface situated in Euclidean space, so also for a space-like surface $r = r(x^1, x^2)$ in Minkowski space, we define the *second fundamental form* of the surface, by

$$b_{ij} = \left\langle \frac{\partial^2 r}{\partial x^i\, \partial x^j}, m \right\rangle, \qquad i, j = 1, 2,$$

where m is the unit normal to the surface (at the point under study). Then by partial analogy with Theorem 8.3.5, we define the *Gaussian curvature K of a space-like surface* in $\mathbb{R}_1^3$ by

$$K = -\frac{\det(b_{ij})}{\det(g_{ij})}. \tag{14}$$

(The minus sign is explained below.) It follows that if the surface is given in the form $t = f(x, y)$, with $1 - f_x^2 - f_y^2 > 0$ (see above), then (cf. §8.3 (27))

$$K = \frac{f_{xy}^2 - f_{xx}f_{yy}}{(1 - f_x^2 - f_y^2)^2}. \tag{15}$$

In particular for the hyperboloid $t^2 - x^2 - y^2 = 1$, we obtain $K \equiv -1$.

Remarks. 1. In Chapter 4 we shall compute the curvature of the Lobachevskian plane in terms of its intrinsic geometry. We shall again arrive at $K \equiv -1$, vindicating the choice of sign in the definition (14).

2. By analogy with the definition given in §8.3, we shall say that a space-like surface in $\mathbb{R}_1^3$ is *convex* if $K < 0$ at all points of the surface.

§11. The Language of Complex Numbers in Geometry

11.1. Complex and Real Co-ordinates

Many geometrical problems are most conveniently formulated and solved in terms of the language of complex numbers. We therefore expound here the simple facts about complex numbers that we shall be needing.

We denote by $\mathbb{C}^n$ the vector space of dimension n over the field of complex numbers $\mathbb{C}$. (A vector space is determined up to isomorphism by its dimension.) If $e_1, \ldots, e_n$ form a basis for this vector space, then every vector $\xi \in \mathbb{C}^n$ has the unique form

$$\xi = z^k e_k, \qquad z^k = x^k + iy^k, \tag{1}$$

where the z^k are the (complex) components.

We can regard $\mathbb{C}^n$ also as a $2n$-dimensional vector space $\mathbb{R}^{2n}$ over the reals; as a basis for this space we might then take

$$e_1, \ldots, e_n, ie_1, \ldots, ie_n. \tag{2}$$

Since

$$\xi = z^k e_k = x^k e_k + y^k(ie_k), \tag{3}$$

the real components of the vector ξ relative to the basis (2), are the numbers $x^1, \ldots, x^n, y^1, \ldots, y^n$. This process of going from a complex vector space to a real one is called *realization*.

The non-singular (i.e. invertible) linear transformations of the vector space $\mathbb{C}^n$ form a group, called the *general linear group of degree n over* $\mathbb{C}$, and denoted by $GL(n, \mathbb{C})$. This group is of course isomorphic to the group of $n \times n$ non-singular matrices with complex entries (so that, as usual, we shall often think of $GL(n, \mathbb{C})$ as consisting of matrices). Upon realization of $\mathbb{C}^n$, each of these transformations becomes a linear transformation of the real vector space $\mathbb{R}^{2n}$. This defines a monomorphism (i.e. one-to-one homomorphism)

$$r: GL(n, \mathbb{C}) \to GL(2n, \mathbb{R}), \tag{4}$$

called the *realization map*.

11.1.1. Example. In the case $n = 1$, our space is just the one-dimensional complex vector space $\mathbb{C}$ co-ordinatized by $z = x + iy$. The invertible linear transformations are just multiplications by a complex number $\lambda \neq 0$ (one such λ for each linear transformation):

$$z \mapsto \lambda z. \tag{5}$$

Writing $\lambda = a + ib$, $a^2 + b^2 \neq 0$, this takes the form

$$z = x + iy \mapsto (a + ib)(x + iy) = (ax - by) + i(bx + ay).$$

Hence the corresponding linear transformation of $\mathbb{R}^2$ has matrix

$$\begin{pmatrix} a & b \\ -b & a \end{pmatrix} = r(\lambda), \qquad a^2 + b^2 \neq 0. \tag{6}$$

Thus we see that certainly not all (invertible) linear transformation of $\mathbb{R}^2$ are obtained in this way from the complex linear transformations of $\mathbb{C}$; in other words, r is not onto.

This example generalizes to the n-dimensional case: It is not difficult to show that if Λ is a matrix in $GL(n, \mathbb{C})$, and $\Lambda = A + iB$ where A and B are real matrices, then

$$r(\Lambda) = \begin{pmatrix} A & B \\ -B & A \end{pmatrix}. \tag{7}$$

11.1.2 Exercise. Prove that $\det(r(\Lambda)) = |\det \Lambda|^2$.

Here is a quick proof of (7) which incidentally yields an alternative characterization of the matrices of the form $r(\Lambda)$, i.e. of the matrices occurring as images under the map $r: GL(n, \mathbb{C}) \to GL(2n, \mathbb{R})$. If Λ is any element of $GL(n, \mathbb{C})$, and ξ is any vector in $\mathbb{C}^n$, then

$$\Lambda(i\xi) = i\Lambda(\xi). \tag{8}$$

Now the linear transformation of $\mathbb{C}^n$ which simply multiplies each vector by the scalar i, goes under r to a linear transformation I of $\mathbb{R}^{2n}$, given by

$$r(i) = I = \begin{pmatrix} 0 & 1 \\ -1 & 0 \end{pmatrix}, \quad \text{where } 1 = n \times n \text{ identity matrix.} \tag{9}$$

(To see this consider the action of i on the basis vectors e_k, ie_k.) It follows from (8) that the matrix I commutes with the matrix $r(\Lambda)$ (since ξ was arbitrary). But then it is straightforward to check that matrices commuting with I must have the form $\begin{pmatrix} A & B \\ -B & A \end{pmatrix}$. We leave to the reader the verification that a matrix of this form does indeed arise from a "complex operator" on $\mathbb{C}^n$.

We conclude by mentioning an important subgroup of $GL(n, \mathbb{C})$, namely that consisting of all matrices of determinant 1, called the *special linear group of degree n over* $\mathbb{C}$, and denoted by $SL(n, \mathbb{C})$.

11.2. The Hermitian Scalar Product

We define a scalar product on the space $\mathbb{C}^n$ as follows:

$$\langle \xi_1, \xi_2 \rangle_{\mathbb{C}} = \sum_{k=1}^{n} z_1^k \bar{z}_2^k, \tag{10}$$

where $\xi_1 = z_1^i e_i$, $\xi_2 = z_2^i e_i$, and the bar indicates the complex conjugate. It follows that if $\xi = z^i e_i$, then

$$\langle \xi, \xi \rangle_{\mathbb{C}} = \sum_{k=1}^{n} |z^k|^2. \tag{11}$$

This scalar product has the following main properties:

$$\begin{aligned} \langle \lambda\xi, \eta \rangle_{\mathbb{C}} &= \lambda \langle \xi, \eta \rangle_{\mathbb{C}}, \\ \langle \xi, \lambda\eta \rangle_{\mathbb{C}} &= \bar{\lambda} \langle \xi, \eta \rangle_{\mathbb{C}}, \\ \langle \xi, \eta \rangle_{\mathbb{C}} &= \overline{\langle \eta, \xi \rangle}_{\mathbb{C}}, \\ \langle \xi_1 + \xi_2, \eta \rangle_{\mathbb{C}} &= \langle \xi_1, \eta \rangle_{\mathbb{C}} + \langle \xi_2, \eta \rangle_{\mathbb{C}}, \\ \langle \xi, \xi \rangle_{\mathbb{C}} &> 0 \quad \text{if } \xi \neq 0. \end{aligned} \tag{12}$$

Any scalar product on $\mathbb{C}^n$ with the properties (12) is called an *Hermitian scalar product*, or *Hermitian form*.

If on the realized space $\mathbb{R}^{2n}$ we introduce the usual Euclidean scalar product, defined for a pair of vectors

$$\xi_1 = (x_1^1, \ldots, x_1^n, y_1^1, \ldots, y_1^n), \ \xi_2 = (x_2^1, \ldots, x_2^n, y_2^1, \ldots, y_2^n)$$

by

$$\langle \xi_1, \xi_2 \rangle_{\mathbb{R}} = \sum_{k=1}^{n} (x_1^k x_2^k + y_1^k y_2^k), \tag{13}$$

then the Hermitian scalar product is related to this one in the following way:

$$\operatorname{Re}\langle \xi_1, \xi_2 \rangle_{\mathbb{C}} = \langle \xi_1, \xi_2 \rangle_{\mathbb{R}}, \tag{14}$$

where Re indicates the real part of a complex number. To see (14), simply compute directly:

$$\operatorname{Re}\langle \xi_1, \xi_2 \rangle_{\mathbb{C}} = \operatorname{Re} \sum_{k=1}^{n} (x_1^k + iy_1^k)(x_2^k - iy_2^k) = \sum_{k=1}^{n} (x_1^k x_2^k + y_1^k y_2^k).$$

It follows in particular from (14) that since $\langle \xi, \xi \rangle_{\mathbb{C}}$ is real (see(11)),

$$\langle \xi, \xi \rangle_{\mathbb{C}} = \langle \xi, \xi \rangle_{\mathbb{R}}. \tag{15}$$

11.2.1. Definition. If $\Lambda \in GL(n, \mathbb{C})$ (i.e. Λ is a non-singular linear transformation of $\mathbb{C}^n$) then we say that Λ is *unitary* if for all vectors ξ_1, ξ_2,

$$\langle \Lambda\xi_1, \Lambda\xi_2 \rangle_{\mathbb{C}} = \langle \xi_1, \xi_2 \rangle_{\mathbb{C}}. \tag{16}$$

Thus, analogously to §4, the unitary linear transformations are just the motions of the Hermitian metric on $\mathbb{C}^n$ which fix the origin, i.e. the origin-fixing isometries of $\mathbb{C}^n$ equipped with the Hermitian scalar product. We denote the group of unitary linear transformations, called the *unitary group*, by $U(n)$.

We now derive a necessary and sufficient condition for a linear transformation Λ to be unitary. Let $e_1, \dots, e_n$ be a basis for $\mathbb{C}^n$, with respect to which the Hermitian scalar product has the form (10). If our linear transformation Λ has matrix (λ_k^l) relative to this basis, then for Λ to be unitary means that

$$\sum_{k=1}^{n} \lambda_k^l \bar{\lambda}_k^{l'} = \delta^{ll'},$$

or, in matrix notation (identifying Λ with its matrix),

$$\Lambda\bar{\Lambda}^{\mathrm{T}} = 1 \Leftrightarrow \bar{\Lambda}^{\mathrm{T}} = \Lambda^{-1}, \tag{17}$$

where as usual T denotes transposition. (Compare (17) with the orthogonality condition $AA^{\mathrm{T}} = 1$ (see §4.4), characterizing origin-fixing isometries of the Euclidean space $\mathbb{R}^n$.) We call matrices satisfying (17) *unitary matrices*, and identify the unitary group $U(n)$ with the group of unitary matrices. We also obtain from (17) that

$$\det(\Lambda\bar{\Lambda}^{\mathrm{T}}) = (\det \Lambda)(\overline{\det \Lambda}) = |\det \Lambda|^2 = 1,$$

whence we have that the determinant of a unitary matrix has modulus 1. Those matrices in $U(n)$ with determinant 1 form a subgroup called the *special unitary group*, denoted by $SU(n)$.

We now characterize the image $r(U(n))$ of $U(n)$, in the group $GL(2n, \mathbb{R})$. It follows from (14) that a linear transformation $\Lambda \in GL(n, \mathbb{C})$ preserves the Hermitian form $\langle \xi, \xi \rangle_{\mathbb{C}}$ if and only if $r(\Lambda)$ preserves the quadratic form

$\langle \xi, \xi \rangle_{\mathbb{R}}$. This and the fact that $\det(r(\Lambda))$ is positive (Exercise 11.1.2), together imply that

$$r(U(n)) = SO(2n) \cap r(GL(n, \mathbb{C})). \tag{18}$$

We conclude this subsection by mentioning the complex analogues of the pseudo-Euclidean spaces. These are the *pseudo-Hermitian spaces* $\mathbb{C}^n_{p,q}$, $p + q = n$, in which the square of the length of a vector $\xi = (z^1, \dots, z^n)$ is given by

$$\langle \xi, \xi \rangle_{p,q} = |z^1|^2 + \cdots + |z^p|^2 - \cdots - |z^n|^2. \tag{19}$$

The group of complex linear transformations of $\mathbb{C}^n$ preserving the form (19) is denoted by $U(p, q)$, and the subgroup of matrices with determinant 1, by $SU(p, q)$.

11.3. Examples of Complex Transformation Groups

(a) As noted above, the group $GL(1, \mathbb{C})$ is (isomorphic to) the group of non-zero complex numbers under multiplication. By (17), the subgroup $U(1)$ consists of all unimodular complex numbers (the "circle group"): $U(1) = \{e^{i\varphi}\}$. Since by (6) $r(e^{i\varphi}) = \begin{pmatrix} \cos\varphi & \sin\varphi \\ -\sin\varphi & \cos\varphi \end{pmatrix}$, we see that the realization map r defines an isomorphism between the groups $U(1)$ and $SO(2)$.

(b) We now consider the group $SL(2, \mathbb{C})$. With each matrix

$$A = \begin{pmatrix} a & b \\ c & d \end{pmatrix} \in SL(2, \mathbb{C})$$

(i.e. such that $ad - bc = 1$), we associate the following linear-fractional transformation of the extended (i.e. with the "point at infinity" adjoined) complex plane $\mathbb{C}$:

$$z' = \frac{az + b}{cz + d}. \tag{20}$$

If $A' = \begin{pmatrix} a' & b' \\ c' & d' \end{pmatrix}$ is another matrix from $SL(2, \mathbb{C})$, then by substituting from (20) we get for the composite of the transformations corresponding to A and A':

$$z'' = \frac{a'z' + b'}{c'z' + d'} = \frac{(a'a + b'c)z + a'b + b'd}{(c'a + d'c)z + c'b + d'd},$$

in other words our association of linear-fractional transformations with elements of $SL(2, \mathbb{C})$, is a homomorphism

$$\varphi\colon SL(2, \mathbb{C}) \to L, \tag{21}$$

where L denotes the group of linear-fractional transformations. It is easy to verify that the kernel of this homomorphism consists of the two matrices $\begin{pmatrix} 1 & 0 \\ 0 & 1 \end{pmatrix}$ and $\begin{pmatrix} -1 & 0 \\ 0 & -1 \end{pmatrix}$, and also that φ is in fact an epimorphism (i.e. is onto L). Thus we have an isomorphism:

$$L \simeq SL(2, \mathbb{C})/\{\pm 1\}. \tag{22}$$

(c) Next we consider $U(2)$. It follows from (17) that a matrix $\begin{pmatrix} a & b \\ c & d \end{pmatrix}$ lies in $U(2)$ precisely if:

$$|a|^2 + |b|^2 = 1, \qquad |c|^2 + |d|^2 = 1, \qquad a\bar{c} + b\bar{d} = 0, \tag{23}$$

while the matrices in the subgroup $SU(2)$ are characterized by these conditions with the further condition $ad - bc = 1$. After a little algebraic manipulation we deduce that $SU(2)$ consists of all matrices of the form

$$\begin{pmatrix} a & b \\ -\bar{b} & \bar{a} \end{pmatrix}, \qquad |a|^2 + |b|^2 = 1. \tag{24}$$

By an analogous argument it can be shown that $SU(1, 1)$ consists of all matrices of the form

$$\begin{pmatrix} c & d \\ \bar{d} & \bar{c} \end{pmatrix}, \qquad |c|^2 - |d|^2 = 1. \tag{25}$$

Note that here $c \neq 0$ (since $|c| \geq 1$). This fact allows us to define a one-to-one map $SU(1, 1) \to SU(2)$ by means of the formula

$$\begin{pmatrix} c & d \\ \bar{d} & \bar{c} \end{pmatrix} \mapsto \begin{pmatrix} \dfrac{1}{c} & \dfrac{d}{\bar{c}} \\ -\dfrac{\bar{d}}{c} & \dfrac{1}{\bar{c}} \end{pmatrix} = \begin{pmatrix} a & b \\ -\bar{b} & \bar{a} \end{pmatrix}. \tag{26}$$

Note that the only matrices of $SU(2)$ which are not images under this map, are those with $a = 0$. This map is not a group homomorphism; its significance will appear later.

§12. Analytic Functions

12.1. Complex Notation for the Element of Length, and for the Differential of a Function

Suppose we are given a curve

$$z^k = z^k(t) = x^k(t) + iy^k(t), \qquad k = 1, \ldots, n, \tag{1}$$

in the space $\mathbb{C}^n$ with co-ordinates z^k. In terms of the real co-ordinates x^k, y^k (where $z^k = x^k + iy^k$), this becomes the curve

$$x^1 = x^1(t), \ldots, x^n = x^n(t), \qquad y^1 = y^1(t), \ldots, y^n = y^n(t),$$

in $\mathbb{R}^{2n}$, and considering $\mathbb{R}^{2n}$ as Euclidean with Euclidean co-ordinates x^k, y^k, the length of an arc of this curve is given by

$$l = \int_{t_0}^{t_1} \sqrt{\sum_{k=1}^{n} (\dot{x}^k)^2 + (\dot{y}^k)^2}\, dt = \int_{t_0}^{t_1} \sqrt{\sum_{k=1}^{n} \dot{z}^k \dot{\bar{z}}^k}\, dt. \tag{2}$$

We write (2) succinctly as

$$dl^2 = \sum_{k=1}^{n} dz^k\, d\bar{z}^k, \tag{3}$$

where dz^k and $d\bar{z}^k$ are defined by

$$dz^k = dx^k + dy^k, \qquad d\bar{z}^k = dx^k - i\, dy^k. \tag{4}$$

We shall in what follows often need to change (as we did above) from the co-ordinates x^k, y^k of $\mathbb{R}^{2n}$ to complex co-ordinates z^k, $\bar{z}^k$, $k = 1, \dots, n$, and back. The equations governing these co-ordinate changes are:

$$\begin{aligned} z^k &= x^k + iy^k, & \bar{z}^k &= x^k - iy^k, \\ x^k &= \tfrac{1}{2}(z^k + \bar{z}^k), & y^k &= \frac{1}{2i}(z^k - \bar{z}^k). \end{aligned} \tag{5}$$

Having introduced complex notation for the element of arc length in $\mathbb{R}^{2n}$, we now introduce similar notational conventions in connexion with the space of complex-valued functions on $\mathbb{R}^{2n}$. We first define two linear operators on that space:

$$\begin{aligned} \frac{\partial}{\partial z^k} &= \frac{1}{2}\left(\frac{\partial}{\partial x^k} - i\frac{\partial}{\partial y^k}\right), \\ \frac{\partial}{\partial \bar{z}^k} &= \frac{1}{2}\left(\frac{\partial}{\partial x^k} + i\frac{\partial}{\partial y^k}\right). \end{aligned} \tag{6}$$

Obviously we then have

$$\begin{aligned} \frac{\partial}{\partial x^k} &= \frac{\partial}{\partial z^k} + \frac{\partial}{\partial \bar{z}^k}, \\ \frac{\partial}{\partial y^k} &= i\left(\frac{\partial}{\partial z^k} - \frac{\partial}{\partial \bar{z}^k}\right). \end{aligned} \tag{7}$$

It is also easy to verify that

$$\begin{aligned} \frac{\partial}{\partial \bar{z}^k}(z^k) &= \frac{\partial}{\partial z^k}(\bar{z}^k) = 0, \\ \frac{\partial}{\partial z^k}(z^k) &= \frac{\partial}{\partial \bar{z}^k}(\bar{z}^k) = 1. \end{aligned} \tag{8}$$

The following lemma is almost immediate from (4) and (7).

12.1.1. Lemma. *The differential of an arbitrary complex-valued function* $f(x^1, \ldots, x^n, y^1, \ldots, y^n)$ is equal to

$$df = \frac{\partial f}{\partial z^1} dz^1 + \cdots + \frac{\partial f}{\partial z^n} dz^n + \frac{\partial f}{\partial \bar{z}^1} d\bar{z}^1 + \cdots + \frac{\partial f}{d\bar{z}^n} d\bar{z}^n. \tag{9}$$

We now consider a special class of complex-valued functions, namely polynomials in the real variables $x^1, \ldots, x^n, y^1, \ldots, y^n$, but with complex coefficients. If $P(x^1, \ldots, x^n, y^1, \ldots y^n)$ is such a polynomial, then on making the change of variables (5), P takes the form of a polynomial

$$Q(z^1, \ldots, z^n, \bar{z}^1, \ldots, \bar{z}^n).$$

We shall say that a polynomial is *independent* of a particular one of its variables, if it has no non-zero terms having a positive power of that variable as a factor (i.e. if the variable does not appear).

12.1.2. Theorem. *A polynomial*

$$Q(z^1, \ldots, z^n, \bar{z}^1, \ldots, \bar{z}^n) = P(x^1, \ldots, x^n, y^1, \ldots, y^n)$$

is independent of a variable $\bar{z}^q$ *if and only if*

$$\frac{\partial P}{\partial \bar{z}^q} \equiv 0. \tag{10}$$

PROOF. It is easy to verify that for our operators $\partial/\partial z^k$, $\partial/\partial \bar{z}^k$, Leibniz' formula holds:

$$\frac{\partial}{\partial z^k}(fg) = \frac{\partial f}{\partial z^k} g + f \frac{\partial g}{\partial z^k},$$

$$\frac{\partial}{\partial \bar{z}^k}(fg) = \frac{\partial f}{\partial \bar{z}^k} g + f \frac{\partial f}{\partial \bar{z}^k} \tag{11}$$

This and (8) together imply that

$$\frac{\partial}{\partial \bar{z}^k}[(z^k)^m] = 0, \qquad \frac{\partial}{\partial \bar{z}^k}[(\bar{z}^k)^m] = m(\bar{z}^k)^{m-1}. \tag{12}$$

From this and the linearity of our operators, we get immediately that if $Q(z, \bar{z})$ is independent of any particular $\bar{z}^q$ then $\partial Q/\partial \bar{z}^q \equiv 0$.

For the converse, suppose that P depends on $\bar{z}^q$, and let $(\bar{z}^q)^m$ be the largest power of $\bar{z}^q$ occurring with non-zero coefficient. We shall show that $\partial P/\partial \bar{z}^q \not\equiv 0$. To this end we write P in the form

$$P = A_0(\bar{z}^q)^m + A_1(\bar{z}^q)^{m-1} + \cdots + A_m,$$

where $A_0, \ldots, A_m$ are polynomials in $z^1, \ldots, z^n$ and those $\bar{z}^k$ other than $\bar{z}^q$. Then by the first part of the proof, we have $\partial A_i/\partial \bar{z}^q \equiv 0$, $i = 0, \ldots, m$. Consequently

$$\frac{\partial P}{\partial \bar{z}^q} = A_0 m(\bar{z}^q)^{m-1} + A_1(m-1)(\bar{z}^q)^{m-2} + \cdots.$$

Therefore since $A_0 \not\equiv 0$, we have that $\partial P/\partial \bar{z}^q \not\equiv 0$, as required. □

Note that this proof is valid not just for polynomials, but also for power series in the $z^k, \bar{z}^k$: thus the condition that such a power series f have no terms with $\bar{z}^q$ as a factor (i.e. that $\bar{z}^q$ not figure in the power series) is equivalent to the condition $\partial f/\partial \bar{z}^q \equiv 0$.

12.1.3. Definition. A complex-valued function $f(x^1, y^1, \ldots, x^n, y^n)$ is said to be *analytic* in a region if the following n identities hold in the region:

$$\frac{\partial f}{\partial \bar{z}^k} \equiv 0, \qquad k = 1, \ldots, n. \tag{13}$$

Thus in the case of a function $f(x, y)$ of two variables, the condition for analyticity (which might alternatively be called "independence of $\bar{z}$") is

$$2\frac{\partial f}{\partial \bar{z}} = \frac{\partial f}{\partial x} + i\frac{\partial f}{\partial y} \equiv 0. \tag{14}$$

If we denote by u and v the real and imaginary parts of f, i.e. $f(x, y) = u(x, y) + iv(x, y)$, then in terms of u, v, the condition (14) becomes

$$\frac{\partial u}{\partial x} = \frac{\partial v}{\partial y}, \qquad \frac{\partial u}{\partial y} = -\frac{\partial v}{\partial x}. \tag{15}$$

These are the familiar "Cauchy–Riemann equations." (Thus we see that our condition (13) for analyticity is equivalent to the traditional one (assuming, as always, that the partial derivatives in (15) are continuous).) From (15) we obtain

$$\frac{\partial^2 u}{\partial x^2} + \frac{\partial^2 u}{\partial y^2} = 0, \qquad \frac{\partial^2 v}{\partial x^2} + \frac{\partial^2 v}{\partial y^2} = 0, \tag{16}$$

from which we draw the significant conclusion that the real and imaginary parts of a complex analytic function are solutions of Laplace's equation (i.e. they are "harmonic" functions). In the conventional notation, (16) appears as $\Delta u = 0$, $\Delta v = 0$, where $\Delta = \partial^2/\partial x^2 + \partial^2/\partial y^2 = 4(\partial^2/\partial z\, \partial \bar{z})$ is the "Laplace operator."

12.2. Complex Co-ordinate Changes

Suppose that some region of n-dimensional complex space $\mathbb{C}^n$ has defined on it two systems of complex co-ordinates

$$\begin{aligned} z^1 &= x^1 + iy^1, \ldots, z^n = x^n + iy^n, \\ w^1 &= u^1 + iv^1, \ldots, w^n = u^n + iv^n. \end{aligned} \tag{17}$$

Then, as usual, we can regard the co-ordinates $w^k = u^k + iv^k$ as functions of the $z^k = x^k + iy^k$:

$$w^k = w^k(x^1, y^1, \ldots, x^n, y^n). \tag{18}$$

12.2.1. Definition. The complex co-ordinate change (18) is said to be *analytic* if the functions $w^k(x^1, y^1, \ldots, x^n, y^n)$ are all analytic, i.e. if

$$\frac{\partial w^k}{\partial \bar{z}^l} \equiv 0, \qquad k, l = 1, \ldots, n. \tag{19}$$

As for real co-ordinate changes in $\mathbb{R}^n$, so also for complex analytic co-ordinate changes (18), do we define the *Jacobian matrix* (a_l^k) by

$$a_l^k = \frac{\partial w^k}{\partial z^l}, \qquad k, l = 1, \ldots, n, \tag{20}$$

and call its determinant the (*complex*) *Jacobian* of the co-ordinate change:

$$J_{\mathbb{C}} = \det(a_l^k). \tag{21}$$

In the realized space $\mathbb{R}^{2n}$, (18) yields the co-ordinate change

$$u^k = u^k(x^1, y^1, \ldots, x^n, y^n), \qquad v^k = v^k(x^1, y^1, \ldots, x^n, y^n), \qquad k = 1, \ldots, n, \tag{22}$$

whose (real) Jacobian $\partial(u, v)/\partial(x, y)$ we denote by $J_{\mathbb{R}}$.

The relation between the complex and real Jacobians $J_{\mathbb{C}}$ and $J_{\mathbb{R}}$ turns out to be very simple.

12.2.2. Lemma. *For a complex analytic co-ordinate change we have*

$$J_{\mathbb{R}} = |J_{\mathbb{C}}|^2.$$

PROOF. Writing $A = (a_l^k) = (\partial w^k/\partial z^l)$ for the complex Jacobian matrix of the co-ordinate change (18), we have $\det A = J_{\mathbb{C}}$. As a first step towards calculating $J_{\mathbb{R}}$, we find the $2n \times 2n$ Jacobian matrix of the change in $\mathbb{R}^{2n}$ from co-ordinates $z^1, \ldots, z^n, \bar{z}^1, \ldots, \bar{z}^n$ to co-ordinates $w^1, \ldots, w^n, \bar{w}^1, \ldots, \bar{w}^n$. Since the complex co-ordinate change from $z^1, \ldots, z^n$ to $w^1, \ldots, w^n$ is, by hypothesis, analytic, we have $\partial w^k/\partial \bar{z}^l = 0 = \partial \bar{w}^k/\partial z^l$. This together with $\partial w^k/\partial z^l = a_l^k$, $\partial \bar{w}^k/\partial \bar{z}^l = \bar{a}_l^k$, implies that the $2n \times 2n$ matrix we are seeking is $\begin{pmatrix} A & 0 \\ 0 & \bar{A} \end{pmatrix}$, which has determinant $|\det A|^2 = |J_{\mathbb{C}}|^2$.

It is clear that the Jacobian matrix of the co-ordinate change in $\mathbb{R}^{2n}$ from the system $x^1, \ldots, x^n, y^1, \ldots, y^n$ to the system $z^1, \ldots, z^n, \bar{z}^1, \ldots, \bar{z}^n$ is

$$B = \left(\begin{array}{ccc|ccc} 1 & & 0 & i & & 0 \\ & \ddots & & & \ddots & \\ 0 & & 1 & 0 & & i \\ \hline 1 & & 0 & -i & & 0 \\ & \ddots & & & \ddots & \\ 0 & & 1 & 0 & & -i \end{array}\right), \qquad \det B = (-2i)^n,$$

and that this matrix B is also the Jacobian matrix of the change from $u^1, \ldots, u^n, v^1, \ldots, v^n$ to $w^1, \ldots, w^n, \bar{w}^1, \ldots, \bar{w}^n$. Since the change from co-ordinates $x^1, \ldots, x^n, y^1, \ldots, y^n$ to co-ordinates $u^1, \ldots, u^n, v^1, \ldots, v^n$ is the composite of the above three changes (from x, y to $z, \bar{z}$ to $w, \bar{w}$ to u, v), we have that the Jacobian matrix of that change of co-ordinates is $B^{-1}\begin{pmatrix} A & 0 \\ 0 & \bar{A} \end{pmatrix} B$, which has determinant $|J_{\mathbb{C}}|^2$, as required. □

12.2.3. Corollary (Inverse Function Theorem for Complex Analytic Co-ordinate Changes). *If the (complex) Jacobian of an analytic co-ordinate change (18) is non-zero at a point, then locally (i.e. in a neighbourhood of the point) the z^k can be expressed as functions of the w^l:*

$$z^k = z^k(w^1, \ldots, w^n), \qquad k = 1, \ldots, n,$$

which are also analytic.

Proof. In the preceding proof we saw that the Jacobian matrix of the co-ordinate change in $\mathbb{R}^{2n}$ from $x^1, \ldots, x^n, y^1, \ldots, y^n$ to $u^1, \ldots, u^n, v^1, \ldots, v^n$ is $B^{-1}\begin{pmatrix} A & 0 \\ 0 & \bar{A} \end{pmatrix} B$. Since this matrix has determinant $|J_{\mathbb{C}}|^2$, which is non-zero by hypothesis, it follows from the Inverse Function Theorem for real co-ordinate changes (1.2.5) that locally this change has an inverse with Jacobian matrix

$$B^{-1}\begin{pmatrix} A^{-1} & 0 \\ 0 & \bar{A}^{-1} \end{pmatrix} B.$$

Since B and B^{-1} correspond to explicit well-behaved co-ordinate changes, it is clear that the change from $z, \bar{z}$ to $w, \bar{w}$ has an inverse locally, with Jacobian matrix

$$\begin{pmatrix} A^{-1} & 0 \\ 0 & \bar{A}^{-1} \end{pmatrix}. \tag{23}$$

Hence the change from z to w has an inverse locally, and from the form of (23), we see that this inverse co-ordinate change is analytic. This completes the proof. □

Thus for example in the case $n = 1$, the change (18) has the form $w = w(z)$, the analyticity condition is $\partial w/\partial \bar{z} \equiv 0$, and the corollary tells us that the function $w = w(z)$ is invertible wherever $\partial w/\partial z \neq 0$.

In the context of a complex space $\mathbb{C}^n$, by a *transformation* we shall understand a one-to-one map from one region of $\mathbb{C}^n$ to another where the functions defining the transformation are analytic.

12.2.4. Examples of Complex Transformations in the Case $n = 1$. (a) *The complex affine transformations*

$$w(z) = az + b, \qquad a \neq 0. \tag{24}$$

For these transformations $dw/dz = a \neq 0$.

Recall that, upon realization, (24) yields a transformation of the real plane which is the composite of a Euclidean isometry and an orientation-preserving dilation (see the conclusion of §4.2).

(b) *The linear-fractional transformations*

$$w(z) = \frac{az + b}{cz + d}, \qquad ad - bc \neq 0. \tag{25}$$

We may suppose that $ad - bc = 1$, since each of the coefficients a, b, c, d can be divided by a square root of $ad - bc$ without altering the transformation. This assumed, we have

$$\frac{dw}{dz} = -\frac{1}{(cz + d)^2} \neq 0. \tag{26}$$

Strictly speaking the transformation (25) is not defined for $z = -d/c$. However this inconvenience can be avoided by regarding (25) as a transformation of the extended complex plane, obtained from $\mathbb{C}$ by adjoining to it (for the time being in a purely formal way) an extra point "at infinity," denoted by ∞, postulated to behave under the transformation (25) as follows:

$$-\frac{d}{c} \mapsto \infty, \qquad \infty \mapsto \frac{a}{c}. \tag{27}$$

As an example of a linear-fractional transformation, consider the one given by

$$w = \frac{1 + iz}{1 - iz}. \tag{28}$$

This maps the upper half-plane $\operatorname{Im} z > 0$ onto the unit disc $|w| < 1$; we used this transformation in §10.1 to construct Klein's model of the Lobachevskian plane.

12.3. Surfaces in Complex Space

We shall consider only the simplest case, namely that of a one-dimensional surface (or complex curve) in 2-dimensional complex space $\mathbb{C}^2$. If the space $\mathbb{C}^2$ is co-ordinatized by co-ordinates w, z, then we shall assume that the curve is given by an equation

$$f(w, z) = 0, \tag{29}$$

where $f(w, z)$ is complex analytic in the arguments w, z. Writing $f = u + iv$ we see that (29) is equivalent to the two real equations $u = 0, v = 0$, so that it defines a 2-dimensional surface in $\mathbb{R}^4(=\mathbb{C}^2)$. By analogy with the real case (see §7.1), we define the *complex gradient*, $\operatorname{grad}_{\mathbb{C}} f$, of the function f, by

$$\operatorname{grad}_{\mathbb{C}} f = \left(\frac{\partial f}{\partial w}, \frac{\partial f}{\partial z}\right), \tag{30}$$

(see (6)), and we say that a point (w_0, z_0) on the curve (29) is *non-singular* if $\operatorname{grad}_{\mathbb{C}} f|_{w_0, z_0} \neq 0$. The following theorem (whose proof we omit) is the complex analogue of the Implicit Function Theorem.

12.3.1. Theorem (The Complex Implicit Function Theorem). *Let $f(w, z)$ be a complex analytic function of the variables w, z and let (w_0, z_0) be a point at which* $\operatorname{grad}_{\mathbb{C}} f \neq 0$. *If for instance $\partial f/\partial w \neq 0$ at the point, then in a sufficiently small neighbourhood (in $\mathbb{C}^2$) of the point (w_0, z_0), the equation $f(w, z) = 0$ has a unique solution $w = w(z)$ (i.e. $f(w(z), z) \equiv 0$ in the neighbourhood); moreover this (unique) solution is analytic, i.e. $\partial w/\partial \bar{z} \equiv 0$ in the neighbourhood (where $\partial/\partial \bar{z}$ is defined in* (6)).

As an (important) example, consider the case that $f(w, z)$ is a polynomial in the variables w, z. We call the totality of all solutions in the form $w = w(z)$, of the equation $f(w, z) = 0$, a *multiple-valued algebraic function*, and the surface (i.e. complex curve) $f(w, z) = 0$ the *graph* or *Riemann surface* of this multiple-valued function.†

An important special case is that of a *hyper-elliptic curve*, defined as the Reimann surface given by an equation of the form

$$f(w, z) = w^2 - P_n(z) = 0, \tag{31}$$

where $P_n(z)$ is a polynomial of degree n. In this case the corresponding multi-valued function may be written as $w = \sqrt{P_n(z)}$, where the square root sign is to be interpreted ambiguously.

12.3.2. Lemma. *The surface* (31) *is free of singularities if and only if the polynomial $P_n(z)$ has no multiple roots.*

PROOF. Calculating the gradient of the function f in (31), we obtain

$$\operatorname{grad}_{\mathbb{C}} f = \left(\frac{\partial f}{\partial w}, \frac{\partial f}{\partial z}\right) = \left(2w, -\frac{dP_n(z)}{dz}\right).$$

Thus a point (w_0, z_0) is a singular point of the curve if and only if

$$2w_0 = 0, \qquad \left.\frac{dP_n(z)}{dz}\right|_{z_0} = 0, \qquad w_0^2 - P_n(z_0) = 0,$$

† This is a simplified variant of the usual definition of a Riemann surface (see, for example, [26]). Our definition is equivalent to the usual one, in the case that the surface $f(w, z) = 0$ has no singular points or self-intersections.

i.e. if and only if $w_0 = 0$, and z_0 is a root both of the polynomial $P_n(z)$ and its derivative. The lemma then follows from the well-known equivalence of this condition on z_0, with the condition that z_0 be a multiple root of $P_n(z)$. (Note that for this equivalence we need the "power rule" (12).) □

The Complex Implicit Function Theorem 12.3.1 allows us to introduce the local co-ordinate z on the surface (31), at (i.e. in neighbourhoods of) those points where $\partial f/\partial w = 2w \neq 0$, i.e. where $P_n(z) \neq 0$. At those points z for which $P_n(z) = 0$, we may take w as local co-ordinate, provided that we do not also have $\partial f/\partial z = -dP_n(z)/dz = 0$, i.e. provided that z is not a multiple root of $P_n(z)$.

We now return to the general situation of an arbitrary complex analytic curve $f(w, z) = 0$. Suppose that (w_0, z_0) is a point on the curve at which $\partial f/\partial w \neq 0$, and let $w = w(z)$ be the analytic function (defined locally around (w_0, z_0)) guaranteed by 12.3.1. By (3) the Hermitian metric in the co-ordinates w, z for $\mathbb{C}^2$ is given by

$$dl^2 = dw\, d\bar{w} + dz\, d\bar{z}. \tag{32}$$

What is the induced metric on the surface $w = w(z)$? To answer this write as usual $w = w(x, y) = w(\frac{1}{2}(z + \bar{z}), (1/2i)(z - \bar{z}))$, i.e. re-co-ordinatize the surface by $z, \bar{z}$; then by Lemma 12.1.1, $dw = (\partial w/\partial z)\, dz + (\partial w/\partial \bar{z})\, d\bar{z} = (\partial w/\partial z)dz$, since by analyticity $\partial w/\partial \bar{z} \equiv 0$. Hence (by (6), (7)) $dw/dz = \partial w/\partial z$. Similarly $d\bar{w} = (\partial \bar{w}/\partial \bar{z})\, d\bar{z}$, and $d\bar{w}/d\bar{z} = \partial \bar{w}/\partial \bar{z}$. Substituting for dw and $d\bar{w}$ in (32) we arrive at the following formula for the metric induced on the surface:

$$dl^2 = \left(1 + \left|\frac{dw}{dz}\right|^2\right) dz\, d\bar{z}, \tag{33}$$

whence we see that the square of an element of arc length on the surface $f(w, z) = 0$, in a neighbourhood of (w_0, z_0), has the form

$$dl^2 = h(z, \bar{z})\, dz\, d\bar{z} = g(x, y)(dx^2 + dy^2), \tag{34}$$

where $h(z, \bar{z}) = g(x, y) = 1 + |dw/dz|^2$, and $z = x + iy$.

12.3.3. Definition. If the metric on a real 2-dimensional surface has the form $dl^2 = g(x, y)(dx^2 + dy^2)$ in terms of real co-ordinates x, y on the surface, then we call these co-ordinates *conformal*.

12.3.4. Lemma. *Suppose we are given a metrized surface on which conformal co-ordinates are defined. Then the co-ordinate changes which preserve the conformal form of the metric (i.e. change it to another conformal form), are precisely those corresponding to complex analytic co-ordinate changes, and the composites of these with complex conjugation.*

PROOF. By hypothesis there are co-ordinates x, y in terms of which the metric is given by

$$dl^2 = g(x, y)(dx^2 + dy^2) = h(z, \bar{z})\, dz\, d\bar{z}.$$

Let $z = z(w)$ define a complex analytic co-ordinate change, so that $\partial z/\partial \bar{w} = 0$. Then as in the derivation of (33) we have

$$dz = \left(\frac{dz}{dw}\right) dw, \qquad d\bar{z} = \overline{\left(\frac{dz}{dw}\right)} d\bar{w},$$

whence

$$dl^2 = h(z, \bar{z})\, dz\, d\bar{z} = h(z(w, \bar{w}), \overline{z(w, \bar{w})}) \left|\frac{dz}{dw}\right|^2 dw\, d\bar{w},$$

which still has conformal form. That complex conjugation preserves the conformal form, is easily verified.

For the converse, suppose that we have a co-ordinate change $z = z(w) = z(w, \bar{w})$, which is neither analytic (so that $\partial z/\partial \bar{w} \not\equiv 0$), nor the composite of an analytic function and complex conjugation (so that $\partial z/\partial w \not\equiv 0$). It follows from our underlying assumption of continuity of the partial derivatives of $z(u, v)$ (where $w = u + iv$), that there exist points at which $\partial z/\partial \bar{w} \neq 0 \neq \partial z/\partial w$ (since otherwise from (6) we should have, at every point, $\partial z/\partial u = \pm i(\partial z/\partial v)$ with neither equality holding everywhere). Now by Lemma 12.1.1, $dz = (\partial z/\partial w)\, dw + (\partial z/\partial \bar{w})\, d\bar{w}$; hence

$$dl^2 = h \left| \frac{\partial z}{\partial w} dw + \frac{\partial z}{\partial \bar{w}} d\bar{w} \right|^2 = h(a_{11}\, du^2 + 2a_{12}\, du\, dv + a_{22}\, dv^2),$$

where $w = u + iv$. Calculation of the a_{ij} shows that the equations $a_{11} = a_{22}$, $a_{12} = 0$ imply that either $\partial z/\partial \bar{w} = 0$ or $\partial z/\partial w = 0$. Since there are points at which neither of these hold, we deduce that our transformation does not preserve the conformal form, completing the proof. □

§13. The Conformal Form of the Metric on a Surface

13.1. Isothermal Co-ordinates. Gaussian Curvature in Terms of Conformal Co-ordinates

Suppose that we have in Euclidean space $\mathbb{R}^3$, a 2-dimensional surface with parametric equations

$$x = x(p, q), \qquad y = y(p, q), \qquad z = z(p, q), \tag{1}$$

where (p, q) ranges over some region of $\mathbb{R}^2$. Then, as we saw in §7.3, the Euclidean metric in $\mathbb{R}^3$ induces on this surface a Riemannian metric

$$dl^2 = E(dp)^2 + 2F\, dp\, dq + G(dq)^2, \tag{2}$$

where $g = EG - F^2 > 0$ (since the form given by (2) is positive definite). It turns out that by means of a change of the local co-ordinates p, q, this metric can be brought into conformal form. (We have already seen this in the case of the sphere (§9).) We state this result precisely as

13.1.1. Theorem. *With the above notation, suppose that E, F, G are real-valued analytic functions of the real variables p, q (i.e. are representable as power series in p, q). Then there exist new (real) local co-ordinates u, v for the surface in terms of which the induced metric (2) takes the form*

$$dl^2 = f(u, v)(du^2 + dv^2). \tag{3}$$

Co-ordinates with this property are called *isothermal* or *conformal* (cf. Definition 12.3.3). Thus for local co-ordinates on a surface to be isothermal the induced metric must have conformal form (i.e. the form (3)) in those co-ordinates.

PROOF OF THEOREM 13.1.1. Factorizing the right-hand side of (2), we obtain

$$dl^2 = \left(\sqrt{E}\,dp + \frac{F + i\sqrt{g}}{\sqrt{E}}\,dq\right)\left(\sqrt{E}\,dp + \frac{F - i\sqrt{g}}{\sqrt{E}}\,dq\right).$$

We are seeking new co-ordinates u, v as functions $u = u(p, q)$, $v = v(p, q)$ of p and q, with the property that in terms of u, v the metric (2) takes the form (3). We shall certainly have achieved this if we can find an "integrating factor," i.e. a complex-valued function $\lambda = \lambda(p, q)$ such that

$$\begin{aligned} \lambda\left(\sqrt{E}\,dp + \frac{F + i\sqrt{g}}{\sqrt{E}}\,dq\right) &= du + i\,dv, \\ \bar{\lambda}\left(\sqrt{E}\,dp + \frac{F - i\sqrt{g}}{\sqrt{E}}\,dq\right) &= du - i\,dv, \end{aligned} \tag{4}$$

since on multiplying these two equations together, we obtain

$$|\lambda|^2\,dl^2 = du^2 + dv^2, \qquad dl^2 = |\lambda|^{-2}(du^2 + dv^2), \tag{5}$$

so that as $f(u, v)$ the function $|\lambda|^{-2}$ will serve. Since the second of the equations (4) is just the complex conjugate of the first, our problem has now become that of finding functions $u(p, q)$, $v(p, q)$, $\lambda(p, q)$ satisfying the equation

$$\lambda\left(\sqrt{E}\,dp + \frac{F + i\sqrt{g}}{\sqrt{E}}\,dq\right) = du + i\,dv = \left(\frac{\partial u}{\partial p} + i\frac{\partial v}{\partial p}\right)dp + \left(\frac{\partial u}{\partial q} + i\frac{\partial v}{\partial q}\right)dq, \tag{6}$$

or, equivalently,

$$\lambda\sqrt{E} = \frac{\partial u}{\partial p} + i\frac{\partial v}{\partial p}, \qquad \lambda\frac{F + i\sqrt{g}}{\sqrt{E}} = \frac{\partial u}{\partial q} + i\frac{\partial v}{\partial q}. \tag{7}$$

Eliminating λ, we get

$$(F + i\sqrt{g})\left(\frac{\partial u}{\partial p} + i\frac{\partial v}{\partial p}\right) = E\left(\frac{\partial u}{\partial q} + i\frac{\partial v}{\partial q}\right), \tag{8}$$

which is equivalent to

$$F\frac{\partial u}{\partial p} - \sqrt{g}\frac{\partial v}{\partial p} = E\frac{\partial u}{\partial q}, \qquad \sqrt{g}\frac{\partial u}{\partial p} + E\frac{\partial v}{\partial p} = E\frac{\partial v}{\partial q}.$$

Hence

$$\begin{aligned} \frac{\partial v}{\partial p} &= \frac{F\dfrac{\partial u}{\partial p} - E\dfrac{\partial u}{\partial q}}{\sqrt{g}}, \qquad & \frac{\partial v}{\partial q} &= \frac{G\dfrac{\partial u}{\partial p} - F\dfrac{\partial u}{\partial q}}{\sqrt{g}}, \\ \frac{\partial u}{\partial p} &= \frac{E\dfrac{\partial v}{\partial q} - F\dfrac{\partial v}{\partial p}}{\sqrt{g}}, \qquad & \frac{\partial u}{\partial q} &= \frac{F\dfrac{\partial v}{\partial q} - G\dfrac{\partial v}{\partial p}}{\sqrt{g}}, \end{aligned} \tag{9}$$

where the two equations on the right-hand side of (9) follow on multiplying equation (8) by $(F - i\sqrt{g})/E$. If we now also demand of the functions u, v that their second partial derivatives be continuous, we shall have $\partial^2 u/\partial p\,\partial q = \partial^2 u/\partial q \partial p$ and $\partial^2 v/\partial p \partial q = \partial^2 v/\partial q \partial p$, and then from (9) we get $Lu = 0$, $Lv = 0$ where the differential operator L is given by

$$L = \frac{\partial}{\partial q}\left[\frac{F\dfrac{\partial}{\partial p} - E\dfrac{\partial}{\partial q}}{\sqrt{EG - F^2}}\right] + \frac{\partial}{\partial p}\left[\frac{F\dfrac{\partial}{\partial q} - G\dfrac{\partial}{\partial p}}{\sqrt{EG - F^2}}\right]. \tag{10}$$

The partial differential equation $Lf = 0$ is known as "Beltrami's equation," and the operator L as the "Beltrami operator." Thus the upshot is that u, v are solutions of Beltrami's equation. Now it is known from the theory of partial differential equations (but we shall not prove it here) that if the functions E, F, G are real analytic, then Beltrami's equation always has solutions of the type we seek, namely functions u, v such that in the region over which (p, q) ranges, the map $(p, q) \mapsto (u(p, q), v(p, q))$ is one-to-one. (Note that once u, say is chosen (as a solution of Betrami's equation) then the equations in the first row of (9) determine grad u, and therefore (essentially) u. The function λ is then determined using (7).) With this we conclude the proof. □

Hence (assuming E, F, G real analytic) isothermal co-ordinates always exist on a surface in Euclidean 3-space. Lemma 12.3.4 then tells us exactly which local co-ordinate changes lead to other isothermal co-ordinates p, q on the surface. For if we write $z = p + iq$, $w = u + iv$, then regarding w as usual as a function of $z, \bar{z}$, i.e. $w = w(z, \bar{z})$, we know from 12.3.4 that p, q are isothermal co-ordinates on the surface if and only if $w(z, \bar{z})$ is analytic in either z or $\bar{z}$.

13.1.2. Example. Consider the complex plane (or line) $\mathbb{C}$ with complex co-ordinate $z = x + iy$. The Euclidean metric $dl^2 = dx^2 + dy^2 = dz\,d\bar{z}$ certainly has conformal form. Since linear-fractional transformations are analytic, we should expect (in view of 12.3.4) that they will preserve the conformal form of the Euclidean metric; and indeed if

$$w = \frac{az + b}{cz + d}, \qquad ad - bc = 1,$$

is any linear-fractional transformation, then $dw = [-1/(cz + d)^2]\,dz$, so that

$$dl^2 = dz\,d\bar{z} = |cz + d|^4\,dw\,d\bar{w},$$

as expected.

We next compute the Gaussian curvature of a surface in Euclidean 3-space, in terms of conformal co-ordinates on the surface.

13.1.3. Theorem. *If u, v are conformal co-ordinates on a surface in Euclidean 3-space, in terms of which the induced metric has the form*

$$dl^2 = g(u, v)(du^2 + dv^2),$$

then the Gaussian curvature of the surface is given by

$$K = -\frac{1}{2g(u, v)}\Delta \ln g(u, v), \tag{11}$$

where $\Delta = \partial^2/\partial u^2 + \partial^2/\partial v^2$ is the Laplace operator.

PROOF. Suppose that in terms of the conformal co-ordinates u, v, the surface is given (locally) by $r = r(u, v)$; $r = (x, y, z)$ (where x, y, z are Euclidean co-ordinates for $\mathbb{R}^3$). Since the metric on the surface in terms of these co-ordinates is given by $dl^2 = g(u, v)(du^2 + dv^2)$, we have (by §7.3 (19)) that

$$\langle r_u, r_u\rangle = \langle r_v, r_v\rangle = g(u, v), \qquad \langle r_u, r_v\rangle = 0. \tag{12}$$

On differentiating these equations with respect to u and v (and, as usual, assuming enough continuity and differentiability), we obtain

$$\begin{aligned} \frac{1}{2}\frac{\partial g(u, v)}{\partial u} &= \langle r_{uu}, r_u\rangle = \langle r_{uv}, r_v\rangle, \\ \frac{1}{2}\frac{\partial g(u, v)}{\partial v} &= \langle r_{vv}, r_v\rangle = \langle r_{uv}, r_u\rangle, \\ \langle r_{uu}, r_v\rangle + \langle r_u, r_{uv}\rangle &= 0 = \langle r_{uv}, r_v\rangle + \langle r_u, r_{vv}\rangle. \end{aligned} \tag{13}$$

Define unit vectors e_1, e_2, n by

$$e_1 = \frac{r_u}{\sqrt{g(u, v)}}, \qquad e_2 = \frac{r_v}{\sqrt{g(u, v)}}, \qquad n = [e_1, e_2]. \tag{14}$$

By (12) and the properties of the vector product the frame e_1, e_2, n is orthonormal at each point of the surface; in addition the vectors e_1, e_2 are tangent to the surface, and the vector n normal to it. By definition (§8.1(3)), the coefficients of the second fundamental form of the surface are

$$b_{11} = L = \langle r_{uu}, n\rangle, \qquad b_{12} = M = \langle r_{uv}, n\rangle,$$
$$b_{22} = \langle r_{vv}, n\rangle = N. \tag{15}$$

It follows from (13) and (15) that relative to the basis e_1, e_2, n, the components of the vectors r_{uu}, r_{uv}, r_{vv} are as follows:

$$\begin{aligned} r_{uu} &= \left(\frac{1}{2\sqrt{g}}\frac{\partial g}{\partial u}, -\frac{1}{2\sqrt{g}}\frac{\partial g}{\partial v}, L\right), \\ r_{uv} &= \left(\frac{1}{2\sqrt{g}}\frac{\partial g}{\partial v}, \frac{1}{2\sqrt{g}}\frac{\partial g}{\partial u}, M\right), \\ r_{vv} &= \left(-\frac{1}{2\sqrt{g}}\frac{\partial g}{\partial u}, \frac{1}{2\sqrt{g}}\frac{\partial g}{\partial v}, N\right), \end{aligned} \tag{16}$$

where of course $g = g(u, v)$. Hence

$$\langle r_{uu}, r_{vv}\rangle - \langle r_{uv}, r_{uv}\rangle = LN - M^2 - \frac{1}{2g}\left[\left(\frac{\partial g}{\partial v}\right)^2 + \left(\frac{\partial g}{\partial u}\right)^2\right].$$

From this and (13) we obtain

$$\begin{aligned} \frac{1}{2}\frac{\partial^2 g}{\partial u^2} &= \langle r_{uuv}, r_v\rangle + \langle r_{uv}, r_{uv}\rangle \\ &= \frac{\partial}{\partial v}\langle r_{uu}, r_v\rangle - \langle r_{uu}, r_{vv}\rangle + \langle r_{uv}, r_{uv}\rangle \\ &= -\frac{1}{2}\frac{\partial^2 g}{\partial v^2} - (LN - M^2) + \frac{1}{2g}\left[\left(\frac{\partial g}{\partial u}\right)^2 + \left(\frac{\partial g}{\partial v}\right)^2\right], \end{aligned}$$

whence it follows (using §8(26) and a little manipulation) that the Gaussian curvature is given by

$$K = \frac{\det(b_{ij})}{\det(g_{ij})} = \frac{LN - M^2}{(g(u, v))^2} = -\frac{1}{2g(u, v)}\Delta \ln g(u, v),$$

concluding the proof of the theorem. □

Note finally that in complex notation the metric $dl^2 = g(u, v)(du^2 + dv^2)$ becomes (setting $z = u + iv$)

$$dl^2 = g(z, \bar{z})\, dz\, d\bar{z},$$

and the formula (11) for the Gaussian curvature of the surface with this metric becomes (again after a little calculation)

$$K = -\frac{2}{g}\frac{\partial^2}{\partial z\,\partial\bar{z}}(\ln g). \tag{17}$$

13.2. Conformal Form of the Metrics on the Sphere and the Lobachevskian Plane

We have already found (in §9) conformal co-ordinates x, y for the sphere: they were obtained by means of the stereographic projection of the sphere onto the plane with Euclidean co-ordinates x, y. Thus for the sphere of radius $R = 1$, the induced metric was found to have the form (see §9(4))

$$dl^2 = 4\frac{dx^2 + dy^2}{(1 + x^2 + y^2)^2}. \tag{18}$$

Setting $z = x + iy$, this takes the form

$$dl^2 = \frac{4}{(1 + |z|^2)^2}\,dz\,d\bar{z}, \tag{19}$$

where z co-ordinatizes the complex plane $\mathbb{C}$.

In §10 we derived (again by means of a stereographic projection) the following formula for the metric of the Lobachevskian plane (with $R = 1$; see §10(10)):

$$dl^2 = \frac{4}{(1 - (x^2 + y^2))^2}(dx^2 + dy^2), \qquad x^2 + y^2 < 1. \tag{20}$$

In complex notation this formula becomes

$$dl^2 = \frac{4}{(1 - |z|^2)^2}\,dz\,d\bar{z}, \qquad |z| < 1, \tag{21}$$

where the complex co-ordinate z ranges over the open unit disc. Recall that the open unit disc with metric (21) (or (20)) is known as the "Poincaré model" of the hyperbolic (i.e. Lobachevskian) plane. If we map the unit disc onto the the upper half-plane $\operatorname{Im} w > 0$ by means of the map $z = (1 + iw)/(1 - iw)$, then we obtain the metric in "Klein's model" of the hyperbolic plane (see §10(13)):

$$dl^2 = -\frac{4}{(w - \bar{w})^2}\,dw\,d\bar{w}, \qquad \operatorname{Im} w > 0. \tag{22}$$

We shall now determine in terms of complex co-ordinates, the isometry groups of the sphere and the Lobachevskian plane. As we saw in the preceding section (Lemma 12.3.4), the conformal form of a metric on a surface is preserved by analytic co-ordinate changes, and therefore certainly by linear-fractional transformations. We shall therefore look for the isometries of our metrics (19) and (21) among the linear-fractional transformations

$$z = \frac{aw + b}{cw + d}, \qquad ad - bc = 1. \tag{23}$$

(We shall prove in Chapter 4 of Part II that in fact all motions of the metrics (19) and (21) are, up to complex conjugation, linear-fractional transformations.)

We begin with the sphere. Under the co-ordinate change (23) the metric (19) becomes

$$\begin{aligned} dl^2 &= \frac{4\,dz\,d\bar{z}}{(1 + |z|^2)^2} = \frac{4\,dw\,d\bar{w}}{[|aw + b|^2 + |cw + d|^2]^2} \\ &= \frac{4\,dw\,d\bar{w}}{[|b|^2 + |d|^2 + w(a\bar{b} + c\bar{d}) + \bar{w}(\bar{a}b + \bar{c}d) + (|a|^2 + |c|^2)|w|^2]^2}. \end{aligned} \tag{24}$$

Hence for the transformation (23) to be an isometry it is necessary and sufficient that

$$|b|^2 + |d|^2 = 1, \qquad a\bar{b} + c\bar{d} = 0, \qquad |a|^2 + |c|^2 = 1. \tag{25}$$

Since these are just the conditions for the matrix $\begin{pmatrix} a & b \\ c & d \end{pmatrix}$ to be unitary (see §11(17)), and since $ad - bc = 1$, we deduce that $\begin{pmatrix} a & b \\ c & d \end{pmatrix}$ belongs to $SU(2)$. This and the fact that the group of linear-fractional transformations is isomorphic to $SL(2, \mathbb{C})/\{\pm 1\}$ (see §11(22)), together give us

13.2.1. Theorem. *The group of all direct motions (i.e. direct isometries) of the sphere S^2, is isomorphic to the group $SU(2)/\{\pm 1\}$.*

(Note that we are assuming here the as yet unproved result that all direct (i.e. proper—cf. §4.2) isometries of the sphere are linear-fractional transformations.)

From this theorem and the remark at the end of §9 concerning the isometry group of the sphere we obtain the following result.

13.2.2. Corollary. *The groups $SO(3)$ and $SU(2)/\{\pm 1\}$ are isomorphic.*

(As an exercise the reader may like to prove this directly using the stereographic projection!)

As mentioned above, the full isometry group of the sphere is obtained from the group of rotations (i.e. of transformations (23) with $\begin{pmatrix} a & b \\ c & d \end{pmatrix} \in SU(2)$)

by supplementing these with the reflection $z \to \bar{z}$ (as a generator). Thus the rotation group has index 2 in the full isometry group.

We now turn to the Lobachevskian plane, beginning with the Poincaré model. Proceeding as in the case of the sphere we calculate that the metric (21) transforms under the linear-fractional transformation (23) to

$$dl^2 = \frac{4\,dz\,d\bar{z}}{(1 - |z|^2)^2}$$
$$= \frac{4\,dw\,d\bar{w}}{[|d|^2 - |b|^2 + (c\bar{d} - a\bar{b})w + (\bar{a}b - \bar{c}d)\bar{w} + (|c|^2 - |a|^2)|w|^2]^2},$$

so that (23) is an isometry precisely if

$$|d|^2 - |b|^2 = 1, \qquad |c|^2 - |a|^2 = -1, \qquad c\bar{d} - a\bar{b} = 0. \tag{26}$$

It is easy to verify that these conditions are exactly those under which the matrix $\begin{pmatrix} a & b \\ c & d \end{pmatrix}$ preserves the quadratic form $\langle \xi, \xi \rangle_{1,1} = (z^1)^2 - (z^2)^2$ (where $\xi = (z^1, z^2)$ is a vector in $\mathbb{C}^2$), i.e. under which $\begin{pmatrix} a & b \\ c & d \end{pmatrix} \in U(1, 1)$ (see the conclusion of §11.2). Since we also have $ad - bc = 1$, it follows that the isometries among the transformations (23), are precisely those for which $\begin{pmatrix} a & b \\ c & d \end{pmatrix} \in SU(1, 1)$. (Note that the conditions (26) also ensure that the transformation (23) maps the open unit disc onto itself, since its boundary $|z| = 1$ goes onto $|w| = 1$, and $w = 0$ is the image of $z = b/d$, which has modulus < 1.)

Finally we consider the Klein model. To begin with we characterize those linear-fractional transformations (23) which map the upper half-plane $\operatorname{Im} z > 0$ onto itself. Since the boundary will then have to be mapped onto itself we shall require that

$$\operatorname{Im} w = 0 \Rightarrow \operatorname{Im} z = 0. \tag{27}$$

A straightforward calculation shows that if w is real ($w \neq -d/c$) and $z = (aw + b)/(cw + d)$, then

$$\operatorname{Im} z = \frac{w^2 \operatorname{Im}(a\bar{c}) + w \operatorname{Im}(b\bar{c} + a\bar{d}) + \operatorname{Im}(b\bar{d})}{|cw + d|^2}.$$

Since w is allowed to be any real number (with the possible unimportant exception of $-d/c$), it follows that for (27) to hold we need

$$\operatorname{Im} a\bar{c} = \operatorname{Im}(b\bar{c} + a\bar{d}) = \operatorname{Im}(b\bar{d}) = 0. \tag{28}$$

Now assume that none of a, b, c, d is zero. (We leave the easier contrary case to the reader.) Since $a\bar{c}$ is real it follows that $a = \alpha c$ where α is real; similarly $b = \beta d$ where β is real. Then since $b\bar{c} + a\bar{d}$ is real, we get that $\beta d\bar{c} + \beta c\bar{d}$ is real. Hence either $\alpha = \beta$, which contradicts $ad - bc = 1$, or $c\bar{d}$ is real, so that $c = \gamma d$ where γ is real. Hence a, b, c, d are all real multiples of some complex number k say. Then from $ad - bc = 1$ it follows that we may take $k = 1$ or i. However if $k = i$, then it is easy to see that, for example, $(ai + b)/(ci + d)$ is in the lower half-plane. The upshot is that the transformation (23) maps the upper half-plane onto itself precisely if a, b, c, d are all real.

It turns out that all transformations (23) with a, b, c, d real are in fact isometries of the Klein model of the Lobachevskian plane, i.e. they all preserve the metric

$$dl^2 = \frac{4\,dz\,d\bar{z}}{-(z - \bar{z})^2}.$$

We shall omit the verification of this, which is similar to that of the preceding two cases. It is noteworthy that in both models of the Lobachevskian plane, all linear-fractional transformations preserving the region of the complex plane on which the metric is defined, turn out to be isometries.

We gather the above results together in the following

13.2.3. Theorem. *The group of direct isometries of the Lobachevskian plane is isomorphic to:*

(i) $SU(1, 1)/\{\pm 1\}$ *(using the Poincaré model)*;
(ii) $SL(2, \mathbb{R})/\{\pm 1\}$ *(using Klein's model)*;
(iii) *the connected component of the identity of* $SO(1, 2)$ *(see* §6.2).

Again we are assuming the result (to be established in Chapter 4 of Part II) that these groups do indeed account for all direct isometries of the Lobachevskian plane. The statement (iii) follows in part from the fact that the connected component of the identity of $SO(1, 2)$, consists of just those isometries of $\mathbb{R}^3_1$ preserving the upper sheet of the pseudo-sphere (see §10.1).

Note also that, analogously to the case of the sphere, the full isometry group of the Lobachevskian plane is generated by all the direct isometries of the Poincaré model (i.e. by $SU(1, 1)/\{\pm 1\}$) together with the map $z \mapsto \bar{z}$, which sends the unit disc onto itself, and is obviously an isometry of that model. In terms of Klein's model, the full isometry group is obtained by including (as a generator) the map $z \mapsto -\bar{z}$, along with the transformations (23) with a, b, c, d real (forming a group isomorphic to $SL(2, \mathbb{R})/\{\pm 1\}$).

13.2.4. Corollary. *The groups* $SU(1, 1)/\{\pm 1\}$, $SL(2, \mathbb{R})/\{\pm 1\}$, *and the connected component of the identity of* $SO(1, 2)$, *are isomorphic.*

As an exercise, prove this directly by using appropriate co-ordinate changes!

13.3. Surfaces of Constant Curvature

Before broaching the theme of this section, we need to answer the following natural question: Are there surfaces in Euclidean 3-space whose induced metrics are hyperbolic (i.e. Lobachevskian)? It turns out that the answer is affirmative: there are surfaces in Euclidean 3-space which are, as they say "locally isometric" to the hyperbolic plane. (However there is no such surface that is globally isometric to the hyperbolic plane; i.e. from which there is a one-to-one metric-preserving map onto, say, the Poincaré model of the hyperbolic plane.) An example is provided by the following hornshaped surface of revolution (to which the name "pseudo-sphere" was first applied):

$$x = \operatorname{sech} u \cos v, \qquad y = \operatorname{sech} u \sin v, \qquad z = u - \tanh u, \qquad u \geqq 0,$$

where x, y, z are Euclidean co-ordinates for $\mathbb{R}^3$. The induced metric is easily calculated to be given by

$$dl^2 = (\tanh^2 u)\, du^2 + (\operatorname{sech}^2 u)\, dv^2,$$

which by a change of local co-ordinates can be brought into one of the usual forms for the Lobachevskian metric (we omit the verification of this). It is also easy to calculate that the Gaussian curvature $K \equiv -1$ (cf. the first of the remarks at the end of §10).

To return to our theme, suppose we are given a surface in Euclidean 3-space whose metric (in terms of certain local co-ordinates) has complex conformal form $dl^2 = g(z, \bar{z})\, dz\, d\bar{z}$. (By 13.1.1 the metrics of many surfaces in Euclidean 3-space can (locally) be brought into this form. By the preceding remarks there are such surfaces locally isometric to the Lobachevskian plane.) Since $g(z, \bar{z}) > 0$, we can define a new function $\varphi(z, \bar{z})$ unambiguously by $g(z, \bar{z}) = e^{\varphi}$. In terms of φ, the formulae (11) and (17) for the Gaussian curvature become

$$K = -\tfrac{1}{2} e^{-\varphi} \Delta\varphi, \qquad K = -2e^{-\varphi} \frac{\partial^2 \varphi}{\partial z\, \partial \bar{z}}. \tag{29}$$

If K is constant, then in the following form these equations are known as "Liouville's equations":

$$\Delta\varphi = -2Ke^{\varphi}, \qquad \frac{\partial^2 \varphi}{\partial z\, \partial \bar{z}} = -\frac{K}{2} e^{\varphi}. \tag{30}$$

13.3.1. Theorem. *A surface in Euclidean 3-space with metric $dl^2 = g(z, \bar{z})\, dz\, d\bar{z}$, and with constant Gaussian curvature K is locally isometric to:*

(i) *the sphere if $K > 0$;*
(ii) *the Euclidean plane if $K = 0$;*
(iii) *the Lobachevskian plane if $K < 0$.*

PROOF. From (29), with K assumed constant, we have

$$0 = \frac{\partial}{\partial z}\left(-\frac{K}{2}\right) = \frac{\partial}{\partial z}\left(e^{-\varphi} \frac{\partial^2 \varphi}{\partial z\, \partial \bar{z}}\right) = e^{-\varphi}\left(\frac{\partial^3 \varphi}{\partial z^2\, \partial \bar{z}} - \frac{\partial \varphi}{\partial z} \frac{\partial^2 \varphi}{\partial z\, \partial \bar{z}}\right)$$

$$= e^{-\varphi} \frac{\partial}{\partial \bar{z}}\left(\frac{\partial^2 \varphi}{\partial z^2} - \frac{1}{2}\left(\frac{\partial \varphi}{\partial z}\right)^2\right),$$

so that the function $\partial^2\varphi/\partial z^2 - \frac{1}{2}(\partial\varphi/\partial z)^2 = \psi(z)$ say, is analytic. Under a complex analytic co-ordinate change $z = f(w)$ the metric takes the form (see Lemma 12.1.1)

$$dl^2 = g(z, \bar{z})\left|\frac{df}{dw}\right|^2 dw\, d\bar{w} = \tilde{g}(w, \bar{w})\, dw\, d\bar{w}.$$

If we then define $\tilde{\varphi}(w, \bar{w})$ by $\tilde{g}(w, \bar{w}) = e^{\tilde{\varphi}}$, it follows that

$$\tilde{\varphi}(w, \bar{w}) = \varphi(z, \bar{z}) + \ln\frac{df}{dw} + \ln\frac{d\bar{f}}{d\bar{w}}. \tag{31}$$

Of course since $\tilde{\varphi}$ was defined completely analogously to φ, the function $\partial^2\tilde{\varphi}/\partial w^2 - \frac{1}{2}(\partial\tilde{\varphi}/\partial w)^2 = \tilde{\psi}(w)$ say, will also be analytic (as a function of w). From this and (31) (and a little computation), we obtain

$$\tilde{\psi}(w) = \psi(z)(f')^2 + \frac{f'''}{f'} - \frac{3}{2}\left(\frac{f''}{f'}\right)^2,$$

where $f' = df/dw$. We wish to choose the function f so that $\tilde{\psi}(w)$ is identically zero, i.e. we seek an analytic function f satisfying the differential equation

$$\frac{f'''}{f'} - \frac{3}{2}\left(\frac{f''}{f'}\right)^2 = -\psi(f(w))(f')^2. \tag{32}$$

It can be shown using the theory of differential equations that (32) has an analytic solution; we shall not give the proof. (In the theory of functions of a complex variable, the left-hand side of (32) is known as the "Schwarz derivative.")

With f chosen to satisfy (32), we have $\tilde{\psi}(w) \equiv 0$, whence

$$\frac{\partial^2 e^{-\tilde{\varphi}/2}}{\partial w^2} = -\tfrac{1}{2} e^{-\tilde{\varphi}/2}\left(\frac{\partial^2\tilde{\varphi}}{\partial w^2} - \frac{1}{2}\left(\frac{\partial\tilde{\varphi}}{\partial w}\right)^2\right) = 0. \tag{33}$$

If we put $w = u + iv$, and analyse the meaning of (33) in terms of the original definition of the operator $\partial^2/\partial w^2$ (see §12(6)), we find that (33) is equivalent to

$$\frac{\partial^2 e^{-\tilde{\varphi}/2}}{\partial u\, \partial v} = 0, \qquad \frac{\partial^2 e^{-\tilde{\varphi}/2}}{\partial u^2} = \frac{\partial^2 e^{-\tilde{\varphi}/2}}{\partial v^2}.$$

It follows easily from these equations that

$$e^{-\tilde{\varphi}/2} = a(u^2 + v^2) + \text{a linear function of } u \text{ and } v, \tag{34}$$

which in complex notation takes the (anticipated) form

$$e^{-\tilde{\varphi}/2} = aw\bar{w} + bw + \bar{b}\bar{w} + c,$$

where a, c are real. Hence in terms of the new co-ordinates w, $\bar{w}$, our metric is given by

$$dl^2 = \frac{dw\, d\bar{w}}{(aw\bar{w} + bw + \bar{b}\bar{w} + c)^2}. \tag{35}$$

From the first of the formulae (29) (with $\tilde{\varphi}$ replacing φ), and (34), it follows that the Gaussian curvature is $K = 4(ac - b\bar{b})$. By means of linear-fractional transformations (which the reader may like to construct), the formula (35) can be brought into the forms:

(i) $\dfrac{4R^4\,dz\,d\bar{z}}{(1 + |z|^2)^2}$ if $K = 4(ac - b\bar{b}) > 0$;

(ii) $dz\,d\bar{z}$ if $K = 4(ac - b\bar{b}) = 0$;

(iii) $\dfrac{4R^4\,dz\,d\bar{z}}{(1 - |z|^2)^2}$ if $K = 4(ac - b\bar{b}) < 0$,

which are the familiar forms of the metrics of the sphere, Euclidean plane, and hyperbolic plane respectively. □

13.4. Exercises

1. Suppose the metric on a surface has the form

$$dl^2 = dx^2 + f(x)\,dy^2,$$

where $f(x)$ is a positive (real-valued) function. Prove that this metric can be brought into conformal form $dl^2 = g(u, v)(du^2 + dv^2)$.

2. Prove that a 2-dimensional pseudo-Riemannian metric (of type (1, 1)) with real analytic coefficients, takes the form

$$dl^2 = \lambda(t, x)(dt^2 - dx^2)$$

after a suitable co-ordinate change.

§14. Transformation Groups as Surfaces in N-Dimensional Space

14.1. Co-ordinates in a Neighbourhood of the Identity

Consider the group $GL(n, \mathbb{R})$ of invertible matrices A, i.e.

$$A = (a^i_j), \qquad \det(a^i_j) \neq 0. \tag{1}$$

Condition (1) defines a region in the space $M(n, \mathbb{R})$ of all $n \times n$ real matrices. If we regard $M(n, \mathbb{R})$ as a vector space (under matrix addition, and multiplication of matrices by scalars), then from this point of view the general linear group is a region of the vector space $\mathbb{R}^{n^2} = M(n, \mathbb{R})$. There is a natural system of co-ordinates for this space, namely the entries a^i_j of the general

$n \times n$ matrix. If $A = (a^i_j)$, $B = (b^i_j)$ are two matrices of degree n over $\mathbb{R}$, then of course their product $C = AB$ is given by $C = (c^i_j)$ where

$$c^i_j = a^i_k b^k_j, \qquad i, j = 1, 2, \ldots, n. \tag{2}$$

For our present purposes the importance of (2) lies in the fact that it expresses each entry in the product C as a smooth (i.e. continuously differentiable) function (in fact as a quadratic polynomial!) of the $2n^2$ entries of A and B. In other words (and restricting attention to $GL(n, \mathbb{R})$) matrix multiplication defines a smooth map of the direct product:

$$GL(n, \mathbb{R}) \times GL(n, \mathbb{R}) \to GL(n, \mathbb{R}),$$

where $(A, B) \mapsto AB$.

If we introduce the Euclidean metric into the space $M(n, \mathbb{R})$, taking the co-ordinates a^i_j as Euclidean, then the square of the norm of a matrix A will be given by

$$|A|^2 = \sum_{i,j} |a^i_j|^2, \qquad A = (a^i_j), \tag{3}$$

and we shall of course have the triangle inequality:

$$|A + B| \leq |A| + |B|. \tag{4}$$

With respect to multiplication of matrices this Euclidean norm behaves analogously:

14.1.1. Lemma. *For all $A, B \in M(n, \mathbb{R})$, we have*

$$|AB| \leq |A||B|. \tag{5}$$

PROOF. It is easy to check (by expanding the right-hand side) that for any $2m$ real numbers $x_1, \ldots, x_m; y_1, \ldots, y_m$,

$$\left(\sum x_i^2\right)\left(\sum y_i^2\right) - \left(\sum x_i y_i\right)^2 = \tfrac{1}{2}\sum (x_i y_j - x_j y_i)^2.$$

Hence

$$\left(\sum x_i y_i\right)^2 \leq \left(\sum x_i^2\right)\left(\sum y_i^2\right), \tag{6}$$

from which (putting $m = n^2$) the lemma follows. □

We shall now construct another useful (local) co-ordinate system in a neighbourhood in $GL(n, \mathbb{R})$ of the identity matrix 1. To this end consider first the open unit ball $|X| < 1$, $X = (x^i_j)$, in the space $M(n, \mathbb{R})$ of all matrices X.

14.1.2. Lemma. *If $|X| < 1$, then the matrix $A = 1 + X$ is invertible, i.e.*

$$A = 1 + X \in GL(n, \mathbb{R}).$$

Proof. We first show that the matrix series

$$B = 1 - X + X^2 - X^3 + \cdots \tag{7}$$

converges (in our Euclidean norm). Using the inequalities (4) and (5) we have

$$\begin{aligned}&|X^m - X^{m+1} + X^{m+2} - \cdots \pm X^{m+k-1}| \\ &\leq |X^m|\cdot|1 + |X| + \cdots + |X|^{k-1}| = |X|^m \frac{1 - |X|^k}{1 - |X|}.\end{aligned}$$

Hence if $|X| < 1$, the sequence of partial sums of the series is a Cauchy sequence, and therefore converges (in our norm). It follows that the n^2 series formed from the entries in the terms of (7) all converge, so that B is well-defined as a matrix. In fact, more generally, the space $M(n, \mathbb{R})$ is complete with respect to our norm: every Cauchy sequence has a limit. Hence

$$AB = (1 + X)(1 - X + X^2 - X^3 + \cdots) = 1,$$

so that $A^{-1} = B$, completing the proof. □

We shall now define co-ordinates on the neighbourhood in $GL(n, \mathbb{R})$ of the identity consisting of all matrices A satisfying

$$|A - 1| < 1.$$

(By Lemma 14.1.2 all such A are in $GL(n, \mathbb{R})$.) Thus if $A = (a^i_j)$, then its n^2 co-ordinates x^i_j are taken to be

$$x^i_j(A) = a^i_j - \delta^i_j,$$

where, as usual,

$$\delta^i_j = \begin{cases} 1 & \text{for } i = j, \\ 0 & \text{for } i \neq j. \end{cases} \tag{8}$$

Thus in particular the co-ordinates of the identity matrix are all zero: $x^i_j(1) = 0$.

Remark. Using this co-ordinatization of a neighbourhood of the identity, we can define co-ordinates in a neighbourhood in $GL(n, \mathbb{R})$ of any point $B_0 \in GL(n, \mathbb{R})$ as follows. The neighbourhood of B_0 in question consists of all matrices B satisfying

$$|B - B_0| < |B_0^{-1}|^{-1}. \tag{9}$$

By multiplying both sides of (9) on the left by $|B_0^{-1}|$ and using Lemma 14.1.1, we see that for such B, $|B_0^{-1}B - 1| < 1$, i.e. $B_0^{-1}B$ is in the above-defined neighbourhood of the identity. We then define the co-ordinates of B to be those of $B_0^{-1}B$ in our above co-ordinatization of the neighbourhood of the identity; in other words, if $B_0^{-1} = (c^i_j)$, $B = (b^i_j)$, then the co-ordinates y^i_j of B are given by

$$y^i_j(B) = c^i_k b^k_j - \delta^i_j. \tag{10}$$

In particular $y^i_j(B_0) = 0$. In this way we can define local co-ordinates around any point.

We now return to our initial co-ordinatization of the matrics in $M(n, \mathbb{R})$ by their entries. In terms of these co-ordinates, $GL(n, \mathbb{R})$ is a region of the space $M(n, \mathbb{R})$. It turns out that there is a natural identification of the tangent space of the group $GL(n, \mathbb{R})$ at the identity, with the space of all matrices of degree n. To see how this arises (and to see what the tangent space *is*), consider an arbitrary curve $A(t)$ in $GL(n, \mathbb{R})$ (i.e. a family of matrices $A(t)$ depending on the parameter t), which passes through the identity when $t = 0$ (i.e. $A(0) = 1$). The tangent or velocity vector to this curve at $t = 0$ is the matrix

$$\dot{A}(t)|_{t=0}, \tag{11}$$

and the tangent space is the vector space spanned by all such matrices. On the other hand, let X be any (constant) matrix of degree n. Then by Lemma 14.1.2 the curve $A(t) = 1 + tX$ lies in $GL(n, \mathbb{R})$, provided we restrict the domain of values of t to a sufficiently small interval about 0. Since, obviously, $A(0) = 1$, $\dot{A}(0) = X$, it follows that every matrix X occurs in the form (11), i.e. occurs as a tangent vector at the identity. Hence the tangent space to $GL(n, \mathbb{R})$ at the identity, coincides with the vector space $M(n, \mathbb{R})$.

Most of the transformation groups which we considered in §§4, 6 were defined by equations which singled out the relevant matrices from within the space $M(n, \mathbb{R})$ of all $n \times n$ matrices over $\mathbb{R}$. From our present point of view therefore, we may regard these transformation groups as surfaces in the space $M(n, \mathbb{R})$.

To begin with, consider the group $SL(n, \mathbb{R})$. It is defined in $M(n, \mathbb{R})$ by the single equation

$$\det A = 1,$$

so that in our present context it is a hypersurface in the n^2-dimensional space $M(n, \mathbb{R})$, lying entirely in the region $GL(n, \mathbb{R})$. (Here we are regarding the matrices in $M(n, \mathbb{R})$ as co-ordinatized by their entries.)

14.1.3. Theorem. *The points of $SL(n, \mathbb{R})$ form a non-singular surface in the space $M(n, \mathbb{R})$.*

Proof. Let x^i_j, $i, j = 1, \ldots, n$, be the entries (or co-ordinates from the other point of view) of a general matrix $A \in M(n, \mathbb{R})$, and write $f(x^i_j) = \det A - 1$. (Hence f is a function of n^2 variables.) Then by definition (see the end of §7.1), non-singularity of the hypersurface $f(x^i_j) = 0$ means that at every point of the hypersurface $\operatorname{grad} f \neq 0$. Now $\partial f/\partial x^i_j$ $(=(\partial \det A)/\partial x^i_j)$ is just the cofactor of x^i_j in the matrix A, so that if $\operatorname{grad} f$ is zero at some point (i.e. matrix), then all cofactors of that matrix will be zero, and the matrix will not be invertible. Hence at any matrix A satisfying $\det A = 1$, we must have $\operatorname{grad} f \neq 0$, as required. □

The truth of this theorem is also revealed by the fact that the tangent space to the hypersurface $SL(n, \mathbb{R})$ at the identity (and therefore at any point), has dimension $n^2 - 1$ (see end of §7.2). This follows from the fact that this tangent space coincides with the subspace of all zero-trace matrices. To see this, let $A(t)$ be a curve in $SL(n, \mathbb{R})$ passing through the identity when $t = 0$; i.e. such that $\det A(t) = 1$, $A(0) = 1$. It is easy to verify that

$$0 = \frac{d}{dt} \det A(t)\bigg|_{t=0} = \operatorname{tr}\left(\frac{d}{dt} A(t)\right)\bigg|_{t=0}.$$

Hence the tangent vector $A(0)$ has zero trace. Conversely if X is any matrix with zero trace, then for the curve $A(t) = 1 + tX$, we have

$$\det A(t) = \det(1 + tX) = 1 + t(\operatorname{tr} X) + o(t) = 1 + o(t).$$

Hence this line is tangent to the hypersurface at the identity (when $t = 0$), so that its direction (i.e. tangent) vector at $t = 0$, namely X, is in the tangent space to $SL(n, \mathbb{R})$ at the identity.

We now consider the group $O(n)$ of orthogonal matrices of degree n. This group is defined as a surface in $\mathbb{R}^{n^2}$ by the equations

$$\sum_k a_k^i a_k^j = \delta^{ij}, \quad \text{or} \quad AA^{\mathrm{T}} = 1, \quad A = (a_j^i). \tag{12}$$

Among these n^2 equations there are duplications, obtained by interchanging the indices i, j. After discarding these we are left with $n(n + 1)/2$ equations. It turns out that $O(n)$ is also a non-singular surface in the space $M(n, \mathbb{R})$. To prove this one might show that if the $n(n + 1)/2$ defining equations of the surface $O(n)$ are written as $f^{ij}(a_l^h) = 0$, $1 \leq i \leq j \leq n$, where

$$f^{ij}(a_l^h) = \sum_k a_k^i a_k^j - \delta^{ij},$$

then the $n^2 \times n(n + 1)/2$ matrix with entry $\partial f^{ij}/\partial a_l^h$ in the $(i, j$th column and (h, l)th row, has rank $n(n + 1)/2$ at every point of $O(n)$ (see the definition immediately following (13) of §7.2). This is feasible since

$$\frac{\partial f^{ij}}{\partial a_l^h} = 0 \quad \text{if } h \neq i, j, \qquad \frac{\partial f^{ij}}{\partial a_l^i} = a_l^j, \qquad \frac{\partial f^{ij}}{\partial a_l^j} = a_l^i,$$

but we shall not pursue it. We shall instead prove that the tangent space to the surface $O(n)$ at the identity, has dimension $n(n - 1)/2 = n^2 - n(n + 1)/2$; by the comment concluding §7.2 this is a geometrical indicator of non-singularity.

We shall in fact show that the tangent space to $O(n)$ at the identity coincides with the space of all skew-symmetric matrices. As before we let

$A(t)$ be a curve in $O(n)$ passing through the identity when $t = 0$; i.e. $A(0) = 1$. By Lemma 5.3.1, under these conditions the matrix

$$X = \frac{d}{dt} A(t)|_{t=0}$$

is skew-symmetric, i.e.

$$X^{\mathrm{T}} + X = 0.$$

Conversely, if X is any skew symmetric matrix, then for the curve $A(t) = 1 + tX$, we have

$$\begin{aligned} A(t)A^{\mathrm{T}}(t) &= (1 + tX)(1 + tX^{\mathrm{T}}) \\ &= 1 + t(X + X^{\mathrm{T}}) + o(t) = 1 + o(t), \end{aligned}$$

so that, as in the case of $SL(2, \mathbb{R})$ above, the line $A(t)$ is tangent to the surface $O(n)$ at the identity, whence X is in the tangent space to $O(n)$ at the identity. It is clear that the subspace of $M(n, \mathbb{R})$ consisting of all skew-symmetric matrices has dimension $n(n - 1)/2$, since each such matrix is specified by its $n(n - 1)/2$ above-diagonal entries.

It follows from the non-singularity of $O(n)$ that its connected component $SO(n)$ is also a non-singular surface in the space $M(n, \mathbb{R})$.

14.1.4. Example. Consider the group $SO(3)$ of rotations of Euclidean 3-space. One system of local co-ordinates for this non-singular surface is provided by the "Euler angles" (perhaps familiar to the reader from analytic geometry), which one associates with a rotation as follows. A rotation moving the co-ordinate frame x, y, z to the position of the frame x', y', z' (as shown in Figure 17), can be achieved by carrying out in succession the following three rotations:

(i) the rotation through angle φ about the z-axis, bringing the x-axis into coincidence with the "nodal line," i.e. the line of intersection of the (x, y)-plane and the (x', y')-plane;

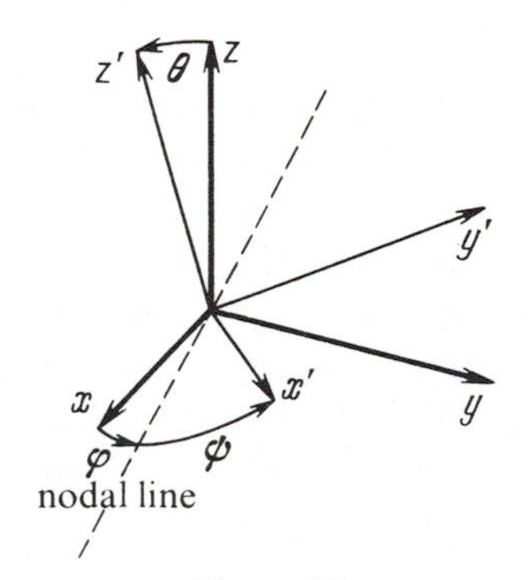

Figure 17

(ii) the rotation through angle θ about the nodal line, which takes the z-axis into the z'-axis;
(iii) the rotation through angle ψ about the z'-axis taking the nodal line into the x'-axis.

Another co-ordinatization of $SO(3)$ is obtained in the following way. Each rotation of Euclidean 3-space is determined by its axis of rotation, and the angle φ, $0 \leq \varphi \leq \pi$, through which the space is rotated about that axis. If from the two possible directions of the axis of rotation, we choose the one for which the rotation through φ is counter-clockwise, then our element of $SO(3)$ is completely specified by the vector $\bar{\varphi}$ whose direction is that chosen for the axis of rotation, and whose magnitude $|\bar{\varphi}|$ is φ. (This is of course essentially the usual way of construing rotations as vectors.) Thus to each point of $SO(3)$ we have associated a point of the (closed) ball of radius π, namely the tip of the vector $\bar{\varphi}$ (when its tail is at the centre of the ball). This correspondence is not one-to-one since diametrically opposite points of the surface of this ball correspond to rotations through π about the same axis but in opposite directions, and these rotations define the same element of $SO(3)$. With this co-ordinatization, the identity of $SO(3)$ corresponds to the origin (i.e. centre of the ball), and the tangent space to $SO(3)$ at the identity, corresponds to the whole 3-dimensional vector space. (We leave the verification of this to the reader.)

We now turn our attention to the complex transformation groups. The group $GL(n, \mathbb{C})$ is a region of the space $M(n, \mathbb{C})$ of all matrices (i.e. in the space $\mathbb{C}^{n^2}$ of dimension n^2), or, from the real point of view, of the space $\mathbb{R}^{2n^2}$. The group $SL(n, \mathbb{C})$ is a complex hypersurface of dimension $n^2 - 1$ (or of dimension $2n^2 - 2$ in $\mathbb{R}^{2n^2}$). As in the real case, the tangent space to $SL(n, \mathbb{C})$ at the identity coincides with the vector space of (complex) zero-trace matrices.

We shall now consider the unitary group $U(n)$. This group is defined in $M(n, \mathbb{C})$ by the equations

$$f^{ij}(A) = \sum_{k=1}^{n} a_k^i \bar{a}_k^j = \delta^{ij}, \quad \text{or} \quad A\bar{A}^{\mathrm{T}} = 1 \quad (\text{where } A = (a_j^i)). \tag{13}$$

It follows that $\partial f^{ij}/\partial \bar{a}_k^j \not\equiv 0$, so that f^{ij} is not analytic. Hence the group $U(n)$ is not a complex surface (see the beginning of §12.3). However as a surface in $\mathbb{R}^{2n^2}$, defined by a set of real equations equivalent to (13) (there remain n^2 such equations after discarding the obviously redundant ones—see below), it turns out to be non-singular. We again omit the details (which the reader may like to attempt to supply, along previous lines).

We shall, however, establish the real dimension of the tangent space to $U(n)$ at the identity, as being $2n^2 - n^2 = n^2$. Note first that each equation $f^{ij} = \delta^{ij}$ gives rise to the two real equations

$$\operatorname{Re} f^{ij} = \delta^{ij}, \qquad \operatorname{Im} f^{ij} = 0. \tag{14}$$

Now since $f^{ij} = \bar{f}^{ji}$ (by inspection of (13)), the equations $f^{ij} = 0$ and $f^{ji} = 0$ are equivalent when $i \neq j$. This and the fact that the equation $\operatorname{Im} f^{ii} = 0$ is automatically fulfilled (since $\sum_k a_k^i \bar{a}_k^i$ is real), together imply that we may discard $[2n(n-1)]/2 + n = n^2$ of the equations (14), leaving $2n^2 - n^2 = n^2$ real equations defining the surface $U(n)$. Let $A(t)$ be a curve in $U(n)$ with $A(0) = 1$. Then by an argument entirely similar to that in Lemma 5.3.1, it follows from the conditions that $A(t)\,\bar{A}^{\mathrm{T}}(t) = 1$ and $A(0) = 1$, that

$$X + \bar{X}^{\mathrm{T}} = 0, \tag{15}$$

where $X = dA/dt|_{t=0}$. We call matrices X satisfying (15) *skew-Hermitian*. It can be shown (much as before) that, conversely, every skew-Hermitian matrix X is in the tangent space to $U(n)$ at the identity, so that the tangent space coincides with the space of all skew-Hermitian matrices. Since a skew-Hermitian matrix (x_j^i) is determined by specifying the n^2 real numbers (Cartesian co-ordinates) $(1/i)x_k^k$, $k = 1, \ldots, n$; $\operatorname{Re} x_j^i$, $\operatorname{Im} x_j^i$, $1 \leq i < j \leq n$, it follows that the tangent space has dimension n^2, as claimed.

It follows from the non-singularity of $U(n)$ that its subgroup $SU(n)$ consisting of the unitary matrices with determinant 1, is also a real non-singular surface in $\mathbb{R}^{2n^2}$. Its dimension is $n^2 - 1$, and its tangent space at the identity coincides with the space of all skew-Hermitian matrices with trace zero.

14.1.5. Example. Consider the group $SU(2)$. In §11.3, we saw that the matrices in $SU(2)$ are just those of the form

$$\begin{pmatrix} a & b \\ -\bar{b} & \bar{a} \end{pmatrix}, \qquad |a|^2 + |b|^2 = 1.$$

If we write $a = x + iy$, $b = u + iv$, it is clear that the equation $|a|^2 + |b|^2 = 1$ is just the equation of the 3-dimensional sphere of radius 1 in the 4-dimensional space co-ordinatized by x, y, u, v.

14.2. The Exponential Function with Matrix Argument

Let T denote the tangent space to the group $GL(n, \mathbb{R})$ at the identity. We define a map

$$\exp\colon T \to GL(n, \mathbb{R}), \qquad \exp(0) = 1, \tag{16}$$

from this tangent space to the group itself, by setting

$$\exp X = 1 + \frac{X}{1!} + \frac{X^2}{2!} + \cdots, \tag{17}$$

for any matrix X. (This definition of $\exp X$ is intended to embrace complex X also.)

14.2.1. Lemma. (i) *The series* (17) *converges for all matrices X (real or complex).*

(ii) *If the matrices X, Y commute, i.e. $XY = YX$, then*

$$\exp(X + Y) = (\exp X)(\exp Y). \tag{18}$$

(iii) *For all matrices X, the matrix $A = \exp X$ is invertible, and*

$$A^{-1} = \exp(-X). \tag{19}$$

(iv) *For all X, $\exp(X^{\mathrm{T}}) = (\exp X)^{\mathrm{T}}$.*

PROOF. Note first that since the realization map is an algebra monomorphism from $M(n, \mathbb{C})$ to $M(2n, \mathbb{R})$, it suffices to prove the lemma for real matrices.

(i) Using the inequalities (4) and (5), we have, for any real matrix X,

$$\left|\frac{X^m}{m!} + \frac{X^{m+1}}{(m+1)!} + \cdots + \frac{X^{m+k-1}}{(m+k-1)!}\right| \leq \frac{|X|^m}{m!} + \cdots + \frac{|X|^{m+k-1}}{(m+k-1)!}.$$

Since the series

$$1 + x + \frac{x^2}{2!} + \cdots$$

converges for all numbers x (to e^x), it follows that the series (17) is Cauchy in the Euclidean norm, so that the norms of the partial sums of the series (17) tend to a limit. As mentioned before, it is then clear that the entries in the (matrix) partial sums of the series (17) each tend to a limit. This proves (i).

(ii) Using the permutability of X and Y we have:

$$\begin{aligned}(\exp X)(\exp Y) &= \left(\sum_{k=0}^{\infty} \frac{X^k}{k!}\right)\left(\sum_{l=0}^{\infty} \frac{Y^l}{l!}\right) \\ &= \sum_{m=0}^{\infty} \frac{1}{m!}\left(\sum_{k+l=m} \frac{m!}{k!\,l!} X^k Y^l\right) = \sum_{m=0}^{\infty} \frac{1}{m!}(X + Y)^m \\ &= \exp(X + Y).\end{aligned}$$

(iii) This follows from (ii) since X and $Y = -X$ commute, and $\exp(0) = 1$. Since (iv) is obvious, the proof of the lemma is complete. □

Now let G be any one of the matrix groups considered in the previous subsection (i.e. $G = SL(n, \mathbb{R})$, $O(n)$, $U(n)$, etc.), and let T be the tangent space to G at the identity, in the appropriate space ($\mathbb{R}^{n^2}$ or $\mathbb{R}^{2n^2}$). We shall show that exp always maps T to G.

14.2.2. Lemma. (i) *If $G = SL(n, \mathbb{R})$ and $X \in T$, i.e.* $\operatorname{tr} X = 0$, *then* $A = \exp X \in SL(n, \mathbb{R})$, *i.e.* $\det A = 1$.

(ii) *If $G = O(n)$ and $X \in T$, i.e. X is skew-symmetric, then $A = \exp X \in O(n)$, i.e. the matrix A is orthogonal.*

(iii) *If $G = U(n)$ and $X \in T$, i.e. X is skew-Hermitian, then* $A = \exp X \in U(n)$, *i.e. the matrix A is unitary.*

Proof. (i) Let X have zero trace, and consider the family of matrices

$$A(t) = \exp(tX),$$

where t is a parameter. By part (ii) of Lemma 14.2.1, we have

$$A(t_1 + t_2) = A(t_1)A(t_2)$$

(since the matrices t_1X and t_2X commute). Write $f(t) = \det A(t)$. Then $f(t_1 + t_2) = f(t_1)f(t_2)$. It is known that the only solution of this functional equation is $f(t) = e^{ct}$ (at least under suitable conditions on f, which are here ensured by the definition of f). We shall show that the constant $c = 0$. Now

$$f(t) = \det \exp(tX) = \det(1 + tX + o(t)) = 1 + t(\operatorname{tr} X) + o(t);$$

thus since $\operatorname{tr}(X) = 0$, we have

$$c = \left.\frac{df}{dt}\right|_{t=0} = \lim_{h\to 0} \frac{f(h) - f(0)}{h} = \lim_{h\to 0} \frac{1 + o(h) - 1}{h} = 0.$$

Hence $\det A(t) = 1$; so, in particular, $\det A = 1$.

(ii) Suppose X is skew-symmetric, i.e. $X^{\mathrm{T}} = -X$. Then X and X^{T} commute so that, putting $A = \exp X$, we have by parts (ii) and (iv) of 14.2.1,

$$AA^T = (\exp X)(\exp X)^T = \exp(X + X^T) = 1,$$

whence $A \in O(n)$, as required.

(iii) Finally, suppose X is skew-Hermitian, i.e. $\bar{X}^{\mathrm{T}} = -X$. Then appealing to parts (ii), (iv) of 14.2.1 we have (setting $A = \exp X$):

$$\begin{aligned} A\bar{A}^{\mathrm{T}} &= (\exp X)(\overline{\exp X})^{\mathrm{T}} = (\exp X)(\exp \bar{X})^{\mathrm{T}} \\ &= (\exp X)\exp(\bar{X}^{\mathrm{T}}) = \exp(X + \bar{X}^{\mathrm{T}}) = 1. \end{aligned}$$

(We have used here also the easy fact that $\overline{\exp X} = \exp \bar{X}$.) Hence A is unitary, and the proof of the lemma is complete. □

As for $SL(n, \mathbb{R})$, so for any group G of unimodular matrices will the tangent space T at the identity consist of zero-trace matrices. Hence by mimicking the proof of part (i) of the above lemma, we shall have that for all $X \in T$, $\det(\exp X) = 1$. This proves the

14.2.3. Corollary. *For $G = SO(n)$, $SU(n)$, we also have* $\exp(X) \in G$ *whenever* $X \in T$.

14.2.4. Lemma. *In some neighbourhood of the origin (the zero matrix of $M(n, \mathbb{R})$), the map* exp *is one-to-one.*

PROOF. By the Inverse Function Theorem (1.2.5), it suffices to show that the Jacobian of the mapping exp (from $\mathbb{R}^{n^2}$ to itself) is non-zero at 0. Write $X = (x^i_j)$, and $A = \exp X = (a^i_j)$. It is easy to see from the definition of exp X, that

$$\frac{\partial a^k_l}{\partial x^i_j} = \delta^{ki}\delta_{jl} + \cdots,$$

where the (infinitely many) terms represented by the dots become zero if we put $X = 0$. (In fact they are all monomials of positive degree in the x^i_j.) Hence the $n^2 \times n^2$ Jacobian matrix of the transformation exp is

$$\left(\frac{\partial a^k_l}{\partial x^i_j}\right) = (\delta^{ki}\delta_{jl}),$$

where the rows are indexed by the pairs (k, l) and the columns by the pairs (i, j). Since this is just the identity matrix of degree n^2, the lemma follows. □

Lemmas 14.2.2 and 14.2.4 (and Corollary 14.2.3) allow us to use the map exp to define another useful local co-ordinate system in a neighbourhood of the identity of each of our matrix groups G, namely the Cartesian co-ordinates (i.e. entries) of the matrices in a neighbourhood of 0 in the tangent plane T. Thus (to be explicit) the co-ordinates x^i_j of a matrix A in a neighbourhood of 1 of G on which the inverse function (ln) of exp exists, are given by

$$x^i_j(A) = (\ln A)^i_j = \left[(A - 1) - \frac{(A - 1)^2}{2} + \frac{(A - 1)^3}{3} - \cdots\right]^i_j. \tag{20}$$

Note that as a mapping from T to G, exp need not be one-to-one. It may also fail to map T *onto* G—see Exercise 3 below.

Remark. Let T be, as usual, the tangent space to the matrix group G at the identity, and consider a straight line in T passing through the origin, i.e. a family of matrices tX, where t is a parameter taking all values in $\mathbb{R}$, and X is fixed. Then the family of matrices $A(t) = \exp(tX)$ forms a *one-parameter subgroup* of G. That this family is indeed a group follows from

$$\begin{gathered} A(s)A(t) = A(s + t), \qquad A(0) = 1, \\ A(-t) = A^{-1}(t). \end{gathered} \tag{21}$$

For instance in the group $SO(3)$, the subgroup of rotations about a fixed axis is a one-parameter subgroup. In this case $A(t + 2\pi) = A(t)$. If the fixed axis is for example the z-axis, then

$$A(t) = \begin{pmatrix} \cos t & \sin t & 0 \\ -\sin t & \cos t & 0 \\ 0 & 0 & 1 \end{pmatrix} = \exp t \begin{pmatrix} 0 & 1 & 0 \\ -1 & 0 & 0 \\ 0 & 0 & 0 \end{pmatrix}.$$

14.3. The Quaternions

The set $\mathbb{H}$ of *quaternions* consists of all linear combinations

$$q \in \mathbb{H}, \qquad q = a + bi + cj + dk, \tag{22}$$

where the coefficients a, b, c, d are real, and $1, i, j, k$ are linearly independent. (Alternatively, write $q = (a, b, c, d)$, $i = (0, 1, 0, 0)$, $j = (0, 0, 1, 0)$, $k = (0, 0, 0, 1)$.) We introduce a multiplication in $\mathbb{H}$ by defining how i, j, k multiply among themselves:

$$\begin{gathered} ij = k = -ji, \qquad jk = i = -kj, \qquad ki = j = -ik, \\ i^2 = j^2 = k^2 = -1, \end{gathered} \tag{23}$$

and then extending the definition to all of $\mathbb{H}$ using the distributive law (assuming that those quaternions with $b, c, d = 0$ are central, i.e. commute with all quaternions). It is easy to see that with this definition of multiplication (and the usual addition of linear combinations), $\mathbb{H}$ is an associative (but not commutative) algebra over the field of real numbers. The following lemma tells us that this algebra is isomorphic to a subalgebra of $M(2, \mathbb{C})$.

14.3.1. Lemma. *For each quaternion* $q = a + bi + cj + dk$, *define* $A(q) \in M(2, \mathbb{C})$ *by*

$$A(q) = \begin{pmatrix} a + bi & c + di \\ -c + di & a - bi \end{pmatrix}. \tag{24}$$

(Note that we may identify the two roles played by the symbol i*!) Then the map* $q \mapsto A(q)$ *is one-to-one, and*

$$A(q_1 q_2) = A(q_1)A(q_2), \tag{25}$$

so that this map is an algebra monomorphism (since it is obviously also linear).

Proof. We shall verify only (25), as the rest of the lemma is trivial. However, once it is pointed out that (25) needs to be verified only for $q_1, q_2 = i, j, k$, this part also becomes obvious. Thus

$$A(i) = \begin{pmatrix} i & 0 \\ 0 & -i \end{pmatrix}, \qquad A(j) = \begin{pmatrix} 0 & 1 \\ -1 & 0 \end{pmatrix}, \qquad A(k) = \begin{pmatrix} 0 & i \\ i & 0 \end{pmatrix},$$

so that $A(i)A(j) = A(k)$, and similarly for the other products. $\square$

Remark. The matrices

$$\sigma_x = -iA(k), \qquad \sigma_y = -iA(j), \qquad \sigma_z = -iA(i) \tag{26}$$

are often called the *Pauli matrices*. They satisfy

$$\sigma_x^2 = \sigma_y^2 = \sigma_z^2 = 1, \qquad \sigma_x \sigma_y = -\sigma_y \sigma_x = i\sigma_z, \ldots. \tag{27}$$

We define an operation of *conjugation* on $\mathbb{H}$ by

$$\bar{q} = a - bi - cj - dk \quad (\text{for } q = a + bi + cj + dk). \tag{28}$$

14.3.2. Lemma. *The map $q \mapsto \bar{q}$ is an anti-isomorphism of the algebra $\mathbb{H}$, i.e. it is linear, and*

$$\overline{q_1 q_2} = \bar{q}_2 \bar{q}_1. \tag{29}$$

PROOF. The lemma follows immediately from 14.3.1, the fact that

$$A(\bar{q}) = \bar{A}^{\mathrm{T}}(q), \tag{30}$$

and the fact that the map $A \mapsto \bar{A}^{\mathrm{T}}$ is an anti-isomorphism of the matrix algebra $M(2, \mathbb{C})$. (Alternatively, it may easily be verified directly.) □

We define the (Euclidean) *norm* $|q|$ of a quaternion by

$$|q|^2 = q\bar{q} = a^2 + b^2 + c^2 + d^2 \quad (\text{for } q = a + bi + cj + dk). \tag{31}$$

From this it follows that if $|q|^2 \neq 0$, i.e. $q \neq 0$, then $q(\bar{q}/|q|^2) = 1$, so that each non-zero quaternion q has multiplicative inverse $q^{-1} = \bar{q}/|q|^2$. We formulate this as a

14.3.3. Lemma. *The algebra $\mathbb{H}$ of quaternions is a division algebra; i.e. for each non-zero quaternion q there is a quaternion q^{-1} inverse to it:*

$$qq^{-1} = 1, \qquad q^{-1}q = 1.$$

From (24) it is immediate that for all quaternions q

$$|q|^2 = \det A(q), \tag{32}$$

whence

$$|q_1 q_2|^2 = |q_1|^2 |q_2|^2. \tag{33}$$

From (33) we deduce that the set of quaternions with norm 1 forms a group under multiplication; we shall denote this group by $\mathbb{H}_1$. From (31) it follows that for $q \in \mathbb{H}_1$, we have $q^{-1} = \bar{q}$.

If we regard $\mathbb{H}$ as a 4-dimensional space with co-ordinates a, b, c, d, then $\mathbb{H}_1$ is the hypersurface defined by the equation

$$a^2 + b^2 + c^2 + d^2 = |q|^2 = 1, \tag{34}$$

i.e. $\mathbb{H}_1$ is just the 3-sphere in $\mathbb{R}^4$. Thus as a surface in $\mathbb{R}^4$, $\mathbb{H}_1$ is familiar to us. From the group-theoretical point of view it is also familiar, since, as we shall now see, as a group $\mathbb{H}_1$ is isomorphic to $SU(2)$. Thus let $q = a + bi + cj + dk \in \mathbb{H}_1$, and put $x = a + bi$, $y = c + di$. Then in terms of x and y, $A(q)$ takes the form

$$A(q) = = \begin{pmatrix} a + bi & c + di \\ -c + di & a - bi \end{pmatrix} = \begin{pmatrix} x & y \\ -\bar{y} & \bar{x} \end{pmatrix}. \tag{35}$$

Moreover, since $|q|^2 = 1$, we have $|x|^2 + |y|^2 = 1$. It follows that (35) defines an isomorphism between $\mathbb{H}_1$ and $SU(2)$ (see §11.3(24)).

We shall now use this fact (that $\mathbb{H}_1 \simeq SU(2)$) to give an alternative proof of Corollary 13.2.2 (which states that $SO(3) \simeq SU(2)/\{\pm 1\}$).

To begin, let $\mathbb{H}_0$ denote the 3-dimensional space consisting of all quaternions x satisfying the condition $\bar{x} = -x$, i.e. with real part zero. We turn this space into Euclidean 3-space by equipping it with the above-defined norm. We then have $|x|^2 = x\bar{x} = -x^2$. We shall need the following fact.

14.3.4. Lemma. *If* $|q|^2 = 1$, *then the transformation defined by*

$$\alpha_q : x \mapsto qxq^{-1}, \qquad x \in \mathbb{H}_0, \tag{36}$$

is a rotation of 3-*dimensional Euclidean space* $\mathbb{H}_0 = \mathbb{R}^3$.

PROOF. Since $\bar{x} = -x$ and $\bar{q} = q^{-1}$, we have (using 14.3.2) that

$$\overline{qxq^{-1}} = \bar{q}^{-1}\bar{x}\bar{q} = -qxq^{-1}.$$

Hence the map α_q sends $\mathbb{H}_0$ to itself. Since $|qxq^{-1}| = |x|$, i.e. the lengths of vectors in $\mathbb{H}_0$ are preserved, it follows that α_q is a isometry of $\mathbb{H}_0$. Finally, writing $q = a + bi + cj + dk$, we see that if $q = a$(i.e. $b = c = d = 0$) then α_q is the identity map, while if $q \neq a$, then $(q - a)$ is fixed by α_q. Hence α_q is a rotation, and the proof is complete. □

It follows easily that the map $q \mapsto \alpha_q$ is a homomorphism from $\mathbb{H}_1 \simeq SU(2)$, to the group of rotations of Euclidean 3-space. It is also easy to verify that α_q is the identity map precisely when q is real, i.e., since $|q|^2 = 1$, when $q = \pm 1$. It is not too difficult to verify also that every rotation has the form (36) (we leave the details to the reader; to begin with knowledge of the axis of rotation may prove useful). The upshot is the promised isomorphism $SO(3) \simeq SU(2)/\{\pm 1\}$.

The isomorphism $\mathbb{H}_1 \simeq SU(2)$ can also be used to prove that

$$SO(4) \simeq (SU(2) \times SU(2))/\{\pm 1\}.$$

We give an indication only of how this is done. Thus suppose $p, q \in \mathbb{H}_1$. Then the map

$$\alpha_{p,q} : x \mapsto pxq^{-1}, \qquad x \in \mathbb{H} = \mathbb{R}^4, \tag{37}$$

clearly preserves the quaternion (i.e. Euclidean) norm, so that it is an isometry of Euclidean 4-space, and therefore, since it fixes the origin, belongs to $O(4)$. It follows that the map $(p, q) \mapsto \alpha_{p,q}$ is a homomorphism from $SU(2) \times SU(2)$ to $O(4)$. It is easy to check that the kernel of this map consists of the pairs $(1, 1)$, $(-1, -1)$ only, so that we have a monomorphism from $(SU(2) \times SU(2))/\{\pm 1\}$ to $O(4)$. The fact that the $\alpha_{p,q}$ actually lie in $SO(4)$ is a consequence of the connectedness of $(SU(2) \times SU(2))/\{\pm 1\}$, (which follows from that of $SO(3)$), and the fact that the map (37) is continuous and so maps connected sets to connected sets. We omit the proof of the fact that every motion in $SO(4)$ occurs as some $\alpha_{p,q}$.

We now turn to the consideration of the n-dimensional *quaternion space* $\mathbb{H}^n$, with basis $e_1, \ldots, e_n$ (e.g. $e_1 = (1, 0, \ldots, 0)$, $e_2 = (0, 1, 0, \ldots,)$, etc. will serve), and co-ordinates $q^1, \ldots, q^n$. For this the following preliminary fact will be important: Any quaternion $q = a + bi + cj + dk$ can be expressed as

$$q = x + yj = x + j\bar{y}, \qquad x = a + bi, \qquad y = c + di, \tag{38}$$

where here x and y can be regarded simply as complex numbers. The following straightforward identity tells us how to multiply two such expressions $x + yj$ and $u + vj$:

$$(x + yj)(u + vj) = (x + yj)(u + j\bar{v}) = (xu - y\bar{v}) + (xv + y\bar{u})j. \tag{39}$$

It follows that the space $\mathbb{H}^n$ can be regarded as the $2n$-dimensional complex space $\mathbb{C}^{2n}$, with basis $e_1, \ldots, e_n, je_1, \ldots, je_n$, and with complex co-ordinates $x^1, \ldots, x^n, y^1, \ldots, y^n$, where $q^k = x^k + y^k j$. This is clear from

$$q^k e_k = x^k e_k + y^k j e_k = x^k e_k + y^k (je_k). \tag{40}$$

We now introduce the group $GL(n, \mathbb{H})$ of invertible transformations of $\mathbb{H}^n$, linear over $\mathbb{H}$. Each transformation $\Lambda \in GL(n, \mathbb{H})$ is determined in the usual way (relative to the basis $e_1, \ldots, e_n$) by an $n \times n$ matrix (λ_l^k), whose entries λ_l^k lie in $\mathbb{H}$; thus the co-ordinates $q^1, \ldots, q^n$ transform according to the rule

$$q^k \to q^l \lambda_l^k = q'^k, \tag{41}$$

where, since $\mathbb{H}$ is non-commutative, the order of the factors is important. From this and (39), it follows that the corresponding complex co-ordinates of $\mathbb{C}^{2n}$ transform according to the rule

$$\begin{aligned} x^k &\to (x^l a_l^k - y^l \bar{b}_l^k) = x'^k, \\ y^k &\to (x^l b_l^k + y^l \bar{a}_l^k) = y'^k, \\ \lambda_l^k &= a_l^k + b_l^k j. \end{aligned} \tag{42}$$

Hence we have that each $\mathbb{H}$-linear transformation of the space $\mathbb{H}^n$ corresponds one-to-one to (or can be regarded as) a $\mathbb{C}$-linear transformation of the corresponding space $\mathbb{C}^{2n}$. Thus, as an analogue of the realization map (see §11.1(4)), we obtain a group monomorphism

$$c: GL(n, \mathbb{H}) \to GL(2n, \mathbb{C}), \tag{43}$$

In view of (42), the explicit action of this map c on an element $\Lambda = A + Bj$ of $GL(n, \mathbb{H})$ is given by

$$c(\Lambda) = \begin{pmatrix} A & B \\ -\bar{B} & \bar{A} \end{pmatrix}. \tag{44}$$

(Here the matrix is taken to act on the right of vectors $(x^1, \ldots, x^n, y^1, \ldots, y^n)$.)

Remark. The matrices occurring as images under c can be characterized in a manner reminiscent of the characterization (in §11.1) of the images of the realization map, namely: The image of $GL(n, \mathbb{H})$ under c consists of just those $2n \times 2n$ complex matrices which commute with the operator $J = c(j)$, i.e. with the linear transformation of $\mathbb{C}^{2n}$ induced by multiplication by j.

We next introduce a form (analogous to the Hermitian form—see §11.2(10)) defined on pairs of vectors in $\mathbb{H}^n$ by

$$\langle \xi_1, \xi_2 \rangle_{\mathbb{H}} = \sum_{k=1}^{n} q_1^k \bar{q}_2^k, \qquad \xi_1 = q_1^k e_k, \quad \xi_2 = q_2^k e_k. \tag{45}$$

Then the norm of a vector in $\mathbb{H}^n$ is defined, as usual, by

$$|\xi|^2 = \langle \xi, \xi \rangle_{\mathbb{H}} = \sum_{k=1}^{n} q^k \bar{q}^k = \sum_{k=1}^{n} |q^k|^2, \qquad \xi = q^k e_k. \tag{46}$$

Writing $q = x + yj$, we have $|q|^2 = |x|^2 + |y|^2$, so that if ξ is regarded as a vector in $\mathbb{C}^{2n}$, with components $x^1, \ldots, x^n, y^1, \ldots, y^n$, then

$$|\xi|^2 = \sum |q^k|^2 = \sum |x^k|^2 + \sum |y^k|^2,$$

i.e. the norm of ξ as a vector in $\mathbb{C}^{2n}$ is the usual Hermitian norm (see §11.2(11)).

14.3.5. Definition. The *symplectic group* $Sp(n)$ is the subgroup of $GL(n, \mathbb{H})$ consisting of all invertible $\mathbb{H}$-linear transformations of $\mathbb{H}^n$ which preserve the form (45).

In terms of complex co-ordinates the form (45) becomes

$$\begin{aligned} \langle \xi_1, \xi_2 \rangle_{\mathbb{H}} &= \sum_{k=1}^{n} (x_1^k + y_1^k j)\overline{(x_2^k + y_2^k j)} \\ &= \sum_k (x_1^k \bar{x}_2^k + y_1^k \bar{y}_2^k) + \sum_k (y_1^k x_2^k - x_1^k y_2^k) j. \end{aligned} \tag{47}$$

From this it follows that the transformations in $Sp(n)$ are just those preserving the Hermitian form on $\mathbb{C}^{2n}$, given by

$$\langle \xi_1, \xi_2 \rangle_{\mathbb{C}} = \sum_{k=1}^{n} (x_1^k \bar{x}_2^k + y_1^k \bar{y}_2^k),$$

and also the skew-symmetric form $\sum_k (y_1^k x_2^k - x_1^k y_2^k)$. In other words, $c(Sp(n))$ consists of those unitary transformations of $\mathbb{C}^{2n}$ (i.e. elements of $U(2n)$), which preserve the above skew-symmetric form.

We end the section by sketching a proof of the fact that $Sp(1)$ is isomorphic to $SU(2)$. From the preceding paragraph we know that $c(Sp(1))$ is that subgroup of $U(2)$ consisting of those matrices preserving the form $y_1 x_2 - x_1 y_2$. But since this is just the area of the complex parallelogram in $\mathbb{C}^2$ having as sides the vectors (x_1, y_1), (x_2, y_2), it follows that $c(Sp(1))$ consists of just the unimodular matrices in $U(2)$.

Alternatively this isomorphism can be inferred directly by simply showing that the unitary matrices which preserve $y_1x_2 - x_1y_2$, are precisely those of the form $\begin{pmatrix} a & b \\ -\bar{b} & \bar{a} \end{pmatrix}$ where $|a|^2 + |b|^2 = 1$; the isomorphism then follows from the fact that this also characterizes the matrices in $SU(2)$ (see §11.2(24)).

14.4. Exercises

1. Prove that the groups $GL(n, \mathbb{C})$, $U(n)$, $Sp(n)$ are connected (see §4.4).

2. Prove the formula

$$\det(\exp X) = e^{tr\, X}$$

where X is any complex matrix.

3. Show that the image under exp of the tangent space to $SL(2, \mathbb{R})$ at the identity, is not the whole of $SL(2, \mathbb{R})$ (see Lemma 14.2.2(i)).

4. Determine the tangent spaces at the identity, of the groups $U(n)$, $SU(n)$, $SO(p, q)$, $SU(p, q)$.

5. Determine all one-parameter subgroups of $SL(2, \mathbb{R})$.

§15. Conformal Transformations of Euclidean and Pseudo-Euclidean Spaces of Several Dimensions

Suppose we have two metrics $g_{\alpha\beta}$ and $g'_{\alpha\beta}$ defined on a region U of the space $\mathbb{R}^n$ with co-ordinates $x^1, \ldots, x^n$. We are interested in the situation where these metrics differ only by a (variable) factor, i.e. $g'_{\alpha\beta}(x) = \lambda(x)g_{\alpha\beta}(x)$, for some real-valued function $\lambda(x) > 0$. In this situation we shall say that the two metrics *define the same conformal structure* on the region U, or are *conformally equivalent*. Thus a metric $g''_{\alpha\beta}(y)$ given to us in terms of co-ordinates $y^1, \ldots, y^n$, is conformally equivalent to $g_{\alpha\beta}(x)$, if after expressing the two metrics in the same co-ordinates (x or y), they differ only by a factor $\lambda(x) = \lambda(x(y))$, i.e. if they are proportional (in an extended sense of that word).

15.1. Definition. A map φ from the region U with metric $g_{\alpha\beta}$ relative to co-ordinates $x^1, \ldots, x^n$, to a region V with co-ordinates $y^1, \ldots, y^n$, is said to be *conformal*, if the metric $g'_{\alpha\beta} = (\partial x^k/\partial y^\alpha)g_{kl}(\partial x^l/\partial y^\beta)$ is proportional to the original:

$$g'_{\alpha\beta} = \lambda g_{\alpha\beta}, \qquad \lambda = \lambda(x) > 0. \tag{1}$$

In this section we shall be concerned only with the Euclidean and pseudo-Euclidean metrics on $\mathbb{R}^n$. We shall moreover suppose that the co-ordinates

$x^1, \ldots, x^n$ for $\mathbb{R}^n$ are Euclidean (i.e. $g_{ij} = \delta_{ij}$), or pseudo-Euclidean (i.e. $g_{ij} = 0$ for $i \neq j$, $g_{ii} = 1$ for $1 \leq i \leq p$, $g_{ii} = -1$ for $p < i \leq n$), as the case may be. Among the latter the case $p = 1$, $q = n - 1$ (i.e. Minkowski space $\mathbb{R}^n_{1,n-1}$) will be of particular interest for us.

We wish to find all conformal transformations of the above spaces. It is clear from our investigations in §§4, 6 of the isometry groups of Euclidean and pseudo-Euclidean spaces, that the only conformal transformations of these spaces that are linear, are as follows:

(i) the isometries, forming the groups $O(n)$ or $O(p, q)$; $\lambda(x) \equiv 1$;
(ii) the dilations $x \mapsto \lambda x$, $\lambda = \text{const.}$;
(iii) more generally, combinations of dilations and isometries: $x \mapsto \lambda A(x)$, $A \in O(n)$ or $O(p, q)$, $\lambda = \text{const.}$

As well as these, we also have the translations (also isometries) $x \mapsto x + x_0$. These (and their composites) account for all the "obvious" conformal transformations. However there are others, namely the inversions

(iv)
$$x^\alpha \to \frac{x^\alpha - x_0^\alpha}{\langle x - x_0, x - x_0 \rangle}, \tag{2}$$

where the denominator is just the scalar square defined by the metric in question, i.e.

$$\langle x - x_0, x - x_0 \rangle = (x^\alpha - x_0^\alpha)(x^\beta - x_0^\beta) g_{\alpha\beta}(x).$$

Note that the map (2) is not defined for $x = x_0$; we can however remove this defect by adjoining to $\mathbb{R}^n$ a point "at infinity" to serve as image of x_0.

To see that (2) defines a conformal transformation, observe first that if we write y for the image of x under (2), then assuming $g_{\alpha\beta} = \delta_{\alpha\beta}$, we have

$$\sum_{k=1}^{n} (y^k)^2 = \langle x - x_0, x - x_0 \rangle^{-1},$$

provided that $x \neq x_0$. Hence

$$x^\alpha = \frac{y^\alpha}{\Sigma (y^k)^2} + x_0^\alpha.$$

From this we can calculate $\partial x^\alpha / \partial y^l$, and thence in turn $g'_{\alpha\beta}$. After this computation, it turns out that indeed the map (2) is conformal, with

$$\lambda(x) = \langle x - x_0, x - x_0 \rangle^2.$$

Let us now consider the conformal transformations of $\mathbb{R}^n$, $\mathbb{R}^n_{p,q}$ for various n. For $n = 1$ all maps trivially qualify as conformal. The case $n = 2$ is covered by Lemma 12.3.4, according to which conformality of a transformation $u + iv = w = f(x, y)$ is essentially equivalent to analyticity. Thus for $n = 1, 2$, we have a large variety of conformal transformations. However for $n \geq 3$, the situation is radically different, as the following theorem shows.

15.2. Theorem (Liouville). *Every sufficiently well-behaved conformal transformation of a region of Euclidean (pseudo-Euclidean) space of dimension ≥ 3, is a composite of isometries, dilations and inversions.*

(For the purposes of this theorem, the transformation is certainly "sufficiently well-behaved" if its fourth-order derivatives are continuous.)

PROOF. We give the proof for $n = 3$ only, indicating at its conclusion how it can be augmented to yield the general case (and to provide for this, we shall throughout use the symbol n rather than 3). We shall assume also that the metric is Euclidean (with Euclidean co-ordinates $x^1, \ldots, x^n$), since the proof in the pseudo-Euclidean case is essentially the same.

Thus let $y^\alpha = y^\alpha(x^1, \ldots, x^n)$ be a (sufficiently well behaved) conformal transformation from a region U to a region V, both in $\mathbb{R}^n$. (We may regard this transformation as a change of co-ordinates for U.) Let ξ, η be any two vectors at the (arbitrary) point $(x^1, \ldots, x^n)$ of U, with components $\xi_x^1, \ldots, \xi_x^n$ and $\eta_x^1, \ldots, \eta_x^n$ respectively, relative to the system x, and components $\xi_y^1, \ldots, \xi_y^n$ and $\eta_y^1, \ldots, \eta_y^n$ respectively, in the co-ordinates y. Then in view of the conformality of our transformation, we have by (1) that

$$\langle \xi, \eta \rangle = g'_{\alpha\beta} \xi_y^\alpha \eta_y^\beta = \lambda(x) g_{\alpha\beta} \xi_y^\alpha \eta_y^\beta. \tag{3}$$

Hence if $g_{\alpha\beta} \xi_x^\alpha \eta_x^\beta = 0$ (i.e. if $\langle \xi, \eta \rangle = 0$), then since $\lambda(x) \neq 0$, we shall have also $g_{\alpha\beta} \xi_y^\alpha \eta_y^\beta = 0$, i.e. (ξ_y^α) and (η_y^α) are othogonal regarded as vectors expressed in the system x (which is in any case the only co-ordinate system for $\mathbb{R}^n$ involved). Since by Definition 2.2.1 (or earlier), $(\xi_y^\alpha) = A(\xi_x^\alpha)$, $(\eta_y^\alpha) = A(\eta_x^\alpha)$, where A is the matrix $(\partial y^i/\partial x^j)$, it follows that if the vectors (ξ_x^α) and (η_x^α) are orthogonal, then so also are the vectors $A(\xi_x^\alpha)$, $A(\eta_x^\alpha)$, regarded as vectors expressed in the system x.

Now let $\eta_1 = (\eta_1^\alpha)$, $\eta_2 = (\eta_2^\alpha)$, $\eta_3 = (\eta_3^\alpha)$ be any three fixed, mutually orthogonal vectors at the (arbitrary) point $(x^1, \ldots, x^n)$ of U. By the above, it follows that then the three vectors $A(\eta_1^\alpha)$, $A(\eta_2^\alpha)$, $A(\eta_3^\alpha)$ are also mutually orthogonal; we write them briefly as $A\eta_1$, $A\eta_2$, $A\eta_3$. Differentiating $\langle A\eta_1, A\eta_2 \rangle = 0$ with respect to x^γ, we obtain (using the fact that the scalar product is the usual Euclidean one):

$$0 = \frac{\partial}{\partial x^\gamma} \langle A\eta_1, A\eta_2 \rangle = \left\langle \frac{\partial^2 y}{\partial x^\beta\, \partial x^\gamma} \eta_1^\beta, A\eta_2 \right\rangle + \left\langle A\eta_1, \frac{\partial^2 y}{\partial x^\beta\, \partial x^\gamma} \eta_2^\beta \right\rangle,$$

whence, multiplying by η_3^γ (and summing over γ), we get

$$\left\langle \frac{\partial^2 y}{\partial x^\beta\, \partial x^\gamma} \eta_1^\beta \eta_3^\gamma, A\eta_2 \right\rangle + \left\langle A\eta_1, \frac{\partial^2 y}{\partial x^\beta\, \partial x^\gamma} \eta_2^\beta \eta_3^\gamma \right\rangle = 0. \tag{4}$$

By permuting the indices 1, 2, 3 in (4), we obtain altogether three such equations. By adding the appropriate two of these, and subtracting the third, we arrive at

$$\left\langle \frac{\partial^2 y}{\partial x^\gamma\, \partial x^\beta} \eta_1^\gamma \eta_2^\beta, A\eta_3 \right\rangle = 0 \tag{5}$$

(and again we obtain from this two further such equations by permuting the symbols 1, 2, 3).

Since $n = 3$, and $A\eta_1$, $A\eta_2$, $A\eta_3$, are mutually perpendicular, it follows from (5) that

$$\frac{\partial^2 y}{\partial x^\gamma \partial y^\beta} \eta_1^\gamma \eta_2^\beta = \mu(x)(A\eta_1) + v(x)(A\eta_2). \tag{6}$$

From this we obtain the following expressions for the coefficients μ and v:

$$\begin{aligned} \mu &= \frac{1}{|A\eta_1|^2} \left\langle \frac{\partial^2 y}{\partial x^\gamma \partial x^\beta} \eta_1^\gamma \eta_2^\beta, A\eta_1 \right\rangle, \\ v &= \frac{1}{|A\eta_2|^2} \left\langle \frac{\partial^2 y}{\partial x^\gamma \partial x^\beta} \eta_1^\gamma \eta_2^\beta, A\eta_2 \right\rangle. \end{aligned} \tag{7}$$

Since by (3) we have $|\eta_1|^2 = \lambda(x)|A\eta_1|^2$, we can rewrite (7) as

$$\begin{aligned} \mu &= \frac{\lambda}{|\eta_1|^2} \left[\frac{1}{2} \eta_2^\alpha \frac{\partial}{\partial x^\alpha} \langle A\eta_1, A\eta_1 \rangle \right] \\ &= \frac{\lambda}{|\eta_1|^2} \left[\frac{1}{2} \eta_2^\alpha \frac{\partial}{\partial x^\alpha} \left(\frac{|\eta_1|^2}{\lambda(x)} \right) \right] = -\frac{1}{2\lambda} \eta_2^\alpha \frac{\partial \lambda}{\partial x^\alpha}, \\ v &= -\frac{1}{2\lambda} \eta_1^\alpha \frac{\partial \lambda}{\partial x^\alpha}. \end{aligned} \tag{8}$$

Substituting these expressions for μ and v in (6) (and rewriting), we obtain

$$\frac{\partial^2 y}{\partial x^\gamma \partial x^\beta} \eta_1^\gamma \eta_2^\beta = -\frac{1}{2\lambda} \left(\frac{\partial \lambda}{\partial x^\gamma} \frac{\partial y}{\partial x^\beta} \eta_1^\beta \eta_2^\gamma + \frac{\partial \lambda}{\partial x^\beta} \frac{\partial y}{\partial x^\gamma} \eta_1^\beta \eta_2^\gamma \right). \tag{9}$$

If we write $\rho = \sqrt{\lambda}$, then

$$\begin{aligned} \frac{\partial^2 (\rho y)}{\partial x^\gamma \partial x^\beta} &= \frac{\partial \rho}{\partial x^\gamma} \frac{\partial y}{\partial x^\beta} + \frac{\partial \rho}{\partial x^\beta} \frac{\partial y}{\partial x^\gamma} + \rho \frac{\partial^2 y}{\partial x^\gamma \partial x^\beta} + y \frac{\partial^2 \rho}{\partial x^\gamma \partial x^\beta} \\ &= \frac{1}{2\sqrt{\lambda}} \left(\frac{\partial \lambda}{\partial x^\gamma} \frac{\partial y}{\partial x^\beta} + \frac{\partial \lambda}{\partial x^\beta} \frac{\partial y}{\partial x^\gamma} + 2\lambda \frac{\partial^2 y}{\partial x^\gamma \partial x^\beta} \right) + y \frac{\partial^2 \rho}{\partial x^\gamma \partial x^\beta}. \end{aligned}$$

From this together with (9) it follows that

$$\frac{\partial^2}{\partial x^\gamma \partial x^\rho} (\rho y \eta_1^\gamma \eta_2^\beta) = \left(\frac{\partial^2 \rho}{\partial x^\gamma \partial x^\beta} \eta_1^\gamma \eta_2^\beta \right) y. \tag{10}$$

Differentiating this equation with respect to x^δ, and then multiplying throughout by η_3^δ and summing over the index δ, we get

$$\frac{\partial^3 (\rho y)}{\partial x^\gamma \partial x^\beta \partial x^\delta} \eta_1^\gamma \eta_2^\beta \eta_3^\delta = \left(\frac{\partial^2 \rho}{\partial x^\gamma \partial x^\beta} \eta_1^\gamma \eta_2^\beta \right) A\eta_3 + \left(\frac{\partial^3 \rho}{\partial x^\gamma \partial x^\beta \partial x^\delta} \eta_1^\gamma \eta_2^\beta \eta_3^\delta \right) y. \tag{11}$$

Since the first and last of the three expressions in (11) are invariant under permutations of the indices 1, 2, 3, so must also the middle expression; hence in particular

$$\left(\frac{\partial^2 \rho}{\partial x^\gamma \, \partial x^\beta} \eta_1^\gamma \eta_2^\beta\right) A\eta_3 = \left(\frac{\partial^2 \rho}{\partial x^\gamma \, \partial x^\beta} \eta_2^\gamma \eta_3^\beta\right) A\eta_1.$$

Since none of $A\eta_1$, $A\eta_2$, $A\eta_3$ is zero (the Jacobian of our transformation being assumed non-zero, at least at the point we are considering), their mutual orthogonality implies that

$$\frac{\partial^2 \rho}{\partial x^\gamma \, \partial x^\delta} \eta_1^\gamma \eta_2^\delta \equiv 0, \tag{12}$$

i.e. the bilinear form (12) is zero for any pair of orthogonal vectors. It follows easily (by considering (12) for various particular orthogonal η_1, η_2) that the coefficient matrix of such a form must be scalar, i.e. a multiple of the identity. Hence

$$\frac{\partial^2 \rho}{\partial x^\gamma \partial x^\delta} = \sigma(x) g_{\gamma\delta}. \tag{13}$$

We shall now show that in fact $\sigma(x) = \text{const}$. Let ξ_1, ξ_2, ξ_3 be any fixed vectors (at the point $(x^1, \ldots, x^n)$). If we differentiate (13) (keeping in mind that the Euclidean metric $g_{\gamma\delta}$ is constant), and then multiply by $\xi_1^\gamma \xi_2^\beta \xi_3^\delta$, and sum over the repeated indices, we obtain

$$\frac{\partial^3 \rho}{\partial x^\gamma \, \partial x^\beta \, \partial x^\delta} \xi_1^\gamma \xi_2^\beta \xi_3^\delta = \left(\frac{\partial \sigma}{\partial x^\gamma} \xi_1^\gamma\right) \langle \xi_2, \xi_3 \rangle.$$

Subtracting this equation from the one obtained by interchanging ξ_1 and ξ_2, we see that

$$\left\langle \left(\frac{\partial \sigma}{\partial x^\alpha} \xi_1^\alpha\right) \xi_2 - \left(\frac{\partial \sigma}{\partial x^\gamma} \xi_2^\gamma\right) \xi_1, \xi_3 \right\rangle = 0. \tag{14}$$

Since (14) holds for all vectors ξ_i, it follows that $\sigma = \text{const.}$, as claimed. Hence

$$\frac{\partial^2 \rho}{\partial x^\beta \partial x^\gamma} = \sigma \delta_{\beta\gamma} = \begin{cases} \sigma & \text{if } \beta = \gamma, \\ 0 & \text{if } \beta \neq \gamma, \end{cases} \qquad \sigma = \text{const.}$$

One then readily deduces that

$$\rho = \sqrt{\lambda} = a_1 |x - x_0|^2 + b_1, \qquad a_1, b_1 \text{ const.} \tag{15}$$

Now since the inverse $y \mapsto x$, from the region V of Euclidean space $\mathbb{R}^n$, to U, is clearly also conformal (with factor $1/\lambda(x)$), we must have that also

$$\frac{1}{\sqrt{\lambda(x)}} = a_2 |y - y_0|^2 + b_2, \qquad a_2, b_2 \text{ const.} \tag{16}$$

From (15) and (16) we obtain

$$(a_1|x - x_0|^2 + b_1)(a_2|y - y_0|^2 + b_2) = 1. \tag{17}$$

From this it follows that our transformation maps (the intersection of U with) any sphere $|x - x_0|^2 = \text{const.}$, to a sphere $|y - y_0|^2 = \text{const.}$ Now take any fixed straight line passing through x_0 and intersecting U, and let x_1, x be two points in U on this line, x_1 fixed, x variable, such that the segment joining x_1 and x lies entirely within U. Since this straight-line segment is perpendicular to all spheres centred at x_0 which intersect it, it follows from the angle-preserving property of conformal transformations (namely that the angle between two curves in U at their point of intersection is the same as the angle between their images in V—this follows immediately from conformality (and is in fact equivalent to it)) that it must go into a curve in V perpendicular to all spheres centred at y_0 which intersect it. We conclude that the image of the segment $x_1 x$ is also a straight-line segment $y_1 y$, say. Now since by (17)

$$\begin{aligned} a_2|y - y_0|^2 + b_2 &= a_2[|y - y_1|^2 + 2|y - y_1||y_1 - y_0| \\ &\quad + |y_1 - y_0|^2] + b_2 \\ &= \frac{1}{a_1|x - x_0|^2 + b_1}, \end{aligned} \tag{18}$$

we have that $|y - y_1|$ is an algebraic function of $|x - x_0|$. On the other hand if we parametrize the segment yy_1 by a parameter τ, $t_1 \leq \tau \leq t$, then, using $\sum_\alpha (dy^\alpha)^2 = (1/\lambda) \sum_\alpha (dx^\alpha)^2$, we have that

$$|y - y_1| = \int_{t_1}^{t} \sqrt{\sum_\alpha \left(\frac{dy^\alpha}{d\tau}\right)^2}\, d\tau = \int_{t_1}^{t} \sqrt{\frac{1}{\lambda} \sum_\alpha \left(\frac{dx^\alpha}{d\tau}\right)^2}\, d\tau. \tag{19}$$

If we choose τ to be arc length along the segment $x_0 x$, with $t_1 = |x_1 - x_0|$, $t = |x - x_0|$, (19) becomes

$$|y - y_1| = \int_{|x_1 - x_0|}^{|x - x_0|} \frac{d\tau}{\sqrt{\lambda}} = \int_{|x_1 - x_0|}^{|x - x_0|} \frac{d\tau}{(a_1 \tau^2 + b_1)}.$$

However if neither a_1 nor b_1 is zero, then this is a transcendental function of $|x - x_0|$, contradicting (18). Hence either $a_1 = 0$ or $b_1 = 0$.

If $a_1 = 0$ then $\lambda = \text{const.}$, and our transformation is an isometry followed by a dilation. If $b_1 = 0$ (in which case $a_1 \neq 0$), then by following the transformation $y = y(x)$ with the inversion

$$x^* = \frac{x - x_0}{|x - x_0|^2},$$

we obtain (since $|x^*| = 1/|x - x_0|$) the equation $a_2|y - y_0|^2 = a_1|x^*|^2$, which brings us back to the preceding case of an isometry followed by a dilation. This completes the proof for the 3-dimensional Euclidean case. □

As mentioned at the beginning of the proof, with the obvious minor changes the proof holds also for the pseudo-Euclidean case. No major changes are required for the proof to go through also for $n > 3$; the condition $n = 3$ was used only in deducing equation (6) for the triple of mutually orthogonal vectors η_1, η_2, η_3. We leave it as an exercise for the reader to establish (6) for arbitrary $n > 3$. This done, we shall have the theorem as stated.

In connexion with this theorem, note finally that it deals with (conformal transformations of) regions (by implication perhaps not the whole of $\mathbb{R}^n$) of Euclidean or pseudo-Euclidean space; as transformations of the whole space, inversions have singular points. Thus for the inversion (2), in the Euclidean case the only singular point is $x = x_0$, while if the metric is pseudo-Euclidean, then all points satisfying $\langle x - x_0, x - x_0 \rangle = 0$, are singular, i.e. all points of the light cone with apex at x_0.

We turn now to the consideration of the group of all conformal transformations of a space with metric $g_{\alpha\beta}$. Clearly this group is the same for all metrics differing from $g_{\alpha\beta}$ by a scalar factor, i.e. for all metrics of the form $\lambda(x)g_{\alpha\beta}$. It follows that we can model the conformal transformations of, in particular, the Euclidean metric $g_{\alpha\beta} = \delta_{\alpha\beta}$, on any space with metric conformally equivalent to the Euclidean metric, i.e. with metric of the form $\lambda(x)\delta_{\alpha\beta}$. For instance the n-dimensional sphere S^n, defined by the equation

$$\sum_{a=0}^{n} (x^a)^2 = 1,$$

can be co-ordinatized in such a way that in terms of these co-ordinates its metric is conformally equivalent to the Euclidean metric; for the 2-sphere, i.e. for $n = 2$, we found such co-ordinates in §9. For larger n conformally Euclidean co-ordinates on S^n (with the "pole" $x^0 = 1, x^1 = 0, \ldots, x^n = 0$ removed) can be defined, similarly to the case $n = 2$, by means of a stereographic projection on the "plane" co-ordinatized by $x^1, \ldots, x^n$ (S^n is situated in $\mathbb{R}^{n+1}$ with co-ordinates $x^0, \ldots, x^n$). In this way the co-ordinates $x^1, \ldots, x^n$ become co-ordinates on the sphere. It follows as in the 2-dimensional case that in terms of these co-ordinates the metric induced on S^n has (to within a constant factor) the form

$$h_{\alpha\beta} = g(x)\delta_{\alpha\beta}, \qquad g(x) = \frac{R^4}{(R^2 + r^2)^2}, \qquad r^2 = (x^1)^2 + \cdots + (x^n)^2, \qquad R = \text{const.} \tag{20}$$

To obtain the group of conformal transformations of the Euclidean metric (in $\mathbb{R}^n$), we may therefore work with the co-ordinates $x^1, \ldots, x^n$ on S^n, i.e. the group can be regarded as consisting of those transformations of S^n generated by the elements of $O(n)$, translations, and dilations, all applied to the co-ordinates $x^1, \ldots, x^n$.

We shall now indicate how the above can be used to show that the group of conformal transformations of the Euclidean metric $\delta_{\alpha\beta}$ in $\mathbb{R}^n$, is isomorphic to $O(1, n + 1)(\simeq O(n + 1, 1))$. To begin, note that the sphere S^n of radius R is the boundary of the $(n + 1)$-dimensional open ball $|x| < R$, of radius R, denoted by D^{n+1}. (In symbols $S^n = \partial D^{n+1}$.) On this ball we can define (analogously to §10) a metric, the Lobachevskian metric, by

$$\tilde{h}_{\alpha\beta} = \frac{R^4}{(R^2 - r^2)^2}\delta_{\alpha\beta}, \qquad r^2 = \sum_{i=0}^{n} (x^i)^2 < R^2. \tag{21}$$

It can be shown that the isometry group of D^{n+1} with this metric (which we call *Lobachevsky space*, and denote by L^{n+1}) is $O(1, n + 1)$ (cf. end of §10.1). Clearly $O(n + 1)$ is a subgroup of $O(1, n + 1)$ (namely the subgroup consisting of those transformations fixing x^0 and orthogonal on $x^1, \ldots, x^{n+1}$). Note that, as in the case $n = 1$, the space L^{n+1} can be realized as the quadric surface defined by the equation

$$(z^0)^2 - \sum_{a=1}^{n+1} (z^a)^2 = 1,$$

situated in the Minkowski space $\mathbb{R}_1^{n+2}$ with co-ordinates $z^0, \ldots, z^{n+1}$, and with metric

$$dl^2 = (dz^0)^2 - \sum_{a=1}^{n+1} (dz^a)^2.$$

(The relationship between the two systems of co-ordinates $z^1, \ldots, z^{n+1}$ and $x^0, \ldots, x^n$, is as follows (cf, §10.1(6) where $x^0 = u$, $x^1 = v$, $z^1 = x$, $z^2 = y$):

$$\begin{aligned} z^\alpha &= \frac{2R^2 x^{\alpha-1}}{R^2 - r^2}, \qquad \alpha = 1, \ldots, n + 1, \\ r^2 &= \sum_{\alpha=1}^{n+1} (x^{\alpha-1})^2.) \end{aligned} \tag{22}$$

With this preamble, we are now able to state the following result (whose proof we omit).

15.3. Theorem. *The group $O(1, n + 1)$ acting on the Lobachevsky space L^{n+1}, induces transformations on the sphere S^n (as for instance in the model of L^{n+1} as the ball D^{n+1} with boundary S^n and metric* (21)*). These induced transformations of S^n are all distinct, conformal in the standard metric* (20) *on S^n, have no singularities, and include the basic conformal transformations (i.e. the elements of $O(n)$, the inversions, dilations, and translations). Consequently the group of conformal transformations of the standard Euclidean metric on $\mathbb{R}^n$ (or the sphere S^n), $n \geq 3$, is isomorphic to $O(1, n + 1)$.*

If we were to prove this theorem, we should need to find a suitable correspondence between the elements of $O(1, n+1)$ or $O(n+1, 1)$ and the translations, dilations and inversions of $\mathbb{R}^n$. In the simplest case, namely $n = 1$, this can be done as follows:

$$\text{translation}(x \mapsto x + a) \leftrightarrow \begin{pmatrix} \frac{a^2}{2} + 1 & a & -\frac{a^2}{2} \\ a & 1 & -a \\ \frac{a^2}{2} & a & 1 - \frac{a^2}{2} \end{pmatrix} \in O(2, 1),$$

$$\text{dilation}(x \mapsto \pm \lambda x, \lambda > 0) \leftrightarrow \begin{pmatrix} \frac{\lambda^2 + 1}{2\lambda} & 0 & \frac{\lambda^2 - 1}{2\lambda} \\ 0 & \pm 1 & 0 \\ \frac{\lambda^2 - 1}{2\lambda} & 0 & \frac{\lambda^2 + 1}{2\lambda} \end{pmatrix} \in O(2, 1),$$

$$\text{inversion}\left(x \mapsto \frac{1}{x}\right) \leftrightarrow \begin{pmatrix} 1 & 0 & 0 \\ 0 & 1 & 0 \\ 0 & 0 & -1 \end{pmatrix} \in O(2, 1).$$

These matrices are conformal on S^1, and generate the whole of $O(2, 1)$.

It can be shown that, analogously to Theorem 15.3, the group of conformal transformations of $\mathbb{R}^n_{p,q}$ is isomorphic to $O(p+1, q+1)$. Thus in particular the group of conformal transformations of Minkowski space $\mathbb{R}^4_1$ is isomorphic to $O(4, 2)$.

We conclude by leaving the reader to conjure with the following remark (also offered without proof): The group $O(4, 2)$ is "locally isomorphic" to $SU(2, 2)$.

CHAPTER 3

Tensors: The Algebraic Theory

§16. Examples of Tensors

The fact that many physical entities find mathematical expression as numerical functions of points in space, will by now be familiar to the reader; the distance from a fixed point to a variable point is one among many examples. If we have several such entities, then their mathematical counterparts form a collection of functions from the points of space to the numbers (or, in other words, a (single) vector-valued function on the points). Thus for instance to fully determine the position of a point in 3-dimensional space, we need the values at that point of (at least) three functions (and of course we call these values "co-ordinates" of the point): each co-ordinate x^i is a function of the point, and together they form an ordered triple (x^1, x^2, x^3) which specifies the point completely. In Chapter 1 we encountered various kinds of co-ordinate systems (i.e. triples of functions); for example in the plane we introduced Cartesian co-ordinates x^1, x^2, and polar co-ordinates r, φ, where $x^1 = r \cos \varphi$, $x^2 = r \sin \varphi$; and in space Cartesian co-ordinates, cylindrical co-ordinates r, z, φ, and spherical co-ordinates r, θ, φ.

Thus a co-ordinate system is a family of numerical functions of points of space, determining precisely the locations of the points. In the same way when we speak of the co-ordinates of a physical system, we mean a family of numerical functions defined on the states of the system, whose values at any particular state fully determine that state. (A "state" of a system is to be thought of as a point in the "space of all possible states" of the system.) For example to specify the state (at an instant), of a moving point-particle we need six numbers, namely three co-ordinates, and the three components of its velocity vector; thus in this case we are dealing with a 6-dimensional space of states.

It turns out, however, that these concepts of a numerical or vector-valued function are by themselves inadequate for many purposes. The point is that many geometrical and physical quantities may only be assigned a specifying collection of numerical functions on points, subsequently to co-ordinates having been assigned to the points; to put it briefly (and familiarly), the specifying collection of functions may be dependent on the particular co-ordinatization of space. To make perfectly clear how this can happen, let us recapitulate an example already considered in §2 (see Definition 2.2.1). Thus suppose we have two co-ordinate systems x^1, x^2, x^3 and z^1, z^2, z^3 linked by

$$x^i = x^i(z^1, z^2, z^3), \qquad i = 1, 2, 3,$$

and let us examine how the components of the velocity vector of a curve

$$z^j = z^j(t), \qquad j = 1, 2, 3,$$

change with the change in co-ordinate systems. In terms of the co-ordinates z^1, z^2, z^3, the components of the velocity vector are given by

$$\left(\frac{dz^1}{dt}, \frac{dz^2}{dt}, \frac{dz^3}{dt}\right)_{t=t_0} = (\eta^1, \eta^2, \eta^3).$$

On the other hand in terms of the co-ordinates x^1, x^2, x^3, the equations for the curve are

$$x^i = x^i(z^1(t), z^2(t), z^3(t)) = x^i(t), \qquad i = 1, 2, 3,$$

so that in those co-ordinates the components of the same velocity vector are given by

$$\left(\frac{dx^1}{dt}, \frac{dx^2}{dt}, \frac{dx^3}{dt}\right)_{t=t_0} = (\xi^1, \xi^2, \xi^3).$$

By the "chain rule", the components in the two systems are linked by

$$\frac{dx^i}{dt} = \sum_{j=1}^{3} \frac{\partial x^i}{\partial z^j} \frac{dz^j}{dt}, \qquad i = 1, 2, 3.$$

Hence we deduce, as before (see §2.2(18)), that the components of a vector attached to a point $(z^1(t_0), z^2(t_0), z^3(t_0))$, transform under a co-ordinate change $x^i = x^i(z^1, z^2, z^3)$ according to the rule

$$\xi^i = \sum_{j=1}^{3} \eta^j \frac{\partial x^i}{\partial z^j}, \qquad i = 1, 2, 3, \tag{1}$$

where ξ^1, ξ^2, ξ^3, and η^1, η^2, η^3 are respectively the components of the vector in the systems x^1, x^2, x^3 and z^1, z^2, z^3.

Among the families of functions (of the points of space) which depend on the particular co-ordinate system in use, and which arise as quantitative representations of the various physical entities, the most frequently encountered, and the most important, are what are known as "tensors". A (point-dependent) vector is the most obvious non-trivial instance of a tensor (the trivial instance being that of a scalar function of the points of space, which does not change under co-ordinate transformations). Before giving the precise mathematical definition of a tensor, we give a few more examples, most of which we have already seen.

(a) *The gradient of a numerical function.* We are accustomed to regarding the gradient of a real-valued function $f(x^1, x^2, x^3)$ of the Cartesian co-ordinates x^1, x^2, x^3, as a vector whose components are given by

$$\operatorname{grad} f = \left(\frac{\partial f}{\partial x^1}, \frac{\partial f}{\partial x^2}, \frac{\partial f}{\partial x^3}\right) = (\xi^1, \xi^2, \xi^3).$$

We wish to see how the components of the gradient of this function (as a function on the points of the Cartesian space) transform under a change to co-ordinates z^1, z^2, z^3, where

$$x^i = x^i(z^1, z^2, z^3) = x^i(z), \qquad i = 1, 2, 3.$$

We have

$$\operatorname{grad} f(x^1(z), x^2(z), x^3(z)) = \left(\frac{\partial f}{\partial z^1}, \frac{\partial f}{\partial z^2}, \frac{\partial f}{\partial z^3}\right) = (\eta_1, \eta_2, \eta_3);$$

$$\frac{\partial f}{\partial z^i} = \sum_{j=1}^{3} \frac{\partial f}{\partial x^j}\frac{\partial x^j}{\partial z^i}, \qquad i = 1, 2, 3.$$

We conclude, therefore, that the components of the gradient change according to the rule

$$\eta_i = \sum_{j=1}^{3} \frac{\partial x^j}{\partial z^i}\xi_j, \tag{2}$$

where ξ_1, ξ_2, ξ_3, and η_1, η_2, η_3, are respectively the components of the gradient computed in terms of the co-ordinates x^1, x^2, x^3, and z^1, z^2, z^3.

Comparing this formula with the transformation rule (1) for the components of the tangent vector to a curve:

$$\begin{aligned} &\text{for the tangent vector:} \quad \xi^i = \sum_{j=1}^{3} \eta^j \frac{\partial x^i}{\partial z^j}, \\ &\text{for the gradient:} \quad \eta_i = \sum_{j=1}^{3} \xi_j \frac{\partial x^j}{\partial z^i}, \end{aligned} \tag{3}$$

we see that the transformation rules are different!

To highlight this difference, and for other reasons which will appear in §17, we shall now see what the conditions are for coincidence of the two rules in (3). To this end we introduce the Jacobian matrix $A = (a^i_j)$, where $a^i_j = \partial x^i/\partial z^j$, and its transpose $A^{\mathrm{T}} = (b^j_k)$ (where $b^j_k = a^k_j$). The transformation rules (3) can then be rewritten briefly as

$$\begin{aligned} \xi &= A\eta \quad \text{(tangent vector)}, \\ \eta &= A^{\mathrm{T}}\xi \quad \text{(gradient)}. \end{aligned} \tag{4}$$

If A is invertible, then so is A^{T}, and therefore in this case the transformation rule for the gradient can be rewritten again as

$$\xi = (A^{\mathrm{T}})^{-1}\eta \qquad \left(\text{or } \xi_i = \sum_{j=1}^{3} \eta_j \frac{\partial z^j}{\partial x^i}\right). \tag{5}$$

From the first equation in (4), together with (5), we see that the transformation rules (3) will be the same at every non-singular point precisely if

$$A = (A^{\mathrm{T}})^{-1}, \quad \text{or} \quad AA^{\mathrm{T}} = 1,$$

i.e. precisely if A is orthogonal at every non-singular point (see §4). (We mention parenthetically, and without proof, the fact that for a sufficiently smooth co-ordinate change $x = x(z)$, if the Jacobian matrix $A = (\partial x^i/\partial z^j)$ is orthogonal at each point of a region, then the co-ordinate change is (up to a translation) a linear transformation, i.e. $A = \text{const.}$)

In any case it is clear that the gradient of a function transforms differently under a general co-ordinate change, than does the tangent vector to a curve. For this reason, in the present context of "tensors" (the gradient being another special kind of tensor) we distinguish the gradient from quantities transforming like tangent vectors, by calling it rather a "covector."

(b) *Riemannian metrics.* We saw in §3 that the "metrical" concepts of length of a curve segment and angle between two curves in a Cartesian space (with co-ordinates $x^1, \ldots, x^n$), or in a region of such a space, are most appropriately defined in terms of the more primitive concept of a "Riemannian metric," i.e. a positive definite quadratic form $g_{ij}(x)$ defined at each point $(x) = (x^1, \ldots, x^n)$ (see Definition 3.1.1 *et seqq.*). Given such a quadratic form, the length of a vector $\xi = (\xi^1, \ldots, \xi^n)$ attached to the point $(x^1, \ldots, x^n)$ was then defined by

$$|\xi|^2 = \sum_{i,j} g_{ij}\xi^i\xi^j. \tag{6}$$

Applying this definition in particular to the tangent vector $(\dot{x}^1, \ldots, \dot{x}^n)$ to a curve $x^i = x^i(t)$, $i = 1, \ldots, n$, and taking arc length along this curve from $t = a$ to $t = b$ to be the integral with respect to the "time" t of the magnitude of the "velocity" or tangent vector, we arrived at the following formula for arc length:

$$l = \int_a^b \sqrt{\sum_{i,j} g_{ij}(x(t))\dot{x}^i\dot{x}^j}\, dt, \tag{7}$$

where $\dot{x}^i = dx^i/dt$.

The transformation rule for the coefficients $g_{ij}(x)$ of the metric under a co-ordinate change

$$x^i = x^i(z^1, \ldots, z^n), \qquad i = 1, \ldots, n,$$

is (see Definition 2.2.2)

$$\tilde{g}_{ij}(z) = \sum_{k,l=1}^{n} g_{kl}(x) \frac{\partial x^k}{\partial z^i} \frac{\partial x^l}{\partial z^j}. \tag{8}$$

Recall that this rule was forced upon us by the form of the transformation rule (1) (with n replacing 3) for the components of a vector, and the very natural requirement that arc length be independent of the co-ordinate system (which serves merely as a means to its calculation). Thus the formula for arc length in terms of the co-ordinates $z^1, \ldots, z^n$, namely

$$l = \int_a^b \sqrt{\sum_{i,j} \tilde{g}_{ij}(z(t)) \dot{z}^i \dot{z}^j}\, dt,$$

where the $z^i(t)$ are given by $x^i(t) = x^i(z^1(t), \ldots, z^n(t))$, is required to yield the same result as the formula (7). In view of the definition of arc length as the integral with respect to time of the length of the velocity vector, this latter requirement comes down to the condition that the square of the length of a vector, given by (6), be invariant under changes of co-ordinate systems.

The point of the rule (8) for us in our present context, is that it makes the quadratic form $g_{ij}(x)$ (defined on vectors) a tensor (more specifically, a particular kind of "tensor of the second rank").

(c) The length of a covector is, as for vectors, most appropriately defined in terms of a quadratic form $g^{ij}(x)$, given at each point (x) of the space. Thus if $\xi = (\xi_1, \ldots, \xi_n)$ is a covector (i.e. if it transforms according to the formula (2) with n in place of 3) at the point $(x^1, \ldots, x^n)$, then we define

$$|\xi|^2 = \sum_{i,j} g^{ij} \xi_i \xi_j.$$

It is not difficult to see that the transformation rule (2) and the natural requirement that the length of a covector be invariant under co-ordinate changes, together imply the following transformation rule for the coefficients $g^{ij}(x)$:

$$\tilde{g}^{ij}(z) = \sum_{k,l=1}^{n} g^{kl} \frac{\partial z^i}{\partial x^k} \frac{\partial z^j}{\partial x^l}, \tag{9}$$

where the co-ordinate change is given by $x^i = x^i(z)$, $i = 1, \ldots, n$. Thus with this transformation rule we shall have

$$|\xi|^2 = \sum_{i,j} \tilde{g}^{ij} \eta_i \eta_j = \sum_{i,j} g^{ij} \xi_i \xi_j,$$

where $\eta_1, \ldots, \eta_n$ are the components of the same covector (and at the same point), in terms of the new co-ordinates $z^1, \ldots, z^n$.

The transformation rule (9) defines $g^{ij}(x)$ as another kind of second-rank tensor.

(d) *Linear operators on vectors* (*or on covectors*) furnish examples of a third (and last) kind of second-rank tensor. Suppose that at each point $(x^1, \ldots, x^n)$ of a space we are given a matrix $(a^i_j(x)) = A(x)$. This matrix then of course determines a linear transformation of the vector space of dimension n (where the vectors are regarded as emanating from $(x) = (x^1, \ldots, x^n)$). Thus if $\xi = (\xi^1, \ldots, \xi^n)$ is a vector at (x) then the linear transformation $A(x)$ maps it to the vector $(\hat{\xi}^1, \ldots, \hat{\xi}^n)$ (also at (x)), where

$$\hat{\xi}^i = \sum_{j=1}^{n} a^i_j(x)\xi^j. \tag{10}$$

It is not difficult to verify that (10) and the rule (1) for transformation of the components of a vector, together imply that the entries of the matrix $A(x)$ transform according to the formula

$$\tilde{a}^i_j = \sum_{k,l} \frac{\partial z^i}{\partial x^k} a^k_l \frac{\partial x^l}{\partial z^j}, \tag{11}$$

under the co-ordinate change $x^i = x^i(z^1, \ldots, z^n)$, $i = 1, \ldots, n$ (which we assume non-singular at the point under consideration).

Similarly, the matrix $A(x)$ determines a linear transformation of the space of covectors at the point (x), defined by

$$\hat{\xi}_j = \sum_{i=1}^{n} a^i_j \xi_i. \tag{12}$$

If the components of the two covectors $(\xi_1, \ldots, \xi_n)$ and $(\hat{\xi}_1, \ldots, \hat{\xi}_n)$ relative to the new co-ordinates $z^1, \ldots, z^n$, are given by $(\eta_1, \ldots, \eta_n)$ and $(\hat{\eta}_1, \ldots, \hat{\eta}_n)$ respectively, then we have from (2) (with n in place of 3), and (12) that

$$\hat{\eta}_j = \sum_k \hat{\xi}_k \frac{\partial x^k}{\partial z^j} = \sum_{k,l} a^l_k \xi_l \frac{\partial x^k}{\partial z^j} = \sum_{i,k,l} a^l_k \eta_i \frac{\partial z^i}{\partial x^l} \frac{\partial x^k}{\partial z^j}.$$

Comparing this with

$$\hat{\eta}_j = \sum_i \tilde{a}^i_j \eta_i,$$

we deduce (since the above equations hold for all covectors) that

$$\tilde{a}^i_j = \sum \frac{\partial z^i}{\partial x^k} a^k_l \frac{\partial x^l}{\partial z^j},$$

which is the same as the rule (11). (Note that once again we are assuming that the co-ordinate change is non-singular at the point under scrutiny.)

Finally we tabulate the above examples of tensors and their rules of transformation.

(1) Scalars (zero-rank tensors) are invariant under co-ordinate changes.

Tensors of rank 1:

(2) Vectors $\xi = (\xi^i)$ (e.g. the tangent vector to a curve) transform according to the formula

$$\tilde{\xi}^j = \sum_{i=1}^{n} \xi^i \frac{\partial z^j}{\partial x^i}.$$

(3) Covectors $\xi = (\xi_i)$ (e.g. the gradient of a function) transform according to the formula

$$\tilde{\xi}_j = \sum_{i=1}^{n} \xi_i \frac{\partial x^i}{\partial z^j}.$$

Second-rank tensors:

(4) The coefficients g_{ij} of a scalar product of vectors transform according to the formula

$$\tilde{g}_{ij} = \sum_{k,l} g_{kl} \frac{\partial x^k}{\partial z^i} \frac{\partial x^l}{\partial z^j}.$$

(5) The coefficients g^{ij} of a scalar product of covectors transform according to the formula

$$\tilde{g}^{ij} = \sum_{k,l} g^{kl} \frac{\partial z^i}{\partial x^k} \frac{\partial z^j}{\partial x^l}.$$

(6) Linear operators $A = (a^i_j)$ on vectors (or covectors) transform according to the formula

$$\tilde{a}^i_j = \sum_{k,l} a^k_l \frac{\partial x^l}{\partial z^j} \frac{\partial z^i}{\partial x^k}.$$

In the above

$$x^i = x^i(z^1, \ldots, z^n), \qquad z^j = z^j(x^1, \ldots, x^n),$$

$$z^j(x^1(z^1, \ldots, z^n), \ldots, x^n(z^1, \ldots, z^n)) = z^j;$$

$$\sum_{k=1}^{n} \frac{\partial z^i}{\partial x^k} \frac{\partial x^k}{\partial z^j} = \delta^i_j.$$

§17. The General Definition of a Tensor

17.1. The Transformation Rule for the Components of a Tensor of Arbitrary Rank

In the preceding section we considered examples of rank-one tensors (vectors and covectors), and of rank-two tensors (quadratic forms g_{ij} on vectors, quadratic forms g^{ij} on covectors, and linear transformations (a^i_j)). These important examples (among others) naturally prompt the following general definition.

17.1.1. Definition. A *tensor* (or *tensor field*) *of type* (p, q) and rank $p + q$, is, relative to a system of co-ordinates $x^1, \ldots, x^n$, a family of numbers $T^{i_1 \ldots i_p}_{j_1 \ldots j_q}$, one such family being given for each point of the space (or region of it) in question; furthermore these numbers or components transform under a co-ordinate change $x^i = x^i(z^1, \ldots, z^n)$ (with inverse $z^j = z^j(x^1, \ldots, x^n)$) according to the formula

$$T^{i_1 \ldots i_p}_{j_1 \ldots j_q} = \sum_{(k), (l)} \tilde{T}^{k_1 \ldots k_p}_{l_1 \ldots l_q} \frac{\partial x^{i_1}}{\partial z^{k_1}} \cdots \frac{\partial x^{i_p}}{\partial z^{k_p}} \frac{\partial z^{l_1}}{\partial x^{j_1}} \cdots \frac{\partial z^{l_q}}{\partial x^{j_q}}, \tag{1}$$

where the $\tilde{T}^{k_1 \ldots k_p}_{l_1 \ldots l_p}$ are the components of the tensor relative to the co-ordinates $z^1, \ldots, z^n$, and where the indices $i_1, \ldots, i_p; j_1, \ldots, j_q$, and $k_1, \ldots, k_p; l_1 \ldots, l_q$, all range from 1 to n (the dimension of the underlying space on which the tensor is defined). Thus the tensor may be regarded as the totality of its representations as point-dependent families of numbers relative to all co-ordinate systems.

Returning to the examples of the previous section, we have, in the light of this definition, that:

a velocity vector is a tensor of type $(1, 0)$;
a covector is a tensor of type $(0, 1)$;
a quadratic form on vectors is a tensor of type $(0, 2)$;
a quadratic form on covectors is a tensor of type $(2, 0)$;
a linear operator on vectors or covectors is a tensor of type $(1, 1)$.

The following theorem tells us that the concept of tensor is well-defined (by 17.1.1).

17.1.2. Theorem. *The components* $\tilde{T}^{k_1 \ldots k_p}_{l_1 \ldots l_q}$ *are given in terms of the* $T^{i_1 \ldots i_p}_{j_1 \ldots j_q}$ *by the formula*

$$\tilde{T}^{k_1 \ldots k_p}_{l_1 \ldots l_q} = \sum_{(i)(j)} T^{i_1 \ldots i_p}_{j_1 \ldots j_p} \frac{\partial z^{k_1}}{\partial x^{i_1}} \cdots \frac{\partial z^{k_p}}{\partial x^{i_p}} \frac{\partial x^{j_1}}{\partial z^{l_1}} \cdots \frac{\partial x^{j_q}}{\partial z^{l_q}}. \tag{2}$$

PROOF. We shall use the equations

$$\sum_j \frac{\partial x^i}{\partial z^j} \frac{\partial z^j}{\partial x^k} = \delta^i_k; \qquad \sum_j \frac{\partial z^k}{\partial x^j} \frac{\partial x^j}{\partial z^q} = \delta^k_q,$$

which are consequences of the fact that the transformations $x = x(z)$, $z = z(x)$ are mutual inverses:

$$x^i(z(x)) = x^i; \qquad z^q(x(z)) = z^q.$$

If we regard (1) as a system of linear equations with constant terms the $T^{i_1 \dots i_p}_{j_1 \dots j_q}$, and unknowns $\tilde{T}^{k_1 \dots k_p}_{l_1 \dots l_q}$, then our problem is to show that this system has the unique solution (2). Now from (1) we have

$$\begin{aligned}
&\sum_{(i),(j)} T^{(i)}_{(j)} \frac{\partial z^{k_1}}{\partial x^{i_1}} \cdots \frac{\partial z^{k_p}}{\partial x^{i_p}} \frac{\partial x^{j_1}}{\partial z^{l_1}} \cdots \frac{\partial x^{j_q}}{\partial z^{l_q}} \\
&\quad = \sum_{(i),(j)} \left(\sum_{(r),(s)} \tilde{T}^{(r)}_{(s)} \frac{\partial x^{i_1}}{\partial z^{r_1}} \cdots \frac{\partial x^{i_p}}{\partial z^{r_p}} \frac{\partial z^{s_1}}{\partial x^{j_1}} \cdots \frac{\partial z^{s_q}}{\partial x^{j_q}} \right) \frac{\partial z^{k_1}}{\partial x^{i_1}} \cdots \frac{\partial z^{k_p}}{\partial x^{i_p}} \frac{\partial x^{j_1}}{\partial z^{l_1}} \cdots \frac{\partial x^{j_q}}{\partial z^{l_q}} \\
&\quad = \sum_{(i),(j),(r),(s)} \tilde{T}^{(r)}_{(s)} \left(\frac{\partial x^{i_1}}{\partial z^{r_1}} \frac{\partial z^{k_1}}{\partial x^{i_1}} \right) \cdots \left(\frac{\partial x^{i_p}}{\partial z^{r_p}} \frac{\partial z^{k_p}}{\partial x^{i_p}} \right) \left(\frac{\partial x^{j_1}}{\partial z^{l_1}} \frac{\partial z^{s_1}}{\partial x^{j_1}} \right) \cdots \left(\frac{\partial x^{j_q}}{\partial z^{l_q}} \frac{\partial z^{s_q}}{\partial x^{j_q}} \right) \\
&\quad = \sum_{(r),(s)} \tilde{T}^{(r)}_{(s)} \delta^{k_1}_{r_1} \cdots \delta^{k_p}_{r_p} \delta^{s_1}_{l_1} \cdots \delta^{s_q}_{l_q} = \tilde{T}^{(k)}_{(l)},
\end{aligned}$$

as required. □

We now indicate some of the simplest properties of tensors. At each point of the underlying space the set of all tensors of the same type forms a linear space: Thus if $T = (T^{i_1 \dots i_p}_{j_1 \dots j_q})$ and $S = (S^{i_1 \dots i_p}_{j_1 \dots j_q})$ are both tensors of type (p, q), then any linear combination $\lambda T + \mu S = U$, is also a tensor of type (p, q), with components $U^{i_1 \dots i_p}_{j_1 \dots j_p} = \lambda T^{i_1 \dots i_p}_{j_1 \dots j_q} + \mu S^{i_1 \dots i_p}_{j_1 \dots j_p}$ (at each point of the underlying space). (We should emphasize that this linear combination is formed from the corresponding components of the respective tensors at the same point of the space; in other words if we regard tensors as (families of) functions defined on the points of the underlying space, then we form linear combinations of them as one normally does for functions.)

The dimension of this linear space of (values of) tensors of type (p, q) at each point, is clearly n^{p+q}. If we denote by $e_1, \dots, e_n$ the usual (canonical) basis vectors in the space co-ordinatized by $x^1, \dots, x^n$, and, regarded as covectors, by $e^1, \dots, e^n$, then the tensors we encountered in the preceding section can be rewritten conveniently as follows:

Vectors: $\xi = \sum_i \xi^i e_i \left(\text{for example } \frac{dx}{dt} = \sum_i \frac{dx^i}{dt} e_i \right);$

Covectors: $\xi = \sum_i \xi_i e^i \left(\text{for example grad } f = \sum_i \frac{\partial f}{\partial x^i} e^i \right);$

Quadratic forms on vectors: $(g_{ij}) = \sum_{i,j} g_{ij} e^i \otimes e^j;$

Quadratic forms on covectors: $(g^{ij}) = \sum_{i,j} g^{ij} e_i \otimes e_j;$

Linear operators: $A = \sum_{i,j} a^i_j e_i \otimes e^j;$

and the general tensor $T = (T^{i_1 \dots i_p}_{j_1 \dots j_q})$ can be rewritten as

$$T = \sum_{i,j} T^{i_1 \dots i_p}_{j_1 \dots j_q} e_{i_1} \otimes \cdots \otimes e_{i_p} \otimes e^{j_1} \otimes \cdots \otimes e^{j_q}. \tag{3}$$

Here the expressions

$$e_{i_1} \otimes \cdots \otimes e_{i_p} \otimes e^{j_1} \otimes \cdots \otimes e^{j_q}, \tag{4}$$

where the i's and j's take (independently) all values in $\{1, \ldots, n\}$, represent basis elements for the linear space of tensors of type (p, q) at the given point (x) of the underlying Cartesian space. It is important to note that in (3) and (4) the order of the indices is essential: interchanging e_{i_1} and e_{i_2} for instance, will yield a different basis element whenever $i_1 \neq i_2$. Thus the basis elements number altogether n^{p+q}, as indeed they should. Under a co-ordinate change $x^i = x^i(z^1, \ldots, z^n)$ we go over to a different basis for the linear space of tensors attached to a given point, namely to that basis expressed as in (4), only in terms of the canonical basis vectors in the system $z^1, \ldots, z^n$. The formulae (1) and (2) allow one to express each of these bases in terms of the other, and thence to re-express in terms of the co-ordinates $z^1, \ldots, z^n$ any tensor written (as in (3)) in terms of the co-ordinates $x^1, \ldots, x^n$. (Note that the basis elements (4) could equally well have been written as ordered $(p + q)$-tuples $(e_{i_1}, \ldots, e^{j_q})$ or even as n^{p+q}-tuples with one entry equal to 1, and all other entries 0. The notation (4) is, however, conventional (in addition to being useful).)

We now consider some examples from the theory of elasticity.

(a) *The stress tensor* (3-dimensional case). At each point of a continuous medium the force on an element of area ΔS, in the direction of the unit normal n to the area element, is given by $\Delta S P(n)$, where $P = (P^i_j)$ is a linear operator. (Clearly P^i_j is the jth component of the force per unit area acting on a small area perpendicular to e_i.) The tensor P^i_j is called the *stress tensor* for the medium. If $n = n^1 e_1 + n^2 e_2 + n^3 e_3$, then

$$P(n) = \sum_{i=1}^{3} \left(\sum_{j=1}^{3} n^j P^i_j \right) e_i,$$

i.e. the ith component of $P(n)$ is $\sum_{j=1}^{3} n^j P^i_j$. In particular if the medium satisfies Pascal's law, i.e. if the force per unit area on an area element in a direction orthogonal to the area element, has the same magnitude p for all directions (at a point), then $P^i_j = \delta^i_j p$; the quantity p is, in this case, called the *pressure* at the point.

(b) *The strain tensor*. Suppose we have a continuous medium on which co-ordinates $x^1, \ldots, x^n$ are defined. We say that the medium has undergone a *deformation* (or *strain*) if each point (x^1, x^2, x^3) is displaced to the nearby point $(x^1 + u^1, x^2 + u^2, x^3 + u^3)$, i.e.

$$x^i \mapsto x^i + u^i.$$

If the co-ordinates x^1, x^2, x^3 are Euclidean, then before the deformation, the distance Δl between two points $(x) = (x^1, x^2, x^3)$ and $(x + \Delta x) = (x^1 + \Delta x^1, x^2 + \Delta x^2, x^3 + \Delta x^3)$ is given by

$$(\Delta l)^2 = \sum_{i=1}^{3} (\Delta x^i)^2,$$

while after the deformation their separation $\Delta l'$ is given by

$$(\Delta l')^2 = \sum_{i=1}^{3} [u^i(x) - \Delta x^i - u^i(x + \Delta x)]^2 = \sum_{i=1}^{3} [\Delta u^i + \Delta x^i]^2.$$

Hence

$$(\Delta l')^2 = (\Delta l)^2 + 2 \sum_{i=1}^{3} \Delta x^i \, \Delta u^i + \sum_{i=1}^{3} (\Delta u^i)^2.$$

Using

$$du^i = \sum_{j=1}^{3} \frac{\partial u^i}{\partial x^j} \Delta x^j,$$

we may formally express the situation in the limit as $\Delta l \to 0$ in terms of differentials as follows:

$$(dl')^2 = (dl)^2 + 2 \sum_{i,j} \frac{\partial u^i}{\partial x^j} dx^i \, dx^j + \sum_{i,j,k} \frac{\partial u^k}{\partial x^i} \frac{\partial u^k}{\partial x^j} dx^i \, dx^j. \tag{5}$$

Since

$$\sum_{i,j} \frac{\partial u^i}{\partial x^j} dx^i \, dx^j = \sum_{i,j} \frac{\partial u^j}{\partial x^i} dx^i \, dx^j,$$

the middle term on the right-hand side of (5) is equal to $\sum \eta_{ij} \, dx^i \, dx^j$, where

$$\eta_{ij} = \frac{\partial u^i}{\partial x^j} + \frac{\partial u^j}{\partial x^i}.$$

Thus rewriting (5), we obtain

$$(dl')^2 - (dl)^2 = \sum_{i,j} \left(\eta_{ij} + \sum_k \frac{\partial u^k}{\partial x^i} \frac{\partial u^k}{\partial x^j} \right) dx^i \, dx^j. \tag{6}$$

17.1.3. Definition. The coefficients $(\eta_{ij} + \sum_k (\partial u^k/\partial x^i)(\partial u^k/\partial x^j))$ in (6), where $\eta_{ij} = \partial u^i/\partial x^j + \partial u^j/\partial x^i$, are the components of the *strain tensor of the medium* (at the point (x^1, x^2, x^3)).

If the change in the u^i per unit change in the x^j is small, i.e. if the displacements of nearby points are similar (this will occur for instance if the medium is "elastic"), then the quadratic terms in the strain tensor are often neglected, yielding the *strain tensor for small deformations*

$$\eta_{ij} = \frac{\partial u^i}{\partial x^j} + \frac{\partial u^j}{\partial x^i}.$$

According to Hooke's law such "small deformations" cause stresses to be set up in the medium, depending linearly on the deformations. More precisely, the stress tensor P and the strain tensor η are linked by the linear system

$$P = U(\eta),$$

or, in index notation,

$$P^i_j = \sum_{k,l} U^{ikl}_j \eta_{kl}, \tag{7}$$

where $P = (P^i_j)$, $\eta = (\eta_{kl})$, $U = (U^{ikl}_j)$. Thus U is a tensor of rank 4, and therefore has 81 components. Surely Hooke's law (for a 3-dimensional continuous medium) does not always require for its specification 81 independent parameters, i.e. 81 numbers at each point of the medium! We shall now indicate how one can in the case of an "isotropic" medium reduce this number to 2.

Recall first the simplifying fact, noted in the preceding section, that under an orthogonal co-ordinate change, vectors and covectors transform in the same way. It is not difficult to show that, more generally, if we restrict ourselves to orthogonal co-ordinate changes, then the distinction between the upper and lower indices of a tensor disappears; i.e. under an orthogonal co-ordinate change the transformation rule for a tensor of type (p, q) is the same as that of a tensor of type $(p + q, 0)$.

The condition that our continuous medium be *isotropic* means that the components U^{ikl}_j of the tensor U at each point are invariant under rotations about that point and reflections in planes through that point, i.e. under orthogonal transformations fixing the point. This condition, which is fulfilled in fluids but by no means always in solid media, allows us to apply the following important result, which we state without proof. (Note that by the above simplifying remark, since the result in question is concerned with orthogonal transformations only, we may ignore the distinction between upper and lower indices.)

17.1.4. Theorem. *A tensor of rank four which is invariant under orthogonal co-ordinate transformations necessarily has the form*

$$U_{ijkl} = \lambda\delta_{ik}\delta_{jl} + \mu\delta_{ij}\delta_{kl} + \nu\delta_{il}\delta_{jk};$$

thus such a tensor is determined by three parameters λ, μ, ν.

Hence assuming isotropy of the continuous medium, (7) simplifies to

$$P_{ij} = \lambda\eta_{ij} + \mu(\operatorname{tr}\eta)\delta_{ij} + \nu\eta_{ji},$$

where $\operatorname{tr}\eta = \sum_i \eta_{ii}$. Since the tensor η is by its very definition symmetric, i.e. $\eta_{ij} = \eta_{ji}$, we see that in fact P depends on only two parameters, namely

$\lambda + \nu$ and μ. We have thus arrived at the conclusion we sought: *In an isotropic medium any linear law linking two (physical) symmetric, second-rank tensors which is given by a fourth-rank tensor, can be specified by two parameters, i.e. by two numbers at each point of the medium.*

In view of Theorem 17.1.4, it is natural to ask what form "isotropic" tensors of ranks 1, 2 and 3 can take.

It is clear that only the null vector or covector is invariant under all rotations.

A tensor (h_{ij}) say, of rank 2, defined on a 2-dimensional space with co-ordinates x, y, goes under the reflection $x \mapsto x$, $y \mapsto -y$ to (h'_{ij}) say, where (by 17.1.1) $h'_{11} = h_{11}$, $h'_{22} = h_{22}$, $h'_{12} = -h_{12}$, $h'_{21} = -h_{21}$, and under the transformation $x \mapsto y$, $y \mapsto x$, to (h''_{ij}) where $h''_{11} = h_{22}$, $h''_{12} = h_{21}$, $h''_{21} = h_{12}$. Invariance of the tensor under these transformations means that $h_{ij} = h'_{ij} = h''_{ij}$, whence $h_{11} = h_{22}$, $h_{12} = h_{21} = 0$, i.e. $h_{ij} = \lambda\delta_{ij}$. This, and the analogous argument for higher dimensions, shows that the only isotropic second-rank tensors are the scalar matrices $(\lambda\delta_{ij})$.

It can be shown that the only isotropic tensor of rank 3 is the zero tensor.

Note that the general concept of isotropy of a medium (or space) involves an isometry group, and hence presupposes the presence of a (Riemannian) metric on the medium. (Above we assumed that the metric was Euclidean.) (Of course the concept of tensor does not in itself rely on the presence of any metric on the underlying space; indeed a metric on a space is itself just one kind of tensor.) For this reason it is worthwhile examining the question of which tensors are invariant not just under orthogonal co-ordinate changes, but under the full linear group.

One can quickly reduce the field of candidates for such invariance as follows. If we subject our space to a dilation with factor $\lambda \neq 1$, then the components of a tensor of type (p, q) will each be multiplied by the factor λ^{q-p}, and so a non-zero tensor can be invariant under such a linear co-ordinate change only if $p = q$. Thus in particular an invariant tensor must have even rank. It follows that the only invariant rank-two tensors are those of the form $h^i_j = \lambda\delta^i_j$, since indeed the only rank-two tensors invariant under orthogonal transformations are of this form (as we showed above). It turns out that the tensors of rank 4 invariant under the full linear group form a two-parameter family: $U^{ij}_{kl} = \lambda\delta^i_k\delta^j_l + \mu\delta^i_l\delta^j_k$.

17.2. Algebraic Operations on Tensors

We begin by introducing some useful (and conventional) notation. Let $T^{i_1 \cdots i_p}_{j_1 \cdots j_q}$ be the components of a tensor of type (p, q) in terms of the co-ordinates $x^1, \ldots, x^n$ for the underlying space, and suppose we have a co-ordinate change

$$x^{k'} = x^{k'}(x^1, \ldots, x^n), \qquad k' = 1', \ldots, n',$$

to new co-ordinates $x^{1'}, \ldots, x^{n'}$. (The primes apply to the indices.) We shall denote the components of our tensor in terms of the primed system by $T^{i'_1 \ldots i'_p}_{j'_1 \ldots j'_q}$. In this notation the transformation rules (2) and (1) take the form

$$T^{i_1 \ldots i_p}_{j_1 \ldots j_q} = \sum_{(i')(j')} T^{i'_1 \ldots i'_p}_{j'_1 \ldots j'_q} \frac{\partial x^{i_1}}{\partial x^{i'_1}} \cdots \frac{\partial x^{i_p}}{\partial x^{i'_p}} \frac{\partial x^{j'_1}}{\partial x^{j_1}} \cdots \frac{\partial x^{j'_q}}{\partial x^{j_q}}, \tag{8}$$

$$T^{i'_1 \ldots i'_p}_{j'_1 \ldots j'_p} = \sum_{(i),(j)} T^{i_1 \ldots i_p}_{j_1 \ldots j_q} \frac{\partial x^{i'_1}}{\partial x^{i_1}} \cdots \frac{\partial x^{i'_p}}{\partial x^{i_p}} \frac{\partial x^{j_1}}{\partial x^{j'_1}} \cdots \frac{\partial x^{j_q}}{\partial x^{j'_q}}. \tag{9}$$

We introduce now also the following convenient notational rule, already employed in §1.2: If in an expression an index occurs twice, once as a lower index and once as an upper index, then summation over that index from 1 to n (where n is the dimension of the underlying space) is assumed implicitly, i.e. the symbol $\sum$ may be omitted. Thus in the formula (8) summation is taken over the indices $i'_1, \ldots, i'_p; j'_1, \ldots, j'_q$, each of which appears twice, once in a lower position, and once in an upper position, while in (9) the indices $i_1, \ldots, i_p; j_1, \ldots, j_q$ are summed over (from 1 to n). This rule, together with the "primed-indices" notation, makes it easier to avoid errors in working with the formulae of tensor analysis.

We now define three very important algebraic operations on tensors. To begin with, we assume that the tensors are expressed in terms of a fixed system of co-ordinates $x^1, \ldots, x^n$.

(i) *Permutation of indices.* Let σ be some permutation of the integers $1, 2, \ldots, q$; in conventional notation we write

$$\sigma = \begin{pmatrix} 1 & \ldots & q \\ \sigma(1) & \ldots & \sigma(q) \end{pmatrix}.$$

We define an action of σ on the ordered n-tuple $(j_1, \ldots, j_q)$ by

$$\sigma(j_1, \ldots, j_q) = (j_{\sigma(1)}, \ldots, j_{\sigma(q)}). \tag{10}$$

We shall say that a tensor $\tilde{T}^{i_1 \ldots i_p}_{j_1 \ldots j_q}$ is obtained from a tensor $T^{i_1 \ldots i_p}_{j_1 \ldots j_q}$ by means of a *permutation σ of the lower indices* if at each point of the underlying space,

$$\tilde{T}^{i_1 \ldots i_p}_{j_1 \ldots j_q} = T^{i_1 \ldots i_p}_{\sigma(j_1, \ldots, j_q)}. \tag{11}$$

Permutations of the upper indices are defined similarly. In general the interchange of upper with lower indices is not permissible since such an operation is not invariant (i.e. does not commute with) co-ordinate changes (see Lemma 17.2.1 and Exercise 1 below). In the case of a second rank tensor (T_{ij}), interchanging i and j is the same as transposing the matrix.

(ii) *Contraction* (or taking "traces"). By the *contraction* of a tensor $T^{i_1 \ldots i_p}_{j_1 \ldots j_q}$ of type (p, q) with respect to the indices i_k, j_l we mean the tensor

$$\tilde{T}^{i_1 \ldots i_{p-1}}_{j_1 \ldots j_{q-1}} = T^{i_1 \ldots i_{k-1} i i_k \ldots i_{p-1}}_{j_1 \ldots j_{l-1} i j_l \ldots j_{q-1}} \tag{12}$$

of type $(p-1, q-1)$. (Note that in (12) summation over the repeated index i from 1 to n is implicit.) Thus contraction essentially amounts to setting an upper and lower index equal and summing over that index. For example the contraction of a tensor T^i_j of type (1, 1) yields a scalar T^i_i, the trace tr T of the matrix T^i_j.

(iii) *Product of tensors.* Given two tensors $T = (T^{i_1 \dots i_p}_{j_1 \dots j_q})$ of type (p, q), and $P = (P^{i_1 \dots i_k}_{j_1 \dots j_l})$ of type (k, l), we define their *product* to be the tensor $S = T \otimes P$ of type $(p+k, q+l)$ with components

$$S^{i_1 \dots i_{p+k}}_{j_1 \dots j_{q+l}} = T^{i_1 \dots i_p}_{j_1 \dots j_q} P^{i_{p+1} \dots i_{p+k}}_{j_{q+1} \dots i_{q+k}}. \tag{13}$$

It is important to note that this multiplication is not in general commutative, since if the components (at any point) are multiplied in the other order, the products will be associated with different values of the indices $i_1, \dots, i_{p+k}$; $j_1, \dots, j_{q+l}$; the multiplication is, however, associative.

17.2.1. Lemma. *The results of applying the operations* (i), (ii), (iii) *above to tensors are again tensors, which moreover are independent of the co-ordinate system in terms of which the operations are performed.*

PROOF. (i) It suffices in this case to prove the lemma for a permutation σ which interchanges k and l, and leaves the remaining integers fixed:

$$\sigma = \begin{pmatrix} 1 \dots k \dots l \dots q \\ 1 \dots l \dots k \dots q \end{pmatrix}.$$

Then

$$\tilde{T}^{i_1 \dots i_p}_{j_1 \dots j_k \dots j_l \dots j_q} = T^{i_1 \dots i_p}_{j_1 \dots j_l \dots j_k \dots j_q}. \tag{14}$$

Changing to another (primed) system of co-ordinates and applying the transformation rule (8), we have

$$\begin{aligned} \tilde{T}^{i_1 \dots i_p}_{j_1 \dots j_q} &= T^{i_1 \dots i_p}_{j_1 \dots j_l \dots j_k \dots j_q} \\ &= T^{i'_1 \dots i'_p}_{j'_1 \dots j'_l \dots j'_k \dots j'_q} \frac{\partial x^{i_1}}{\partial x^{i'_1}} \cdots \frac{\partial x^{i_p}}{\partial x^{i'_p}} \frac{\partial x^{j'_1}}{\partial x^{j_1}} \cdots \frac{\partial x^{j'_l}}{\partial x^{j_l}} \cdots \frac{\partial x^{j'_k}}{\partial x^{j_k}} \cdots \frac{\partial x^{j'_q}}{\partial x^{j_q}} \\ &= \tilde{T}^{i'_1 \dots i'_p}_{j'_1 \dots j'_q} \frac{\partial x^{i_1}}{\partial x^{i'_1}} \cdots \frac{\partial x^{i_p}}{\partial x^{i'_p}} \frac{\partial x^{j'_1}}{\partial x^{j_1}} \cdots \frac{\partial x^{j'_q}}{\partial x^{j_q}}, \end{aligned}$$

where to obtain the last expression from the preceding one we have simply rearranged the factors $\partial x^{j'}/\partial x^j$. Hence $(\tilde{T}^{i_1 \dots i_p}_{j_1 \dots j_q})$ is a tensor of rank (p, q) (by 17.1.1).

(ii) The contraction of the tensor $T^{i_1 \dots i_p}_{j_1 \dots j_q}$ with respect to the indices i_k and j_l transforms under the usual co-ordinate change as follows:

$$
\begin{aligned}
T^{i_1 \dots i_p}_{j_1 \dots j_q}\Big|_{i_k = j_l = i} &= T^{i'_1 \dots i'_p}_{j'_1 \dots j'_q} \frac{\partial x^{i_1}}{\partial x^{i'_1}} \cdots \frac{\partial x^{i_p}}{\partial x^{i'_p}} \frac{\partial x^{j'_1}}{\partial x^{j_1}} \cdots \frac{\partial x^{j'_q}}{\partial x^{j_q}}\Bigg|_{i_k = j_l = i} \\
&= T^{i'_1 \dots i'_p}_{j'_1 \dots j'_q} \cdot \delta^{j'_l}_{i'_k} \left(\frac{\partial x^{i_1}}{\partial x^{i'_1}} \cdots \frac{\partial x^{i_{k-1}}}{\partial x^{i'_{k-1}}} \frac{\partial x^{i_{k+1}}}{\partial x^{i'_{k+1}}} \cdots \frac{\partial x^{i_p}}{\partial x^{i'_p}} \right) \\
&\quad \times \left(\frac{\partial x^{j'_1}}{\partial x^{j_1}} \cdots \frac{\partial x^{j'_{l-1}}}{\partial x^{j_{l-1}}} \frac{\partial x^{j'_{l+1}}}{\partial x^{j_{l+1}}} \cdots \frac{\partial x^{j'_q}}{\partial x^{j_q}} \right) \qquad (15) \\
&= T^{i'_1 \dots i'_p}_{j'_1 \dots j'_q} \left(\frac{\partial x^{i_1}}{\partial x^{i'_1}} \cdots \frac{\partial x^{i_{k-1}}}{\partial x^{i'_{k-1}}} \frac{\partial x^{i_{k+1}}}{\partial x^{i'_{k+1}}} \cdots \frac{\partial x^{i_p}}{\partial x^{i'_p}} \right) \\
&\quad \times \left(\frac{\partial x^{j'_1}}{\partial x^{j_1}} \cdots \frac{\partial x^{j'_{l-1}}}{\partial x^{j_{l-1}}} \frac{\partial x^{j'_{l+1}}}{\partial x^{j_{l+1}}} \cdots \frac{\partial x^{j'_q}}{\partial x^{j_q}} \right)\Bigg|_{i'_k = j'_l = i'},
\end{aligned}
$$

which gives the desired conclusion. (To arrive at (15) we used the equalities $(\partial x^i/\partial x^{i'_k})(\partial x^{j'_l}/\partial x^i) = \delta^{j'_l}_{i'_k}$.)

Since (iii) is immediate, the proof of the lemma is complete. □

We now give examples of the above tensor operations.

17.2.2. Examples. (a) Given a vector ξ^i and a covector η_j, we can form their tensor product $T^i_j = \xi^i \eta_j$, which is a tensor of type (1, 1). We can then perform the contraction, obtaining T^i_i, the trace of T^i_j. The trace T^i_i is a scalar, the "scalar product" of the vector and covector.

(b) Given a vector ξ^i and a linear operator A^k_l, we can form their tensor product $T^{ik}_l = A^k_l \xi^i$, a tensor of type (2, 1). The contraction

$$\eta^k = T^{ik}_i = A^k_i \xi^i$$

is again a vector, which we of course recognize as the result of applying the linear transformation A^k_l to the vector ξ^i.

Remark. Using the scalar product of a vector with a covector, defined in Example (a) above, we can associate with each vector $\xi = (\xi^i)$, a linear differential operator (on functions defined on the points of the underlying space) as follows: Since the gradient $(\partial f/\partial x^i)$ of a function f is a covector, the quantity

$$\partial_\xi f = \xi^i \frac{\partial f}{\partial x^i} \qquad (16)$$

will be a scalar, called the *directional derivative* of f in the direction ξ. In particular, if $e_1, \ldots, e_n$ are the canonical basis vectors (which is to say that the ith component of e_k is δ^i_k), then from (16) we obtain

$$\partial_{e_k}(f) = \frac{\partial f}{\partial x^k}.$$

Thus under this correspondence between vectors and differential operators, the canonical basis vectors $e_1, \ldots, e_n$ correspond to the operators $\partial/\partial x^1, \ldots, \partial/\partial x^n$ respectively, while an arbitrary vector ξ corresponds to the operator

$$\partial_\xi = \xi^i \frac{\partial}{\partial x^i}.$$

17.3. Exercises

1. Produce an example to show that in general the interchange of an upper with a lower index of a tensor is not an operation on tensors (i.e. the result depends on the co-ordinate system in which the interchange is carried out).
2. A second-rank tensor is called *non-singular* if as a matrix it is non-singular (at every point of a region). Show that for such a tensor, the inverse of the matrix is also a tensor (i.e. that taking the inverse is a tensor operation).

§18. Tensors of Type (0, k)

18.1. Differential Notation for Tensors with Lower Indices Only

We consider to begin with tensors of type (0, 1) (i.e. covectors) of which our standard example is, of course, the gradient $(\partial f/\partial x^i)$ of a function f. Recall that in analysis the differential of a function of $x^1, \ldots, x^n$, corresponding to increments dx^i in the x^i, is defined by

$$df = \frac{\partial f}{\partial x^i} dx^i. \tag{1}$$

Let $x^i = x^i(x^{1'}, \ldots, x^{n'})$ be a co-ordinate change. The differentials $dx^{i'}$ corresponding to the dx^i satisfy

$$dx^i = \frac{\partial x^i}{\partial x^{i'}} dx^{i'}, \tag{2}$$

whence

$$df = \frac{\partial f}{\partial x^i} dx^i = \left(\frac{\partial f}{\partial x^i}\frac{\partial x^i}{\partial x^{i'}}\right) dx^{i'} = \frac{\partial f}{\partial x^{i'}} dx^{i'}, \tag{3}$$

i.e. the expression df is invariant under co-ordinate changes. More generally, given any covector (T_i) (in place of the particular covector $(\partial f/\partial x^i)$) it follows in the same way that the *differential form* $T_i\, dx^i$ is invariant under co-ordinate changes. This suggests that it might be convenient to write the covector in the differential notation $T_i\, dx^i$. (Note that the above formulae (1), (2), (3) are meant to be suggestive of the relationship between the change in f and the dx^i, $dx^{i'}$, for "vanishingly small" dx^i.)

We pursue further the representation of a covector (T_i) by the differential form $T_i\, dx^i$. Thus let $e^1, \ldots, e^n$ be the canonical basis covectors at the point under scrutiny (i.e. the kth component of e^i is δ^i_k). Then for any covector (T_i) we may write

$$T_i e^i = T_{i'} e^{i'},$$

by which we mean simply that the components of the covector in the unprimed and primed systems are respectively T_i and $T_{i'}$. From the transformation rule for covectors, it is clear that in terms of the primed co-ordinate system, the components of e^i are $\partial x^i/\partial x^{1'}, \ldots, \partial x^i/\partial x^{n'}$, which we may express by

$$e^i = \frac{\partial x^i}{\partial x^{i'}} e^{i'}. \tag{4}$$

The similarity between this formula and the transformation rule (2) for the dx^i, shows the appropriateness of the differential notation $T_i\, dx^i$ for the covector $T_i e^i$; in particular we shall use the symbols dx^i to denote the basis covectors e^i. As we saw in Example 17.2.2(a) it is natural to regard a covector as a linear form on vectors. From this point of view, the value taken for instance by the linear form $df = (\partial f/\partial x^i)\, dx^i$ (representing the gradient covector) on the vector $\Delta\xi = \Delta x^i e_i$ is defined to be

$$\left(\frac{\partial f}{\partial x^i} dx^i, \Delta\xi\right) = \frac{\partial f}{\partial x^i} \Delta x^i. \tag{5}$$

Note that the latter expression is just the linear part of Taylor's formula for the change in f in traversing the vector $\Delta\xi$.

We now examine various aspects of a second important case, namely that of tensors of type (0, 2). As in §17.1 we take as a basis for the space of such tensors (at a given point) the products

$$e^i \otimes e^j.$$

In terms of this basis an arbitrary tensor T_{ij} has the form

$$T_{ij} e^i \otimes e^j. \tag{6}$$

A tensor T_{ij} of type (0, 2) can be regarded as a bilinear form on vectors, since if ξ, η are vectors then the scalar

$$T_{ij} \xi^i \eta^j$$

can be considered as the value of the bilinear form on those vectors. Thus does the tensor T_{ij} become at each point a bilinear function on pairs of tangent vectors at the point.

Any tensor T_{ij} of type (0, 2) can be expressed as the sum of a symmetric and a skew-symmetric tensor; indeed if we set

$$T_{ij}^{\text{sym}} = \tfrac{1}{2}(T_{ij} + T_{ji}), \qquad T_{ij}^{\text{alt}} = \tfrac{1}{2}(T_{ij} - T_{ji}), \tag{7}$$

then

$$\begin{gathered} T_{ij} = T_{ij}^{\text{sym}} + T_{ij}^{\text{alt}}; \\ T_{ij}^{\text{sym}} = T_{ji}^{\text{sym}}, \qquad T_{ij}^{\text{alt}} = -T_{ji}^{\text{alt}}. \end{gathered} \tag{8}$$

From (6) and (7) we obtain as a basis for the space of symmetric tensors of type (0, 2) at a point the set of tensors of the form

$$\frac{e^i \otimes e^j + e^j \otimes e^i}{2}, \qquad i \leq j, \tag{9}$$

and for the space of skew-symmetric tensors of type (0, 2) the basis consisting of the tensors

$$e^i \otimes e^j - e^j \otimes e^i, \qquad i < j. \tag{10}$$

Thus in terms of the basis elements (9), a symmetric tensor T_{ij} has the form

$$\begin{aligned} T_{ij} e^i \otimes e^j &= \sum_{i \leq j} T_{ij} e^i \otimes e^j + \sum_{i > j} T_{ij} e^i \otimes e^j \\ &= \sum_{i} T_{ij} e^i \otimes e^i + \sum_{i < j} 2T_{ij} \left(\frac{e^i \otimes e^j + e^j \otimes e^i}{2} \right), \end{aligned} \tag{11}$$

while if T_{ij} is skew-symmetric, then in terms of the basis elements (10) it has the form

$$\begin{aligned} T_{ij} e^i \otimes e^j &= \sum_{i < j} T_{ij} e^i \otimes e^j + \sum_{i > j} T_{ij} e^i \otimes e^j \\ &= \sum_{i < j} T_{ij} (e^i \otimes e^j - e^j \otimes e^i). \end{aligned} \tag{12}$$

In differential notation we write the basis elements (9) in the form $dx^i \, dx^j = dx^j \, dx^i$, and the basis elements (10) in the form

$$dx^i \wedge dx^j = -dx^j \wedge dx^i.$$

This notation ties in with our previous notation

$$dl^2 = g_{ij} \, dx^i \, dx^j \tag{13}$$

for the square of the element of length with respect to a Riemannian metric g_{ij}. Such a metric is a symmetric tensor of type (0, 2), and (13) gives its decomposition relative to the "basis" $dx^i \, dx^j$.

18.2. Skew-Symmetric Tensors of Type (0, k)

We first define such objects.

18.2.1. Definition. A *skew-symmetric tensor* of type (0, k) is a tensor $T_{i_1 \ldots i_k}$ satisfying

$$T_{\sigma(i_1, \ldots, i_k)} = (\operatorname{sgn} \sigma) T_{i_1 \ldots i_k}, \tag{14}$$

for all permutations σ. (Here sgn σ is $+1$ or -1 according as σ is an even or odd permutation.)

In view of the fact that permutation of the indices is a tensor operation (Lemma 17.2.1), this definition is independent of co-ordinate changes. Thus if $(T_{i_1 \ldots i_k})$ is a skew-symmetric tensor (relative to a particular co-ordinate system), then given a component $T_{i_1 \ldots i_k}$, the components corresponding to even permutations of $i_1, \ldots, i_k$, are the same, while those corresponding to odd permutations are just the negative of $T_{i_1 \ldots i_k}$. Hence in particular if k exceeds n, the dimension of the underlying space, then the skew-symmetric tensor $(T_{i_1 \ldots i_k})$ must be the zero tensor (since for each component at least two of the indices $i_1, \ldots, i_k$ will then coincide). In what follows we assume $k \leq n$.

For skew-symmetric tensors it is convenient to use the notation of differential forms. Thus we shall denote the elements of the standard basis for the space of all skew-symmetric tensors of type (0, k) at a given point, by

$$dx^{i_1} \wedge \cdots \wedge dx^{i_k}, \qquad i_1 < \cdots < i_k, \tag{15}$$

where

$$dx^{i_1} \wedge \cdots \wedge dx^{i_k} = \sum_{\sigma \in S_k} (\operatorname{sgn} \sigma) e^{\sigma(i_1)} \otimes \cdots \otimes e^{\sigma(i_k)}. \tag{16}$$

Here S_k is the symmetric group on $1, \ldots, k$, i.e. the group of all permutations of $1, \ldots, k$. (Recall that by $\sigma(i_l)$ we mean $i_{\sigma(l)}$.) Then, as in the special case $k = 2$ treated in the preceding subsection (see (12)), we obtain the following representation of the skew-symmetric tensor $(T_{i_1 \ldots i_k})$ as a differential form:

$$T_{i_1 \ldots i_k} e^{i_1} \otimes \cdots \otimes e^{i_k} = \sum_{i_1 < \cdots < i_k} T_{i_1 \ldots i_k} dx^{i_1} \wedge \cdots \wedge dx^{i_k}. \tag{17}$$

From (16) it is clear that the expression $dx^{i_1} \wedge \cdots \wedge dx^{i_k}$ is skew-symmetric, i.e. for any permutation σ

$$dx^{\sigma(i_1)} \wedge \cdots \wedge dx^{\sigma(i_k)} = (\operatorname{sgn} \sigma)\, dx^{i_1} \wedge \cdots \wedge dx^{i_k}. \tag{18}$$

18.2.2. Examples. (a) A skew-symmetric tensor $T_{i_1 \ldots i_n}$ of type (0, n) in n-dimensional space, is determined by the single component $T_{12 \ldots n}$, since those components with repeated indices are zero, while the remainder are given by

$$T_{\sigma(1, \ldots, n)} = (\operatorname{sgn} \sigma) T_{1 \ldots n}.$$

It follows that the space of skew-symmetric tensors of type (0, n) at a given point is one-dimensional, since the basis (15) contains just the single element $dx^1 \wedge \cdots \wedge dx^n$. In the physics literature, the components of the basic skew-symmetric tensor of type (0, n) given by $T_{12\dots n} = 1$ (i.e. of the differential form (16) with $k = n$) are denoted by $\varepsilon_{i_1 \dots i_n}$. Thus $\varepsilon_{i_1 \dots i_n} = 0$ whenever two or more of the indices coincide, and otherwise is given by

$$\varepsilon_{i_1 \dots i_n} = \begin{cases} +1 & \text{if } \operatorname{sgn}(i_1, \dots, i_n) = +1, \\ -1 & \text{if } \operatorname{sgn}(i_1, \dots, i_n) = -1. \end{cases} \tag{19}$$

This tensor is often called the "Levi-Civita tensor" of rank n. Clearly $T_{i_1 \dots i_n} = T_{12\dots n}\varepsilon_{i_1 \dots i_n}$.

We include as part of this example the following result, which tells us how skew-symmetric tensors of type (0, n) transform.

18.2.3. Theorem. *Skew-symmetric tensors of type* (0, n), *where n is the dimension of the underlying space, transform under co-ordinate changes $x^{i'} = x^{i'}(x^1, \dots, x^n)$ according to the rule*

$$T_{12\dots n} = T_{1'2'\dots n'} J, \tag{20}$$

where, as usual, J is the Jacobian of the co-ordinate change, i.e. $J = \det(\partial x^{i'}/\partial x^j)$.

Proof. It clearly suffices to prove (20) for the basis tensor $\varepsilon_{i_1 \dots i_n}$. From the general transformation rule for tensors (17.1.1) we have

$$\varepsilon_{12\dots n} = \varepsilon_{i'_1 \dots i'_n} \frac{\partial x^{i'_1}}{\partial x^1} \cdots \frac{\partial x^{i'_n}}{\partial x^n} = \sum_{\sigma \in S_n} (\operatorname{sgn} \sigma) \frac{\partial x^{i'_1}}{\partial x^1} \cdots \frac{\partial x^{i'_n}}{\partial x^n}, \tag{21}$$

where σ is the permutation $\begin{pmatrix} 1 & 2 \dots n \\ i'_1 & i'_2 \dots i'_n \end{pmatrix}$. The last expression in (21) is just J, as required. □

(b) Suppose that we have a tensor g_{ij} of type (0, 2), which, regarded as a quadratic form, is non-singular, i.e. $g = \det(g_{ij}) \neq 0$, at the point under scrutiny. Under a co-ordinate change $x^{i'} = x^{i'}(x^1, \dots, x^n)$, the tensor transforms according to the rule

$$g_{i'j'} = g_{ij} \frac{\partial x^i}{\partial x^{i'}} \frac{\partial x^j}{\partial x^{j'}},$$

or, in matrix notation, $G' = A^{\mathrm{T}} G A$, where $A = (\partial x^i/\partial x^{i'})$, $G = (g_{ij})$, $G' = (g_{i'j'})$. Hence the determinant $g = \det G$ transforms according to the rule

$$g' = \det G' = \det(A^{\mathrm{T}} G A) = (\det A)^2 \det G. \tag{22}$$

Thus if $\det A > 0$, we shall have

$$\sqrt{|g'|} = \sqrt{|g|} \det A.$$

Comparison of this rule with (20), yields the following corollary of 18.2.3:

18.2.4. Corollary. *The expression* $\sqrt{|g|}\,dx^1 \wedge \ldots \wedge dx^n$ *behaves like a tensor under co-ordinate changes for which the Jacobian* $J = \det A = \det(\partial x^i/\partial x^{i'})$ *is positive.*

The differential form $\sqrt{|g|}\,dx^1 \wedge \ldots \wedge dx^n$ is called the *volume element for the metric* g_{ij}. (This name is justified intuitively by §7.4 and the fact that the form $dx^1 \wedge dx^2$ evaluated at a pair (ξ, η) of tangent vectors at any given point, gives the Euclidean area of the parallelogram determined by ξ and η (and analogously for higher rank forms $dx^1 \wedge dx^2 \wedge dx^3$, etc.).)

We now consider briefly the complex case. Thus suppose the underlying space to be complex n-space with co-ordinates $z^1, \ldots, z^n; \bar{z}^1, \ldots, \bar{z}^n$, where $z^\alpha = x^\alpha + iy^\alpha$, $\bar{z}^\alpha = x^\alpha - iy^\alpha$ (see §11 *et seqq.*). A tensor on this space is then defined, in terms of these co-ordinates, analogously to 17.1.1. In complex notation a skew-symmetric differential form corresponding to a tensor of type $(0, k)$ may be written as

$$T = \sum_{p+q=k} T^{(p,q)},$$

where the summands are given by

$$T^{(p,q)} = \sum_{\substack{i_1 < \ldots < i_p \\ j_1 < \ldots < j_q}} T_{i_1 \ldots i_p; j_1 \ldots j_q} dz^{i_1} \wedge \cdots \wedge dz^{i_p} \wedge d\bar{z}^{j_1} \wedge \cdots \wedge d\bar{z}^{j_q}, \quad (23)$$

the $T_{i_1 \ldots i_p; j_1 \ldots j_q}$ being the (complex) components of a tensor in terms of the co-ordinates $z^\alpha, \bar{z}^\alpha$, with appropriate properties with respect to permutations of $i_1, \ldots, i_p$ and of $j_1, \ldots, j_q$. These $T^{(p,q)}$ are called "forms of type (p, q)."

For example consider a form of type (1, 1):

$$\Omega = T_{\alpha\beta}\, dz^\alpha \wedge d\bar{z}^\beta.$$

If the components $T_{\alpha\beta}$ satisfy $T_{\beta\alpha} = -\overline{T}_{\alpha\beta}$, then we shall have

$$iT_{\beta\alpha} = \overline{iT_{\alpha\beta}},$$

so that the $iT_{\alpha\beta}$ are the coefficients of an Hermitian form $\sum iT_{\alpha\beta}\, dz^\alpha\, d\bar{z}^\beta$ (see §11.1). Thus we can regard an Hermitian metric as a form of type (1, 1), in the above sense.

18.3. The Exterior Product of Differential Forms. The Exterior Algebra

As an application of the algebraic tensor operations introduced in §17.2, we now define the "exterior product" of two skew-symmetric tensors of types $(0, p)$ and $(0, q)$ (or, equivalently, of two differential forms of ranks p and q). Let the two forms in question be

$$\begin{aligned} \omega_1 &= \sum_{i_1 < \ldots < i_p} T_{i_1 \ldots i_p}\, dx^{i_1} \wedge \cdots \wedge dx^{i_p}, \\ \omega_2 &= \sum_{j_1 < \ldots < j_q} S_{j_1 \ldots j_q}\, dx^{j_1} \wedge \cdots \wedge dx^{j_q}. \end{aligned} \quad (24)$$

Their *exterior product* (or *wedge product*) $\omega = \omega_1 \wedge \omega_2$ is then defined to be the form of rank $(p + q)$ given by

$$\omega = \omega_1 \wedge \omega_2 = \sum_{k_1 < \dots < k_{p+q}} R_{k_1 \dots k_{p+q}} dx^{k_1} \wedge \cdots \wedge dx^{k_{p+q}}, \tag{25}$$

where

$$R_{k_1 \dots k_{p+q}} = \sum_{\sigma \in S_{p+q}} \frac{\operatorname{sgn}(\sigma)}{p!\,q!} T_{\sigma(k_1 \dots k_p} S_{k_{p+1} \dots k_{p+q})}. \tag{26}$$

Thus the $R_{k_1 \dots k_{p+q}}$ are the components of a tensor obtained from the tensors $T_{i_1 \dots i_p}$ and $S_{j_1 \dots j_q}$ by means of the operations of tensor product and permutation of indices (followed by the taking of a linear combination). Hence the definition (26) is independent of the co-ordinate system, i.e. yields the same tensor whatever co-ordinate system is used as context for carrying out the operation. It is clear from (26) that this tensor is again skew-symmetric.

18.3.1. Lemma. *The exterior product of skew differential forms is a bilinear associative operation. Moreover if ω_1, ω_2 are skew forms of ranks p and q, then*

$$\omega_2 \wedge \omega_1 = (-1)^{pq} \omega_1 \wedge \omega_2. \tag{27}$$

PROOF. The bilinearity is obvious from (26). The associativity, though not quite so obvious, follows easily from (26). The formula (27) is immediate from the fact that the sign of the permutation

$$\begin{pmatrix} 1 & \dots q & q+1 \dots p+q \\ p+1 \dots p+q & 1 & \dots p \end{pmatrix}$$

is $(-1)^{pq}$. (Verify it!) □

We conclude with the remark that the exterior product of the basic forms $\omega_1 = dx^i$ and $\omega_2 = dx^j$ as given by (26), coincides with the previous definition of the symbols $dx^i \wedge dx^j$ (at the end of §18.1). More generally if $\omega_1 = dx^{i_1} \wedge \cdots \wedge dx^{i_p}$ and $\omega_2 = dx^{j_1} \wedge \cdots \wedge dx^{j_q}$, then (cf. (16)):

$$\omega_1 \wedge \omega_2 = dx^{i_1} \wedge \cdots \wedge dx^{i_p} \wedge dx^{j_1} \wedge \cdots \wedge dx^{j_q}. \tag{28}$$

The *exterior algebra* at a given point of the underlying space is the linear space of all skew forms at the point, equipped with the operation of exterior multiplication.

18.4. Exercises

1. Let $\omega^j = a_i^j\, dx^i$. Establish the formula

$$\omega^{i_1} \wedge \cdots \wedge \omega^{i_k} = J^{i_1 \dots i_k}_{j_1 \dots j_k} dx^{j_1} \wedge \cdots \wedge dx^{j_k},$$

where $J^{i_1 \dots i_k}_{j_1 \dots j_k}$ is the $k \times k$ minor of (a_i^j) formed by the intersections of the rows numbered $i_1, \dots, i_k$ with the columns numbered $j_1, \dots, j_k$. Thus in particular

$$\omega^1 \wedge \cdots \wedge \omega^n = \det(a_i^j)\, dx^1 \wedge \cdots \wedge dx^n.$$

2. Calculate the dimension of the space of (skew-symmetric) k-forms (at a given point).
3. Prove that

$$\sum_k a^j_{i_k} \varepsilon_{i_1 \dots i_{k-1} j i_{k+1} \dots i_n} = \varepsilon_{i_1 \dots i_n} \operatorname{tr}(a^j_i),$$

where $\operatorname{tr}(a^j_i) = a^i_i$ (i.e. the trace of the matrix (a^j_i)).

§19. Tensors in Riemannian and Pseudo-Riemannian Spaces

19.1. Raising and Lowering Indices

Let g_{ij} be a tensor of type (0, 2) defining either a Riemannian or a pseudo-Riemannian metric on the underlying space. Recall that relative to such a metric we define the scalar product of two vectors (ξ^i), (η^j) by

$$\langle \xi, \eta \rangle = \xi^i \eta^j g_{ij}. \tag{1}$$

Note that the resulting scalar tensor may be regarded as being obtained from the three tensors entering into it, by means of the operations of forming the tensor product and of contraction.

Similarly, given a tensor g^{ij} of type (2, 0), we define the scalar product (relative to this tensor) of covectors (ξ_j), (η_i) by

$$\langle \xi, \eta \rangle = g^{ij} \xi_j \eta_i. \tag{2}$$

In terms of a given metric g_{ij} one defines the fundamentally important tensor operation of "lowering" of indices as follows: If $T^{i_1 \dots i_p}_{j_1 \dots j_q}$ is a tensor of type (p, q), then by the operation of *lowering the index i_1 using the metric g_{ij}* we obtain the tensor of type $(p - 1, q + 1)$ given by

$$T^{i_2 \dots i_p}_{i_1 j_1 \dots j_q} = g_{i_1 k} T^{k i_2 \dots i_p}_{j_1 \dots j_q}. \tag{3}$$

Clearly the result is indeed a tensor, and is co-ordinate-independent, as it is obtained by forming the product of two tensors, and then performing a contraction.

For example if we lower the index of a vector ξ^i, we obtain the covector

$$\xi_i = g_{ij} \xi^j. \tag{4}$$

Thus the operation of lowering the index of vectors (using a fixed metric g_{ij}) yields, for each point, a linear map from the tangent space of vectors attached to the point, to the space of covectors at the point. This correspondence between vectors and covectors can be described alternatively as follows: If $\langle\ ,\ \rangle$ denotes the scalar product of vectors in the metric g_{ij}, then to the vector η corresponds that covector which, regarded as a linear form on vectors, takes the value $\langle \xi, \eta \rangle$ on each vector ξ.

We now turn to the related operation of "raising" of indices. Again the operation is performed in the context of a given metric g_{ij}. To define the operation we need the inverse of the matrix (g_{ij}) (which exists by virtue of the definitions of a Riemannian and a pseudo-Riemann metric, since in either case $\det(g_{ij}) \neq 0$). Denote the inverse by g^{ij}; thus

$$g^{ij}g_{jk} = \delta^i_k. \tag{5}$$

Then by *raising the index* j_1 *of the tensor* $T^{i_1 \dots i_p}_{j_1 \dots j_p}$ *using the metric* g_{ij}, we obtain the tensor

$$T^{j_1 i_1 \dots i_p}_{j_2 \dots j_q} = g^{j_1 k} T^{i_1 \dots i_p}_{k j_2 \dots j_q}. \tag{6}$$

19.1.1. Lemma. *If we lower an index and then raise it, we obtain the original tensor.*

PROOF. Lowering the index i_1 in the tensor $T^{i_1 \dots i_p}_{j_1 \dots j_q}$, we obtain the tensor $g_{i_1 k} T^{k i_2 \dots i_p}_{j_1 \dots j_q}$. If we now raise the same index i_1 in the latter tensor we obtain

$$g^{i_1 l} g_{lk} T^{k i_2 \dots i_p}_{j_1 \dots j_q} = \delta^{i_1}_k T^{k i_2 \dots i_p}_{j_1 \dots j_q} = T^{i_1 \dots i_p}_{j_1 \dots j_q},$$

where we have used (5). This completes the proof. □

As we have just seen (in (2)), given a matrix (g^{ij}), we can define a scalar product of covectors. If the matrix (g^{ij}) arises as the inverse of the matrix of a metric g_{ij}, then we can regard this scalar product as proceeding from the metric. The following simple lemma tells us that the value of this scalar product on any two covectors, is the same as the scalar product of the two vectors obtained by lowering the indices of those covectors.

19.1.2. Lemma. *The scalar products of a pair of vectors* $\xi = (\xi^i)$, $\eta = (\eta^i)$, *and of the corresponding pair of covectors* $\hat{\xi} = (\xi_i) = (g_{ij}\xi^j)$, $\hat{\eta} = (\eta_i) = (g_{ij}\eta^j)$, *coincide; i.e.* $\langle \hat{\xi}, \hat{\eta} \rangle = \langle \xi, \eta \rangle$.

PROOF. Since $\langle \xi, \eta \rangle = g_{ij}\xi^i\eta^j$ and $\langle \hat{\xi}, \hat{\eta} \rangle = g^{ij}\xi_i\eta_j$, We have

$$\begin{aligned} \langle \hat{\xi}, \hat{\eta} \rangle &= g^{ij}\xi_j\eta_i = g^{ij}g_{jk}\xi^k g_{il}\eta^l \\ &= \delta^i_k \xi^k g_{il}\eta^l = \xi^i g_{il}\eta^l = \langle \xi, \eta \rangle, \end{aligned}$$

as required. □

Remark. In the above we have used only the non-singularity of the matrix of the metric g_{ij}; neither the positive definiteness (which holds in the Riemannian case), nor the symmetry (which holds for both kinds of metric), was involved. In Chapter 5 we shall consider skew-symmetric metrics in connexion with "Hamiltonian formalism."

19.2. The Eigenvalues of a Quadratic Form

Suppose that we have a linear operator T^i_j, or, in other words, a tensor of type (1, 1). We cannot *ab initio* say of such a tensor that it is symmetric or skew-symmetric, since interchanging the indices i, j is not permitted, i.e. such an interchange is not in general a tensor operation. If however our underlying space comes equipped with a metric g_{ij}, then we can define symmetry and skew-symmetry of the tensor T^i_j relative to this metric, by first using the metric to lower the index i, thereby obtaining the tensor $T_{ij} = g_{ik} T^k_j$ of type (0, 2). The latter tensor gives rise in the usual way to a bilinear form $\{\ ,\ \}_T$ on pairs of vectors, defined by

$$\{\xi, \eta\}_T = \xi^i T_{ij} \eta^j,$$

where $\xi = (\xi^i)$, $\eta = (\eta^i)$. If we denote by $\langle\ ,\ \rangle$ the scalar product afforded by the metric g_{ij}, then

$$\{\xi, \eta\}_T = \xi^i g_{ij} T^j_k \eta^k = \langle \xi, T\eta \rangle. \tag{7}$$

19.2.1. Definition. A linear operator T^j_i defined on a space with a metric g_{ij}, is said to be *symmetric* if the bilinear form $T_{ij} = g_{ik} T^k_j$ is symmetric, i.e. if $T_{ij} = T_{ji}$; and *skew-symmetric* if $T_{ij} = -T_{ji}$.

19.2.2. Theorem. *A linear operator $T = (T^i_j)$ defined on a space with a Riemannian or pseudo-Riemannian metric g_{ij}, is symmetric if and only if for all vectors ξ, η we have*

$$\langle T\xi, \eta \rangle = \langle \xi, T\eta \rangle, \tag{8}$$

and is skew-symmetric if and only if

$$\langle T\xi, \eta \rangle = -\langle \xi, T\eta \rangle, \tag{9}$$

again for all vectors ξ, η.

This follows easily from (7) and the symmetry (skew-symmetry) of T.

Suppose now that we are given a bilinear form T_{ij} defined on a Riemannian or pseudo-Riemannian space with metric g_{ij}. In order to define in a coherent manner the "eigenvalues" of such a bilinear form we need to turn it into a linear operator, or, more precisely, to raise its index i, obtaining thereby the operator $T^i_j = g^{ik} T_{kj}$ (where, as before, $g^{ij} g_{jk} = \delta^i_k$).

19.2.3. Definition. By the *eigenvalues of a quadratic form* T_{ij} relative to the metric g_{ij}, we shall mean the eigenvalues of the linear operator $T^i_j = g^{ik} T_{kj}$, where $(g^{ij}) = (g_{ij})^{-1}$.

Let λ be an eigenvalue of the above linear transformation T^i_j, and (ξ^i) a corresponding eigenvector, i.e. $T^i_j \xi^j = \lambda \xi^i$, $(\xi^i) \neq 0$. By the definition of T^i_j,

$$T^i_j \xi^j = \lambda \xi^i \Leftrightarrow g^{ik} T_{kj} \xi^j = \lambda \xi^i, \tag{10}$$

so that the eigenvector (ξ^i) is a (non-zero) solution of the linear system of equations

$$T_{kj}\xi^j = \lambda g_{ki}\xi^i, \qquad k = 1, \ldots, n. \tag{11}$$

It is clear that the trace tr $T^i_j = T^i_i = g^{ik}T_{ik}$, and determinant det T^i_j of the linear operator $T^i_j = g^{ik}T_{kj}$, are metrical invariants of the form T_{ij}, i.e. they are invariant under co-ordinate changes, but of course depend on the metric g_{ij}. As the following example will recall to the reader, we have met these invariants before, albeit in a slightly different guise.

19.2.4. Example. In Chapter 2 we associated with (each point of) a surface $r = r(u, v)$; $r = (x, y, z)$, $x^1 = u$, $x^2 = v$, in Euclidean 3-space, two quadratic forms:

(i) the induced metric $g_{ij}\,dx^i\,dx^j$, or tensor g_{ij};
(ii) the second fundamental form $b_{ij}\,dx^i\,dx^j$, or tensor b_{ij}.

(Here the summation is from 1 to 2, since the surface is 2-dimensional.)

In §8 we obtained the following formulae expressing the Gaussian and mean curvatures of the surface in terms of these two tensors:

$$\text{Gaussian curvature } K = \frac{\det(b_{ij})}{\det(g_{ij})};$$

$$\text{mean curvature } H = b^i_i = g^{ij}b_{ij}.$$

Thus we see that, in our present terminology, the mean curvature is just the trace of the tensor b_{ij} relative to the metric g_{ij}, while since

$$K = (\det(g_{ij}))^{-1}\det(b_{ij}) = \det(g^{ik}b_{kj}) = \det(b^i_j),$$

the Gaussian curvature is the second of the above-mentioned metrical invariants of the second fundamental form b_{ij}.

19.3. The Operator $*$

The presence of a metric g_{ij} allows us to derive from any skew-symmetric tensor T of type $(0, k)$ a skew-symmetric tensor $*T$ of type $(0, n - k)$, defined as follows.

19.3.1. Definition. Given a skew-symmetric tensor T of type $(0, k)$, with components $T_{i_1 \ldots i_k}$, we define the skew-symmetric tensor $*T$ of type $(0, n - k)$ by the formula

$$(*T)_{i_{k+1} \ldots i_n} = \frac{1}{k!}\sqrt{|g|}\;\varepsilon_{i_1 \ldots i_n} T^{i_1 \ldots i_k}, \tag{12}$$

where $g = \det(g_{ij})$, and $T^{i_1 \dots i_k}$ is obtained from $T_{i_1 \dots i_k}$ by raising the indices, i.e.

$$T^{i_1 \dots i_k} = g^{i_1 j_1} \dots g^{i_k j_k} T_{j_1 \dots j_k}. \tag{13}$$

By Corollary 18.2.4 the expression $\sqrt{|g|}\varepsilon_{i_1 \dots i_n}$, defined on n-dimensional space, behaves like a tensor under co-ordinate changes with positive Jacobian. It follows that $* T$ also behaves like a tensor under such co-ordinate changes. Its skew-symmetry is clear from the definition.

Note finally that if we apply the operator $*$ twice, we obtain

$$*(*T) = (-1)^{k(n-k)} \operatorname{sgn}(g) T. \tag{14}$$

We leave this to the reader to verify.

19.4. Tensors in Euclidean Space

In a Euclidean space with Euclidean co-ordinates, the metric is given by $g_{ij} = \delta_{ij}$. It follows that if the co-ordinates of the underlying space are Euclidean, then raising or lowering of indices does not affect the components of an arbitrary tensor $T^{i_1 \dots i_p}_{j_1 \dots j_q}$:

$$T^{i_2 \dots i_p}_{i_1 j_1 \dots j_q} = \delta_{i_1 i} T^{i i_2 \dots i_p}_{j_1 \dots j_q} = T^{i_1 i_2 \dots i_p}_{j_1 \dots i_p}. \tag{15}$$

Hence in this case the distinction between upper and lower indices disappears, so that they may all be regarded as lower indices, provided of course that we restrict ourselves to co-ordinate changes preserving the metric, i.e. to combinations of orthogonal transformations and translations.

In particular, therefore, in terms of Euclidean co-ordinates the entries of the matrix of a linear operator coincide with the coefficients of the corresponding quadratic form (see §19.2), and under Euclidean isometries the components of the gradient of a function will transform like those of a vector, and so on.

We conclude by considering the effect of the operator $*$ when the underlying space is Euclidean, with Euclidean co-ordinates x, y, z. The reader will easily verify that

$$*dx = dy \wedge dz, \qquad *dy = -dx \wedge dz, \qquad dz = dx \wedge dy.$$

It follows that on an arbitrary 1-form (i.e. covector) $\omega = P\,dx + Q\,dy + R\,dz$, the effect of the operator $*$ is as follows:

$$*\omega = P\,dy \wedge dz + Q\,dz \wedge dx + R\,dx \wedge dy, \qquad *(*\omega) = \omega.$$

If f is a scalar then $*f$ is a 3-form, namely

$$*f = f\,dx \wedge dy \wedge dz,$$

and $*(*f) = f$.

19.5. Exercises

1. Let $\varepsilon_{\alpha\beta\gamma}$ be the skew-symmetric tensor defined in §18.2(19), in the case where the underlying space is 3-dimensional Euclidean. Prove the following formulae:

(i) $\varepsilon_{\alpha\beta\gamma}\varepsilon_{\lambda\mu\nu} = \begin{vmatrix} \delta_{\alpha\lambda} & \delta_{\alpha\mu} & \delta_{\alpha\nu} \\ \delta_{\beta\lambda} & \delta_{\beta\mu} & \delta_{\beta\nu} \\ \delta_{\gamma\lambda} & \delta_{\gamma\mu} & \delta_{\gamma\nu} \end{vmatrix}$;

(ii) $\varepsilon_{\alpha\beta\gamma}\varepsilon_{\lambda\mu\gamma} = \delta_{\alpha\lambda}\delta_{\beta\mu} - \delta_{\alpha\mu}\delta_{\beta\lambda}$;

(iii) $\varepsilon_{\alpha\beta\gamma}\varepsilon_{\lambda\beta\gamma} = 2\delta_{\alpha\lambda}$;

(iv) $\varepsilon_{\alpha\beta\gamma}\varepsilon_{\alpha\beta\gamma} = 6$.

(Here $\delta_{\alpha\beta}$ is the Kronecker delta, and summation is to be taken over all repeated indices.)

2. For

$$\omega_1 = \sum_{i_1 < \dots < i_p} T_{i_1 \dots i_p}\, dx^{i_1} \wedge \cdots \wedge dx^{i_p},$$

$$\omega_2 = \sum_{j_1 < \dots < j_p} S_{j_1 \dots j_p}\, dx^{j_1} \wedge \cdots \wedge dx^{j_p},$$

define

$$\{\omega_1, \omega_2\} = \frac{1}{p!} g^{i_1 j_1} \cdots g^{i_p j_p} T_{i_1 \dots i_p} S_{j_1 \dots j_p}.$$

Show that $\omega_1 \wedge *\omega_2 = \{\omega_1, \omega_2\} \sqrt{|g|}\, dx^1 \wedge \cdots \wedge dx^n$.

§20. The Crystallographic Groups and the Finite Subgroups of the Rotation Group of Euclidean 3-Space. Examples of Invariant Tensors

In this section we shall first study the crystallographic groups, and shall then classify the finite subgroups of the rotation group $SO(3)$. (Tensors will make a brief appearance at the very end of the section.)

We shall consider a "crystal lattice" of points spaced apart in some regular (i.e. finitely defined) fashion throughout the whole of the Euclidean plane or Euclidean 3-space. This model for a crystal is a standard one: the crystal is regarded as consisting of a few types of atoms fixed rigidly in space (or in the plane), and distributed throughout space (or the plane) in a regular fashion. Of course a real crystal has boundaries; however by imposing the condition of "periodicity" on the crystal (which condition we shall explain below), it becomes clear that the study of such a "finite" crystal is equivalent to the study of infinite lattices of points, i.e. of "infinite" crystals. Thus it is appropriate to define a *crystal lattice* as the totality of points (in Euclidean 3-space or in the Euclidean plane) giving the locations of the atoms of an infinite crystal.

We shall confine our attention to a rather narrow class of such lattices, namely those invariant under certain translations. (See below for the precise conditions.) This restriction is justified by the fact that most real crystals correspond (approximately) to just such lattices. We shall suppose that our crystal lattice always contains as a subset the set of all points with position vectors of the form $\alpha = n_1\alpha_1 + n_2\alpha_2 + n_3\alpha_3$ (or, if the lattice is planar, then all points with position vectors of the form $\alpha = n_1\alpha_1 + n_2\alpha_2$), where n_1, n_2, n_3 are arbitrary integers, and the vectors $\alpha_1, \alpha_2, \alpha_3$ are the vectors corresponding to the "basic translations." The vectors α_1, α_2, α_3 are called the *primitive vectors* of the lattice, and are always assumed to be linearly independent.

We now impose on our lattice (L say) the fundamental requirement foreshadowed above, that it go into itself under translations along the vectors α_1, α_2, α_3, and consequently under the translations corresponding to all linear combinations of $\alpha_1, \alpha_2, \alpha_3$, with integer coefficients, and that the latter are the only translations preserving the lattice. In other words we require that our lattice L be invariant under all translations generated (over the integers) by the vectors α_1, α_2, α_3, and under these only. As we mentioned above, this property is satisfied (essentially) by most real "infinite" crystals.

If we denote the translations along α_1, α_2, α_3 by τ_1, τ_2, τ_3 respectively, then the general translation of the type we have been considering has the form

$$T = n_1\tau_1 + n_2\tau_2 + n_3\tau_3,$$

where n_1, n_2, n_3 are integers.

20.1. Definition. The (spatial) lattice L is called *translation-invariant* if there exist primitive translations τ_1, τ_2, τ_3 such that L is sent to itself by all translations of the form $T = n_1\tau_1 + n_2\tau_2 + n_3\tau_3$, and by no other translation. (Translation-invariant planar lattices are defined analogously.)

Thus, to repeat, we shall henceforth consider only translation-invariant lattices (in the plane or in space).

Remark. Many expositions of the mathematical theory of crystals begin by defining a crystallographic group to be a discrete subgroup Γ of the group G_3 of all motions of Euclidean 3-space, with compact quotient G_3/Γ. It is then proved that the subgroup of translations in Γ has finite index in Γ. As atoms one then takes the points in one of the orbits of Γ; for this purpose one of the most symmetrical of the orbits is chosen. A detailed account along these lines may be found in the book [39]. (Alternatively, see Chapter 4 of [38].)

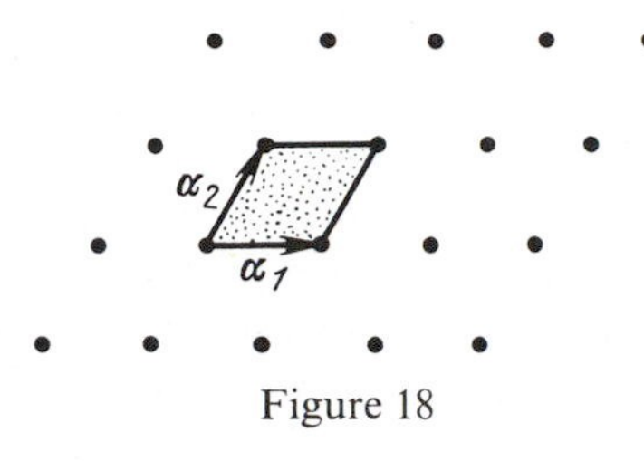

Figure 18

20.2. Definition. A parallelepiped with sides the vectors $\alpha_1, \alpha_2, \alpha_3$, is called a *primitive cell* or *fundamental region* of the lattice L.

We remark that it is often assumed that the primitive translation vectors $\alpha_1, \alpha_2, \alpha_3$ (or α_1, α_2 in the planar case) are such that the volume of the parallelepiped formed from them is smallest possible.

In Figure 18 the simplest kind of 2-dimensional lattice is depicted, with one of its primitive cells shaded in. It is clear that in this case the condition of translation-invariance entails that the whole crystal lattice is simply the union of the primitive cells, which are all obtained by applying translations to any particular one of them. In fact the lattice of Figure 18 has the property that any one of its atoms (i.e. of its points) can be obtained from any other atom by means of one of the permitted translations $T = n_1\tau_1 + n_2\tau_2$. This is usually expressed by saying that the set (in fact it is a group) of all permitted translations is *transitive* on the lattice. However this happy situation by no means obtains for all lattices, in particular for the following reason. In general a crystal lattice is comprised of several different types of atoms, so that it is natural to require (at least as far as our model is concerned) that under basic translations atoms of one type are sent to (the former positions of) atoms of the same type, and not to points occupied by atoms of a different type. Thus it is clear that if there is more than one type of atom in the crystal we are modelling, then it is entirely possible that the set of all translations generated by the basic ones will not be transitive on the lattice. Such a lattice is shown in Figure 19, where atoms of type A may not be translated to the

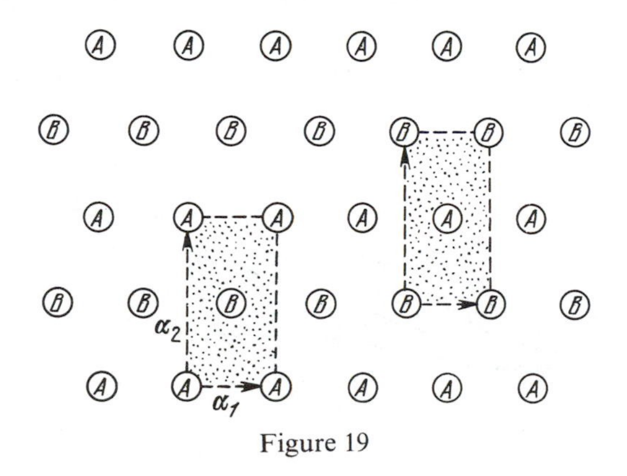

Figure 19

positions of type B atoms, and vice versa. While the set of all permitted translations is transitive on each of the two sets of atoms of the same type, it is not transitive on their union, the set of all atoms of the lattice. (In other, more technical, terminology, the set of type A atoms and the set of type B atoms are the orbits of the group of permitted translations of the lattice.)

It follows that in order to specify a crystal lattice completely it is in general not enough to give only the basic translations. On the other hand it is clear that if in addition to the basic translations we know the positions of the atoms of the lattice in any particular primitive cell, then since the lattice as a whole is the union of its primitive cells, we shall be able to reconstruct it. Thus from amongst the class of all translation-invariant lattices it is natural to single out those which *are* determined by their basic translations, i.e. all points of which are translates, under permitted translations, of a single point.

20.3. Definition. A lattice L (in the plane or in space) is termed a *Bravais lattice*, if all of its atoms have position vectors of the form $n_1\alpha_1 + n_2\alpha_2 + n_3\alpha_3$ ($n_1\alpha_1 + n_2\alpha_2$ in the planar case), where n_1, n_2, n_3 are integers.

In other words, a Bravais lattice is one on which the set of permitted translations is transitive. It is clear from the definition that different Bravais lattices will differ only in the shapes of their primitive cells, so that any Bravais lattice can be transformed into any other (of the same dimension) by means of a suitable affine transformation. In other words, from the point of view of affine geometry there is only one Bravais lattice. Bravais lattices which differ metrically (i.e. are not transformable one into the other by means of combinations of orthogonal transformations and translations) will in general have primitive translation vectors of different lengths with different angles between them.

20.4. Definition. Let $X_1, \ldots, X_n$ be all the atoms of a lattice L lying inside a primitive cell with one boundary atom at the origin of co-ordinates 0 (as in Figure 20). Then the vectors from 0 to $X_1, \ldots, X_n$ together form a *basis* for the lattice (relative to the given set of primitive translation vectors).

20.5. Proposition. A lattice is completely determined by a set of primitive translation vectors $\alpha_1, \alpha_2, \alpha_3$ (or α_1, α_2 in the planar case), together with the corresponding basis for the lattice.

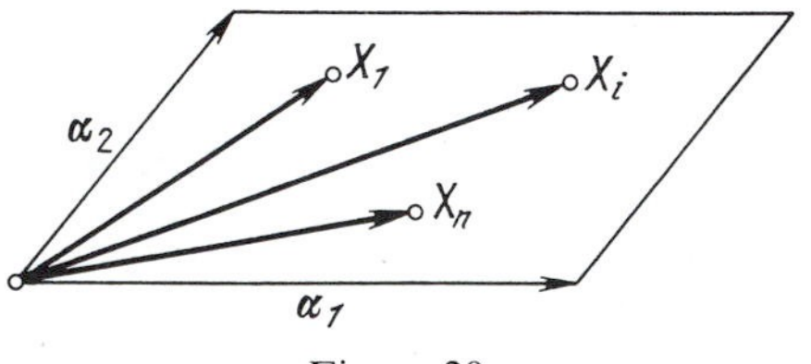

Figure 20

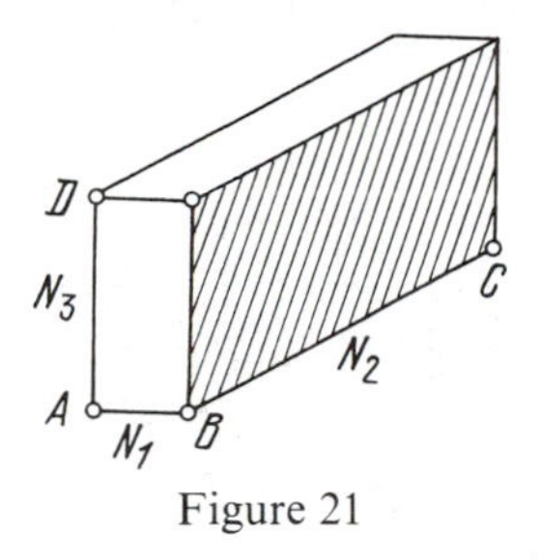

Figure 21

This proposition is (at least intuitively) clear from our definition of basis and of translation invariance.

We digress for a moment to give a more detailed justification for modelling a finite, real crystal by means of translation-invariant infinite lattices. Figure 21 represents a (somewhat idealized) real, perfect, 3-dimensional crystal with boundary. The positive integers N_1, N_2, N_3 count the numbers of primitive cells stacked along the corresponding edges of the parallelepiped (i.e. of the crystal); thus $AB = N_1\alpha_1$, $BC = N_2\alpha_2$, $AD = N_3\alpha_3$.

We now apply to our real crystal the permitted translations (in particular translations along the primitive translation vectors $\alpha_1, \alpha_2, \alpha_3$), but with the following modification: If for instance we translate the crystal along α_1, we then slice off the layer protruding beyond the original parallelepiped, and glue it to the opposite (i.e. the left) face of the crystal which has now moved into the parallelepiped along the vector α_1; by this procedure we have made the crystal "periodic" in the direction of α_1. Essentially the same effect can be obtained by imagining the opposite faces of the crystal identified. The introduction of such a periodic model is vindicated by the fact that most physical constructions and calculations involving crystals are not altered by the imposition of periodicity. Moreover, it is clear that the periodic model (where the same procedure is applied in all three directions) is equivalent to our infinite crystal lattice.

The condition of periodicity may be elucidated by considering, in geometrical terms, what the basic translations become when applied to the periodic crystal. In the case of a 1-dimensional crystal, consisting of N aligned atoms, with the two atoms numbered 1 and N comprising the boundary, the imposition of periodicity amounts to glueing the two ends to form a circular chain, so that a permissible translation, as applied to the "glued" crystal (i.e. to the 1-dimensional crystal with the condition of periodicity on its ends), becomes a rotation of the circle through an integral multiple of $2\pi/N$. In the planar case periodicity is equivalent to the formation of the surface T^2, the 2-dimensional torus, by identifying the opposite edges of the parallelogram occupied by the crystal. Finally the 3-dimensional periodic crystal corresponds to the 3-dimensional torus T^3.

We now return to the mainstream of our development. With each (translation-invariant) lattice there is associated in a natural way a "symmetric cell" (not to be confused with a primitive cell), having an atom at its centre.

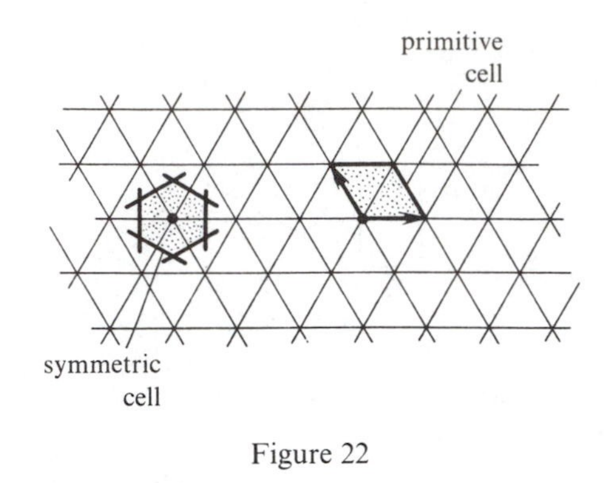

Figure 22

20.6. Definition. For each atom of a lattice L, the *symmetric cell* with that atom at its centre, is the set of points of space (or of the plane if the lattice is planar) nearer to that atom than to any other atom of the lattice.

Symmetric cells are sometimes called also "Wigner-Seitz cells", and in the theory of discrete groups are known as "Dirichlet regions". In Figure 22, which represents the so-called plane "hexagonal" lattice, a primitive cell and a symmetric cell are indicated. (The reader will notice that they are different!) In general the boundary of a symmetric cell of a planar lattice is formed, as in Figure 22, from segments of the perpendicular bisectors of the "edges" of the lattice joining the central atom of the cell to its nearest neighbours.

We now transfer our attention to the Euclidean motions preserving a given lattice, i.e. mapping the lattice onto itself. As before we shall denote by G_3 the group of all motions (i.e. isometries) of Euclidean 3-space. (The isometry group of the Euclidean plane will be denoted by G_2.)

20.7. Proposition. Every element g of the group G_3 (and also of G_2) can be expressed in exactly one way as a product $g = T \circ \alpha$, Where T is a translation, and α is an orthogonal transformation.

PROOF. In §4.3 it was shown that every motion g of Euclidean 3-space is of one of the following two types:

(1) a screw-displacement $g = T\alpha$, where α is a rotation (so that $\det \alpha = 1$) and T is a translation along the axis of rotation;
(2) a rotatory reflection $g = \alpha$, where $\alpha \in O(3)$ and $\det \alpha = -1$.

The uniqueness of the expression follows easily from the fact that the elements of $O(3)$ fix the origin, while a non-trivial translation does not. The proof for the planar case, i.e. for G_2, follows similarly (from §4.2). □

Remark. In general the translations do not commute with rotations, i.e. it is not true that $T\alpha = \alpha T$ for all rotations α and translations T. (We leave it to the reader to give an example.)

The set of all translations of $\mathbb{R}^3$ is a subgroup of G_3, the *translation subgroup*. It is easy to verify that in fact it is a normal subgroup of G_3, i.e. for all translations T and all $g \in G_3$, gTg^{-1} is again a translation. (In view of Proposition 20.7 it suffices to verify this for $g \in O(3)$.) On the other hand the group $O(3)$ is not a normal subgroup of G_3. (Analogous statements hold for the translation subgroup of G_2, and for $O(2)$.)

In our present context we are interested in those transformations in G_3 (or G_2 as the case may be) which preserve a given lattice. It is easy to see that for a given lattice the set of all such transformations is a group.

20.8. Definition. The subgroup consisting of all the motions in G_3 (resp. G_2) preserving a given lattice L, is called the *crystallographic space group* of the lattice, and is denoted by $G_3(L)$ (resp. $G_2(L)$).

From now on we shall state definitions and propositions for the 3-dimensional case only, with the understanding that the corresponding definitions and facts hold good in the planar case.

20.9. Definition. The subgroup of $G_3(L)$ consisting of all translations of the lattice L, is called the *translation group* of L, and is denoted by $T_3(L)$.

Thus $T_3(L)$ consists of all translations T of the form $T = n_1\tau_1 + n_2\tau_2 + n_3\tau_3$, where n_1, n_2, n_3 are integers, and τ_1, τ_2, τ_3 are the primitive, or basic, translations of our translation-invariant lattice L.

It is easily verified that the subgroup $T_3(L)$ is normal in the group $G_3(L)$. For it is easy to show that if t is any translation and g is any element of $G_3(L)$, then the element gtg^{-1} is again a translation (in $G_3(L)$), and therefore lies in $T_3(L)$.

We may obviously assume (and shall do so from now on) that the origin of co-ordinates, denoted by O, is a vertex of a primitive cell of our lattice L, and we shall imagine the primitive vectors emanating from O. Then of course all orthogonal transformations will fix the atom O of the lattice.

By Proposition 20.7 each transformation $g \in G_3$ can be expressed in exactly one way as a product $g = T\alpha$ where T is a translation and $\alpha \in O(3)$. It is sometimes useful to frame this decomposition of the elements of G_3 (or more generally of G_n) in matrix terms, as follows. Recall first that an arbitrary motion g of the Euclidean space $\mathbb{R}^n$ may be uniquely written in the form $y = Ax + b$, where y, x are vectors in $\mathbb{R}^n$, $A \in O(n)$, and b is a fixed vector in $\mathbb{R}^n$. (Here b is the vector along which the translation T above moves the space; thus Proposition 20.7 and the normality of the translation subgroup hold, more generally, for G_n.) Using this we can represent such a transformation

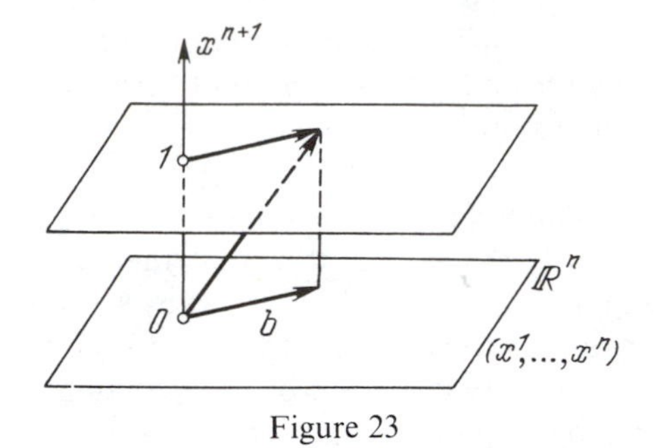

Figure 23

g of $\mathbb{R}^n$ uniquely by the following non-singular transformation of $\mathbb{R}^{n+1}$ (cf. Exercise 3 in §4.5):

$$\hat{g} = \left(\begin{array}{c|c} A & \begin{array}{c} b^1 \\ \vdots \\ b^n \end{array} \\ \hline 0 \ldots 0 & 1 \end{array}\right).$$

The linear transformation $\hat{g}$ acting on $\mathbb{R}^{n+1}$, induces on the hyperplane $\mathbb{R}^n$ through the point $(0, \ldots, 0, 1)$ parallel to the co-ordinate hyperplane $x^{n+1} = 0$, a non-singular transformation which coincides with the original motion g of $\mathbb{R}^n$ (see Figure 23). This matrix representation of g makes it particularly evident how it is that G_n is a semidirect product of the group of translations, which correspond to matrices

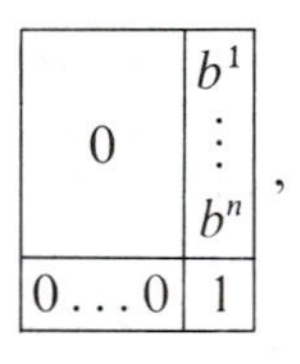

by the subgroup of orthogonal transformations, which correspond to matrices of the form

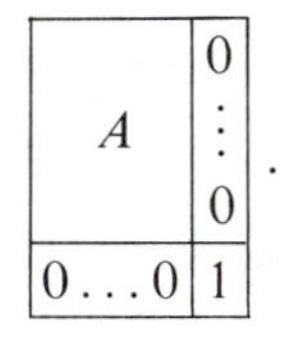

We now make a further remark in connexion with our earlier one concerning the usual abstract mathematical definition of a crystallographic group. In §4.4 we showed in particular that the group $O(3)$ has two connected components, namely the rotation subgroup $SO(3)$ whose transformations correspond to matrices of determinant 1, i.e. the subgroup of proper (or direct) motions, and the coset consisting of transformations with determinant -1, the improper (or opposite) orthogonal transformations of Euclidean

3-space. The translation subgroup of G_3 is also connected, since clearly it may be identified in a natural way with $\mathbb{R}^3$ itself. In contrast with this it is not difficult to see that the space group $G_3(L)$ of a crystal lattice L is discrete. (A transformation group is defined to be *discrete* if it does not contain transformations arbitrarily close to, but different from, the identity transformation. (Cf. §§18.4, 20 of Part II.))

Narrowing our sights to the subgroup $G_3(L)$ of G_3, where L is the translation-invariant lattice we are investigating, we see that by Proposition 20.7 every element g of $G_3(L)$ is expressible in exactly one way as a product $g = T\alpha$, where T is a translation and $\alpha \in O(3)$. It is important to emphasize that since $G_3(L)$ is a proper subgroup of G_3 it may (and does) happen that for some $g \in G_3(L)$ the corresponding T and α do not lie in $G_3(L)$, i.e. that T and α do not preserve the lattice L, though their product $g = T\alpha$ does. An example of this (in the planar case) is given below. As in the case of the full group of motions G_3, the order of composition of the T and α composing an element $g \in G_3(L)$ is important, i.e. in general $T\alpha \neq \alpha T$.

20.10. Definition. The set of all orthogonal transformations $\alpha \in O(3)$ such that for some translation T the product $T\alpha$ lies in $G_3(L)$, is called the *crystallographic point group* of the crystal lattice L, and is denoted by $S_3(L)$.

(We shall show below (Proposition 20.11) that this set is indeed a group.) One sometimes hears the point group of a lattice referred to as the "symmetry group" of the crystal, and its elements as "symmetry operations" on the crystal. To recapitulate the definition: An orthogonal transformation α belongs to the point group precisely if on following it by some translation T (i.e. on forming $T\alpha$) we obtain a lattice-preserving motion, i.e. an element of $G_3(L)$.

20.11. Proposition. *The set $S_3(L)$ of transformations is a group.*

PROOF. It follows from the normality of the translation group T_3 in G_3 and the uniqueness of the expression $T\alpha$ for each element g of G_3, that the projection map $\pi: G_3 \to O(3)$ defined by $\pi: g = T\alpha \to \alpha$, is a homomorphism. Hence $\pi G_3(L)$, the image under π of the subgroup $G_3(L)$ of G_3, is a subgroup of $O(3)$. But from their definitions it is clear that the image is just the set $S_3(L)$. Hence $S_3(L)$ is indeed a group, as required.

Remark. By the "homomorphism theorems" of group theory, the group $\pi G_3(L)$ is isomorphic to the quotient group $G_3(L)/(G_3(L) \cap \operatorname{Ker} \pi)$, where $\operatorname{Ker} \pi$ denotes the kernel of the homomorphism π (i.e. the subgroup of G_3 consisting of those elements sent under π to the identity). Since clearly $\operatorname{Ker} \pi = T_3$, and since $T_3(L) = G_3(L) \cap T_3$, we have

$$S_3(L) \simeq \pi G_3(L) \simeq G_3(L)/G_3(L) \cap T_3 = G_3(L)/T_3(L),$$

i.e. the point group of the crystal is isomorphic to the quotient $G_3(L)/T_3(L)$.

Proposition 20.11 prompts the following question (whose answer is sometimes useful): Given two elements α_1, α_2 of the point group of a crystal lattice L, and translations T_1, T_2 guaranteed by the definition of the point group, i.e. satisfying $T_1\alpha_1, T_2\alpha_2 \in G_3(L)$, what explicit translation T will serve for the composite $\alpha = \alpha_1\alpha_2$, i.e. will have the property that $T\alpha \in G_3(L)$? The answer is not difficult to find: since $T_1\alpha_1$ and $T_2\alpha_2$ belong to $G_3(L)$, so does $T_1\alpha_1 T_2\alpha_2 = T_1\alpha_1 T_2\alpha_1^{-1}\cdot\alpha_1\alpha_2$, so that for T we may take (among other possible candidates) the translation $T_1\alpha_1 T_2\alpha_1^{-1}$. In more explicit terms, if the orthogonal transformations α_1, α_2 have matrices A_1, A_2 respectively, and if T_1, T_2 are translations along vectors b_1, b_2 respectively, then for any vector r in $\mathbb{R}^3$, we have

$$(T_1\alpha_1)(r) = A_1 r + b_1, \qquad (T_2\alpha_2)(r) = A_2 r + b_2$$

(where it is clear that in applying $T_i\alpha_i$ to r, the orthogonal transformation has been applied first, followed by the translation). It is easy to calculate that then

$$(T_1\alpha_1 T_2\alpha_1^{-1})(r) = r + (A_1 b_2 + b_1),$$

so that for T we may take the translation along the vector $A_1 b_2 + b_1$.

We now give the promised example of a 2-dimensional crystal lattice L and an element $g \in G_2(L)$ whose (unique) expression in the form $T\alpha$ has the property that both $T, \alpha \notin G_2(L)$. The lattice in question is depicted in Figure 24, where the primitive vectors α_1, α_2 are indicated. It is clear that neither the reflection α ($\in O(2)$) in the straight line l, nor the translation T along the vector β (which is half the primitive vector α_1) preserves the lattice, i.e. neither α nor T belongs to $G_2(L)$. On the other hand it is easy to see that the composite transformation $T\alpha$ (a glide-reflection) does preserve the lattice. Thus in particular α is in $S_3(L)$, the point group of L, even though it does not preserve L, a situation which serves to underline the fact that, generally speaking, the point group is not a subgroup of the group of motions of a crystal.

The point group is of first importance in the theory of crystalline structures. It is not for nothing that it bears the alternative name of "symmetry group" of the lattice, since as well as containing the "genuine" symmetries of the lattice, it includes those motions of Euclidean space which preserve

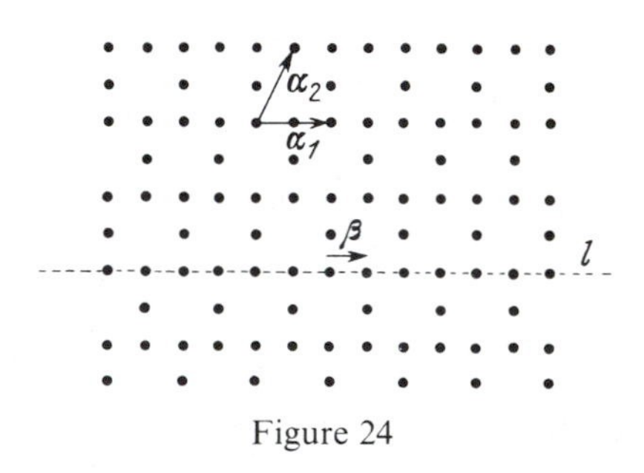

Figure 24

the lattice only after the application of translations (also not lattice-preserving).

As we have seen, the lattice of Figure 24 has a glide reflection as one of its "basic" symmetries. In three dimensions, one frequently meets also with crystals preserved under screw-displacements. (Recall from §4.3 that a screw-displacement is a rotation followed by a translation along the axis of rotation.) We recommend to the reader that he construct an example of a 3-dimensional lattice having a screw-displacement as one of its "basic" symmetries.

In crystallography, the following natural group of transformations of a lattice, consisting of the above-mentioned "genuine" symmetries, is also commonly considered.

20.12. Definition. The *stationary group* $H_3(L)$ of a lattice L is the subgroup of $G_3(L)$ consisting of those motions of the lattice fixing the origin O. In other words $H_3(L) = G_3(L) \cap O(3)$.

Thus $H_3(L)$ is a subgroup of $S_3(L)$ (since in the unique expression of $\alpha \in H_3(L)$ as a translation T following an element of $O(3)$, we shall have $T = 1$, the identity transformation). (As the lattice of Figure 24 shows, $H_3(L)$ will in general be a proper subgroup, i.e. will not be equal to $S_3(L)$. However it is easy to see that if the lattice is Bravais then $H_3(L) = S_3(L)$.) We therefore obtain immediately the following

20.13. Proposition. *The stationary group $H_3(L)$ is the intersection of the space group $G_3(L)$ with the point group $S_3(L)$, i.e. $H_3(L) = S_3(L) \cap G_3(L)$. (The groups involved are all regarded as subgroups of G_3.)*

By no means all subgroups of $O(3)$ can occur as point groups of lattices. It turns out that the requirement that the lattices be translation-invariant greatly narrows the field of possible candidates for the groups $G_3(L)$, $S_3(L)$ and $H_3(L)$. This is borne out (in the case of $H_3(L)$ at least) by the following theorem, which is fundamental in the theory of crystal lattices. Before stating the theorem, we introduce the notation $H_3(L)_{(0)}$ for the group consisting of the proper motions in $H_3(L)$, i.e. we define $H_3(L)_{(0)} = SO(3) \cap G_3(L)$.

20.14. Theorem. *If L is a translation-invariant lattice, then the group $H_3(L)_{(0)}$ is finite, and each of its (finitely many) elements is a rotation about an axis through O, through an angle which is an integer multiple of $\pi/3$ or of $\pi/2$.*

PROOF. (i) We consider first the case that L is a Bravais lattice, i.e. that every atom of L is a translate of O by means of an integral linear combination of the primitive translations $\alpha_1, \alpha_2, \alpha_3$. By §4.3 a transformation $\Phi \in H_3(L)_{(0)}$, being a proper orthogonal transformation, is a rotation through some angle φ say, about an axis l passing through the point O (which we are assuming,

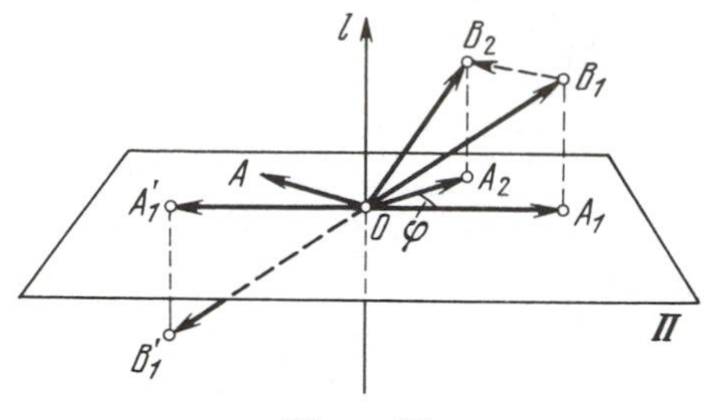

Figure 25

as always, to be an atom of the lattice). Denote by Π the plane through O perpendicular to the straight line l. Our lattice L, being Bravais, consists precisely of the points with position vectors of the form $\alpha = n_1\alpha_1 + n_2\alpha_2 + n_3\alpha_3$. We now consider the projections parallel to the axis l of all of these points onto the plane Π, and choose one point A_1 say, from among those of the projected points closest to, but different from, the point O (see Figure 25). Since the lattice $L$is symmetric about O, together with each of its points B_1 it will have as an atom the diametrically opposite point B_1', the reflection of B_1 in the point O. Upon rotating the space through the angle φ about the axis l, the point B_1 will go into a point B_2 of the lattice, and the line segment OA_1 will swing through the angle φ to take up the position of the segment OA_2. (See Figure 25, where B_1 is taken to be the atom of which A_1 is the projection.) Since the vectors OB_1 and OB_2 correspond to permitted translations of the lattice, so will their difference B_1B_2 correspond to a permitted translation. Since the latter vector is parallel to the plane Π, on translating along this vector, the point O will, as in Figure 25, go into a point A in the plane Π, and this point A will again be an atom of the lattice. Since A_1 and A_2 are closest (among all projections of atoms of the lattice) to O, the length of OA cannot be less than the length of OA_1 (which is the same as that of OA_2). Since $|OA| = |B_1B_2| = |A_1A_2|$, it follows that the base of the isosceles triangle A_1OA_2 is at least as long as the sides OA_1 and OA_2 of equal length, whence $\varphi \geq \pi/3$, i.e. any lattice-preserving rotation must rotate the lattice through at least $\pi/3$. It follows that if the angle of rotation is less than π, then it must be $2\pi/3$, $\pi/3$ or $\pi/2$, since otherwise some power α^n of the rotation α would correspond to a rotation through an angle less than $\pi/3$. Hence the only possible angles of rotation are integer multiples of $\pi/2$ or $\pi/3$. This completes the proof for the special case of a Bravais lattice.

(ii) Having established the theorem for Bravais lattices we might now prove it in general by showing that every translation-invariant lattice contains as a sublattice a Bravais lattice preserved by any particular rotation Φ. We shall however not follow this avenue of proof, preferring a different method which illustrates another idea of considerable importance in the study of crystal lattices.

We first consider the case of a planar lattice, with primitive translation vectors α_1, α_2. As before let Φ denote a lattice-preserving rotation of the

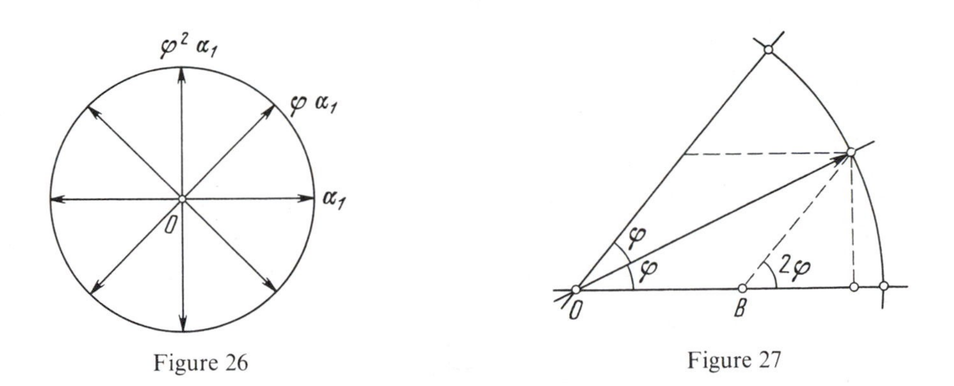

Figure 26

Figure 27

plane about O through an angle φ. To the vector α_1 we apply successive powers of Φ, obtaining $\alpha_1, \Phi\alpha_1, \Phi^2\alpha_1, \ldots$. Since a primitive cell contains only a finite number of atoms, this sequence of images of α_1 under successive rotations through the angle φ must eventually repeat, i.e. $\Phi^n\alpha_1 = \alpha_1$ for some n, or, what amounts to the same thing, $\varphi = 2m\pi/n$ for some integral m, n (see Figure 26). It is easy to check that if τ_1 denotes the translation along the vector α_1, then the translations along the vectors $\Phi\alpha_1, \Phi^2\alpha_1, \ldots$, are given respectively by $\Phi\tau_1\Phi^{-1}, \Phi^2\tau_1\Phi^{-2}, \ldots$, which clearly all lie in the translation subgroup of the lattice. It follows that the vectors $\alpha_1, \Phi\alpha_1, \Phi^2\alpha_1, \ldots$, must all be integral linear combinations of the primitive vectors α_1 and α_2. If α_1 and $\Phi^2\alpha_1$ are linearly dependent (over the reals) then since Φ merely rotates α_1 through the angle φ, this angle must be an integer multiple of $\pi/2$. Thus we may assume from now on that α_1 and $\Phi^2\alpha_1$ are linearly independent over the reals. Since in this case these two vectors span the whole of 2-dimensional space, in particular $\Phi\alpha_1$ will be a linear combination of them (over the reals). However since all three of the vectors $\alpha_1, \Phi\alpha_1, \Phi^2\alpha_1$ are integral linear combinations of α_1 and α_2, it follows that in fact $\Phi\alpha_1$ is a rational linear combination of α_1 and $\Phi^2\alpha_1$. (Alternatively this follows from the fact that if either of the coefficients in the expression of $\Phi\alpha_1$ as a linear combination of α_1 and $\Phi^2\alpha_1$, is irrational, then by repeated translations of the lattice along the vector $\Phi\alpha_1$, we should obtain infinitely many atoms in a primitive cell, contradicting the assumed discreteness of the space group of a crystal lattice.)

The fact, established above, that the vector $\Phi\alpha_1$ is a rational linear combination of α_1 and $\Phi^2\alpha_1$, implies that $\cos\varphi$ is rational. To see this we express $\Phi\alpha_1$ explicitly in terms of α_1 and $\Phi^2\alpha_1$; thus, referring to Figure 27, we see that

$$\Phi\alpha_1 = \frac{OB}{|\alpha_1|}(\alpha_1 + \Phi^2\alpha_1),$$

so that $OB/|\alpha_1|$ is rational. However from Figure 27 it is clear that

$$OB = |\alpha_1|\cos\varphi - OB\cos 2\varphi,$$

whence

$$\frac{OB}{|\alpha_1|} = \frac{1}{2\cos\varphi}.$$

Hence $\cos\varphi$ is rational, so that φ is restricted to being an integer multiple of $\pi/2$ or $\pi/3$.

(iii) Finally we drop all restrictions, and consider a general 3-dimensional translation-invariant lattice. As before we let Φ be an arbitrary lattice-preserving rotation through an angle φ, about an axis l, and denote by Π the plane through O orthogonal to l. As in (ii) above we concentrate our attention on (any) one of the three primitive vectors, say α_1. If this vector lies in the plane Π then the argument given in (ii) applies without change. If on the other hand α_1 does not lie in Π, then application of Φ moves it along the cone with apex at O and axis l. If we then project this cone (together with the various vectors $\Phi^n\alpha_1$ lying on it) onto the plane Π, then the argument in (ii) can be applied to the projected vectors to give the desired conclusion. This completes the proof of the theorem. □

Theorem 20.14, which of course holds also for planar lattices, allows us to give immediately a complete list of all the possibilities for the group $H_2(L)_{(0)}$, where $L$is any planar, translation-invariant lattice.

20.15. Classification Theorem for the Groups $\mathbf{H_2(L)_{(0)}}$. *For $n = 1, 2, 3, 4, 6$, let C_n denote the group consisting of the n elements*

$$\begin{pmatrix} \cos\dfrac{2\pi k}{n} & \sin\dfrac{2\pi k}{n} \\ -\sin\dfrac{2\pi k}{n} & \cos\dfrac{2\pi k}{n} \end{pmatrix}, \qquad 0 \le k \le n-1,$$

i.e. C_n is the cyclic group generated by the rotation of the plane about O through the angle $2\pi/n$. Then for any planar, translation-invariant lattice L, the group $H_2(L)_{(0)}$ is one of the C_n, $n = 1, 2, 3, 4, 6$.

Thus if a group of rotations preserves a plane lattice, then it must be one of these five groups. (Note that C_1 is the trivial group.)

We may easily derive from this list a complete list of all possibilities for $H_2(L)$, i.e. for the full group of othogonal transformations preserving a plane lattice L, by allowing for the possibility that some origin-fixing reflection preserves L, i.e. by adjoining to $H_2(L)_{(0)}$ such a reflection (if there is one), together with all of its composites with elements of $H_2(L)_{(0)}$. In this way we obtain the following complete list of possible stationary groups $H_2(L)$:

$$C_1, C_2, C_3, C_4, C_6; D_1, D_2, D_3, D_4, D_6.$$

Here D_i is obtained from C_i by forming the semi-direct product of C_i with the cyclic group of order 2 generated by the lattice-preserving reflection; in each case the effect of conjugation of the elements of C_i by the reflection is to invert them. Thus $D_i/C_i \simeq \mathbb{Z}_2$, the cyclic group of order 2.; in particular it follows that $D_1 \simeq C_2$. (The D_i are called "dihedral" groups.)

It is not difficult to construct for each of these 10 groups a plane lattice having that group as its stationary group. We leave this to the reader as an exercise.

It is noteworthy that C_5 does not appear in the above list of possible orthogonal symmetry groups of a plane lattice. Since plane translation-invariant lattices correspond closely to the possible ornamental repeating patterns, they have been for centuries, in one guise or another, the objects of study by specialists in ornamental design. In Arabic artwork, in particular, one can observe the results of attempts to make a repeating pattern based on the number 5 (i.e. with rotation group isomorphic to C_5). Though in the strictest sense these attempts proved vain, they resulted in so-called "compromise variants," based on the number 5, but inevitably having their symmetry flawed in places.

Having completely classified the stationary groups $H_2(L)$ of a plane translation-invariant lattice, it is natural to ask next what the possibilities are for the full group $G_2(L)$ of motions of such a lattice. It turns out that such a group $G_2(L)$ is isomorphic to one of 17 (pairwise non-isomorphic) possible groups (corresponding to each of which there is a lattice having that group as its space group). It is interesting to note that all of the corresponding 17 types of ornamentation occur in antique decorative patterns (chiefly Egyptian).

We now return to 3-dimensional lattices. As might be expected, the problem of classifying the groups $H_3(L)$ and $G_3(L)$ is markedly more complex than in the planar case. We shall therefore not carry out this classification in full detail, but shall confine ourselves to drawing up a complete list of all the possible finite subgroups of rotations of Euclidean 3-space, i.e. the finite subgroups of $SO(3)$. Since the stationary group (as also the symmetry group $S_3(L)$) of an arbitrary translation-invariant 3-dimensional lattice is discrete, and therefore, as is not too difficult to see, finite, it follows that once having obtained a complete list of the finite subgroups of $SO(3)$, we shall have at least significantly narrowed the field of candidates for the groups $H_3(L)_{(0)}$ and $S_3(L)_{(0)}$.

Before actually listing the finite subgroups of $SO(3)$ and then proving that there are no omissions, we need to describe the finite subgroups in question. To this end let l be some straight line through the origin O of $\mathbb{R}^3$, and denote by Π the plane through O orthogonal to l. For each $n = 1, 2, 3, \ldots$, there is an obvious (and unique) rotation of 3-space about l as axis, which induces on the plane Π the rotation about O (in one direction or the other) through the angle $2\pi/n$. In keeping with our previous notation, we denote by C_n the cyclic subgroup generated by this rotation of the plane Π, and then by C_n'

the cyclic subgroup generated by the corresponding rotation of 3-space about l; thus C_n' induces the group C_n on Π. (In particular C_1' consists of just the identity transformation.) As well as the C_n, the groups D_n (see above) act on the plane Π. Recall that D_n is obtained from C_n by adjoining to it a reflection of Π in some straight line q lying in Π and passing through O, together with the composites of that reflection with the elements of C_n. (As noted before, D_n is a semi-direct product of C_n by a 2-cycle with an inverting action.) Such a reflection of the plane Π is induced by the rotation of 3-space through the angle π about q as axis of rotation. Thus the improper orthogonal transformations of the plane Π are induced by proper orthogonal transformations, i.e. by rotations, of 3-dimensional space. We denote by D_n' the resulting group of rotations of 3-space inducing the group D_n on Π. It is easy to see that the group D_n' consists of the following rotations: those in C_n, and a further n rotations through the angle π with their n respective axes of rotation lying in the plane Π and passing through O, each such axis making an angle of $2\pi/2n = \pi/n$ with its immediate neighbours. In particular D_1' has just two elements (namely the identity transformation, and a rotation of space about a straight line lying in Π), and is therefore isomorphic to C_2. Hence if we wish our list to have on it just one group from each isomorphism class of finite subgroups of $SO(3)$, then we should exclude the duplicate D_1'. Thus so far our (incomplete) list contains the following groups: C_n, $n = 1, 2, 3, \ldots$; D_n', $n = 2, 3, 4, \ldots$.

Apart from these two infinite sequences of subgroups, there are a few rather more exotic discrete subgroups of $SO(3)$. These arise in connexion with the five regular (or "platonic") polyhedra (the tetrahedron, cube, octahedron, dodecahedron and icosahedron). With each of these polyhedra (imagined with its centre at O) there is associated the finite group of motions of 3-space sending the polyhedron to itself (i.e. the symmetry group of the polyhedron). Among the five finite groups obtained in this way there are however isomorphic ones; in fact only three of them are distinct (in the sense of being non-isomorphic), since the symmetry groups of the cube and octahedron are isomorphic, as are those of the dodecahedron and icosahedron. This is due to the fact that the cube and octahedron on the one hand, and the dodecahedron and icosahedron on the other, are "dual" pairs of polyhedra. To see what this means, consider a sphere inscribed in a cube, and inside the sphere an octahedron whose vertices touch the sphere at those points where the sphere touches the cube, i.e. at the centres of the faces of the cube (see Figure 28). It is clear that with these two polyhedra in this relative position, any motion which moves the cube into itself is also a symmetry of the octahedron, and conversely; hence the symmetry groups of the cube and octahedron coincide when they are as in Figure 28, and they will therefore be in general isomorphic. Entirely similar considerations show that the icosahedral and dodecahedral groups are isomorphic.

We denote by T, W, P the respective groups of proper (i.e. direct) symmetries of the tetrahedron, cube (and octahedron), dodecahedron (and icosa-

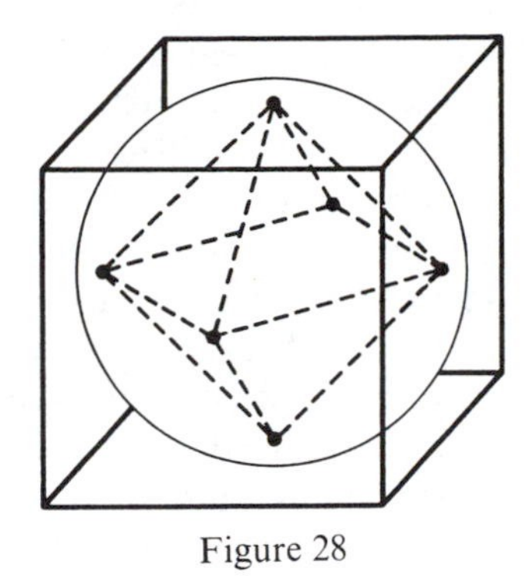

Figure 28

hedron). We leave it to the reader to verify that the orders of these three groups are 12, 24, 60 respectively, and that they are distinct from the groups C'_n and D'_n. By adjoining to each of the groups T, W, P a reflection preserving the corresponding polyhedron, together with the composites of that reflection with the elements of the group (i.e. by including the improper symmetries), we obtain the full symmetry groups $\tilde{T}, \tilde{W}, \tilde{P}$ of the polyhedra, having orders 24, 48, 120 respectively.

It turns out that the above finite groups of rotations of 3-space exhaust the possibilities.

20.16. Theorem. *Every finite subgroup of the rotation group $SO(3)$ of 3-dimensional Euclidean space is isomorphic to one of the following groups:*

$$C_n \quad (n = 1, 2, 3, \ldots),$$

$$D'_n \quad (n = 2, 3, \ldots),$$

$$T, W, P;$$

where C_n is the (cyclic) group consisting of the n rotations through the angles $2k\pi/n$, $k = 0, 1, \ldots, n-1$ about a fixed axis l passing through O; D'_n consists of the rotations in C'_n, together with n rotations through π about n axes passing through O, perpendicular to l, with a uniform angular spacing of π/n; and T, W, P are respectively the groups of rotations preserving the regular tetrahedron, cube (or octahedron), and dodecahedron (or icosahedron).

Proof. We saw in §4.3 that every transformation in $SO(3)$ is a rotation of 3-space about some axis through the origin O. We associate with each rotation of 3-space the pair of diametrically opposite points of the unit sphere in which the axis of rotation meets that sphere, calling the pair of points of the unit sphere associated with a given rotation in this way, the *poles* of the rotation.

Suppose that Γ is a finite group consisting of N rotations of 3-space. Consider the poles of the $N - 1$ non-identity transformations in Γ. It is easy to see that Γ acts (faithfully) as a permutation group on this set of poles.

We call the number ν of transformations in Γ which fix a pole P the *multiplicity* (relative to Γ) of that pole. (In other words ν is the order of the stabilizer in Γ of the point P.) It is clear that the stabilizer of P is just the cyclic group of order ν generated by the rotation through the angle $2\pi/\nu$ about the axis through O and P; we denote this subgroup of Γ by Γ_P. We shall show that the orbit of the point P under the group Γ has size N/ν. (This is a particular case of a general result in the theory of permutation groups, i.e. the statement holds for any permutation group Γ whatever, as indeed is shown by the general nature of the proof which follows.) Thus with each point Q of the orbit in question (i.e. the orbit containing P) we associate the set of all motions in Γ which map P to Q. It is easy to see that this set is simply the coset $g\Gamma_P$ where g is any motion in Γ sending P to Q, and that this association of orbit elements with left cosets of Γ_P in Γ is one-to-one and onto. Since the cosets $g\Gamma_P$ are pairwise disjoint, and they all have $|\Gamma_P|$ elements, we have the desired conclusion. (We have here also indicated the proof of Lagrange's theorem, according to which the order of a subgroup of a finite group always divides the order of the whole group.) We conclude that for any pole P of the finite group Γ, if ν_P is the multiplicity of the pole and n_P the size of the pole's orbit, then $|\Gamma| = N = n_P \nu_P$.

We now calculate in two different ways the number of pairs (S, P) where P is a pole and S is a non-identity transformation in Γ, fixing P. In view of the fact that each of the $N - 1$ non-identity motions in Γ fixes just two poles, the number of such pairs is $2(N - 1)$. On the other hand since the stabilizer of a pole P has order ν_P, there are exactly $\nu_P - 1$ non-identity motions in Γ fixing P; hence the number of pairs (S, P) is also given by $\sum (\nu_P - 1)$, where the summation is over the set of all poles P. Since $\nu_Q = \nu_P$ for all poles Q in the same orbit as P (the stabilizer of Q being simply the conjugate of the stabilizer of P by any motion g in Γ which sends P to Q), after numbering the orbits (from 1 to however many there are), we obtain the equation

$$2(N - 1) = \sum_i (\nu_i - 1)n_i, \tag{1}$$

where n_i is the size of the ith orbit, and ν_i is the multiplicity of an element in the ith orbit. Since, as we have already shown, $N = n_i \nu_i$ for all i, on dividing (1) by N, we obtain

$$2 - \frac{2}{N} = \sum_i \left(1 - \frac{1}{\nu_i}\right). \tag{2}$$

We shall now see that (2) places severe restrictions on the possible values for the ν_i, N, and the number of orbits. Since the trivial group is on our list, we may assume that Γ is non-trivial, i.e. that the left-hand side of (2) lies between 1 and 2 (and cannot equal 2), so that on the right-hand side there must clearly be no fewer than 2 orbits. On the other hand since by definition each pole is stabilized by at least one non-trivial transformation in Γ, i.e. since $\nu_i > 1$, we cannot have more than 3 summands on the right-hand side of (2). Thus there are either 2 or 3 orbits.

In the case that there are 2 orbits equation (2) takes the form

$$2 = \frac{N}{v_1} + \frac{N}{v_2} = n_1 + n_2.$$

This forces $n_1 = n_2 = 1$, which means that there are but two poles each of which is stabilized by Γ. Hence Γ is a cyclic group of order N consisting of rotations about the axis through these poles, i.e. Γ is isomorphic to C'_N.

If there are 3 orbits, equation (2) can be rewritten as

$$\frac{1}{v_1} + \frac{1}{v_2} + \frac{1}{v_3} = 1 + \frac{2}{N}. \tag{3}$$

We may assume (by renumbering the orbits if necessary) that $v_1 \leq v_2 \leq v_3$. Not all of v_1, v_2, v_3 can exceed 2, since if they did then the left-hand side of (3) would be less than or equal to $\frac{1}{3} + \frac{1}{3} + \frac{1}{3} = 1$, whereas the right-hand side is greater than 1. Hence $v_1 = 2$, and from (3) we obtain

$$\frac{1}{v_2} + \frac{1}{v_3} = \frac{1}{2} + \frac{2}{N}. \tag{4}$$

Again we cannot have both $v_2, v_3 \geqq 4$ since the left-hand side of (4) would otherwise be at most $\frac{1}{2}$. Hence $v_2 = 2$ or 3.

Consider first the case $v_1 = v_2 = 2$, $N = 2v_3$. Put $n = v_3$. In this case we have 2 orbits of size n, the stabilizers of whose elements have order 2, and a further orbit of size 2 with stabilizer of order n. The latter stabilizer is clearly isomorphic to C'_n, and it is easy to see (from the definition of the dihedral groups given above) that in fact Γ in this case is isomorphic to D'_n.

Suppose next that $v_1 = 2$, $v_2 = 3$, in which case $1/v_3 = \frac{1}{6} + 2/N$. Since $v_3 \geq v_2 = 3$, we have only three possibilities: $v_3 = 3$, $N = 12$; $v_3 = 4$, $N = 24$; $v_3 = 5$, $N = 60$. We shall now see that corresponding to these three cases we have $\Gamma \simeq T, W, P$ respectively.

In the case $v_3 = 3$, $N = 12$, two of the orbits each have 4 poles all of multiplicity 3. It is clear that the poles in either one of these orbits are the vertices of a regular tetrahedron, with those in the other orbit diametrically opposed to them. Thus in this case Γ is the group of proper motions preserving a tetrahedron, i.e. $\Gamma \simeq T$. (The 6 poles of multiplicity 2 comprising the remaining orbit will be the projections onto the unit sphere of the midpoints of the 6 edges of the tetrahedron.)

In the case $v_3 = 4$, $N = 24$, the third orbit contains 6 poles of multiplicity 4, which are clearly the vertices of a regular octahedron inscribed in the unit sphere, so that Γ is the group of proper motions preserving that octahedron, i.e. $\Gamma \simeq W$. (The 8 poles of multiplicity 3 of the second orbit are the projections of the centres of the faces of this octahedron onto the unit sphere, while the 12 poles of multiplicity 2 of the first orbit are the projections onto the unit sphere of the midpoints of the 12 edges of the octahedron.)

In the final case $v_3 = 5$, $N = 60$, the 12 poles, each of multiplicity 5, of the third orbit are the vertices of a regular icosahedron inscribed in the unit sphere. The first and second orbits contain respectively 30 poles of order 2 (the projections of the midpoints of the 30 edges of the icosahedron), and 20 poles of order 3 (the projections of the centres of the 20 faces of the icosahedron).

Since we have enumerated all possibilities, the theorem is proved. □

This theorem can be used for compiling a complete list of all the possible stationary groups of 3-dimensional crystal lattices. As a preliminary to giving this list (without the details of the proof) we need to define the new groups that will appear in it.

To begin with let B denote the reflection of 3-space in the origin O, i.e. the transformation sending each point X to $-X$. Since the matrix of B is just the negative of the identity matrix, it follows both that B is an improper orthogonal transformation, and that it commutes with every element of $O(3)$; in fact it is not difficult to show that B generates the centre of $O(3)$ (essentially because it is the only non-identity scalar matrix in $O(3)$). Hence if H is any subgroup of $SO(3)$, then the subgroup $\tilde{H}$ of $O(3)$, generated by H together with the element B, is the direct product of H with the cyclic group of order 2 generated by B; thus H has index 2 in $\tilde{H}$.

We next describe another way in which finite subgroups of $O(3)$ are built up from certain subgroups of $SO(3)$ by the adjunction of plane reflections. Let Γ be a subgroup of index 2 in a subgroup Φ of $SO(3)$. Denote the coset $\Phi\backslash\Gamma$ by S; thus S is just the set of elements of Φ outside Γ. if we replace in Φ the subset S by the set BS, i.e. if we form the union $\Gamma \cup BS$, it is easy to see that we obtain a group different from, but isomorphic to, the group Φ. (We are making heavy use here of the centrality of B.) This new group, which we denote by $\Phi\Gamma$, contains Γ, and has its other half made up of plane reflections. For example we can form a group WT since the group T of proper motions of the tetrahedron occurs as a subgroup of index 2 in the group W of proper motions of the octahedron.

20.17. Theorem. *The following is a complete list of all possible stationary groups of 3-dimensional, translation-invariant lattices:* $C'_1, C'_2, C'_3, C'_4, C'_6$; $\tilde{C}'_1, \tilde{C}'_2, \tilde{C}'_3, \tilde{C}'_4, \tilde{C}'_6$; D'_2, D'_3, D'_4, D'_6; $\tilde{D}'_2, \tilde{D}'_3, \tilde{D}'_4, \tilde{D}'_6$; $C'_2C'_1, C'_4C'_2, C'_6C'_3$; $D'_4D'_2, D'_6D'_3$; $D'_2C'_2, D'_3C'_3, D'_4C'_4, D'_6C'_6$; $T, W, \tilde{T}, \tilde{W}, WT$. *Each of these groups actually arises as the stationary group of some translation-invariant, 3-dimensional lattice. (Note that the 32 groups in the list are not all isomorphically distinct.)*

The proof of this theorem involves combining Theorems 20.16, which gives (up to isomorphism) a complete list of the finite subgroups of $SO(3)$, with the above-described procedure for "incorporating" plane reflections with groups of rotations, and by exploiting the fact that poles can have only the multiplicities 2, 3, 4 or 6. We omit the details.

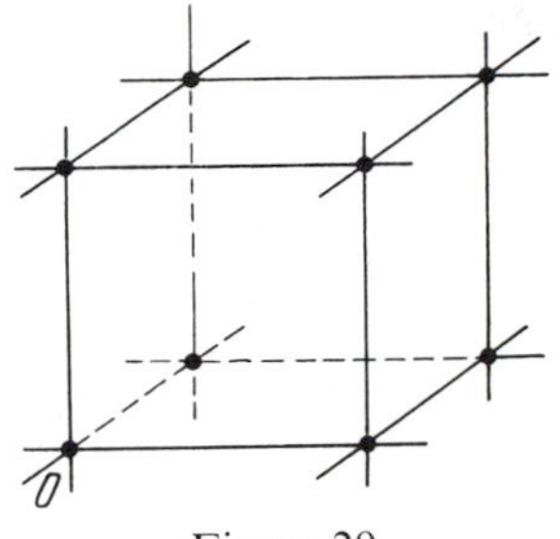

Figure 29

We conclude this section by considering crystallographic groups in connexion with concrete tensor fields. We shall limit ourselves to one simple example of a macroscopic property possessed by crystals, namely their electrical conductivity, which determines the precise relationship between the strength ε of the ambient electric field and the current density vector j in the crystals. This relationship is given by the empirical equation

$$j_k = \sigma_k^s \varepsilon_s,$$

where $j = (j_k)$, $\varepsilon = (\varepsilon_s)$, and (σ_k^s) is the electrical conductivity tensor of the medium. If the medium is isotropic then $\sigma_k^s = \sigma \delta_k^s$, i.e. in this simplest of cases the electrical conductivity is determined by the single scalar σ. In the general case (σ_k^s) will be more complicated.

In our present context we wish to consider as medium a cubic crystal, represented by the cubic lattice in $\mathbb{R}^3$ depicted in Figure 29. We shall assume that the crystal is perfect and fills the whole of 3-space. It is clear that the symmetry group of this crystal contains the following three orthogonal transformations:

$$\beta_1 = \begin{pmatrix} 0 & 1 & 0 \\ -1 & 0 & 0 \\ 0 & 0 & 1 \end{pmatrix}, \qquad \beta_2 = \begin{pmatrix} 0 & 0 & 1 \\ 0 & 1 & 0 \\ -1 & 0 & 0 \end{pmatrix},$$

$$\beta_3 = \begin{pmatrix} 1 & 0 & 0 \\ 0 & 0 & 1 \\ 0 & -1 & 0 \end{pmatrix};$$

these are rotations through $\pi/2$ about the z-axis, y-axis and x-axis respectively.

Since these three rotations preserve the lattice, they must also preserve the electrical conductivity tensor (σ_k^s) of the crystal. If we denote by A the matrix (σ_k^s) then under the transformation β_i, A goes into $A_i = \beta_i A \beta_i^{-1}$ which, since the tensor is preserved, is again equal to A. If we calculate A_1 explicitly, we obtain

$$A_1 = \begin{pmatrix} \sigma_2^2 & -\sigma_1^2 & \sigma_3^2 \\ -\sigma_2^1 & \sigma_1^1 & -\sigma_3^1 \\ \sigma_2^3 & -\sigma_1^3 & \sigma_3^3 \end{pmatrix} = A = \begin{pmatrix} \sigma_1^1 & \sigma_2^1 & \sigma_3^1 \\ \sigma_1^2 & \sigma_2^2 & \sigma_3^2 \\ \sigma_1^3 & \sigma_2^3 & \sigma_3^3 \end{pmatrix}.$$

Hence $\sigma_1^1 = \sigma_2^2$. Explicit calculation of A_2 and A_3 yields similarly $\sigma_2^2 = \sigma_3^3$. The three equations $A = A_1 = A_2 = A_3$ also give that $\sigma_j^i = 0$ for $i \neq j$. Alternatively this follows by calculating explicitly the matrices $\beta_i^2 A \beta_i^{-2}$ and setting them equal to A; for instance

$$\beta_1^2 A \beta_1^{-2} = \begin{pmatrix} \sigma_1^1 & \sigma_2^1 & -\sigma_3^1 \\ \sigma_1^2 & \sigma_2^2 & -\sigma_3^2 \\ -\sigma_1^3 & -\sigma_2^3 & \sigma_3^3 \end{pmatrix} = A = \begin{pmatrix} \sigma_1^1 & \sigma_2^1 & \sigma_3^1 \\ \sigma_1^2 & \sigma_2^2 & \sigma_3^2 \\ \sigma_1^3 & \sigma_2^3 & \sigma_3^3 \end{pmatrix}$$

gives $\sigma_3^1 = \sigma_3^2 = \sigma_1^3 = \sigma_2^3 = 0$. We conclude that

$$A = (\sigma_k^s) = \sigma \begin{pmatrix} 1 & 0 & 0 \\ 0 & 1 & 0 \\ 0 & 0 & 1 \end{pmatrix},$$

i.e. $\sigma_k^s = \sigma \delta_k^s$, where σ is a scalar. We have thus shown that the electrical conductivity of a cubic crystal is like that of an isotropic medium, i.e. is independent of direction. It is not physically obvious that this should be the case, since one might expect that for instance the conductivity of a cubic crystal in directions parallel to the edges would differ from that in the diagonal direction. Thus we have used our knowledge of the symmetry group of the crystal in a significant way to establish the number of independent components of the tensor (σ_k^s).

§21. Rank 2 Tensors in Pseudo-Euclidean Space, and Their Eigenvalues

21.1. Skew-Symmetric Tensors. The Invariants of an Electromagnetic Field

Of great importance in physics are the skew-symmetric rank 2 tensors defined on Minkowski space. In particular the electromagnetic field F_{ik} is a tensor of this kind. (We shall in this physical context adopt the concomitant terminology, calling tensors "fields," etc.)

21.1.1. Definition. The *invariants* of a field F_{ik} (defined on $\mathbb{R}_{1,3}^4$) are the coefficients of the characteristic polynomial

$$P(\lambda) = \det(F_{ik} - \lambda g_{ik}), \tag{1}$$

where g_{ik} is the Minkowski metric.

Given any skew-symmetric rank 2 tensor F_{ik}, we define the (*electric* and *magnetic*) *vector fields* **E** and **H** by

$$E_\alpha = F_{0\alpha}, \qquad \alpha = 1, 2, 3;$$
$$H^1 = F_{23}, \quad H^2 = F_{31}, \quad H^3 = F_{12}, \tag{2}$$

where $\mathbf{E} = (E_1, E_2, E_3)$, $\mathbf{H} = (H^1, H^2, H^3)$ (at each point of 3-space). Hence in terms of these the matrix (F_{ik}) takes the form

$$(F_{ik}) = \begin{pmatrix} 0 & E_1 & E_2 & E_3 \\ -E_1 & 0 & H^3 & -H^2 \\ -E_2 & -H^3 & 0 & H^1 \\ -E_3 & H^2 & -H^1 & 0 \end{pmatrix}. \tag{3}$$

21.1.2. Lemma. *The characteristic polynomial of a skew-symmetric, rank 2 tensor F_{ik} has the form*

$$P(\lambda) = -\lambda^4 + (|\mathbf{E}|^2 - |\mathbf{H}|^2)\lambda^2 + \langle \mathbf{E}, \mathbf{H}\rangle^2, \tag{4}$$

where $|\mathbf{E}|^2 = E_1^2 + E_2^2 + E_3^2, |\mathbf{H}|^2 = (H^1)^2 + (H^2)^2 + (H^3)^2$ and $\langle \mathbf{E}, \mathbf{H}\rangle = E_i H^i$.

This can be proved by simply expanding the right-hand side of (1), and recalling that in terms of Minkowski co-ordinates for $\mathbb{R}^4_{1,3}$ (which we are assuming), the metric g_{ij} is given by: $g_{00} = 1$, $g_{11} = g_{22} = g_{33} = -1$, $g_{ij} = 0$ for $i \neq j$ (see §3.2). Alternatively, we may appeal to Theorem 21.1.5 below, which gives the canonical form of a rank 2 skew-symmetric tensor in pseudo-Euclidean space.

Remark. There is a theorem of linear algebra to the effect that a skew-symmetric matrix can be brought into a standard block form by a suitable orthogonal co-ordinate change; however in the present context the Lorenz transformations are the pertinent ones, and they will not in general serve for bringing the matrix into this particular canonical form.

In the language of differential forms (see §18.1), (2) becomes

$$\begin{aligned} F &= \sum_{i<j} F_{ij}\, dx^i \wedge dx^j \\ &= E_\alpha\, dx^0 \wedge dx^\alpha + H^1\, dx^2 \wedge dx^3 + H^2\, dx^3 \wedge dx^1 + H^3\, dx^1 \wedge dx^2. \end{aligned} \tag{5}$$

In saying above that the entities **E** and **H** are "vectors," we meant only that they behave like vectors under orthogonal changes of the co-ordinates x^1, x^2, x^3. Since F_{ik} is a tensor, and therefore must transform as stipulated in the definition (17.1.1), we calculate that in particular under the Lorentz transformation

$$x^1 = \frac{x^{1'} + vt'}{\sqrt{1 - \frac{v^2}{c^2}}}; \qquad x^2 = x^{2'}; \qquad x^3 = x^{3'}; \qquad t = \frac{t' + \frac{v}{c^2} x^{1'}}{\sqrt{1 - \frac{v^2}{c^2}}} \tag{6}$$

where $x^0 = ct$, $x^{0'} = ct'$, the components of $\mathbf{E}$ and $\mathbf{H}$ transform as follows:

$$E_1 = E'_1; \qquad E_2 = \frac{E'_2 + \frac{v}{c} H^{3'}}{\sqrt{1 - \frac{v^2}{c^2}}}; \qquad E_3 = \frac{E'_3 - \frac{v}{c} H^{2'}}{\sqrt{1 - \frac{v^2}{c^2}}};$$

$$H^1 = H^{1'}; \qquad H^2 = \frac{H^{2'} - \frac{v}{c} E'_3}{\sqrt{1 - \frac{v^2}{c^2}}}; \qquad H^3 = \frac{H^{3'} + \frac{v}{c} E'_2}{\sqrt{1 - \frac{v^2}{c^2}}}. \tag{7}$$

The set of all skew-symmetric tensors F_{ik} of type (0, 2) forms at each point a 6-dimensional linear space. The following lemma describes the action of the operator $*$ (see 19.3.1) on each such linear space.

21.1.3. Lemma. *Given a skew-symmetric tensor F of type* (0, 2) *which as a differential form is given by*

$$F = E_\alpha \, dx^0 \wedge dx^\alpha + H^1 \, dx^2 \wedge dx^3 + H^2 \, dx^3 \wedge dx^1 + H^3 \, dx^1 \wedge dx^2,$$

we have

$$* F = - \sum_{\alpha=1}^{3} H^\alpha \, dx^0 \wedge dx^\alpha + E_1 \, dx^2 \wedge dx^3 + E_2 \, dx^3 \wedge dx^1 + E_3 \, dx^1 \wedge dx^2. \tag{8}$$

PROOF. By Definition 19.3.1, the operator $*$ acts on F as follows:

$$(* F)_{ik} = \tfrac{1}{2} \varepsilon_{iklm} F^{lm}; \qquad F^{lm} = g^{lp} g^{mq} F_{pq}. \tag{9}$$

Here g_{ij} is the Minkowski metric, i.e. in terms of our assumed pseudo-Euclidean co-ordinates,

$$g_{ij} = \begin{pmatrix} 1 & & & 0 \\ & -1 & & \\ & & -1 & \\ 0 & & & -1 \end{pmatrix},$$

and also $(g^{ij}) = (g_{ij})^{-1}$. Hence the tensor F^{ik} is given by

$$F^{0\alpha} = -F_{0\alpha} = -E_\alpha; \qquad F^{\alpha\beta} = F_{\alpha\beta}, \qquad \alpha, \beta = 1, 2, 3. \tag{10}$$

The lemma now follows by direct calculation from this, (9) and the definition of the tensor ε_{iklm}. □

21.1.4. Corollary (cf. §19.3(14)). *With F as in the lemma, we have*

$$*(*F) = -F,$$

i.e. the square of the operator $*$ *as applied to skew-symmetric rank* 2 *tensors, is just multiplication by* -1.

This corollary, together with the fact that $*$ preserves addition of tensors, allows us to introduce on the set of skew-symmetric rank 2 tensors, the structure of a complex linear space. We do this by defining

$$(a + bi)F = aF + b * F, \tag{11}$$

for any such tensor F and any complex number $a + bi$. In this way we obtain at each point a 3-dimensional complex space $\mathbb{C}^3$.

Remark. We may then regard **E** and **H** as the "real" and "imaginary" parts of the tensor F, i.e. we may write formally $F = \mathbf{E} + i\mathbf{H}$. This fits in with (11) since multiplication by i gives

$$i(\mathbf{E} + i\mathbf{H}) = -\mathbf{H} + i\mathbf{E},$$

which accords with the action of the operator $*$ as described in Lemma 21.1.3. This leads naturally to the complex co-ordinates

$$z^\alpha = E_\alpha + iH^\alpha, \qquad \alpha = 1, 2, 3, \tag{12}$$

for the space $\mathbb{C}^3$ attached to each point of the underlying space. Since the transformations in $SO(1, 3)$ have Jacobian 1, it follows from 18.2.4 that they preserve the operator $*$, i.e. with respect to these transformations, the operator $*$ is a tensor operation. Hence $SO(1, 3)$ acts on $\mathbb{C}^3$ as a group of complex linear transformations. We now define a quadratic form $\langle F, F\rangle$ on skew-symmetric tensors $F = (F_{ik})$, which is preserved by $SO(1, 3)$. To this end note first that $F \wedge F$ and $F \wedge (*F)$ are skew-symmetric tensors of type (0, 4). Since the underlying space has dimension 4, it follows (as in Example 18.2.2(a)) that each of these tensors has all its non-zero components equal up to sign. It is a straightforward calculation from the definitions to show that

$$(F \wedge (*F))_{0123} = \tfrac{1}{2}F_{ik}F^{ki}; \qquad (F \wedge F)_{0123} = -\tfrac{1}{2}\varepsilon^{ijkl}F_{ij}F_{kl}.$$

It follows from the definition of the operator $*$ that $*(F \wedge F)$ and $*(F \wedge (*F))$ are scalars, so that any linear combination of them will also be a scalar. We take one such linear combination as defining our quadratic form, namely:

$$\langle F, F\rangle = -*(F \wedge (*F) + i(F \wedge F)) = -\tfrac{1}{2}(F_{ik}F^{ki} + i\varepsilon^{ijkl}F_{ij}F_{kl}). \tag{13}$$

(There would appear to be no possibility here of confusing the complex number i with the index i.) Since this is a scalar, it is invariant under co-ordinate changes, and so in particular is preserved by $SO(1, 3)$. Using (2) and (10) to express this quadratic form in terms of the co-ordinates z^1, z^2, z^3 given by (12), we obtain

$$\langle F, F\rangle = -|\mathbf{H}|^2 + |\mathbf{E}|^2 + 2i\langle \mathbf{E}, \mathbf{H}\rangle = \sum_{\alpha=1}^{3} (z^\alpha)^2. \tag{14}$$

Thus the form defined in (13) is just the "scalar square" (i.e. the sum of the squares of the components) of the complex 3-vector $\mathbf{E} + i\mathbf{H}$. The complex linear transformations of $\mathbb{C}^3$ preserving this scalar square of vectors are called *complex orthogonal transformations* by analogy with the real case, where this scalar square defines the Euclidean metric. The group of all complex orthogonal transformations of $\mathbb{C}^3$ is denoted by $O(3, \mathbb{C})$. (It must not be confused with $U(3)$.) We now come to the point of our remark, namely that since the transformations in $SO(1, 3)$ act like complex linear transformations on the space of skew-symmetric rank 2 tensors at a point, i.e. on $\mathbb{C}^3$, our introduction of a complex linear structure on that space yields a monomorphism from $SO(1, 3)$ to $O(3, \mathbb{C})$. Note finally the additional consequence that the quantities $\mathrm{Re}\langle F, F\rangle = |\mathbf{E}|^2 - |\mathbf{H}|^2$, and $\frac{1}{2}\mathrm{Im}\langle F, F\rangle = \langle \mathbf{E}, \mathbf{H}\rangle$, are invariants of the electromagnetic field, i.e. are preserved by Lorentz transformations. (This also follows from Lemma 21.1.2.)

We turn now to the question of reducing a skew-symmetric tensor F of rank 2 to canonical form by means of Lorentz transformations.

21.1.5. Theorem. *Let $\langle F, F\rangle$ be as in* (14), *i.e.*

$$\langle F, F\rangle = |\mathbf{E}|^2 - |\mathbf{H}|^2 + 2i\langle \mathbf{E}, \mathbf{H}\rangle.$$

(i) (a) *If $\langle F, F\rangle \neq 0$ and $\langle \mathbf{E}, \mathbf{H}\rangle \neq 0$, then there is a Lorentz transformation of co-ordinates such that the vectors* $\mathbf{E}$ *and* $\mathbf{H}$ *defined in terms of the transformed components of F are parallel and both non-zero.*

(b) *If $\langle F, F\rangle \neq 0$ and $\langle \mathbf{E}, \mathbf{H}\rangle = 0$, then $|\mathbf{E}|^2 - |\mathbf{H}|^2 \neq 0$, and there is a Lorentz transformation yielding co-ordinates in terms of which: if $|\mathbf{E}|^2 - |\mathbf{H}|^2 > 0$ then $\mathbf{E} \neq 0$, $\mathbf{H} = 0$; while if $|\mathbf{E}|^2 - |\mathbf{H}|^2 < 0$, then $\mathbf{E} = 0$, $\mathbf{H} \neq 0$.*

In both cases (a), (b), *the canonical form of the tensor F is*

$$F = \begin{pmatrix} 0 & E' & 0 & 0 \\ -E' & 0 & 0 & 0 \\ 0 & 0 & 0 & H' \\ 0 & 0 & -H' & 0 \end{pmatrix}, \tag{15}$$

so that $E'^2 - H'^2 = |\mathbf{E}|^2 - |\mathbf{H}|^2$, $E'H' = \langle \mathbf{E}, \mathbf{H}\rangle$.

(ii) *If $\langle F, F\rangle = 0$, then $|\mathbf{E}|^2 - |\mathbf{H}|^2 = 0$, $\langle \mathbf{E}, \mathbf{H}\rangle = 0$, and these equations are preserved by Lorentz transformations, i.e. after any Lorentz transformation of co-ordinates, the vectors* $\mathbf{E}$ *and* $\mathbf{H}$ *will remain perpendicular* *and of equal length. In this case the tensor F can, by a suitable Lorentz transformation of co-ordinates, be brought into the form*

$$F = \begin{pmatrix} 0 & E & 0 & 0 \\ -E & 0 & E & 0 \\ 0 & -E & 0 & 0 \\ 0 & 0 & 0 & 0 \end{pmatrix}. \tag{16}$$

PROOF. (i) Suppose $\langle F, F\rangle \neq 0$. Treating F as a vector in $\mathbb{C}^3$, as in the above remark, we write $F = dn$ where $d = \sqrt{\langle F, F\rangle}$ and $\langle n, n\rangle = 1$. It is easy to see that, analogously to the real case, the unit vector n is taken to any other unit vector, and so in particular to the unit vector $n' = (1, 0, 0)$ in the direction of the z^1-axis, by a suitable transformation from $SO(3, \mathbb{C})$. Writing $d = E' + iH'$, it is clear that in terms of the transformed co-ordinates F has the form (15) (since $F = (E' + iH', 0, 0)$). It is also obvious that $\mathbf{E}' = (E', 0, 0)$ and $\mathbf{H}' = (H', 0, 0)$ are parallel. Since $\langle F, F\rangle$ is preserved by $SO(3, \mathbb{C})$ we have that

$$d^2 = |\mathbf{E}|^2 - |\mathbf{H}|^2 + 2i\langle \mathbf{E}, \mathbf{H}\rangle = \langle F, F\rangle = E'^2 - H'^2 + 2i\langle \mathbf{E}', \mathbf{H}'\rangle, \tag{17}$$

whence if $\langle \mathbf{E}, \mathbf{H}\rangle \neq 0$, then $\langle \mathbf{E}', \mathbf{H}'\rangle = E'H' \neq 0$, and neither $\mathbf{E}'$ nor $\mathbf{H}'$ can be zero. This proves (i)(a). On the other hand if $\langle \mathbf{E}, \mathbf{H}\rangle = 0$, then again by (17) we must have $E'H' = 0$ while $E'^2 - H'^2 = |\mathbf{E}|^2 - |\mathbf{H}|^2 \neq 0$. Hence if $E'^2 - H'^2 > 0$ we shall have $E' \neq 0$, $H' = 0$, while if $E'^2 - H'^2 < 0$ then $E' = 0$, $H' \neq 0$, which is the claim of (i)(b). To complete the proof of (i) we need to know that the transformations in $SO(3, \mathbb{C})$ are Lorentz transformations. We omit the verification of this; the above remark makes it at least plausible.

(ii) By means of a (real) transformation from $SO(3) \subset SO(1, 3)$, we can arrange the co-ordinates z^1, z^2, z^3 so that $\mathbf{E}$ has the form $(E, 0, 0)$, i.e. is directed along the z^1-axis. Then $\mathbf{H}$ will lie in the (z^2, z^3)-plane. By rotating this plane, if necessary, we may then assume that $\mathbf{H}$ has the direction of the z^3-axis, i.e. $\mathbf{H} = (0, 0, H)$. Then since $|\mathbf{H}| = |\mathbf{E}|$, we must have $H = E$, whence we obtain the form (16) for F. □

21.2. Symmetric Tensors and Their Eigenvalues. The Energy-Momentum Tensor of an Electromagnetic Field

Let T_{ik} be a symmetric tensor of type (0, 2) defined on Minkowski space $\mathbb{R}^4_{1,3}$ with the Minkowski metric g_{ik}.

21.2.1. Definition. The *eigenvalues of the tensor* T_{ik} are the roots of the characteristic equation

$$P(\lambda) = \det(T_{ik} - \lambda g_{ik}) = 0. \tag{18}$$

As noted previously, it is clear that the eigenvalues of T_{ik} are just the eigenvalues, in the usual sense, of the linear operator $T^i_k = g^{ij}T_{jk}$, where $(g^{ij}) = (g_{ij})^{-1}$. The eigenvectors corresponding to an eigenvalue λ are the non-zero vectors (ξ^k) satisfying

$$T_{ik}\xi^k = \lambda g_{ik}\xi^k. \tag{19}$$

It is a well-known result of linear algebra that a symmetric matrix can always be brought into diagonal form by a suitable orthogonal transformation of co-ordinates; however, since our underlying space is not Euclidean, but pseudo-Euclidean, this is not relevant to our present situation. We should instead like to know what canonical forms a symmetric matrix T_{ik} can be brought into under Lorentz transformations of co-ordinates. The following theorem gives us the answer to this question in the case of the 2-dimensional Minkowski space $\mathbb{R}^2_1$.

21.2.2. Theorem. *Let T_{ik} be a symmetric tensor defined on 2-dimensional Minkowski space $\mathbb{R}^2_1$.*

(i) *If the roots λ_0, λ_1 of the characteristic equation $P(\lambda) = 0$ are real and distinct, then after a suitable Lorentz transformation of co-ordinates, the matrix T_{ik} transforms to*

$$\begin{pmatrix} \lambda_0 & 0 \\ 0 & -\lambda_1 \end{pmatrix}. \tag{20}$$

(ii) *If the roots λ_0, λ_1 of $P(\lambda) = 0$ are complex conjugates, say $\lambda_0 = \alpha + i\beta$, $\lambda_1 = \alpha - i\beta$, $\beta \neq 0$, then by means of a Lorentz transformation of co-ordinates, the matrix T_{ik} can be brought into the form*

$$\begin{pmatrix} \alpha & \beta \\ \beta & -\alpha \end{pmatrix}. \tag{21}$$

(iii) *If $\lambda_0 = \lambda_1 = \lambda$, then in any co-ordinate system the matrix T_{ik} has the form*

$$\begin{pmatrix} \lambda + \mu & -\mu \\ -\mu & -\lambda + \mu \end{pmatrix}, \tag{22}$$

where if $\mu \neq 0$ then μ is not an invariant of the tensor T_{ik}, and cannot be made 0 by applying a Lorentz transformation.

PROOF. Suppose first that $\lambda_0 \neq \lambda_1$, and let ξ_0, ξ_1 be corresponding eigenvectors defined as in (19), necessarily linearly independent since the eigenvalues λ_0, λ_1 are distinct. Since

$$\xi^i_0 T_{ik}\xi^k_1 = \lambda_0 g_{ik}\xi^i_0\xi^k_1 = \lambda_1 g_{ik}\xi^i_0\xi^k_1,$$

it follows from the assumption $\lambda_0 \neq \lambda_1$ that $\langle \xi_0, \xi_1 \rangle = g_{ik}\xi^i_0\xi^k_1 = 0$, i.e. that the vectors ξ_0, ξ_1 are orthogonal.

Suppose now that in addition to being distinct λ_0 and λ_1 are real. A little explicit calculation then shows that since ξ_0, ξ_1 are orthogonal and (because $\lambda_0 \neq \lambda_1$) independent, one of them will be time-like and the other space-like (see §6.1). It follows that we can take these as the directions of the co-ordinate axes, i.e. there is a co-ordinate change making these the co-ordinate axes, and preserving the metric (see §6.2). This completes the proof of (i).

We now prove (ii). (Note that this case can arise, since although (g^{ij}) and (T_{ik}) are symmetric, the matrix $(g^{ij}T_{jk})$ in general need not be symmetric, and may therefore have complex eigenvalues.) Thus suppose $\lambda_0 = \alpha + i\beta$, $\lambda_1 = \alpha - i\beta$. From (19) it follows that if $\xi_0 = a + ib$ is an eigenvector corresponding to λ_0, then $\xi_1 = \bar{\xi}_0 = a - ib$ will be an eigenvector corresponding to the eigenvalue λ_1. Since any scalar multiple of $\xi_0 = a + ib$ is also an eigenvector we may assume that its pseudo-Euclidean norm satisfies

$$\langle a + ib, a + ib \rangle = 2,$$

i.e. $(a^1 + ib^1)^2 - (a^2 + ib^2)^2 = 2$. From this condition and the orthogonality of ξ_0 and ξ_1 we deduce that

$$\langle a, b \rangle = 0, \qquad \langle a, a \rangle + \langle b, b \rangle = 0, \qquad \langle a, a \rangle - \langle b, b \rangle = 2.$$

It follows that $|a|^2 = 1$, $|b|^2 = -1$. In terms of the basis vectors a, b, the matrix (T_{ik}) takes the form (21). Since these are real, orthogonal vectors, the same reasoning as in the proof of (i) shows that the change from the original co-ordinate system to one based on a and b, can be effected by a Lorentz transformation. This completes the proof of (ii).

Finally, suppose $\lambda_0 = \lambda_1 = \lambda$. The characteristic polynomial is given by

$$\det(T_{ik} - \lambda g_{ik}) = \begin{vmatrix} T_{00} - \lambda & T_{01} \\ T_{01} & T_{11} + \lambda \end{vmatrix}$$
$$= -\{\lambda^2 + \lambda(T_{11} - T_{00}) + T_{01}^2 - T_{00}T_{11}\}.$$

Since this polynomial has only one root, its discriminant must be zero; i.e.

$$(T_{11} - T_{00})^2 - 4(T_{01}^2 - T_{00}T_{11}) = (T_{11} + T_{00})^2 - 4T_{01}^2 = 0,$$

whence

$$T_{11} + T_{00} = \pm 4T_{01}. \tag{23}$$

The unique root is

$$\lambda = \frac{T_{00} - T_{11}}{2}. \tag{24}$$

Solving (23) and (24) for T_{11} and T_{00} in terms of λ and $\mu = T_{01}$ we see that the matrix (T_{ik}) has the form (22). The case $\mu = 0$ corresponds to the situation that the space spanned by the eigenvectors of λ has dimension 2. We leave it to the reader to show that μ can be altered by suitable Lorentz transformations of co-ordinates. This completes the proof of (iii), and therefore of the theorem. □

Suppose now that F_{ik} is a skew-symmetric tensor defined on the Minkowski space $\mathbb{R}^4_1$. From F_{ik} we construct a symmetric tensor (T_{ik}), defining (see §37.3(24))

$$T_{ik} = \frac{1}{4\pi}(-g^{lm}F_{il}F_{km} + \tfrac{1}{4}F_{lm}F^{lm}g_{ik}). \tag{25}$$

If F_{ik} is an electromagnetic field tensor, then T_{ik} is called the *energy-momentum tensor* of that electromagnetic field. In terms of the real 3-vectors **E**, **H** defined in (2), the components of (T_{ik}) are given by

$$T_{00} = \frac{E^2 + H^2}{8\pi}, \qquad T_{0\alpha} = -\frac{c}{4\pi}[\mathbf{E}, \mathbf{H}]_\alpha,$$

$$T_{\alpha\beta} = \frac{1}{4\pi}\{-E_\alpha E_\beta - H_\alpha H_\beta + \tfrac{1}{2}\delta_{\alpha\beta}(E^2 + H^2)\}, \qquad \alpha, \beta = 1, 2, 3, \tag{26}$$

where $E = |\mathbf{E}|$, $H = |\mathbf{H}|$, and $[\mathbf{E}, \mathbf{H}]$ denotes the cross product of **E** and **H**.

The quantity $W = T_{00}$ is called the *energy density* of the electromagnetic field F_{ik}, the vector $(S_\alpha) = -(T_{0\alpha})$ is *Poynting's vector*, and the 3-dimensional tensor $T_{\alpha\beta}(1 \leq \alpha \leq 3, 1 \leq \beta \leq 3)$ is *Maxwell's electromagnetic stress tensor*.

We conclude this section by exhibiting the canonical forms of the energy-momentum tensor (25), i.e. the forms taken by the energy-momentum tensor when the electromagnetic tensor F_{ik} is brought into one or another of the canonical forms (15), (16) given in Theorem 21.1.5. Thus when F_{ik} has the form (15), i.e. when $\mathbf{E} = (E, 0, 0)$, $\mathbf{H} = (H, 0, 0)$, we have from (26) that

$$(T_{ik}) = \begin{pmatrix} +W & & & 0 \\ & -W & & \\ & & +W & \\ 0 & & & +W \end{pmatrix}, \qquad W = \frac{1}{8\pi}(E^2 + H^2), \tag{27}$$

whence we see that its eigenvalues are all $\pm W$. When F_{ik} has the form (16), i.e. when

$$\mathbf{E} = (E, 0, 0), \qquad \mathbf{H} = (0, 0, H), \qquad E = H,$$

then from (26) we obtain

$$(T_{ik}) = \begin{pmatrix} W & 0 & -W & 0 \\ 0 & 0 & 0 & 0 \\ -W & 0 & W & 0 \\ 0 & 0 & 0 & 0 \end{pmatrix}, \qquad W = \frac{E^2}{4\pi} = \frac{H^2}{4\pi}. \tag{28}$$

This is analogous to the third case in Theorem 21.2.2 (see (22)), with $\mu = W$, and corresponds to the situation of electromagnetic waves of the form $f(x - ct)$ propagated in a single direction—see §37.3. The energy-momentum tensor in this case does not reduce to diagonal form; its eigenvalues are all zero.

§22. The Behaviour of Tensors Under Mappings

22.1. The General Operation of Restriction of Tensors with Lower Indices

Suppose we are given a mapping F from a region of an m-dimensional Cartesian space with co-ordinates $x^{1'}, \ldots, x^{m'}$, to a region of an n-dimensional space with co-ordinates $x^1, \ldots, x^n$:

$$x^i = x^i(x^{1'}, \ldots, x^{m'}), \qquad i = 1, \ldots, n. \tag{1}$$

Then with each tensor $T_{i_1 \ldots i_k}$ of type $(0, k)$ defined on the range-space (i.e. the region of the space co-ordinatized by $x^1, \ldots, x^n$), we can associate the tensor F^*T defined on the domain of F (i.e. on the primed space) whose components are given by

$$(F^*T)_{i'_1 \ldots i'_k}(x^{1'}, \ldots, x^{m'}) = \left[T_{i_1 \ldots i_k} \frac{\partial x^{i_1}}{\partial x^{i'_1}} \cdots \frac{\partial x^{i_k}}{\partial x^{i'_k}}\right](x^i(x^{1'}, \ldots, x^{m'})). \tag{2}$$

We leave to the reader the verification that as defined by (2), F^*T is indeed a tensor of type $(0, k)$. Thus the operation F^* maps tensors of type $(0, k)$ in the opposite direction, so to speak, to that of the map F of the underlying spaces. We call such an operation a *restriction* (or *pullback*) of the tensor T.

22.1.1. Example. Suppose that in an n-dimensional space with metric g_{ij} we are given an m-dimensional surface

$$x^i = x^i(x^{1'}, \ldots, x^{m'}), \qquad i = 1, \ldots, n. \tag{3}$$

Then restriction of the tensor (g_{ij}) to this surface yields the tensor

$$g_{i'j'} = g_{ij} \frac{\partial x^i}{\partial x^{i'}} \frac{\partial x^j}{\partial x^{j'}}, \qquad i', j' = 1, \ldots, m,$$

which is just the metric induced on the surface by the metric g_{ij} of the space containing it (see §7.3).

We shall now consider the restriction of a skew-symmetric tensor $T_{i_1 \ldots i_k}$ of type $(0, k)$ to a k-dimensional surface $x^i = x^i(x^{1'}, \ldots, x^{k'})$ in an n-dimensional space. The following explicit formula for such a restriction of such a tensor will be useful to us later on in the theory of integration of skew-symmetric forms.

22.1.2. Theorem. *The restriction of the skew-symmetric form*

$$\sum_{i_1 < \cdots < i_k} T_{i_1 \ldots i_k} dx^{i_1} \wedge \cdots \wedge dx^{i_k}$$

to the k-dimensional surface $x^i = x^i(x^{1'}, \ldots, x^{k'})$ *is given by*

$$\sum_{i_1 < \cdots < i_k} J^{i_1 \ldots i_k} T_{i_1 \ldots i_k}\, dx^{1'} \wedge \cdots \wedge dx^{k'},$$

where $J^{i_1 \ldots i_k}$ *is the* $k \times k$ *minor of the matrix* $(\partial x^i/\partial x^{i'})$ *formed from the columns numbered* $i_1, \ldots, i_k$. *It follows that on the surface we have*

$$\sum_{i_1 < \cdots < i_k} T_{i_1 \ldots i_k}\, dx^{i_1} \wedge \cdots \wedge dx^{i_k} = \sum_{i_1 < \cdots < i_k} J^{i_1 \ldots i_k} T_{i_1 \ldots i_k}\, dx^{1'} \wedge \cdots \wedge dx^{k'}. \quad (4)$$

PROOF. By the definition of restriction (see (2)),

$$T_{1' \ldots k'} = T_{i_1 \ldots i_k} \frac{\partial x^{i_1}}{\partial x^{1'}} \cdots \frac{\partial x^{i_k}}{\partial x^{k'}}. \quad (5)$$

In view of the skew-symmetry of $T_{i_1 \ldots i_k}$, the right-hand side of (5) can be rewritten as

$$T_{i_1 \ldots i_k} \frac{\partial x^{i_1}}{\partial x^{1'}} \cdots \frac{\partial x^{i_k}}{\partial x^{k'}} = \sum_{i_1 < \cdots < i_k} T_{i_1 \ldots i_k} \left(\sum_{\sigma \in S_k} \operatorname{sgn}(\sigma) \frac{\partial x^{i_{\sigma(1)}}}{\partial x^{1'}} \cdots \frac{\partial x^{i_{\sigma(k)}}}{\partial x^{k'}} \right)$$
$$= \sum_{i_1 < \cdots < i_k} T_{i_1 \ldots i_k} J^{i_1 \ldots i_k}.$$

Alternatively, (4) can be proved by expressing the dx^i in terms of the $dx^{i'}$. □

22.2. Mappings of Tangent Spaces

It is in general not possible to define a strict analogue of the above operation of restriction for tensors with upper indices. However, given as before a map $F: x^i = x^i(x^{1'}, \ldots, x^{m'})$, $i = 1, \ldots, n$, we can define a map F_* from the space of vectors T (i.e. tangent space) at a particular point $(x^{1'}, \ldots, x^{m'})$ to the space of vectors at the point $x^i(x^{1'}, \ldots, x^{m'})$, $i = 1, \ldots, n$, by setting

$$(F_* T)^i \Big|_{x^k = x^k(x^{1'}, \ldots, x^{m'})} = T^{i'} \frac{\partial x^i}{\partial x^{i'}} \Big|_{(x^{1'}, \ldots, x^{m'})}. \quad (6)$$

(One can clearly extend this definition to obtain an analogous map F_* of the space of tensors of type $(k, 0)$ at any particular point $(x^{1'}, \ldots, x^{m'})$.) Thus the tangent spaces are mapped by F_* in the "same direction" as the original map F. This map F_* of the tangent spaces at each point is often called the *differential of the mapping F*.

For general F the map F_* cannot be extended to a map of vector fields (as wholes), since if F maps distinct points P_1, P_2 to the same point P, then the right-hand side of (6) may yield different vectors according as $x^{1'}, \ldots, x^{m'}$ are the co-ordinates of P_1 or P_2; thus in this case (6) may define two vectors at the point P for each vector field T. This difficulty disappears if F belongs to the important class of functions called "diffeomorphisms": a smooth mapping F is a *diffeomorphism* if it is one-to-one and onto, and its inverse F^{-1} is also smooth (i.e. has continuous partial derivatives). It is well known that in this case $m = n$. If F is a diffeomorphism then F_* is an isomorphism between the tangent spaces at corresponding points.

If F is a smooth real-valued function on $\mathbb{R}^n$ (i.e. if F is a continuously differentiable function from $\mathbb{R}^n$ to $\mathbb{R}$), then F_* defines for each point a linear transformation from the space of vectors attached to the point (i.e. from the tangent space at the point), to the real line $\mathbb{R}$, i.e. a linear functional, or linear form, on vectors. From the definition (6) the reader will recognize this as the differential $dF = (\partial F/\partial x_i)\, dx^i$ in the more familiar sense of that word. (Recall also that, as noted previously, we often consider dx^i as the map on the tangent space at the point (x^i) which picks out the ith component of each tangent vector.)

EXERCISE

Show that if ω_1, ω_2 are differential forms, and F, as defined above, is smooth, then

$$F^*(\omega_1 \wedge \omega_2) = F^*(\omega_1) \wedge F^*(\omega_2).$$

§23. Vector Fields

23.1. One-Parameter Groups of Diffeomorphisms

Let $x^1, \ldots, x^n$ be co-ordinates for an n-dimensional Cartesian space. With each vector field $\xi^i = \xi^i(x^1, \ldots, x^n)$ defined on a region of that space, there is associated the following autonomous system of differential equations:

$$\dot{x}^i(t) = \xi^i(x^1(t), \ldots, x^n(t)), \qquad i = 1, \ldots, n,$$
$$\dot{x}^i = \frac{dx^i}{dt}. \tag{1}$$

The solutions $x^i = x^i(t)$ of this system are called the *integral curves* of the vector field ξ^i; the latter is comprised of tangent vectors to the integral curves. We denote by

$$F_t^i(x_0^1, \ldots, x_0^n) = x^i = x^i(t, x_0^1, \ldots, x_0^n), \tag{2}$$

the integral curve of the vector field ξ^i, satisfying the initial conditions

$$x^i|_{t=t_0} = x_0^i. \tag{3}$$

(It is known from the theory of ordinary differential equations that if the functions $\xi^i(x^1, \ldots, x^n)$ are smooth, then there is exactly one solution (2) satisfying (3).)

Formula (2) defines a self-map

$$F_t\colon (x_0^1, \ldots, x_0^n) \mapsto (x^1(t, x_0^1, \ldots, x_0^n), \ldots, x^n(t, x_0^1, \ldots, x_0^n)) \tag{4}$$

of our region, depending on the parameter t. (In visual terms, F_t applied to a point $P(x_0^1, \ldots, x_0^n)$ gives the new position of a particle at that point after a time interval t, as the point-particle moves along the integral curve through P.) The theory of ordinary differential equations tells us that given any point $(x_0^1, \ldots, x_0^n)$ at which $(\xi^i) \neq 0$, there is a neighbourhood of that point on which, for all sufficiently small t, the map F_t is a diffeomorphism (see §22.2). (In other, briefer, words F_t is locally a diffeomorphism.) We express the fact that F_0 is the identity map of our region, and that for sufficiently small values of the parameter the functions F_t are, locally, diffeomorphisms satisfying

$$F_{t+s} = F_t \circ F_s, \qquad F_{-t} = (F_t)^{-1}, \tag{5}$$

by saying that the diffeomorphisms F_t define a *local group*. In this way our original vector field ξ^i gives rise to a local one-parameter group of diffeomorphisms.

We note parenthetically, for later use, that for small t, Taylor's theorem gives us the following more explicit form for the maps F_t:

$$x^i(t, x_0^1, \ldots, x_0^n) = x_0^i + t\xi^i(x_0^1, \ldots, x_0^n) + o(t). \tag{6}$$

Hence the entries in the Jacobian matrix of F_t satisfy

$$\frac{\partial x^i(t)}{\partial x_0^j} = \delta_j^i + t\frac{\partial \xi^i}{\partial x_0^j} + o(t), \tag{7}$$

so that the entries in the inverse matrix satisfy

$$\frac{\partial x_0^i}{\partial x^j} = \delta_j^i - t\frac{\partial \xi^i}{\partial x^j} + o(t). \tag{8}$$

The above construction can be reversed: Given a one-parameter local group of diffeomorphisms $F_t = (F_t^1, \ldots, F_t^n)$ we define its *velocity field* to be the vector field

$$\xi^i = \left(\frac{d}{dt}F_t^i\right)_{t=0}, \qquad i = 1, \ldots, n. \tag{9}$$

23.1.1. Example. Consider the one-parameter group of rotations through the angle t about the origin of the Euclidean plane with Euclidean coordinates x, y. In this case F_t is given by

$$x = x_0 \cos t + y_0 \sin t,$$

$$y = -x_0 \sin t + y_0 \cos t,$$

whence

$$\left.\frac{dx}{dt}\right|_{t=0} = y_0, \qquad \left.\frac{dy}{dt}\right|_{t=0} = -x_0.$$

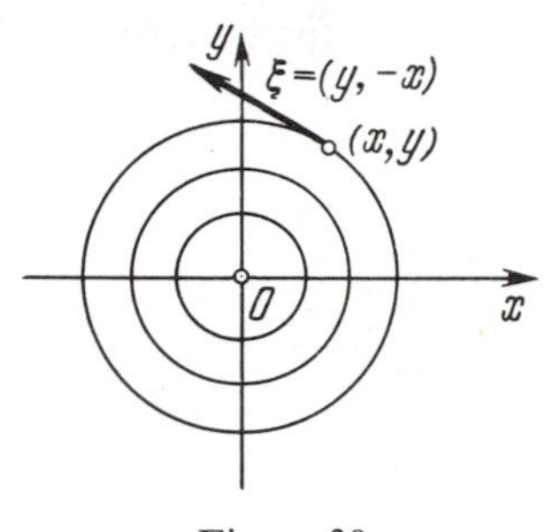

Figure 30

Hence in terms of the co-ordinates x, y, the velocity field ξ^i, $i = 1, 2$, has the form

$$\xi(x, y) = (y, -x),$$

and the integral curves of this vector field are the circles $x^2 + y^2 = \text{const.}$ (see Figure 30).

23.2. The Lie Derivative

Let $\xi = (\xi^i)$ denote a vector, with associated one-parameter local group of diffeomorphisms F_t, and let $T^{i_1 \dots i_p}_{j_1 \dots j_q}$ be a tensor of type (p, q). We restrict our attention to a region of the underlying space on which the F_t are one-to-one (for all appropriately small t). In view of this one-to-oneness, each F_t defines on the region a change from the co-ordinates $x^i(t)$ to the co-ordinates x^i_0. If $T^{i_1 \dots i_p}_{j_1 \dots j_q}$ are the components of our tensor relative to the co-ordinates $x^i(t)$, then by the definition of tensor (17.1.1), its components relative to the co-ordinates x^i_0 are given by

$$(F_t T)^{i_1 \dots i_p}_{j_1 \dots j_q} = T^{l_1 \dots l_p}_{k_1 \dots k_q} \frac{\partial x^{k_1}}{\partial x^{j_1}_0} \cdots \frac{\partial x^{k_q}}{\partial x^{j_q}_0} \frac{\partial x^{i_1}_0}{\partial x^{l_1}} \cdots \frac{\partial x^{i_p}_0}{\partial x^{l_p}}. \tag{10}$$

23.2.1. Definition.† The *Lie derivative* of a tensor $(T^{i_1 \dots i_p}_{j_1 \dots j_q})$ along a vector field ξ is the tensor $L_\xi T$ given by

$$L_\xi T^{i_1 \dots i_p}_{j_1 \dots j_q} = \left[\frac{d}{dt} (F_t T)^{i_1 \dots i_p}_{j_1 \dots j_q}\right]_{t=0}. \tag{11}$$

Thus if we regard the F_t as defining a time-dependent deformation of the underlying space, then the Lie derivative measures the rate of change of the tensor $T^{i_1 \dots i_p}_{j_1 \dots j_q}$ resulting from this deformation. It is clear that $L_\xi T$ is a tensor (of the same type (p, q) as T).

† In the theory of the mechanics of continuous media the expression $dT/d\tau = \partial T/\partial \tau + L_\xi T$, where $T = T(\tau, x)$ is an arbitrary tensor field, is called the "full derivative" of T along the velocity field ξ.

We shall now obtain an explicit formula for the Lie derivative. Using (7) and (8) to rewrite (10) we get

$$
\begin{aligned}
(F_t T)^{i_1 \dots i_p}_{j_1 \dots j_q} &= T^{l_1 \dots l_p}_{k_1 \dots k_q}\left(\delta^{k_1}_{j_1} + t\frac{\partial \xi^{k_1}}{\partial x_0^{j_1}}\right)\cdots\left(\delta^{k_q}_{j_q} + t\frac{\partial \xi^{k_q}}{\partial x_0^{j_q}}\right)\\
&\quad \times \left(\delta^{i_1}_{l_1} - t\frac{\partial \xi^{i_1}}{\partial x^{l_1}}\right)\cdots\left(\delta^{i_p}_{l_p} - t\frac{\partial \xi^{i_p}}{\partial x^{l_p}}\right)\\
&= T^{i_1 \dots i_p}_{j_1 \dots j_q} + t\left[T^{i_1 \dots i_p}_{k j_2 \dots j_q}\frac{\partial \xi^k}{\partial x_0^{j_1}} + \cdots + T^{i_1 \dots i_p}_{j_1 \dots j_{q-1} k}\frac{\partial \xi^k}{\partial x_0^{j_q}}\right.\\
&\quad \left. - T^{l i_2 \dots i_p}_{j_1 \dots j_q}\frac{\partial \xi^{i_1}}{\partial x^l} - \cdots - T^{i_1 \dots i_{p-1} l}_{j_1 \dots j_q}\frac{\partial \xi^{i_p}}{\partial x^l}\right] + o(t). \qquad (12)
\end{aligned}
$$

Differentiating equation (12) with respect to t and setting $t = 0$ (so that $x_0^i = x^i$) we obtain the desired formula:

$$
\begin{aligned}
L_\xi T^{i_1 \dots i_p}_{j_1 \dots j_q} &= \xi^s \frac{\partial T^{i_1 \dots i_p}_{j_1 \dots j_q}}{\partial x^s} + T^{i_1 \dots i_p}_{k j_2 \dots j_q}\frac{\partial \xi^k}{\partial x^{j_1}} + \cdots + T^{i_1 \dots i_p}_{j_1 \dots j_{q-1} k}\frac{\partial \xi^k}{\partial x^{j_q}}\\
&\quad - T^{l i_2 \dots i_p}_{j_1 \dots j_q}\frac{\partial \xi^{i_1}}{\partial x^l} - \cdots - T^{i_1 \dots i_{p-1} l}_{j_1 \dots j_q}\frac{\partial \xi^{i_p}}{\partial x^l}. \qquad (13)
\end{aligned}
$$

We now give some important (and illustrative) examples.

23.2.2. Examples. (a) In the case of a zero-rank tensor f (i.e. a scalar) we have from (11)

$$L_\xi f = \xi^i \frac{\partial f}{\partial x^i} = \partial_\xi f, \qquad (14)$$

the directional derivative of the function f in the direction ξ (i.e. along the vector field). Thus if $L_\xi f \equiv 0$, then the function f is constant on the integral curves of the field ξ. Such a function is called an *integral of the field* (or of the corresponding system of differential equations $\dot{x}^i = \xi^i(x^1, \dots, x^n)$). Thus in Example 23.1.1 the function $f(x, y) = x^2 + y^2$ is an integral.

If f is an integral of the vector field ξ, then the integral curves of the field all lie entirely in one or another of the surfaces with equations of the form $f(x^1, \dots, x^n) = \text{const.}$; clearly the field itself is tangent to these surfaces. We can exploit this to reduce from n to $n - 1$ the order of the original system of differential equations $\dot{x}^i = \xi^i(x^1, \dots, x^n)$, since if we have such an integral, then it suffices (for finding the general solution) to consider instead the restriction of this system (i.e. the restriction of the vector field ξ) to any particular hypersurface $f(x^1, \dots, x^n) = \text{const.}$, which has dimension $n - 1$. This brings out the connection between the simple geometric procedure of forming the restriction of a vector field, and the well-known result concerning the general solution of a system of differential equations, to the effect that knowledge of a solution leads to a reduction by 1 of the order of the system.

(b) We next consider tensors of type (1, 0), i.e. vectors. Thus let $\eta = (\eta^i)$ be another vector field. From (13) we obtain

$$L_\xi \eta^i = \xi^j \frac{\partial \eta^i}{\partial x^j} - \eta^j \frac{\partial \xi^i}{\partial x^j}, \tag{15}$$

whence

$$L_\xi \eta = -L_\eta \xi. \tag{16}$$

23.2.3. Definition. If ξ, η are two vector fields, then we call the vector field $L_\xi \eta$ the *commutator* of ξ with η, and denote it by $[\xi, \eta]$. (Thus from (16) we have $[\xi, \eta] = -[\eta, \xi]$.) Similarly the *commutator* $[\partial_\xi, \partial_\eta]$ *of the operators* $\partial_\xi, \partial_\eta$ on functions f (i.e. on scalars) is given by

$$[\partial_\xi, \partial_\eta]f = \partial_\xi(\partial_\eta f) - \partial_\eta(\partial_\xi f). \tag{17}$$

The following important theorem tells us that the action of the vector field $L_\xi \eta = [\xi, \eta]$ on functions (defined by the forming of the directional derivative along this vector field) is the same as that of $[\partial_\xi, \partial_\eta]$.

23.2.4. Theorem. *Given two vector fields* ξ, η, *we have*

$$\partial_{L_\xi \eta} f = \partial_{[\xi, \eta]} f = [\partial_\xi, \partial_\eta] f$$

for all (twice continuously differentiable) functions f. *Thus the commutator of the operators* ∂_ξ *and* ∂_η *is again a first order operator, namely directional differentiation along the vector field* $L_\xi \eta = -L_\eta \xi = [\xi, \eta]$.

PROOF. Explicit evaluation of the commutator $[\partial_\xi, \partial_\eta]$ yields:

$$\begin{aligned}
[\partial_\xi, \partial_\eta]f &= \partial_\xi(\partial_\eta f) - \partial_\eta(\partial_\xi f) = \partial_\xi\left(\eta^i \frac{\partial f}{\partial x^i}\right) - \partial_\eta\left(\xi^i \frac{\partial f}{\partial x^i}\right) \\
&= \xi^j \eta^i \frac{\partial^2 f}{\partial x^i \partial x^j} + \xi^j \frac{\partial \eta^i}{\partial x^j} \frac{\partial f}{\partial x^i} - \eta^j \xi^i \frac{\partial^2 f}{\partial x^i \partial x^j} - \eta^j \frac{\partial \xi^i}{\partial x^j} \frac{\partial f}{\partial x^i} \\
&= \left(\xi^j \frac{\partial \eta^i}{\partial x^j} - \eta^j \frac{\partial \xi^i}{\partial x^j}\right) \frac{\partial f}{\partial x^i} = \partial_{L_\xi \eta} f,
\end{aligned}$$

where $\xi^j \eta^i (\partial^2 f/\partial x^i \partial x^j) - \xi^i \eta^j (\partial^2 f/\partial x^i \partial x^j) = 0$ since, by the smoothness of f, $\partial^2 f/\partial x^i \partial x^j = \partial^2 f/\partial x^j \partial x^i$. This completes the proof. □

By direct computation using (15) one can verify the following analogue of Leibniz' formula:

$$L_\xi(f\eta) = f L_\xi \eta + \eta(\partial_\xi f). \tag{18}$$

From this formula the following theorem readily follows.

23.2.5. Theorem. *Let $\xi_1, \ldots, \xi_m$ be m vector fields in $\mathbb{R}^n$. In order for there to exist a system of co-ordinates $y^1, \ldots, y^n$ such that at every point the vector ξ_j is tangent to the co-ordinate axis determined by y^j, $j = 1, \ldots, m$, the vector fields must satisfy the condition*

$$[\xi_j, \xi_k] = f_{jk}^{(1)}\xi_j + f_{jk}^{(2)}\xi_k, \tag{19}$$

where the $f_{jk}^{(i)}$ are scalars (i.e. real-valued functions on the points of $\mathbb{R}^n$).

PROOF. As usual we write $e_1 = (1, 0, \ldots, 0), \ldots, e_n = (0, \ldots, 0, 1)$ for the standard basis vectors in the system of co-ordinates $y^1, \ldots, y^n$. Clearly if the components of the vectors ξ_j, ξ_k are constant relative to this system of co-ordinates then by (15) (with y^j in place of x^j) we have $[\xi_j, \xi_k] \equiv 0$. Hence in particular if $\xi_j = e_j$, $\xi_k = e_k$ at every point $(y^1, \ldots, y^n)$, then $[\xi_j, \xi_k] \equiv 0$. Now tangency of the ξ_j to the corresponding co-ordinate axes of the system $y^1, \ldots, y^n$, means simply that $\xi_j = f_j(y)e_j$, $j = 1, \ldots, m$, where the $f_j(y)$ are scalars. Hence in this case since $\partial_{e_i} = \partial/\partial y^i$, we have from (18), (16) and the fact that $[e_j, e_k] = 0$, that

$$[\xi_j, \xi_k] = [f_j e_j, f_k e_k] = f_j \frac{\partial f_k}{\partial y^j} e_k - f_k \frac{\partial f_j}{\partial y^k} e_j.$$

The desired equation (19) now follows by setting $f_{jk}^{(1)} = -(f_k/f_j)(\partial f_j/\partial y^k)$ and $f_{jk}^{(2)} = (f_j/f_k)(\partial f_k/\partial y^j)$. □

(c) For a tensor T_j of type (0, 1), i.e. for covectors, the formula (13) yields

$$(L_\xi T)_j = \xi^k \frac{\partial T_j}{\partial x^k} + T_k \frac{\partial \xi^k}{\partial x^j}. \tag{20}$$

In particular if we take T to be the gradient of a sufficiently smooth function f, i.e. $T_j = \partial f/\partial x^j$, we obtain

$$(L_\xi T)_j = \xi^k \frac{\partial^2 f}{\partial x^k \partial x^j} + \frac{\partial f}{\partial x^k} \frac{\partial \xi^k}{\partial x^j}. \tag{21}$$

It is easy to see that the right-hand side of (21) is also the ith coefficient in $d(L_\xi f)$. We conclude therefore that:

The operation of taking the Lie derivative commutes with that of forming the differential:

$$L_\xi(df) = d(L_\xi f). \tag{22}$$

(d) For a tensor g_{ij} of type (0, 2), i.e. for a bilinear form, we have (from (13))

$$L_\xi g_{ij} = \xi^s \frac{\partial g_{ij}}{\partial x^s} + g_{kj} \frac{\partial \xi^k}{\partial x^i} + g_{ik} \frac{\partial \xi^k}{\partial x^j} \equiv u_{ij}. \tag{23}$$

The tensor u_{ij} is called the *strain tensor* (for small deformations). It describes how the metric g_{ij} of the underlying space changes under the small deformation F_t defined by the vector field ξ. In particular if the space is Euclidean, so that $g_{ij} = \delta_{ij}$ (relative to suitable co-ordinates) then the tensor u_{ij} takes the form (cf. 17.1.3 *et seqq.*)

$$u_{ij} = \frac{\partial \xi^i}{\partial x^j} + \frac{\partial \xi^j}{\partial x^i}. \tag{24}$$

(e) Finally, we calculate the Lie derivative of the volume element $\sqrt{|g|}\varepsilon_{i_1 \ldots i_n}$ (or $\sqrt{|g|}\, dx^1 \wedge \cdots \wedge dx^n$), where $g = \det g_{ij}$. From formula (13) we obtain

$$\begin{aligned} L_\xi \sqrt{|g|}\varepsilon_{i_1 \ldots i_n} &= \xi^k \frac{\partial \sqrt{|g|}}{\partial x^k} \varepsilon_{i_1 \ldots i_n} \\ &\quad + \sqrt{|g|}\left(\varepsilon_{k i_2 \ldots i_n} \frac{\partial \xi^k}{\partial x^{i_1}} + \cdots + \varepsilon_{i_1 \ldots i_{n-1} k} \frac{\partial \xi^k}{\partial x^{i_n}}\right). \end{aligned} \tag{25}$$

It is not difficult to see that the bracketed expression in (25) is just $(\partial \xi^i / \partial x^i)\varepsilon_{i_1 \ldots i_n}$ (this was Exercise 3 in §18.4). Using this, equation (25) becomes

$$\begin{aligned} L_\xi \sqrt{|g|}\varepsilon_{i_1 \ldots i_n} &= \sqrt{|g|}\varepsilon_{i_1 \ldots i_n}\left(\xi^k \frac{\partial \ln \sqrt{|g|}}{\partial x^k} + \frac{\partial \xi^i}{\partial x^i}\right) \\ &= \sqrt{|g|}\varepsilon_{i_1 \ldots i_n}\left(\xi^k \tfrac{1}{2} g^{im} \frac{\partial g_{im}}{\partial x^k} + \frac{\partial \xi^i}{\partial x^i}\right). \end{aligned} \tag{26}$$

(Here, as usual, $(g^{ij}) = (g_{ij})^{-1}$, and we are using the fact that $\partial g / \partial x^k = A^{im}(\partial g_{im} / \partial x^k)$, where A^{im} is the cofactor of g_{im} in $\det(g_{ij})$.) However by comparison with (23) we see that the last bracketed expression in (26) is just the trace of the strain tensor (relative to the given metric), i.e. $\operatorname{tr} u_{im} = g^{im} u_{im}$; hence our final expression for the Lie derivative of the volume element is

$$L_\xi(\sqrt{|g|}\varepsilon_{i_1 \ldots i_n}) = \tfrac{1}{2} g^{im} u_{im} \sqrt{|g|}\varepsilon_{i_1 \ldots i_n}. \tag{27}$$

In the Euclidean case, when $g_{ij} = \delta_{ij}$, the trace of the strain tensor is

$$g^{im} u_{im} = 2 \frac{\partial \xi^i}{\partial x^i},$$

and (27) simplifies accordingly.

23.3. Exercises

1. Prove Leibniz' formula for the Lie derivative:

$$L_\xi(T \otimes R) = (L_\xi T) \otimes R + T \otimes L_\xi R,$$

where T, R are arbitrary tensors.

2. Let ω_1, ω_2 be differential forms. Show that

$$L_\xi(\omega_1 \wedge \omega_2) = L_\xi\omega_1 \wedge \omega_2 + \omega_1 \wedge L_\xi\omega_2.$$

3. Let F be a diffeomorphism from the region U to the region V, let X_1, X_2 be vector fields defined on U, and let $Y_i = F_* X_i$, $i = 1, 2$, be the corresponding vector fields on V. Prove that $F_*[X_1, X_2] = [Y_1, Y_2]$.

4. Let F_t, G_s be elements of one-parameter groups of diffeomorphisms with corresponding vector fields X, Y respectively. Show that F_t and G_s commute for all t, s if and only if the commutator of the fields X and Y is (identically) zero.

5. Let $X_1, \ldots, X_n$ be linearly independent vector fields on a region of n-dimensional space, satisfying $[X_i, X_j] \equiv 0$. Show that there is a (local) system of co-ordinates $x^1, \ldots, x^n$ such that for all i the field X_i is tangent to the ith co-ordinate axis, i.e. such that $\partial_{X_i}(x^k) = \delta_i^k$.

§24. Lie Algebras

24.1. Lie Algebras and Vector Fields

We begin with the definition.

24.1.1. Definition. A linear space V on which there is defined a skew-symmetric bilinear operation [,], is called a *Lie algebra* if Jacobi's identity holds, i.e. if

$$[\xi, [\eta, \zeta]] + [\eta, [\zeta, \xi]] + [\zeta, [\xi, \eta]] = 0 \tag{1}$$

for all $\xi, \eta, \zeta \in V$. The bracket operation [,] is called *commutation.*

Remark. For each $\xi \in V$ we define a linear transformation $\operatorname{ad} \xi: V \to V$ by $\operatorname{ad} \xi(\eta) = [\xi, \eta]$. Considering commutation from this point of view, Jacobi's identity signifies that each map $\operatorname{ad} \xi$ is, as they say in algebra, a "derivation" of the Lie algebra V, meaning that the following analogue of Leibniz' formula holds:

$$\operatorname{ad} \xi([\eta, \zeta]) = [\operatorname{ad} \xi(\eta), \zeta] + [\eta, \operatorname{ad} \xi(\zeta)]. \tag{2}$$

24.1.2. Examples. (a) Three-dimensional Euclidean space is a Lie algebra under the operation of the vector (or cross) product.

(b) An algebra V of linear operators can be made into a Lie algebra by defining

$$[A, B] = AB - BA. \tag{3}$$

To see that Jacobi's identity holds for this bracket operation we calculate:

$$\begin{aligned}[A, [B, C]] &= A[B, C] - [B, C]A \\ &= ABC - ACB - BCA + CBA, \\ [C, [A, B]] &= CAB - CBA - ABC + BAC, \\ [B, [C, A]] &= BCA - BAC - CAB + ACB.\end{aligned}$$

On adding the three right-hand side expressions we obtain zero, as required.

24.1.3. Corollary. *The space $M(n, \mathbb{R})$ of all real $n \times n$ matrices forms a Lie algebra under the bracket operation*

$$[A, B] = AB - BA. \tag{4}$$

In the light of our present example, Theorem 23.2.4 implies the following

24.1.4. Corollary. *The space of all vector fields on $\mathbb{R}^n$ is a Lie algebra with respect to the bracket operation given by*

$$[\xi, \eta]^i = \xi^j \frac{\partial \eta^i}{\partial x^j} - \eta^j \frac{\partial \xi^i}{\partial x^j}. \tag{5}$$

(Such Lie algebras of vector fields will of course be infinite-dimensional.)

24.1.5. Theorem. *If vector fields ξ and η are both tangent to a given smooth surface, then their commutator is also tangent to the surface.*

PROOF. Since an arbitrary surface (of dimension $< n$) is locally the intersection of hypersurfaces, it suffices to prove the theorem when the surface in question is a hypersurface $f(x^1, \ldots, x^n) = 0$. We may without loss of generality further suppose that our hypersurface has as its equation

$$x^n = 0, \tag{6}$$

since locally (in a neighbourhood of a non-singular point of the surface) this can be arranged by choosing suitable co-ordinates. Tangency of the fields ξ, η to the surface (6) means that at each point of the surface

$$\xi^n|_{x^n=0} = 0, \qquad \eta^n|_{x^n=0} = 0. \tag{7}$$

The nth component of $[\xi, \eta]$ is given by

$$[\xi, \eta]^n = \xi^k \frac{\partial \eta^n}{\partial x^k} - \eta^k \frac{\partial \xi^n}{\partial x^k}. \tag{8}$$

From (7) it follows that on the surface $x^n = 0$, we have for $k \neq n$ that $\partial\xi^n/\partial x^k = \partial\eta^n/\partial x^k = 0$. Substituting in (8) with $x^n = 0$ we then obtain $[\xi, \eta]^n|_{x^n=0} = 0$, completing the proof. □

24.1.6. Corollary. *The linear space consisting of all vector fields tangent to a given smooth surface, is a subalgebra of the Lie algebra of all vector fields.*

24.2. The Fundamental Matrix Lie Algebras

With each of the linear groups previously considered (see especially §§14.1, 14.2), there is associated a Lie algebra of matrices, namely the tangent space at the identity of the group, with the usual (as in (4)) commutation of matrices.

We begin by listing the most important matrix groups, together with their tangent spaces at the identity (which we shall denote by the same letters, but in lower case).

(1) The special linear groups $\mathrm{SL}(n, \mathbb{R})$, $\mathrm{SL}(n, \mathbb{C})$ are the groups of all $n \times n$ (respectively real or complex) matrices of determinant 1. We saw in §14.1 that the tangent spaces $sl(n, \mathbb{R})$, $sl(n, \mathbb{C})$ at the identity are just the spaces of all $n \times n$ matrices with zero trace (in $M(n, \mathbb{R})$, $M(n, \mathbb{C})$ respectively).

(2) The rotation groups $SO(n, \mathbb{R})$, $SO(n, \mathbb{C})$ are the groups of (respectively real or complex) $n \times n$ orthogonal matrices A with determinant 1:

$$AA^{\mathrm{T}} = 1, \qquad \det A = 1, \qquad A \in SO(n, \mathbb{R}) \text{ or } SO(n, \mathbb{C}). \tag{9}$$

As indicated in §§14.1, 14.2, the tangent spaces $so(n, \mathbb{R})$, $so(n, \mathbb{C})$ at the identity of these groups are the algebras of all (real or complex as the case may be) skew-symmetric matrices X:

$$X^{\mathrm{T}} = -X, \qquad X \in so(n, \mathbb{R}), so(n, \mathbb{C}). \tag{10}$$

(3) If $G = (g_{ij})$ is the pseudo-Euclidean metric in the space $\mathbb{R}^n_{p,q}$, then $SO(p, q)$ is the group of all real $n \times n$ matrices with determinant 1, preserving the metric $G = (g_{ij})$, or, as we saw (essentially) in §6.2, $SO(p, q)$ is the group of all $n \times n$ real matrices A such that

$$AGA^{\mathrm{T}} = G, \qquad \det A = 1. \tag{11}$$

It follows without too much difficulty that $so(p, q)$ consists of all real $n \times n$ matrices $X = (x^i_j)$ satisfying.

$$XG + GX^{\mathrm{T}} = 0, \tag{12}$$

or in terms of entries,

$$g_{ij}x^i_k + x^j_i g_{jk} = 0. \tag{13}$$

This is equivalent to the condition that the matrix

$$GX = (g_{ij}x^j_k) = (u_{ik}) \tag{14}$$

be skew-symmetric. Hence the map $X \mapsto GX$ defines a one-to-one and onto linear map (i.e. a vector space isomorphism) from $so(p, q)$ to the space of all real $n \times n$ skew-symmetric matrices.

(4) The unitary group $U(n)$ is the group of all unitary matrices of degree n, i.e. of all complex $n \times n$ matrices A preserving the Hermitian form, or, in other words, satisfying

$$A\bar{A}^{\mathrm{T}} = 1. \tag{15}$$

In §14.1 we saw that the tangent space $u(n)$ at the identity of $U(n)$, is the algebra of all skew-Hermitian matrices, i.e. complex $n \times n$ matrices X satisfying

$$\overline{X}^{\mathrm{T}} = -X. \tag{16}$$

(5) The special unitary group $SU(n)$ is the subgroup of $U(n)$ formed by the unitary matrices of determinant 1. As indicated in §14.1, the algebra $su(n)$ consists of all zero-trace skew-Hermitian matrices, i.e. of all $n \times n$ matrices X such that

$$\overline{X}^{\mathrm{T}} = -X, \qquad \operatorname{tr} X = 0. \tag{17}$$

(6) The pseudo-unitary group $U(p, q)$ is the group of linear transformations of complex n-dimensional space (where $n = p + q$), which preserve the pseudo-Hermitian scalar product

$$\begin{gathered}\langle \xi, \eta \rangle = \sum_{i=1}^{p} x^i \bar{y}^i - \sum_{i=p+1}^{n} x^i \bar{y}^i = g_{ij} x^i \bar{y}^j,\\ \xi = (x^1, \ldots, x^n), \quad \eta = (y^1, \ldots, y^n); \quad n = p + q.\end{gathered} \tag{18}$$

Writing as before $G = (g_{ij})$ (the usual pseudo-Euclidean metric of type (p, q)), one sees in the usual way that $U(p, q)$ consists of those $n \times n$ complex matrices A satisfying

$$AG\overline{A}^{\mathrm{T}} = G, \tag{19}$$

and that the tangent space $u(p, q)$ consists of all $n \times n$ complex matrices X satisfying

$$XG + G\overline{X}^{\mathrm{T}} = 0. \tag{20}$$

Since G is real and symmetric, it follows from (20) that GX is skew-Hermitian, so that the map $X \mapsto GX$ defines a vector space isomorphism between $u(p, q)$ and the space of all skew-Hermitian matrices.

(7) The group $SU(p, q)$ is the subgroup of $U(p, q)$ consisting of those matrices of the latter having determinant 1. The algebra $su(p, q)$ consists of those matrices X in $u(p, q)$ with trace zero. If we write $X = (x^i_j)$ and $GX = (u_{ik}) = g_{ij} x^j_k$, then the condition that $\operatorname{tr} X = 0$, i.e. that $x^i_i = 0$, is equivalent to the condition $g^{ik} u_{ik} = 0$, i.e. that the corresponding skew-Hermitian matrix (u_{ik}) have zero trace relative to the metric g_{ij}.

24.2.1. Theorem. *The spaces $sl(n, \mathbb{R})$, $sl(n, \mathbb{C})$, $so(n, \mathbb{R})$, $so(n, \mathbb{C})$, $so(p, q)$, $u(n)$, $su(n)$, $u(p, q)$, $su(p, q)$ are Lie algebras under the usual commutation of matrices.*

PROOF. Since each of these sets of matrices certainly forms a linear space, it remains to show only that each of them is closed under commutation. For this it clearly suffices to prove the following three assertions concerning arbitrary matrices X, Y and arbitrary real symmetric matrix G:

(i) for any matrices X, Y, $\operatorname{tr}[X, Y] = 0$;

(ii) if X, Y both satisfy condition (12) (for arbitrary real symmetric G) then so does their commutator $[X, Y]$;
(iii) if X, Y both satisfy (20) (with arbitrary real symmetric G), then so does $[X, Y]$.

Assertion (i) follows from the easy fact that for any matrices X, Y, we have $\operatorname{tr}(XY) = \operatorname{tr}(YX)$. To see (ii), suppose that X, Y both satisfy (12), i.e.

$$GX^{\mathrm{T}} = -XG, \qquad GY^{\mathrm{T}} = -YG.$$

Then

$$G[X, Y]^{\mathrm{T}} = GY^{\mathrm{T}}X^{\mathrm{T}} - GX^{\mathrm{T}}Y^{\mathrm{T}} = YXG - XYG = -[X, Y]G.$$

The proof of (iii) being similar, we omit the details. □

24.2.2. Definition. Let G be any of the transformation groups (1) through (7) above. The tangent space at the identity of G, equipped with the operation of commutation of matrices, is called the *Lie algebra of the group G.*

24.2.3. Examples. (a) As we have seen, the Lie algebra $so(3, \mathbb{R})$ of the rotation group $SO(3, \mathbb{R})$, is the algebra of all 3×3 real skew-symmetric matrices. If we take as a basis for $so(3, \mathbb{R})$ the matrices X_1, X_2, X_3 given by

$$X_1 = \begin{pmatrix} 0 & 0 & 0 \\ 0 & 0 & -1 \\ 0 & 1 & 0 \end{pmatrix}, \qquad X_2 = \begin{pmatrix} 0 & 0 & 1 \\ 0 & 0 & 0 \\ -1 & 0 & 0 \end{pmatrix},$$
$$X_3 = \begin{pmatrix} 0 & -1 & 0 \\ 1 & 0 & 0 \\ 0 & 0 & 0 \end{pmatrix}, \tag{21}$$

then we find that

$$[X_1, X_2] = X_3, \qquad [X_2, X_3] = X_1, \qquad [X_3, X_1] = X_2, \tag{22}$$

which brings us immediately to the following conclusion:

The Lie algebra of the group $SO(3, \mathbb{R})$ is isomorphic to the Lie algebra of vectors in Euclidean 3-space with commutation taken to be the vector (or cross) product.

(This justifies our earlier use of the bracket notation for the cross product.)

(b) Consider the Lie algebra $so(p, q)$. As usual we assume a co-ordinate system in terms of which the pseudo-Euclidean metric has the standard form

$$g_{ij} = \varepsilon_i \delta_{ij}, \qquad \varepsilon_i = \pm 1.$$

In (3) (see in particular (14)) we saw that the map defined by $(x_k^i) = X \mapsto GX = (g_{ij}x_k^j) = (u_{ik}) = u$, is a vector space isomorphism from $so(p, q)$ to the space of all skew-symmetric matrices. If we compute the effect of this map on commutators, we find that the corresponding commutation on skew-symmetric matrices is not the usual commutation of matrices, but is given instead by

$$[u, v]_{ij} = \sum_k \varepsilon_k (u_{ik} v_{kj} - v_{ik} u_{kj}),$$
$$u = (u_{ij}), \quad v = (v_{ij}), \quad u_{ji} = -u_{ij}, \quad v_{ji} = -v_{ij}. \tag{23}$$

(c) The Lie algebra $su(2)$ consists of all 2×2 skew-Hermitian matrices with trace zero. It is easy to see that the matrices

$$s_1 = \begin{pmatrix} i & 0 \\ 0 & -i \end{pmatrix}, \quad s_2 = \begin{pmatrix} 0 & 1 \\ -1 & 0 \end{pmatrix}, \quad s_3 = \begin{pmatrix} 0 & i \\ i & 0 \end{pmatrix}, \tag{24}$$

form a basis for the space $su(2)$. The pairwise commutators of these elements are found to be

$$[s_1, s_2] = 2s_3; \qquad [s_2, s_3] = 2s_1; \qquad [s_3, s_1] = 2s_2. \tag{25}$$

It follows from Lemma 14.3.1 that this Lie algebra is isomorphic (as Lie algebra) to the Lie algebra of purely imaginary quaternions (i.e. those quaternions x such that $\bar{x} = -x$) under the commutation $[x, y] = xy - yx$. Under the isomorphism given in that lemma,

$$i \leftrightarrow s_1, \qquad j \leftrightarrow s_2, \qquad k \leftrightarrow s_3. \tag{26}$$

The following fact is obvious from (22) and (25); however we give an alternative proof with a geometric flavour.

24.2.4. Theorem. *The Lie algebras $su(2)$ and $so(3, \mathbb{R})$ are isomorphic.*

Proof. With each matrix $X \in su(2)$ we associate the linear transformation $\operatorname{ad} X$ of the 3-dimensional space $su(2)$, defined by

$$\operatorname{ad} X: Z \mapsto \operatorname{ad} X(Z) = [X, Z], \qquad X, Z \in su(2). \tag{27}$$

It follows from Jacobi's identity that

$$[\operatorname{ad} X, \operatorname{ad} Y] = \operatorname{ad}[X, Y], \tag{28}$$

which means that the map $X \mapsto \operatorname{ad} X$ from $su(2)$ to the algebra of linear transformations of the vector space $su(2)$, preserves commutation; since (as is easily seen) this map is also linear, it follows that it is a Lie algebra homomorphism. In order to obtain from this map a homomorphism from $su(2)$ to $so(3, \mathbb{R})$, we need to realize each transformation $\operatorname{ad} X$ as an element of the tangent space at the identity of $SO(3, \mathbb{R})$.

To this end we first make $su(2)$ into Euclidean 3-space by defining the square of the length of a vector $Z = bs_1 + cs_2 + ds_3$ ($\leftrightarrow bi + cj + dk$ under the map defined by (26)) to be $b^2 + c^2 + d^2$. Then since $b^2 + c^2 + d^2 = \det Z$, and $\det Z = \det(AZA^{-1})$, it follows that for each $A \in SU(2)$, the self-map of the Euclidean 3-space $su(2)$ defined by

$$Z \mapsto AZA^{-1}, \tag{29}$$

is orthogonal. Now let $A = A(t)$ be a smooth family of transformations in $SU(2)$ satisfying

$$\left.\frac{dA(t)}{dt}\right|_{t=0} = X, \qquad A(0) = 1,$$

where $X \in su(2)$. Then the derivative with respect to t of the family of transformations

$$Z \mapsto A(t)ZA(t)^{-1}, \qquad Z \in su(2), \tag{30}$$

evaluated at $t = 0$, is given by

$$Z \mapsto XZ - ZX = \operatorname{ad} X(Z),$$

i.e. ad X is tangent to the curve $A = A(t)$ at $A(0) = 1$. Since by (29) this can be regarded as a curve in $SO(3, \mathbb{R})$, it follows that we have realized ad X as an element of $so(3, \mathbb{R})$.

Hence the map

$$su(2) \to so(3, \mathbb{R}); \qquad X \mapsto \operatorname{ad} X, \tag{31}$$

is a homomorphism. Its kernel is zero (since if ad $X(Z) = 0$ for all Z then $X = 0$), and therefore, the dimensions of the linear spaces $su(2)$ and $so(3, \mathbb{R})$ being the same (namely 3), it follows that the homomorphism (31) is in fact an isomorphism. This completes the proof. □

Remark. One can calculate that the matrices of the transformations ad s_1, ad s_2, ad s_3 (where s_1, s_2, s_3 are given by (24)) are

$$\operatorname{ad} s_1 = 2X_1, \qquad \operatorname{ad} s_2 = 2X_2, \qquad \operatorname{ad} s_3 = 2X_3, \tag{32}$$

where the basis elements X_1, X_2, X_3 for the space $so(3, \mathbb{R})$ are as in (21).

(d) As a basis for the Lie algebra $sl(2, \mathbb{R})$ we take the matrices

$$Y_0 = \begin{pmatrix} 0 & 1 \\ -1 & 0 \end{pmatrix}, \qquad Y_1 = \begin{pmatrix} 1 & 0 \\ 0 & -1 \end{pmatrix}, \qquad Y_2 = \begin{pmatrix} 0 & 1 \\ 1 & 0 \end{pmatrix}. \tag{33}$$

The commutators of these matrices are

$$[Y_0, Y_1] = -2Y_2, \qquad [Y_0, Y_2] = 2Y_1, \qquad [Y_1, Y_2] = 2Y_0. \tag{34}$$

24.2.5. Theorem. *The Lie algebras $sl(2, \mathbb{R})$ and $so(1, 2)$ are isomorphic.*

PROOF. As in the proof of the preceding theorem, we associate with each matrix Y in $sl(2, \mathbb{R})$ the linear transformation ad Y of the 3-dimensional space $sl(2, \mathbb{R})$. As before this defines a Lie algebra homomorphism. The self-transformations of the space $sl(2, \mathbb{R})$ of the form

$$Y \mapsto AYA^{-1} \tag{35}$$

preserve the quadratic form

$$|Y|^2 = \det Y. \tag{36}$$

By imitating the proof of the preceding theorem (24.2.4), it can be seen that ad Y is tangent at the identity to the group of transformations of $sl(2, \mathbb{R})$ preserving the metric (36). However this metric is the Minkowski metric since

$$\det Y = \det(y^0 Y_0 + y^1 Y_1 + y^2 Y_2)$$
$$= \det\begin{pmatrix} y_1 & y_2 + y_0 \\ y_2 - y_0 & -y_1 \end{pmatrix} = y_0^2 - y_1^2 - y_2^2. \tag{37}$$

Hence we have realized ad Y as an element of $so(1, 2)$, and the theorem follows as before. □

24.3. Linear Vector Fields

Let $X = (X_k^i)$ be a fixed real (or complex) $n \times n$ matrix. For each such X we construct a vector field T_X on $\mathbb{R}^n$ (or $\mathbb{C}^n$) by taking its value at each point $x \in \mathbb{R}^n$ (or $\mathbb{C}^n$) to be the negative of the result of applying the matrix X to the vector x, i.e.

$$T_X(x) = -Xx,$$

or, in terms of components,

$$T_X^i(x) = -X_k^i x^k. \tag{38}$$

We call such a field T_X a *linear vector field.*

The following theorem tells us what the integral curves of a linear vector field are, i.e. it gives the solutions of the system of equations

$$\dot{x} = -Xx. \tag{39}$$

24.3.1. Theorem. *The integral curve $x(t)$ of the vector field* (38), *satisfying the initial condition $x(0) = x_0$, is given by*

$$x(t) = \exp(-tX)x_0. \tag{40}$$

(Note that when $x_0 = 0$, the integral curve is just a point.)

PROOF. Since the functions on the right-hand side of our system (39) are as well behaved as could be desired, we know from the appropriate uniqueness theorem of the theory of ordinary differential equations that there is at most one solution of (39) satisfying the given initial condition. Hence we need merely show that (40) is indeed a solution of the system (39).

Recall (from §14.2(17)) that the exponential function applied to matrices is defined as the sum of a series:

$$\exp(-tX) = 1 - \frac{tX}{1!} + \frac{t^2X^2}{2!} - \cdots. \tag{41}$$

Since this series converges for all t (see Lemma 14.2.1) we can differentiate it term-by-term to obtain

$$\frac{d}{dt}\exp(-tX) = -X + \frac{tX^2}{1!} - \cdots = -X\exp(-tX). \tag{42}$$

Hence on differentiating both sides of (40) we have that

$$\frac{dx}{dt} = \frac{d}{dt}(\exp(-tX)x_0) = -X\exp(-tX)x_0 = -Xx,$$

which shows that (40) is a solution of (39), as required. □

It follows from this theorem that the one-parameter group of diffeomorphisms associated with the linear field T_X consists simply of applications of the matrices $\exp(-tX)$.

24.3.2. Example. The matrices X_1, X_2, X_3 given in (21), which form a basis for the Lie algebra $so(3)$, give rise, in the above way, to three linear vector fields defined on (Euclidean) 3-space. These three linear fields are usually denoted respectively by L_x, L_y, L_z. The respective values of these vector fields at the point (x, y, z) are given by

$$L_x = (0, +z, -y), \qquad L_y = (-z, 0, +x), \qquad L_z = (+y, -x, 0). \tag{43}$$

By calculating the matrices $\exp(-tX_i)$, $i = 1, 2, 3$, we see that, as might be expected, the three one-parameter groups corresponding to the vector fields L_x, L_y, L_z are the groups of rotations of $\mathbb{R}^3$ about the x-, y-, z-axes respectively.

Returning to the general situation, we take an arbitrary pair of $n \times n$ matrices X and Y, and calculate the commutator of the two linear vector fields T_X and T_Y.

24.3.3. Theorem. *The commutator of two linear vector fields T_X and T_Y has the form*

$$[T_X, T_Y] = T_{[X,Y]}. \tag{44}$$

PROOF. From the formula §23 (15) for the components of the commutator of a pair of vector fields we have

$$\begin{aligned}[T_X, T_Y]^i &= X^k_l x^l \frac{\partial(Y^i_m x^m)}{\partial x^k} - Y^k_l x^l \frac{\partial(X^i_m x^m)}{\partial x^k} = X^k_l x^l Y^i_k - Y^k_l x^l X^i_k \\ &= (-[X, Y]x)^i.\end{aligned}$$

Since the last expression here is the ith component of the linear vector field $T_{[X,Y]}$, the theorem is proved. □

24.3.4. Corollary. *A Lie algebra of $n \times n$ matrices (under the usual commutation of matrices) is isomorphic to the Lie algebra of linear vector fields defined by those matrices.*

Thus in particular if G is any one of the "classical" groups of transformations of (real or complex) n-dimensional space, considered in §24.2 above, and $\mathfrak{g}$ is its Lie algebra, then the linear vector fields T_X, $X \in \mathfrak{g}$ form a Lie algebra isomorphic to $\mathfrak{g}$. Since by Lemma 14.2.2 (and its extension to the cases it does not include) $X \in \mathfrak{g}$ entails $\exp(-tX) \in G$, it follows that the one-parameter group of diffeomorphisms corresponding to each of the vector fields T_X, consists simply of applications of the elements of some one-parameter subgroup of G.

Remark. Let G and $\mathfrak{g}$ be as in the preceding paragraph. Corresponding to each linear vector field T_X, $X \in \mathfrak{g}$, we have the differential operator ∂_{T_X} on smooth functions defined on $\mathbb{R}^n$. These differential operators are called *generators of the action* of the group G. If we know these generators, we can reconstruct the action of G on functions, i.e. the action given by

$$f(x_0) \to f(x(t)) = f(\exp(-tX)x_0) \equiv F(t, x_0), \tag{45}$$

by solving the differential equation

$$\dot{F} = \partial_{T_X} F. \tag{46}$$

(To see this use $\dot{x} = T_X(x)$ to get

$$\dot{F} = \frac{d}{dt} f(x(t)) = \frac{\partial f(x(t))}{\partial x^i} \dot{x}^i = \frac{\partial F}{\partial x^i} T_X^i = \partial_{T_X} F,$$

which gives (46). Alternatively, use geometric intuition!) The solution of (46) can be expressed formally as

$$F(t, x) = \exp(t\,\partial_{T_X}) f(x), \tag{47}$$

where $\exp(t\,\partial_{T_X})$ is the following formal power series in the differential operator $t\,\partial_{T_X}$:

$$\exp(t\,\partial_{T_X}) = 1 + \frac{t\,\partial_{T_X}}{1!} + \frac{t^2\,\partial_{T_X}^2}{2!} + \cdots. \tag{48}$$

Thus the right-hand side of (47) is

$$\exp(t\,\partial_{T_X}) f(x) = f(x) + t\,\partial_{T_X} f(x) + \tfrac{1}{2} t^2 \partial_{T_X}^2 f(x) + \cdots. \tag{49}$$

It is clear that this will indeed satisfy the original differential equation (46) whenever it is defined, i.e. for all functions $f(x)$ for which the right-hand side of (49) converges (for some interval of values of t).

We now give two examples illustrating this. (Note that clearly formula (47) for the solution of equation (46) is valid for all vector fields, not just the linear ones (cf. the first example below).)

24.3.5. Examples. (a) The generator of the action $f(x) \to f(x + t)$ of the group of translations of the real line is the operator d/dx. In this case the formula (47) takes the familiar form

$$f(x + t) = e^{t(d/dx)} f(x) = f(x) + \frac{t}{1!} f'(x) + \frac{t^2}{2!} f''(x) + \cdots$$

of Taylor's formula for the function $f(x)$.

(b) The generators corresponding to the matrices X_1, X_2, X_3 (see (21)), of the action of the group $SO(3, \mathbb{R})$ of rotations of Euclidean 3-space, are the following differential operators (cf. (43) where the same symbols are used for the corresponding linear vector fields):

$$L_x = z\frac{\partial}{\partial y} - y\frac{\partial}{\partial z}, \qquad L_y = x\frac{\partial}{\partial z} - z\frac{\partial}{\partial x}, \qquad L_z = y\frac{\partial}{\partial x} - x\frac{\partial}{\partial y}. \tag{50}$$

By Theorem 24.3.3 and (22), the commutators of these differential operators are given by

$$[L_x, L_y] = L_z, \qquad [L_y, L_z] = L_x, \qquad [L_z, L_x] = L_y. \tag{51}$$

24.3.6. Application (Left-invariant Fields Defined on Transformation Groups). Let X be a fixed $n \times n$ matrix (for definiteness assumed to be real). To each such X there corresponds the linear transformation

$$A \mapsto AX \tag{52}$$

of the space $\mathbb{R}^{n^2}$ of $n \times n$ real matrices A. We denote by L_X the linear vector field on the space $\mathbb{R}^{n^2}$ which at the point (i.e. $n \times n$ matrix) A, takes the value

$$L_X(A) = AX. \tag{53}$$

(Note that this differs from the previous definition (see (38)) in that the point A is on the left of the matrix X.) The integral curves of the field L_X are the solutions of the system of differential equations (or single matrix equation)

$$\dot{A} = L_X(A) = AX. \tag{54}$$

By closely imitating the proof of Theorem 24.3.1, we find that the unique solution of (54) satisfying the initial condition $A|_{t=0} = A_0$ is

$$A = A_0 \exp(tX). \tag{55}$$

Thus the one-parameter group of diffeomorphisms determined by the vector field L_X, consists of right multiplications by the matrices $\exp(tX)$ (i.e. of right translations by $\exp(tX)$).

A computation similar to that occurring in the proof of Theorem 24.3.3 shows that the conclusion of that theorem holds also for commutators of vector fields of the form L_X:

$$[L_X, L_Y] = L_{[X,Y]}. \tag{56}$$

From their definition (see (53)) it is clear that fields of the form L_X possess the important property of *left invariance* (or *invariance under left translations*):

$$BL_X(A) = L_X(BA). \tag{57}$$

As before, let G be any one of the "classical" groups which we considered in §24.2, regarded, as usual, as a smooth surface in the space $\mathbb{R}^{n^2}$ of all $n \times n$ real matrices, and let $\mathfrak{g}$ be the tangent space to G at the identity. Recall that if $X \in \mathfrak{g}$, then the matrices $\exp(tX)$ belong to G for all t (forming therefore a one-parameter subgroup of G). (See 14.2.2 where this was verified for several of the groups G in question.)

24.3.7. Lemma. *For each $X \in \mathfrak{g}$, the vector field L_X is tangent to the surface G; hence its restriction to G is a vector field on G.*

Proof. From (53) and (54) we see that the vector $L_X(A_0)$ where $A_0 \in G$, is the velocity vector at $t = 0$ of the curve $A_0 \exp(tX)$, which lies in the surface G. This gives the lemma. □

We shall use the same symbol L_X (where $X \in \mathfrak{g}$) to denote also the restriction of the field L_X to the surface G. Note that by the definition of L_X, its value at the identity is X, and also that the vector field L_X on the space G is left-invariant under left translations by elements of G.

24.3.8. Definition. A vector field of the form L_X on a (classical) group G, where X is an element of the Lie algebra $\mathfrak{g}$ of G, is called a *left-invariant field on the group G*.

The following theorem follows readily from Theorem 24.2.1 and the formula (56).

24.3.9. Theorem. *The left-invariant vector fields on a (classical) group G form a Lie algebra isomorphic to the Lie algebra $\mathfrak{g}$ of the group G.*

The following result will also be of use.

24.3.10. Lemma. *The values of all the left-invariant vector fields at an arbitrary point of the group G together comprise the whole of the tangent space to G at that point.*

Proof. If $X_1, \ldots, X_N$ form a basis for the Lie algebra $\mathfrak{g}$, then the vectors $L_{X_1}(A), \ldots, L_{X_N}(A)$ are linearly independent and tangent to the surface G at the point $A \in G$. Hence the space which they span, having the right dimension, must be the whole of the tangent space at A.

24.4. The Killing Metric

We first define a Killing metric on an arbitrary Lie algebra $\mathfrak{g}$.

24.4.1. Definition. A Euclidean or pseudo-Euclidean scalar product $\langle\ ,\ \rangle_0$ on a Lie algebra $\mathfrak{g}$ is called a *Killing metric* (or *Killing form*) if all the operators $\operatorname{ad} X$, $X \in \mathfrak{g}$, are skew-symmetric in that metric, i.e. if

$$\langle \operatorname{ad} X(Y), Z\rangle_0 = -\langle Y, \operatorname{ad} X(Z)\rangle_0. \tag{58}$$

We have already seen examples of Killing metrics on the Lie algebras $su(2)$ (where in the proof of Theorem 24.2.4 we defined the length-squared of a vector Z in $su(2)$ to be $\det Z$) and $sl(2, \mathbb{R})$ (see (36) above). (These metrics were used in the proofs of Theorems 24.2.4 and 24.2.5.) Note that the Killing metric on $su(2)$ was Euclidean, while the Killing metric (36) on $sl(2, \mathbb{R})$ was pseudo-Euclidean (of type (1, 2)).

Now let $\mathfrak{g}$ be the Lie algebra of a transformation group G, and suppose a Killing metric $\langle\ ,\ \rangle_0$ is given on $\mathfrak{g}$. We shall now use this Killing metric and the left-invariant fields on G defined in the preceding subsection, to introduce a metric on the surface G itself. Thus let A be any point (i.e. matrix) in G; then by Lemma 24.3.10 every vector tangent to G at A has the form $L_X(A)$ for some (unique) $X \in \mathfrak{g}$. It follows that in setting

$$\langle L_X(A), L_Y(A)\rangle = \langle X, Y\rangle_0, \tag{59}$$

for all $X, Y \in \mathfrak{g}$, we shall have defined the scalar product of any pair of vectors tangent to G at the arbitrary point A, i.e. we shall have defined a metric on G. This metric is called the *Killing metric on the group* G (determined by the given Killing form on $\mathfrak{g}$). Thus relative to the Killing metric on G, the scalar product of a pair of vectors $L_X(A)$, $L_Y(A)$ tangent to G at A, is the scalar product (afforded by the Killing form on $\mathfrak{g}$) of the corresponding left translates (by A^{-1}) X, Y in the tangent space $\mathfrak{g}$ at the identity of G, i.e. the scalar product of the values of the vector fields L_X, L_Y at the identity.

24.4.2. Example. Consider the case $G = SO(n, \mathbb{R})$. We shall show that the Euclidean metric on the space $\mathbb{R}^{n^2}$ of all $n \times n$ real matrices induces a Killing metric on the group $SO(n, \mathbb{R})$. The Euclidean metric on the space of $n \times n$ matrices is, of course, defined by

$$\langle X, Y\rangle = \sum_{i,j} x^i_j y^i_j, \qquad X = (x^i_j), \quad Y = (y^i_j), \tag{60}$$

which may be rewritten as†

$$\langle X, Y\rangle = \operatorname{tr}(XY^{\mathrm{T}}),$$

† Cf. the usual definition of the "Killing form" on a finite-dimensional Lie algebra as $\langle A, B\rangle = -\operatorname{tr}(\operatorname{ad} A \operatorname{ad} B)$. (See Exercise 9 of §24.7 below, and §3.1 of Part II.)

where tr denotes the trace. Note incidentally that with respect to this metric the surface $SO(n, \mathbb{R})$ is contained in the sphere of radius $\sqrt{n}$, since for orthogonal $n \times n$ matrices A we have

$$\langle A, A \rangle = \operatorname{tr}(AA^{\mathrm{T}}) = \operatorname{tr}(1) = n.$$

Now (as noted before) by Lemma 24.3.10 any vector tangent to $SO(n, \mathbb{R})$ at A has the form $L_X(A)$ for some (unique) $X \in so(n, \mathbb{R})$. Hence the metric induced on $SO(n, \mathbb{R})$ by the Euclidean metric (60) is completely determined by the values of $\langle L_X(A), L_Y(A) \rangle$, $X, Y \in so(n, \mathbb{R})$, $A \in SO(n, \mathbb{R})$. Now

$$\begin{aligned} \langle L_X(A), L_Y(A) \rangle &= \operatorname{tr}(AX(AY)^{\mathrm{T}}) = \operatorname{tr}(AXY^{\mathrm{T}}A^{\mathrm{T}}) \\ &= \operatorname{tr}(A^{\mathrm{T}}AXY^{\mathrm{T}}) = \operatorname{tr}(XY^{\mathrm{T}}), \end{aligned}$$

where we have used the invariance of the trace under conjugation, and so under cyclic permutation of the factors in a product of matrices, and also the fact that $A^{\mathrm{T}}A = 1$. Hence

$$\langle L_X(A), L_Y(A) \rangle = \langle L_X(1), L_Y(1) \rangle,$$

which accords with the definition (see (59)) of a Killing metric on $SO(n, \mathbb{R})$, assuming the yet-to-be-established fact that the Euclidean metric on $so(n, \mathbb{R})$ is a Killing form.

Thus it remains to show that the operators ad X, $X \in so(n, \mathbb{R})$, are skew-symmetric in the Euclidean metric on $so(n, \mathbb{R})$. To this end let $Y, Z \in so(n, \mathbb{R})$, i.e. let Y, Z be such that $Y^{\mathrm{T}} = -Y$, $Z^{\mathrm{T}} = -Z$. Then

$$\langle Y, Z \rangle = \operatorname{tr}(YZ^{\mathrm{T}}) = -\operatorname{tr}(YZ).$$

Hence

$$\begin{aligned} \langle \operatorname{ad} X(Y), Z \rangle &= \operatorname{tr}(YXZ) - \operatorname{tr}(XYZ), \\ \langle Y, \operatorname{ad} X(Z) \rangle &= \operatorname{tr}(YZX) - \operatorname{tr}(YXZ). \end{aligned} \tag{61}$$

Since the right-hand sides in (61) differ only in sign, it follows (from Definition 24.4.1) that the Euclidean metric on $so(n, \mathbb{R})$ is a Killing form, so that the induced metric on $SO(n, \mathbb{R})$ is a Killing metric on that group (and therefore also on each of its subgroups).

If a group G is given as a subgroup of the unitary group $U(n)$, then a Killing metric on G may be obtained by exploiting the embedding of $U(n)$ in $SO(2n, \mathbb{R})$ given by the operation of realization (see §11.2), and the method of the foregoing example. The explicit form of the resulting Killing metric can be calculated to be given by

$$\langle X, Y \rangle = \operatorname{Re} \operatorname{tr}(X\bar{Y}^{\mathrm{T}}) = -\operatorname{Re} \operatorname{tr}(XY), \qquad X, Y \in u(n).$$

24.5. The Classification of the 3-Dimensional Lie Algebras

Let L denote a 3-dimensional Lie algebra, with basis elements e_1, e_2, e_3 say. The operation of commutation in L determines (and is determined by) "structural constants" c_{ij}^k defined by

$$[e_i, e_j] = c_{ij}^k e_k, \qquad i, j = 1, 2, 3. \tag{62}$$

(We note parenthetically that if L is the Lie algebra of a classical group G, then the c_{ij}^k behave like the components of a tensor on G.)

From the skew-symmetry of commutation we have

$$c_{ij}^k = -c_{ji}^k, \tag{63}$$

and from Jacobi's identity

$$c_{ij}^k c_{kl}^m + c_{jl}^k c_{ki}^m + c_{li}^k c_{kj}^m = 0. \tag{64}$$

We wish now to choose a basis e_1, e_2, e_3 in terms of which the "tensor" (c_{ij}^k) takes its simplest form. From (63) we see that the commutation operation on L is determined by the 9 constants c_{ij}^k with $i < j$. Consider the following system of 9 linear equations in the 9 unknowns a_1, a_2, a_3 and $b^{kl} = b^{lk}$, $k, l = 1, 2, 3$:

$$c_{ij}^k = \varepsilon_{ijl} b^{lk} + \delta_j^k a_i - \delta_i^k a_j, \qquad i < j. \tag{65}$$

It is easily seen that this system always has solutions; thus we regard (65) as expressing the tensor c_{ij}^k in terms of the symmetric tensor (b^{ij}) and the vector (a_i). We have already taken the condition (63) into account in (65). On writing out the condition (64) in terms of the right-hand sides in (65), it follows without difficulty that

$$b^{ij} a_j = 0, \tag{66}$$

i.e. that the vector (a_i) is either the zero vector, or else an eigenvector of the matrix (b^{ij}) corresponding to the eigenvalue 0. Since the matrix (b^{ij}) is symmetric it can be brought into diagonal form $(b^{(i)}\delta^{ij})$ by means of a suitable change of basis. Moreover since (a_i) is an eigenvector, this change of basis can be arranged that in terms of the new basis (a_i) takes the form $(a, 0, 0)$. Then from (66) it follows that $b^{(1)}a = 0$, so that either a or $b^{(1)}$ is zero. In terms of such a (new) basis e_1, e_2, e_3, the commutation operation is given by

$$[e_1, e_2] = ae_2 + b^{(3)}e_3,$$

$$[e_2, e_3] = b^{(1)}e_1, \qquad [e_3, e_1] = b^{(2)}e_2 - ae_3.$$

If we take into account the latitude still remaining us of replacing e_1, e_2, e_3 by any (non-zero) scalar multiples, we end up with the following table of possible (isomorphism classes of) 3-dimensional Lie algebras ("Bianchi's classification"):

Type	a	$b^{(1)}$	$b^{(2)}$	$b^{(3)}$	Type	a	$b^{(1)}$	$b^{(2)}$	$b^{(3)}$
I	0	0	0	0	V	1	0	0	0
II	0	1	0	0	VI	1	0	0	1
VII	0	1	1	0	VII	a	0	1	1
VI	0	1	-1	0	III ($a = 1$) VI ($a \neq 1$)	a	0	1	-1
IX	0	1	1	1					
VIII	0	1	1	-1					

Note that the "abelian" Lie algebra of type I is the Lie algebra of the group of translations of 3-dimensional space, while the Lie algebra of $SO(3, \mathbb{R})$ is of type IX.

24.6. The Lie Algebras of the Conformal Groups

We shall now investigate the vector fields associated with conformal transformations of Euclidean and pseudo-Euclidean space. By Theorem 15.2, in dimensions ≥ 3 the conformal transformations are, in contrast to the 2-dimensional case, of a very restricted nature, being combinations of isometries, dilations and inversions. Each of the three groups G of (pseudo-) rotations, of translations, and of dilations, gives rise to vector fields. By the remark in §24.3 (see in particular the parenthetical observation preceding 24.3.5) each such field is determined by a differential operator (the generator of the action of G), which in the above cases, for suitable basis vectors X of the space $\mathfrak{g}$, is given by the first three of the following four first-order differential operators (the fourth will be explained subsequently):

(pseudo-) rotations (cf. Example 24.3.5(b)):

$$\Omega_{ab} = g_{ac}x^c \frac{\partial}{\partial x^b} - g_{bc}x^c \frac{\partial}{\partial x^a}, \qquad a, b = 1, \dots, n;$$

translations (cf. Example 24.3.5(a)): $P_a = \dfrac{\partial}{\partial x^a}$;

$$\text{dilations: } D = x^a \frac{\partial}{\partial x^a}; \tag{67}$$

inversions: $K_a = 2g_{ac}x^c x^b \dfrac{\partial}{\partial x^b} - g_{bc}x^b x^c \dfrac{\partial}{\partial x^a}$.

If the metric is Euclidean, i.e. if $g_{ac} = \delta_{ac}$, then the operator $\exp(t\Omega_{ab})$ (see the remark in §24.3) defines a rotation fixing the (x^a, x^b)-plane. If the metric is pseudo-Euclidean (so that $g_{ac} = \lambda_a \delta_{ac}$, $\lambda_a = \pm 1$), the operator $\exp(t\Omega_{ab})$ defines either a rotation (if $\lambda_a = \lambda_b$) or an elementary Lorentz transformation (if $\lambda_a = -\lambda_b$), of the (x^a, x^b)-plane. The transformations $\exp(t(\partial/\partial x^a))$ define (or, with a slight abuse of language, "are") translations along the x^a-axis, and the transformations $\exp(tx^\alpha(\partial/\partial x^\alpha))$ "are" the dilations $D(x) = tx$. It is a more complicated matter to identify the one-parameter group of transformations of the form $\exp(tK_a)$, since the field corresponding to K_a is non-linear; however we shall later (see Theorem 24.6.2) at least show that they are conformal on $\mathbb{R}^n$. Conversely, it can be shown, (using Liouville's theorem (15.2) among other things) that for $n \geq 3$ any conformal transformation of $\mathbb{R}^n$ (or, equivalently, of S^n—see §15) sufficiently close to the identity transformation, can be put into the form $\exp(tA)$ where

$$A = \sum_{a,b=1}^{n} \lambda^{ab}\Omega_{ab} + \sum_{a=1}^{n} \mu^a P_a + \gamma D + \sum_{a=1}^{n} \delta^a K_a, \tag{68}$$

(with the analogous result holding for $\mathbb{R}^n_{p,q}$).

The linear space spanned by the vector fields (67) is a Lie algebra under the usual bracket operation on vector fields (see 24.1.4 and 23.2.4). It is not difficult to verify the following commutator relations:

$$\begin{aligned} &[\Omega_{ab}, \Omega_{cd}] = g_{ac}\Omega_{bd} - g_{bc}\Omega_{ad} + g_{ad}\Omega_{cb} - g_{bd}\Omega_{ca};\\ &[\Omega_{ab}, P_c] = g_{ac}P_b - g_{bc}P_a; \qquad [\Omega_{ab}, K_c] = g_{ac}K_b - g_{bc}K_a;\\ &[\Omega_{ab}, D] = [P_a, P_b] = [K_a, K_b] = 0;\\ &[P_a, K_b] = 2(g_{ab}D + \Omega_{ab}); \qquad [P_a, D] = P_a; \qquad [K_a, D] = -K_a. \end{aligned} \tag{69}$$

Consider the correspondence

$$\begin{aligned} &\Omega_{a,b} \mapsto \Omega_{\mu,\nu}, \qquad \mu = a = 1, \dots, n; \qquad \nu = b = 1, \dots, n,\\ &P_a \mapsto \Omega_{a,n+1} - \Omega_{a,n+2},\\ &K_a \mapsto \Omega_{a,n+1} + \Omega_{a,n+2},\\ &D \mapsto \Omega_{n+1,n+2}, \end{aligned} \tag{70}$$

where the Ω_{ab}, P_a, K_a, D on the left-hand sides are given by (67) with the metric g_{ab} assumed Euclidean (i.e. $g_{ab} = \delta_{ab}$), while on the right-hand sides the $\Omega_{\mu\nu}$, μ, $\nu = 1, \dots, n+2$, correspond to the basic pseudo-rotations of $\mathbb{R}^{n+2}_{n+1,1}$, so that $g_{\mu\nu} = \lambda_\mu \delta_{\mu\nu}$ with $\lambda_\mu = 1$ for $\mu = 1, \dots, n+1$ and $\lambda_{n+2} = -1$. It is not difficult to show that (70) extends to a Lie algebra isomorphism. (The bijectivity follows by comparing dimensions.) From (68) we can deduce therefore the following important

24.6.1. Proposition. *The correspondence* (70) *extends to an isomorphism between the Lie algebra defined by* (69) *with the metric* g_{ab} *Euclidean* (*i.e. the Lie algebra of the group of conformal transformations of* $\mathbb{R}^n$, $n \geq 3$), *and the Lie algebra of the group* $SO(n+1, 1)$.

Thus the transformations of Euclidean space $\mathbb{R}^n$ of the form $\exp(tA)$ with A in the Lie algebra defined by (69) can (in virtue of the map defined by (70)) be identified with the pseudo-rotations of $\mathbb{R}^{n+2}_{n+1,1}$.

Remarks. 1. From among the multifarious conformal transformations of the Euclidean plane (or rather of the 2-sphere S^2) we previously singled out for special attention the (subgroup of) linear fractional transformations

$$z \mapsto \frac{az+b}{cz+d}, \qquad ad - bc \neq 0.$$

(We used various subgroups of this group in our investigations of the motions of the Euclidean plane $\mathbb{R}^2$, the Lobachevskian plane L^2, and the sphere S^2 (see §§9, 10.1, 13.2).) We saw earlier that this group is isomorphic to $SL(2, \mathbb{C})/\{\pm 1\}$, and is generated precisely by the rotations, translations, dilations and inversions of $\mathbb{R}^2$ (after projecting S^2 onto $\mathbb{R}^2$ by means of the stereographic projection). It follows that in this case (70) defines an isomorphism between the Lie algebra of complex, zero-trace, 2×2 matrices (the tangent space at the identity of $SL(2, \mathbb{C})$), and the Lie algebra of the Lorentz group $SO(3, 1)$, the group of pseudo-rotations, given in the form $\exp(tA)$. This isomorphism is called the "semi-spinor" representation of the Lie algebra of the Lorentz group $SO(3, 1)$ by complex 2×2 matrices. Complex conjugation of the entries in the matrices yields of course another such representation.

2. Proposition 24.6.1 can be generalized as follows: The Lie algebra of the group of conformal transformations of the pseudo-Euclidean space $\mathbb{R}^n_{p,q}$ is isomorphic to $so(p+1, q+1)$. (Prove it!)

We end the chapter by fulfilling our earlier promise to indicate the proof in the Euclidean case of the conformality of the transformations $\exp(tA)$ with A as in (67) (or, equivalently, as in (68)).

24.6.2. Theorem. *If $g_{ab} = \delta_{ab}$, then any transformation of the form* $\exp(tA)$, *where A has the form*

$$A = \lambda^{ab}\Omega_{ab} + \mu^a P_a + \gamma D + \delta^a K_a,$$

is (locally) a conformal transformation of Euclidean space $\mathbb{R}^n$.

(The need for the adverb "locally" here is a consequence of the fact that inversions are not defined at every point of $\mathbb{R}^n$. Note also that the theorem clearly holds if the δ^a are all zero, since, as we already know, $\exp(tA)$ is conformal for $A = \Omega_{ab}, P_a, D$.)

Proof of Theorem 24.6.2. We wish to prove that the group of transformations $S_t = \exp(tA)$ is locally conformal. Observe first that if a vector field $u = (u^a)$ defines a local group of Euclidean isometries, then the strain tensor, which measures the distortion of distance under the transformation (see Example 23.2.2(d)), must be zero, i.e.

$$\frac{\partial u^a}{\partial x^b} + \frac{\partial u^b}{\partial x^a} = 0$$

(where in this case there is no summation over repeated indices). If, more generally, the local group of the vector field $u = (u^a)$ is conformal, then for similar reasons the strain tensor must be proportional to the metric, i.e.

$$\frac{\partial u^a}{\partial x^b} + \frac{\partial u^b}{\partial x^a} = \gamma(x)\delta_{ab}, \tag{71}$$

where $\gamma(x)$ is a smooth function. We note parenthetically that in this case the effect on the tensor $g_{ab} = \delta_{ab}$, of the local group of the vector field u is given by

$$S_t^*(g_{ab})_{ij} = [1 + t\mu(x)]\delta_{ij} + O(t^2). \tag{72}$$

(To see this note first that by §22.1(2)

$$S_t^*(\delta_{ab})_{ij} = \delta_{ab}\frac{\partial x^a}{\partial y^i}\frac{\partial x^b}{\partial y^j} = \sum_{a=1}^{n}\frac{\partial x^a}{\partial y^i}\frac{\partial x^a}{\partial y^j}. \tag{73}$$

By §23.1(6) the transformed co-ordinates y^i satisfy

$$y^i = x^i + tu^i(x^1, \ldots, x^n) + o(t),$$

whence

$$\frac{\partial y^i}{\partial x^a} = \delta_a^i + t\frac{\partial u^i}{\partial x^a} + o(t).$$

This and the condition (71) together imply that

$$\sum_{i=1}^{n}\frac{\partial y^i}{\partial x^a}\frac{\partial y^i}{\partial x^b} = [1 + t\gamma(x)]\delta_{ab} + O(t^2).$$

By inverting this and substituting in (73) we arrive at the desired (72).)

By direct verification it can be seen that all four of the vector fields (67) satisfy the condition (71). (For instance setting $\partial_u \ (= u^i(\partial/\partial x^i)) = K_a$ yields

$$u^j = 2x^a x^j, \qquad j \neq a; \qquad u^a = 2(x^a)^2 - \sum_i (x^i)^2,$$

from which (71) is immediate.) Hence the theorem will follow if we can show that condition (71) implies conformality of the transformations S_t defined by the vector field $u = (u^\alpha)$. The proof of this, which is based on the analogous, though simpler, argument for the more restricted case of isometries, is as follows.

A co-ordinate shift along the integral curves of the system

$$\dot{x}^\alpha = u^\alpha(x^1, \ldots, x^n), \qquad \alpha = 1, \ldots, n, \tag{74}$$

through a time interval Δt, may be broken up into N consecutive shifts through intervals $\Delta t/N$:

$$S_{\Delta t} = S_{(\Delta t/N)} \circ \cdots \circ S_{(\Delta t/N)} = S^N_{(\Delta t/N)}.$$

By §23.1(7), the Jacobian of the transformation $S_\tau(\tau = \Delta t/N)$ satisfies (for small τ)

$$\frac{\partial x^i(\tau)}{\partial x^j} = \delta^i_j + \tau \frac{\partial u^i}{\partial x^j} + O(\tau^2).$$

Hence if $\xi = (\xi^a_0)$ is a vector at the point $x_0 = (x^1_0, \ldots, x^n_0)$, then by the transformation rule for vectors, its components in the shifted co-ordinates will be given for small $\Delta t/N$ by

$$\begin{aligned} \xi^a_0 = \xi^a \to \xi^a + \left(\frac{\Delta t}{N}\right) \frac{\partial u^a}{\partial x^b}(x_0)\xi^b + O\left(\frac{\Delta t}{N}\right)^2 &= \xi^a_1, \\ \xi^a_1 \to \xi^a_2 = \xi^a_1 + \left(\frac{\Delta t}{N}\right) \frac{\partial u^a}{\partial x^b}(x_1)\xi^b_1 + O\left(\frac{\Delta t}{N}\right)^2&, \\ \cdots\cdots\cdots\cdots\cdots\cdots\cdots\cdots&\cdots \\ \xi^a_{N-1} \to \xi^a_N = \xi^a_{N-1} + \left(\frac{\Delta t}{N}\right) \frac{\partial u^a}{\partial x^b}(x_{N-1})\xi^b_{N-1} + O\left(\frac{\Delta t}{N}\right)^2&. \end{aligned} \tag{75}$$

Here the points x_q, $q = 1, \ldots, N - 1$, which lie on the integral curve $(x^a(t))$ of the system (74) satisfying $x^a(0) = x^a_0$, are defined by

$$\begin{gathered} x^a\left(\frac{\Delta t}{N}\right) = x^a_1, \\ \cdots\cdots\cdots\cdots\cdots\cdots \\ x^a\left((N-1)\frac{\Delta t}{N}\right) = x^a_{N-1}. \end{gathered}$$

From (75) together with condition (71) we obtain the following expressions for the scalar squares of the vectors ξ_i:

$$\begin{aligned} \langle \xi_1, \xi_1 \rangle &= \left(1 + \mu(x_0)\frac{\Delta t}{N}\right)\langle \xi_0, \xi_0 \rangle + O\left(\frac{\Delta t}{N}\right)^2, \\ \langle \xi_2, \xi_2 \rangle &= \left(1 + \mu(x_1)\frac{\Delta t}{N}\right)\langle \xi_1, \xi_1 \rangle + O\left(\frac{\Delta t}{N}\right)^2, \\ &\cdots\cdots\cdots\cdots\cdots\cdots\cdots\cdots \\ \langle \xi_N, \xi_N \rangle &= \left(1 + \mu(x_{N-1})\frac{\Delta t}{N}\right)\langle \xi_{N-1}, \xi_{N-1} \rangle + O\left(\frac{\Delta t}{N}\right)^2. \end{aligned} \tag{76}$$

Hence

$$\langle \xi_N, \xi_N \rangle = \left[1 + \left(\sum_{i=0}^{N-1} \mu(x_i)\right)\frac{\Delta t}{N}\right]\langle \xi_0, \xi_0 \rangle + O\left(\frac{\Delta t}{N}\right)^2.$$

In the limit as $N \to \infty$ the right-hand side of this becomes

$$\left[1 + \int_{\Delta t} \mu(x(t))\, dt\right]\langle \xi_0, \xi_0 \rangle = \rho(x_0)\langle \xi_0, \xi_0 \rangle,$$

so that the transformation $S_{\Delta t}$ is conformal, as claimed. □

Remarks. 1. It follows from this proof that if the strain tensor is zero then S_t is an isometry (since in (76) we shall have $\mu(x) \equiv 0$). Hence for each t, S_t is, modulo a translation, a constant linear transformation $\exp(tB)$ where $B = B_t$, and $u = Bx$ (see §24.3). From this it follows that the matrix B_t (which may be written as $S_t^{-1}(dS_t/dt)$) is $(\partial u^a/\partial x^b)$. Clearly the condition that S_t be an isometry is equivalent to skew-symmetry of the matrix $(\partial u^a/\partial x^b) = B(x)$ at all points x.

2. It can happen in some cases that the subspace spanned by the set of all $B(x) = (\partial u^a/\partial x^b)$ (with x ranging over $\mathbb{R}^n$) is a subalgebra of the Lie algebra of all $n \times n$ real matrices. In particular in certain cases with n even, it turns out to be the subalgebra of all $n/2 \times n/2$ complex matrices.

24.7. Exercises

1. Prove that if the situation of the last sentence of the preceding remark obtains, then S_t defines a holomorphic transformation $\mathbb{C}^{n/2} \to \mathbb{C}^{n/2}$. (In general for $n/2 > 1$ such transformations need not be conformal.)
2. Prove that if all $B(x)$ (see the first of the preceding remarks) have zero trace, then the transformation S_t is volume-preserving.
3. Establish the following Lie algebra isomorphisms:

   (i) $su(1, 1) \simeq sl(2, \mathbb{R})$;
   (ii) $su(2) \times su(2) \simeq so(4)$;
   (iii) $sl(2, \mathbb{C}) \simeq so(1, 3)$;
   (iv) $so(1, 2) \simeq$ algebra of vectors in $\mathbb{R}^3_{1,2}$ with the "vector product" bracket operation defined in Exercise 1, §6.3.
4. Compute the left-invariant vector fields on the group of unimodular quaternions.
5. Express the Killing metric on the group $SO(3, \mathbb{R})$ in terms of the Euler angles (see Example 14.1.4).
6. Let g^0_{ij} be a Killing metric on a Lie algebra $\mathfrak{g}$ with basis $X_1, \ldots, X_n$. Write $[X_i, X_j] = c^k_{ij}X_k$, $c_{kij} = g^0_{kl}c^l_{ij}$. Show that the tensor c_{kij} is skew-symmetric.
7. The inner antomorphisms $B \mapsto ABA^{-1}$ of the group $SO(n, \mathbb{R})$ are motions of the Killing metric on that group.

8. A Killing metric on the group $SO(p, q)$ can be obtained as the restriction to this group of the pseudo-Euclidean metric

$$\langle X,Y\rangle = \operatorname{tr}(GXGY^T),$$

where G is the matrix of the pseudo-Euclidean metric of type (p, q). What is the type of the resulting pseudo-Riemannian metric on $SO(p, q)$?

9. All of the examples of Killing metrics mentioned in the text (see §24.4) are given to within a constant factor, by $\operatorname{tr}(\operatorname{ad} X \cdot \operatorname{ad} Y)$.

10. We saw in §13.2 that the group $PSL(2, \mathbb{R})$ is the (direct) isometry group of (the upper half-plane (or Kleinian) model of) the Lobachevskian plane. Hence to each one-parameter subgroup of $SL(2, \mathbb{R})$ (see Exercise 5 of §14.4) there corresponds a one-parameter group of diffeomorphisms of the Lobachevskian plane. Find the corresponding vector fields in (Klein's model of) the Lobachevskian plane. Calculate their commutators.

11. Compute the Lie algebras of:
    (i) the affine group on $\mathbb{R}^n$ (see Exercise 3, §4.5);
    (ii) the isometry group of Euclidean n-space;
    (iii) the isometry group of the Minkowski space $\mathbb{R}^3_{1,2}$
    Calculate the respective generators of the actions of these groups.

12. Show that the linear vector fields L_x, L_y, L_z in $\mathbb{R}^3$, corresponding to the action of $SO(3, \mathbb{R})$ (see §24.3(43)) are tangent to every sphere with centre at the origin. Find the form taken by the corresponding first-order differential operators on the unit sphere in terms of spherical co-ordinates.

13. Define a right-invariant field on a matrix group G to be the restriction to G of a vector field of the form $R_X(A) = -XA$. Prove that $[R_X, R_Y] = R_{[X,Y]}, [L_X, R_Y] \equiv 0$.

14. Identify the types of the Lie algebras $so(1, 2)$ and $sl(2, \mathbb{R})$ in the classification given in §24.5.

CHAPTER 4

The Differential Calculus of Tensors

§25. The Differential Calculus of Skew-Symmetric Tensors

25.1. The Gradient of a Skew-Symmetric Tensor

Most physical laws find mathematical expression as relationships between derivatives of one kind or another of various physical quantities. Many of these physical quantities are most appropriately represented as tensor (in particular vector) fields throughout space, or a region of space. We are therefore naturally interested in the question of what differential operations on tensors there exist, which are (in a sense to be made precise) independent of the co-ordinate system in terms of which we calculate their effect. The simplest example of such an operation is the following one: If we have a function $f(x, \alpha)$, or a tensor field $T^{i_1 \dots i_p}_{j_1 \dots j_q}(x, \alpha)$, depending as usual on the point $x = (x^1, x^2, x^3)$ of space, but also on some parameter α independent of the points, then we can take the partial derivative with respect to that parameter:

$$\frac{\partial f}{\partial \alpha}(x, \alpha) \quad \text{or} \quad \frac{\partial T^{i_1 \dots i_p}_{j_1 \dots j_q}(x, \alpha)}{\partial \alpha}$$

at each point x. (In classical mechanics the time t plays the role of an independent parameter.) This operation is not connected with the geometry of the underlying space co-ordinatized by x^1, x^2, x^3, being carried out separately, in effect, at each point of that space. Another familiar example of a differential operation not related in any way to whatever Riemannian metric the underlying space may have on it, is that of forming the gradient of a function (i.e. of a scalar field):

$$\left(\frac{\partial f}{\partial x^1}, \frac{\partial f}{\partial x^2}, \frac{\partial f}{\partial x^3}\right) = \operatorname{grad} f.$$

The result is a covector constructed from the function f, which is invariant in the sense that under co-ordinate changes its components transform according to the transformation rule for tensors:

$$x = x(z), \qquad \frac{\partial f}{\partial z^j} = \frac{\partial f}{\partial x^i}\frac{\partial x^i}{\partial z^j}.$$

The reader will probably be familiar with special cases of the following generalization of the concept of gradient to arbitrary skew-symmetric tensors defined on n-dimensional Cartesian space.

25.1.1. Definition. Let $T_{i_1 \ldots i_k}$ be a skew-symmetric tensor defined on an n-dimensional space with co-ordinates $x^1, \ldots, x^n$ (so that the i_q range from 1 to n). By the *gradient* or *differential* $dT_{j_1 \ldots j_k j_{k+1}}$ of this tensor we shall mean the skew-symmetric tensor of type $(0, k + 1)$ with components

$$(dT)_{j_1 \ldots j_{k+1}} = \sum_{q=1}^{k+1} \frac{\partial T_{j_1 \ldots j_{q-1} j_{q+1} \ldots j_{k+1}}}{\partial x^{j_q}} \cdot (-1)^{q-1}. \tag{1}$$

Before verifying that dT is indeed a tensor we consider some special (and perhaps familiar) cases of it.

25.1.2. Examples. (a) If $T = f(x)$ is a function, so that $k + 1 = 1$, then by definition

$$(dT)_i = \frac{\partial T}{\partial x^i},$$

which is just the usual gradient of f.

(b) If $T = (T_i)$ is a covector, then

$$(dT)_{ij} = \frac{\partial T_j}{\partial x^i} - \frac{\partial T_i}{\partial x^j} = -(dT)_{ji}.$$

This tensor is called the *curl* (or sometimes *rotation*) of the covector field (T_i), and is denoted also by curl T (or sometimes rot(T)). Thus the curl is a skew-symmetric tensor of type $(0, 2)$.

Remark. When the ambient space is 3-dimensional Euclidean, with Euclidean co-ordinates x^1, x^2, x^3, it is customary to use the notation curl T to refer to the vector $(\eta^k) = * \, dT$ (see §19.3); thus $\eta^i = \frac{1}{2}\varepsilon^{ijk}(dT)_{jk}$, so that

$$\eta^1 = \frac{\partial T_3}{\partial x^2} - \frac{\partial T_2}{\partial x^3} = (dT)_{23} = -(dT)_{32},$$

$$\eta^2 = \frac{\partial T_1}{\partial x^3} - \frac{\partial T_3}{\partial x^1} = (dT)_{31} = -(dT)_{13},$$

$$\eta^3 = \frac{\partial T_2}{\partial x^1} - \frac{\partial T_1}{\partial x^2} = (dT)_{12} = -(dT)_{21}.$$

(c) Let $n = 3$, and let $T_{ij} = - T_{ji}$ be a skew-symmetric tensor of type (0, 2). Then the skew-symmetric tensor dT of rank 3 is given by

$$(dT)_{123} = \frac{\partial T_{12}}{\partial x^3} - \frac{\partial T_{13}}{\partial x^2} + \frac{\partial T_{23}}{\partial x^1}.$$

Remark. If the co-ordinates x^1, x^2, x^3 are Euclidean, and if (in the manner indicated in the preceding remark) we define the vector (η^i) in terms of the skew-symmetric tensor T by setting $\eta^1 = T_{23}$, $\eta^2 = - T_{13}$, $\eta^3 = T_{12}$ (i.e. $\eta^i = \frac{1}{2}\varepsilon^{ijk}T_{jk}$), then we obtain

$$(dT)_{123} = \frac{\partial \eta^1}{\partial x^1} + \frac{\partial \eta^2}{\partial x^2} + \frac{\partial \eta^3}{\partial x^3} = \frac{\partial \eta^i}{\partial x^i}.$$

In the present context (i.e. when the co-ordinates x^1, x^2, x^3 are Euclidean) the scalar $\partial\eta^i/\partial x^i$ is called the *divergence* of the vector field $(\eta^i) = \eta$, and is denoted by div η.

We now turn to the justification of Definition 25.1.1, i.e. to the verification that dT is a tensor.

25.1.3. Theorem. *The gradient dT of a skew-symmetric tensor of type* (0, k) *is a skew-symmetric tensor of type* (0, $k + 1$).

PROOF (for $k = 0, 1$ only). Suppose we are given a co-ordinate change

$$x^i = x^i(x^{1'}, \ldots, x^{n'}), \qquad i = 1, \ldots, n. \tag{2}$$

Write $T_{i_1 \ldots i_k}$ for the components of the tensor T relative to the co-ordinate system x, and $T_{i'_1 \ldots i'_k}$ for the components relative to the primed system x'. Since T is a tensor we have by the very definition that

$$T_{i'_1 \ldots i'_k} = T_{i_1 \ldots i_k} \frac{\partial x^{i_1}}{\partial x^{i'_1}} \cdots \frac{\partial x^{i_k}}{\partial x^{i'_k}}. \tag{3}$$

In the definition of gradient (25.1.1) the co-ordinate system was arbitrary; thus

$$(dT)_{i_1 \dots i_{k+1}} = \sum_q (-1)^{q-1} \frac{\partial T_{i_1 \dots i_{q-1} i_{q+1} \dots i_{k+1}}}{\partial x^{i_q}}, \tag{4}$$

$$(dT)_{i'_1 \dots i'_{k+1}} = \sum_p (-1)^{p-1} \frac{\partial T_{i'_1 \dots i'_{p-1} i'_{p+1} \dots i'_{k+1}}}{\partial x^{i'_p}}. \tag{5}$$

From the formulae (3), (4) and (5) we wish to deduce that the $dT_{(i')}$ and $dT_{(i)}$ are related according to the transformation rule for tensors of type $(0, k + 1)$. In view of the notational complexity of this deduction, we shall content ourselves with carrying it out for the case $k = 1$ only. (The case $k = 0$ was verified earlier—see for instance §16.)

Thus let T_i be a covector; in this case, the formulae (3), (4), (5) simplify to:

$$T_{i'} = T_i \frac{\partial x^i}{\partial x^{i'}};$$

$$(dT)_{ij} = \frac{\partial T_j}{\partial x^i} - \frac{\partial T_i}{\partial x^j}, \qquad (dT)_{k'l'} = \frac{\partial T_{l'}}{\partial x^{k'}} - \frac{\partial T_{k'}}{\partial x^{l'}}.$$

Hence

$$\begin{aligned}
(dT)_{k'l'} &= \frac{\partial T_{l'}}{\partial x^{k'}} - \frac{\partial T_{k'}}{\partial x^{l'}} = \frac{\partial}{\partial x^{k'}}\left(T_j \frac{\partial x^j}{\partial x^{l'}}\right) - \frac{\partial}{\partial x^{l'}}\left(T_i \frac{\partial x^i}{\partial x^{k'}}\right) \\
&= \frac{\partial T_j}{\partial x^{k'}} \frac{\partial x^j}{\partial x^{l'}} + T_j \frac{\partial^2 x^j}{\partial x^{k'} \partial x^{l'}} - \frac{\partial T_i}{\partial x^{l'}} \frac{\partial x^i}{\partial x^{k'}} - T_i \frac{\partial^2 x^i}{\partial x^{k'} \partial x^{l'}} \\
&= \left(\frac{\partial T_j}{\partial x^i} \frac{\partial x^i}{\partial x^{k'}}\right) \frac{\partial x^j}{\partial x^{l'}} - \left(\frac{\partial T_i}{\partial x^j} \frac{\partial x^j}{\partial x^{l'}}\right) \frac{\partial x^i}{\partial x^{k'}} \\
&= \left(\frac{\partial T_j}{\partial x^i} - \frac{\partial T_i}{\partial x^j}\right) \frac{\partial x^i}{\partial x^{k'}} \frac{\partial x^j}{\partial x^{l'}} = (dT)_{ij} \frac{\partial x^i}{\partial x^{k'}} \frac{\partial x^j}{\partial x^{l'}},
\end{aligned}$$

completing the proof for this case. □

25.2. The Exterior Derivative of a Form

We now give an alternative definition of the gradient of a skew-symmetric tensor, in terms of its associated differential form. Corresponding to a skew-symmetric tensor $T_{i_1 \dots i_k}$, we have as usual the form

$$\omega = \sum_{i_1 < \dots < i_k} T_{i_1 \dots i_k} dx^{i_1} \wedge \cdots \wedge dx^{i_k}. \tag{6}$$

The *exterior derivative* (or *differential*) of this form is the form of rank $k+1$ given by

$$d\omega = \sum_{\substack{i_0 \\ i_1 < \dots < i_k}} \frac{\partial T_{i_1 \dots i_k}}{\partial x^{i_0}} dx^{i_0} \wedge dx^{i_1} \wedge \cdots \wedge dx^{i_k}. \tag{7}$$

In particular if $\omega = f$, a scalar, then $d\omega = df = (\partial f/\partial x^i)\, dx^i$ is just the familiar differential of the function f.

The connexion with our earlier concept of gradient (or differential) is revealed in the following

25.2.1. Theorem. *For any skew-symmetric tensor* $T_{i_1 \dots i_k}$, *we have*

$$d\omega = \sum_{j_1 < \dots < j_{k+1}} (dT)_{j_1 \dots j_{k+1}} dx^{j_1} \wedge \cdots \wedge dx^{j_{k+1}}. \tag{8}$$

PROOF. From the definition (25.1.1) of dT it follows that

$$\sum_{j_1 < \dots < j_{k+1}} (dT)_{j_1 \dots j_{k+1}} dx^{j_1} \wedge \cdots \wedge dx^{j_{k+1}}$$

$$= \sum_{j_1 < \dots < j_{k+1}} \sum_q (-1)^{q-1} \frac{\partial T_{j_1 \dots j_{q-1} j_{q+1} \dots j_{k+1}}}{\partial x^{j_q}} dx^{j_1} \wedge \cdots \wedge dx^{j_{k+1}}.$$

Using

$$dx^{j_1} \wedge \cdots \wedge dx^{j_{k+1}}$$

$$= (-1)^{q-1} dx^{j_q} \wedge dx^{j_1} \wedge \cdots \wedge dx^{j_{q-1}} \wedge dx^{j_{q+1}} \wedge \cdots \wedge dx^{j_{k+1}},$$

the last expression becomes

$$\sum_q \sum_{j_1 < \dots < j_{k+1}} \frac{\partial T_{j_1 \dots j_{q-1} j_{q+1} \dots j_{k+1}}}{\partial x^{j_q}}$$

$$\times\, dx^{j_q} \wedge dx^{j_1} \wedge \cdots \wedge dx^{j_{q-1}} \wedge dx^{j_{q+1}} \wedge \cdots \wedge dx^{j_{k+1}}.$$

If in the qth summand we put $j_1 = i_1, \dots, j_{q-1} = i_{q-1}, j_{q+1} = i_q, \dots, j_{k+1} = i_k$, this sum becomes

$$\sum_q \sum_{\substack{i_1 < \dots < i_k \\ i_{q-1} < j_q < i_q}} \frac{\partial T_{i_1 \dots i_k}}{\partial x^{j_q}} dx^{j_q} \wedge dx^{i_1} \wedge \cdots \wedge dx^{i_k},$$

which can be seen without much difficulty to be the same as the defining expression for $d\omega$ given in (7). □

25.2.2. Theorem. *Two successive applications of the gradient operation to a skew-symmetric tensor T result in the zero tensor*:

$$d(dT) = 0 \quad or \quad d(d\omega) = 0. \tag{9}$$

PROOF. If

$$\omega = \sum_{i_1 < \dots < i_k} T_{i_1 \dots i_k} \, dx^{i_1} \wedge \cdots \wedge dx^{i_k},$$

then we have

$$d\omega = \sum_{\substack{p \\ i_1 < \dots < i_k}} \frac{\partial T_{i_1 \dots i_k}}{\partial x^p} \, dx^p \wedge dx^{i_1} \wedge \cdots \wedge dx^{i_k},$$

$$d(d\omega) = \sum_{\substack{p, q \\ i_1 < \dots < i_k}} \frac{\partial^2 T_{i_1 \dots i_k}}{\partial x^q \, \partial x^p} \, dx^q \wedge dx^p \wedge dx^{i_1} \wedge \cdots \wedge dx^{i_k}. \tag{10}$$

Since under our blanket smoothness assumptions the expression $\partial^2 T_{i_1 \dots i_k}/\partial x^q \, \partial x^p$ is symmetric in the indices p and q, while $dx^q \wedge dx^p$ is skew-symmetric, it follows that the sum in (10) is zero, as required. □

Remark. The formula (7) for the exterior derivative of a form may be re-written as

$$d\left(\sum_{i_1 < \dots < i_k} T_{i_1 \dots i_k} \, dx^{i_1} \wedge \cdots \wedge dx^{i_k}\right) = \sum_{i_1 < \dots < i_k} (dT_{i_1 \dots i_k}) \wedge dx^{i_1} \wedge \cdots \wedge dx^{i_k}, \tag{11}$$

where by $dT_{i_1 \dots i_k}$ we mean the differential of the component $T_{i_1 \dots i_k}$ regarded as a scalar function. This yields the following alternative proof of the fact that dT is a tensor: Since $dT_{i_1 \dots i_k}$ is a tensor (corresponding as it does to the gradient of a scalar function), it follows that its exterior product (see §18.3) with $dx^{i_1} \wedge \cdots \wedge dx^{i_k}$ is also a tensor; however this exterior product is just the right-hand side of (11).

In terms of the operation of commutation of vector fields (see Definition 23.2.3) yet another expression can be given for the exterior derivative of a form.

25.2.3. Theorem (Cartan's Formula). *Let ω be a differential form of rank k, and let $X_1, \dots, X_{k+1}$ be smooth vector fields. Then the value of the form $d\omega$ on the fields $X_1, \dots, X_{k+1}$ is given by the following formula:*

$$\begin{aligned} (k+1) \, d\omega(X_1, \dots, X_{k+1}) &= \sum_i (-1)^i \, \partial_{X_i} \omega(X_1, \dots, \hat{X}_i, \dots, X_{k+1}) \\ &\quad + \sum_{i<j} (-1)^{i+j} \omega([X_i, X_j], X_1, \dots, \hat{X}_i, \dots, \hat{X}_j, \dots, X_{k+1}). \end{aligned} \tag{12}$$

(Here the hat over a symbol indicates that that symbol is omitted. Note also that of course the value of a form $T_{i_1 \ldots i_k}\, dx^{i_1} \wedge \cdots \wedge dx^{i_k}$ on vectors $X_1, \ldots, X_k$, is by definition just $T_{i_1 \ldots i_k} X_1^{i_1} \cdots X_k^{i_k}$.)

We again give the proof for the case $k = 1$ only. (In the case $k = 0$ it is immediate.) Thus suppose $\omega = T_i\, dx^i$. Then from (8) we have that

$$d\omega = \sum_{i<j} \left(\frac{\partial T_j}{\partial x^i} - \frac{\partial T_i}{\partial x^j} \right) dx^i \wedge dx^j,$$

whence

$$2\, d\omega = \sum_{i,j} \left(\frac{\partial T_j}{\partial x^i} - \frac{\partial T_i}{\partial x^j} \right) dx^i \wedge dx^j.$$

Hence the value taken by the form $d\omega$ on the vector fields $X = (X^i)$, $Y = (Y^i)$, is given by

$$2\, d\omega(X, Y) = X^i Y^j \left(\frac{\partial T_j}{\partial x^i} - \frac{\partial T_i}{\partial x^j} \right). \tag{13}$$

Now in the case $k = 1$, the formula (12) becomes

$$\begin{aligned} 2\, d\omega(X, Y) &= \partial_X \omega(Y) - \partial_Y \omega(X) - \omega([X, Y]) \\ &= \partial_X(T_i Y^i) - \partial_Y(T_i X^i) - T_i \left(X^k \frac{\partial Y^i}{\partial x^k} - Y^k \frac{\partial X^i}{\partial x^k} \right) \\ &= X^i Y^j \left(\frac{\partial T_j}{\partial x^i} - \frac{\partial T_i}{\partial x^j} \right), \end{aligned}$$

which is the same as the right-hand side of (13). This completes the proof. □

How does the differential operator d act on the exterior product of two differential forms? The following theorem supplies the answer.

25.2.4. Theorem. *Let ω_1, ω_2 be differential forms of degrees p and q respectively. Then*

$$d(\omega_1 \wedge \omega_2) = d\omega_1 \wedge \omega_2 + (-1)^p \omega_1 \wedge d\omega_2. \tag{14}$$

PROOF. It suffices to carry out the proof in the case where ω_1 and ω_2 are "monomials":

$$\omega_1 = f\, dx^{i_1} \wedge \cdots \wedge dx^{i_p}, \qquad \omega_2 = g\, dx^{j_1} \wedge \cdots \wedge dx^{j_q}.$$

Then we have

$$\omega_1 \wedge \omega_2 = fg\, dx^{i_1} \wedge \cdots \wedge dx^{i_p} \wedge dx^{j_1} \wedge \cdots \wedge dx^{j_q},$$

whence by the definition of d (see (7))

$$\begin{aligned} d(\omega_1 \wedge \omega_2) &= \frac{\partial f}{\partial x^k} g\, dx^k \wedge dx^{i_1} \wedge \cdots \wedge dx^{j_q} + f \frac{\partial g}{\partial x^k} dx^k \wedge dx^{j_1} \wedge \cdots \wedge dx^{j_q} \\ &= \left(\frac{\partial f}{\partial x^k} dx^k \wedge dx^{i_1} \wedge \cdots \wedge dx^{i_p}\right) \wedge (g\, dx^{j_1} \wedge \cdots \wedge dx^{j_q}) \\ &\quad + (-1)^p (f\, dx^{i_1} \wedge \cdots \wedge dx^{i_p}) \wedge \left(\frac{\partial g}{\partial x^k} dx^k \wedge \cdots \wedge dx^{j_q}\right) \\ &= d\omega_1 \wedge \omega_2 + (-1)^p\, \omega_1 \wedge d\omega_2, \end{aligned}$$

where we have used the fact that

$$\begin{aligned} &dx^k \wedge dx^{i_1} \wedge \cdots \wedge dx^{i_p} \wedge dx^{j_1} \wedge \cdots \wedge dx^{j_q} \\ &\qquad = (-1)^p\, dx^{i_1} \wedge \cdots \wedge dx^{i_p} \wedge dx^k \wedge dx^{j_1} \wedge \cdots \wedge dx^{j_q}. \end{aligned}$$

This completes the proof. □

The presence of a metric g_{ij} on the underlying space, allows us to define another important differential operation on forms which reduces their ranks by 1. The operation we have in mind is called the *divergence* (of a skew-symmetric tensor), and is denoted by δ; it is defined by

$$\delta = *^{-1}\, d\, *, \tag{15}$$

where $*$ is the metric-dependent operation on skew-symmetric tensors, defined in §19.3. A more explicit formula for the operation δ will be derived in §29. Since the operation $*$ is a tensor operation in general only if we restrict the permitted co-ordinate changes to those with positive Jacobian, it follows that this also is the most we can claim (on the face of it at least) for the operation δ.

25.2.5. Examples. In the first three of these examples the underlying space is assumed to be 3-dimensional Euclidean, with Euclidean co-ordinates x^1, x^2, x^3.

(a) In the case of a scalar field $f(x)$ the exterior derivative is the usual differential df, which is a covector. If we restrict the co-ordinate changes to being orthogonal then we need not distinguish upper and lower indices, so that df becomes the vector $\operatorname{grad} f = (\partial f/\partial x^1, \partial f/\partial x^2, \partial f/\partial x^3)$.

(b) Suppose $\omega = T_1\, dx^1 + T_2\, dx^2 + T_3\, dx^3$. Then by (8)

$$\begin{aligned} d\omega &= \left(\frac{\partial T_2}{\partial x^1} - \frac{\partial T_1}{dx^2}\right) dx^1 \wedge dx^2 \\ &\quad + \left(\frac{\partial T_3}{\partial x^2} - \frac{\partial T_2}{\partial x^3}\right) dx^2 \wedge dx^3 + \left(\frac{\partial T_1}{\partial x^3} - \frac{\partial T_3}{\partial x^1}\right) dx^3 \wedge dx^1. \end{aligned}$$

If we apply the operation $*$ to the 2-form $d\omega$ (see §19.3), we obtain the co-vector (or 1- form)

$$* \, d\omega = \left(\frac{\partial T_3}{\partial x^2} - \frac{\partial T_2}{\partial x^3}\right) dx^1 + \left(\frac{\partial T_1}{\partial x^3} - \frac{\partial T_3}{\partial x^1}\right) dx^2 + \left(\frac{\partial T_2}{\partial x^1} - \frac{\partial T_1}{\partial x^2}\right) dx^3. \quad (16)$$

If, as in (a), we identify vectors and covectors, then the operation $* \, d$ may be described as sending each vector T to the vector $* \, dT = \text{curl } T$, the curl of the vector field T.

(c) Suppose again that $\omega = T_1 \, dx^1 + T_2 \, dx^2 + T_3 \, dx^3$, and let us compute $\delta\omega = *^{-1} \, d * \omega$. This, as we shall see, is a form of rank zero (i.e. a scalar) (which is to be expected since δ reduces the rank by 1). Thus

$$\begin{aligned} \omega &= T_1 \, dx^1 + T_2 \, dx^2 + T_3 \, dx^3 \\ &\xrightarrow{*} T_1 \, dx^2 \wedge dx^3 + T_2 \, dx^3 \wedge dx^1 + T_3 \, dx^1 \wedge dx^2 \\ &\xrightarrow{d} \left(\frac{\partial T_1}{\partial x^1} + \frac{\partial T_2}{\partial x^2} + \frac{\partial T_3}{\partial x^3}\right) dx^1 \wedge dx^2 \wedge dx^3 \xrightarrow{*^{-1}} \frac{\partial T_1}{\partial x^1} + \frac{\partial T_2}{\partial x^2} + \frac{\partial T_3}{\partial x^3}. \end{aligned}$$

We conclude that

$$\delta(T) = \text{div } T = \frac{\partial T_1}{\partial x^1} + \frac{\partial T_2}{\partial x^2} + \frac{\partial T_3}{\partial x^3}. \quad (17)$$

Note that in deriving this formula for the divergence of a vector field we have made heavy use of our assumption that the co-ordinates are Euclidean; the formula (17) does not hold in general in other co-ordinates.

(d) We now take the underlying space to be 4-dimensional space-time with co-ordinates $x^0 = ct, x^1, x^2, x^3$ (where c is the speed of light), equipped with the usual pseudo-Euclidean metric

$$dl^2 = (dx^0)^2 - (dx^1)^2 - (dx^2)^2 - (dx^3)^2,$$

or, equivalently,

$$g_{\alpha\beta} = \begin{pmatrix} 1 & & & 0 \\ & -1 & & \\ & & -1 & \\ 0 & & & -1 \end{pmatrix}$$

Recall (from §21.1) that an electromagnetic field is a rank 2 skew-symmetric tensor $F_{\alpha\beta}$, where $\alpha, \beta = 0, 1, 2, 3$. From the physics of electromagnetic fields we know that the tensor $F_{\alpha\beta}$ must satisfy Maxwell's equations. The first "pair", one might say, of Maxwell's equations are given by

$$(dF)_{ijk} = \frac{\partial F_{jk}}{\partial x^i} - \frac{\partial F_{ik}}{\partial x^j} + \frac{\partial F_{ij}}{\partial x^k} = 0, \quad (18)$$

or, more briefly, $dF = 0$, where $F = \sum_{\alpha<\beta} F_{\alpha\beta}\, dx^\alpha \wedge dx^\beta$. In terms of the electric field $\mathbf{E} = (E_\alpha)$, $E_\alpha = F_{0\alpha}$, and the magnetic field $\mathbf{H} = (H^\alpha)$, $H^1 = F_{23}$, $H^2 = -F_{13}$, $H^3 = F_{12}$, those of the equations (18) with $i, j, k = 1, 2, 3$ in some order, are equivalent to the equation

$$\frac{\partial F_{12}}{\partial x^3} - \frac{\partial F_{13}}{\partial x^2} + \frac{\partial F_{23}}{\partial x^1} = 0 \quad \text{or} \quad \operatorname{div} \mathbf{H} = 0, \tag{19}$$

and the remaining equations in (18) are equivalent to

$$\operatorname{curl} \mathbf{E} + \frac{\partial \mathbf{H}}{\partial x^0} = 0 \quad \text{or} \quad \operatorname{curl} \mathbf{E} = -\frac{1}{c}\frac{\partial \mathbf{H}}{\partial t}. \tag{20}$$

Thus the system (18) is equivalent to the system consisting of the two equations (19) and (20), the first a scalar equation, and the second a vector equation. It is for this reason that (18) is referred to as the first "pair" of Maxwell's equations. Note that there is no connexion between these equations and the pseudo-Euclidean metric.

On the other hand, as we shall now see, the second pair of Maxwell's equations involve the metric in an essential way. In terms of the electromagnetic field tensor F this second "pair" takes the form

$$\delta F = *^{-1}\, d * F = \frac{4\pi}{c} j_{(4)}, \tag{21}$$

where $j_{(4)}$ denotes the 4-dimensional current vector, $j_{(4)} = (\rho c, \rho v^1, \rho v^2, \rho v^3)$, (where in turn ρ is the electric charge density at each point of 3-dimensional space, and $v = (v^1, v^2, v^3)$ is the usual velocity of charge at each point of the 3-dimensional space with co-ordinates x^1, x^2, x^3). By using Lemma 21.1.3, where $* F$ is expressed in terms of $\mathbf{E}$ and $\mathbf{H}$, we can rewrite (21) as the pair of equations (writing $\mathbf{j} = \rho v$, the current)

$$\operatorname{div} \mathbf{E} = 4\pi\rho, \tag{22}$$

$$\operatorname{curl} \mathbf{H} + \frac{1}{c}\frac{\partial \mathbf{E}}{\partial t} = \frac{4\pi}{c}\mathbf{j}, \tag{23}$$

again consisting of a scalar equation and a vector equation.

25.3. Exercises

1. Let $X_1, \ldots, X_n$ be vector fields defined on an n-dimensional space, linearly independent at each point of a region of that space, and let $\omega^1, \ldots, \omega^n$ be (at each point) the dual basis of 1-forms, i.e. linear functionals defined by $\omega^i(X_j) = \delta^i_j$. Show that

$$d\omega^k = -\tfrac{1}{2} c^k_{ij} \omega^i \wedge \omega^j,$$

where the quantities c^k_{ij} are defined by

$$[X_i, X_j] = c^k_{ij} X_k.$$

2. The operations of taking the differential and taking the Lie derivative of forms, commute; i.e. $L_\xi\, d\omega = d(L_\xi \omega)$.

3. For each vector field X define a linear operator $i(X)$ on forms by

$$[i(X)\omega](X_1, \ldots, X_{k-1}) = \omega(X, X_1, \ldots, X_{k-1}),$$

where ω is any form of rank k, and $X_1, \ldots, X_{k-1}$ are arbitrary vector fields.

(i) Prove that $i(X)$ is "anti-differentiation", i.e.

$$i(X)(\omega_1 \wedge \omega_2) = (i(X)\omega_1) \wedge \omega_2 + (-1)^k \omega_1 \wedge i(X)\omega_2,$$

where k is the rank of the form ω_1.

(ii) Establish the formula

$$i(X)\, d + d i(X) = L_X,$$

where L_X denotes the operation of taking the Lie derivative along the field X.

§26. Skew-Symmetric Tensors and the Theory of Integration

26.1. Integration of Differential Forms

From a second course in calculus the reader will recall the definition (analogous to that for the case $n = 2$ sketched in §7.4) of the multiple integral

$$\int \underset{U}{\cdots} \int f(z)\, dz^1 \cdots dz^n \tag{1}$$

of a (suitably well-behaved) function $f(z^1, \ldots, z^n)$ over a (suitable) region U of n-dimensional Cartesian space. For reasons soon to be made apparent (see also the remark following Corollary 18.2.4), we shall use the alternative notation

$$\int \underset{U}{\cdots} \int f(z)\, dz^1 \wedge \cdots \wedge dz^n \tag{2}$$

for the integral (1). The reader may also recall that, given a co-ordinate change $z = z(y)$, the integral (1) can be evaluated using the new co-ordinates $y^1, \ldots, y^n$ via the formula

$$\int \underset{U}{\cdots} \int f(z)\, dz^1 \wedge \cdots \wedge dz^n = \int \underset{V}{\cdots} \int J f(z(y))\, dy^1 \wedge \cdots \wedge dy^n, \tag{3}$$

where $J = \det(\partial z^i/\partial y^i)$ is the Jacobian of the co-ordinate change, and V is the same region U, but co-ordinatized by the new co-ordinates $y^1, \ldots, y^n$. (We shall confine ourselves throughout the remainder of this section to co-ordinate changes with positive Jacobian.)

By Theorem 18.2.3 the transformation rule for a skew-symmetric tensor T of rank n is given by $T'_{1\ldots n} = J \cdot T_{1\ldots n}$. From this and (3) it follows that we may regard the expression following the integral signs in (2) as a skew-symmetric tensor T of rank n, where $T_{1\ldots n} = f(z)$ (whence the appropriateness of the notation in (2)).

26.1.1. Example. If g_{ij} is a (pseudo-) Riemannian metric, then under a co-ordinate change $z = z(y)$, its determinant $g = \det(g_{ij})$ transforms as follows:

$$g' = \det(g'_{ij}) = \det\left(g_{kl}\frac{\partial z^k}{\partial y^i}\frac{\partial z^l}{\partial y^j}\right) = J^2 g. \tag{4}$$

Thus, as indeed we saw in 18.2.4, under co-ordinate changes with positive Jacobian, the expression $\sqrt{|g|}$ comports itself like a skew-symmetric tensor of rank n. Recall that by Definition 7.4.1 the area of a region U of a surface is

$$\sigma(U) = \iint\limits_U \sqrt{|g|}\, du\, dv; \qquad u = z^1, \quad v = z^2, \quad n = 2 \tag{5}$$

(where g_{ij} is the metric on the surface). Thus we now see that this definition of surface area is independent of the co-ordinates on the surface.

Suppose we are given a surface $x^i = x^i(z^1, z^2)$ in Euclidean 3-space with Euclidean co-ordinates x^1, x^2, x^3, and a function $f(x(z))$ (which might typically be in some essential way related to the surface—such as for instance its Gaussian curvature). The *surface integral* of this function over a region U of the surface is defined to be

$$\iint\limits_U f(x(z))\sqrt{|g|}\, dz^1 \wedge dz^2, \tag{6}$$

where $\iint_U \varphi(z)\, dz^1 \wedge dz^2$ is the usual double integral. (The motivation for this definition of surface integral is given in §7.4.) The expression $\sqrt{|g|}\, dz^1\, dz^2$ is sometimes called the *element of area* (or *element of measure*) on the surface.

The following 7 conclusions summarize (and extend) the above.

(i) For any bounded region U (with nice enough boundary) of n-dimensional Cartesian space, and any skew-symmetric tensor $T = (T_{i_1\ldots i_n})$ of type $(0, n)$, the multiple integral $\int \cdots \int_U T$ is defined.

(ii) In the notation of differential forms the tensor T can be written as

$$T = T_{1\ldots n}\, dz^1 \wedge \cdots \wedge dz^n$$

(or, without the wedges, as $T_{1\ldots n}\, dz^1 \cdots dz^n$).

(iii) Although $f(z)$ is given initially as a scalar function on points, for the purposes of integration, in view of the equality

$$\int \cdots \int\limits_U f(z)\, dz^1 \wedge \cdots \wedge dz^n = \int \cdots \int\limits_V J f(z)\, dy^1 \wedge \cdots \wedge dy^n,$$

we change our point of view of $f(z)$, regarding it instead as the component $T_{1 \dots n} = f(z)$ of a skew-symmetric tensor of rank n.

(iv) To integrate a function $\varphi(z)$ over a region U of a space (i.e. to form what might be called, by analogy with the term "surface integral," a "space integral") we need to be given beforehand the volume element (or element of measure), which is a skew-symmetric tensor (at least under certain relevant co-ordinate changes) of rank equal to the dimension of the space. If we denote this tensor by T, then the integral of $\varphi(z)$ over U is defined to be

$$\int_U \cdots \int \varphi(z)\, T = \int_U \cdots \int \varphi(z)\, T_{1 \dots n}\, dz^1 \wedge \cdots \wedge dz^n. \tag{7}$$

(v) If in (iv) the space is (pseudo-) Riemannian with metric $g_{ij}(z)$, then the tensor T (the *volume element*) is given by

$$T = d\sigma = \sqrt{|g|}\, dz^1 \wedge \cdots \wedge dz^n,$$

so that in this case, by (7) the integral of a function $\varphi(z)$ over a region U is

$$\int_U \cdots \int \varphi(z) T = \int_U \cdots \int \varphi(z) \sqrt{|g|}\, dz^1 \wedge \cdots \wedge dz^n.$$

(Note once again that here T is a tensor relative only to co-ordinate changes with positive Jacobian.)

(vi) Given a co-ordinate change $z^i = z^i(y)$, it follows from the equalities $dz^i = (\partial z^i / \partial y^j)\, dy^j$ and $dy^i \wedge dy^j = -dy^j \wedge dy^i$, that

$$dz^{i_1} \wedge \cdots \wedge dz^{i_k} = \sum_{j_1 \dots j_k} J^{i_1 \dots i_k}_{j_1 \dots j_k}\, dy^{j_1} \wedge \cdots \wedge dy^{j_k}, \tag{8}$$

where $J^{(i)}_{(j)}$ denotes the appropriate minor of the Jacobian matrix $(\partial z^i / \partial y^j)$. In particular when $k = n$ we get the expected formula

$$dz^1 \wedge \cdots \wedge dz^n = J\, dy^1 \wedge \cdots \wedge dy^n, \tag{9}$$

where J is, as usual, the Jacobian.

(vii) If the co-ordinates $z^1, \dots, z^n$ are Euclidean, then $\sqrt{|g|} = 1$, so that the volume element is given by $d\sigma = dz^1 \wedge \cdots \wedge dz^n$.

This completes our list of conclusions. We emphasize in particular the need to distinguish between the two kinds of integrals of a function arising above:

1. *The integral of the second kind* of a function (or, from the other point of view, of a rank n skew-symmetric tensor) over a region. This the integral occurring in (i), (ii) and (iii) above; it is just the multiple integral of the function and is independent of any metric which may be defined on the n-dimensional Cartesian space.

2. *The integral of the first kind* of a function over a region. This is the integral in (iv) and (v), for whose evaluation one needs to know the volume element or measure with respect to which the integration is to be carried out. It is the ordinary multiple (n-fold) integral of the product of the given function with the volume element, taken over the specified region. (In this sense this integral reduces to an integral of the second kind.)

Having given appropriate definitions(s) of the integral of a tensor of type $(0, n)$ over regions of n-dimensional Cartesian space, we now address the analogous problem for tensors of type $(0, k)$, $k < n$.

To begin with we consider the case $k = 1$; thus let T_j be a covector defined on $\mathbb{R}^n$. As noted in §18.1, we can associate with each such tensor the differential form $\omega = T_j\,dz^j$, of degree 1. The phrase "differential form" as applied to the expression $T_j\,dz^j$, suggests that we have in mind some useful interpretation of the symbol $\int T_j\,dz^j$. Indeed we do! Given a segment of a curve: $z^i = z^i(t)$, $a \leq t \leq b$, then we define

$$\int_a^b T_j\,dz^j = \int_a^b T_j \xi^j\,dt = \int_a^b T_j\,\dot{z}^j\,dt \tag{10}$$

(where $\xi^j = \dot{z}^j = dz^j/dt$ is the velocity vector), and we call this integral the *line integral* of the form ω along the curve segment. (This is another instance of an "integral of the second kind" (to borrow once again from the terminology of analysis).)

As previously, we now consider the corresponding "integral of the first kind", which involves whatever metric the underlying space may carry. This integral arises typically when we wish to integrate (along the curve segment $z^i = z^i(t)$, $a \leq t \leq b$) a scalar function $f(z)$ bearing some relation to the curve, in such a way that we need to introduce the element of measure (or, equivalently, element of length) $dl = |\dot{z}|\,dt$, on the curve. Thus our line integral of the second kind is defined to be

$$\int_a^b f(z(t))\,dl = \int_a^b f(z(t))|\dot{z}|\,dt. \tag{11}$$

Note that the element of length on the curve, namely $dl = |\dot{z}|\,dt$, may be regarded as just a 1-dimensional version of the n-dimensional volume element $d\sigma = \sqrt{|g|}\,dz^1 \wedge \cdots \wedge dz^n$ which we met with earlier, since when $n = 1$ we have $|g| = |g_{11}|$ and $\sqrt{|g_{11}|}\,dt = dl$, where $g_{11} = |\dot{z}|^2$, $t = z^1$.

The line integral of the second kind of a covector (or first-degree differential form) along a curve segment, has the following invariance properties:

(i) it is independent of the particular parametrization of the curve; i.e. if $t = t(\tau)$, where τ varies from a' to b' as t varies from a to b, then

$$\int_a^b T_\alpha \frac{dz^\alpha}{dt}\,dt = \int_{a'}^{b'} T_\alpha \frac{dz^\alpha}{d\tau}\,d\tau; \tag{12}$$

(ii) it is also independent of the co-ordinatization of the ambient space; i.e. if $z = z(y)$ is a co-ordinate change, so that $T'_\beta = T_\alpha(\partial z^\alpha/\partial y^\beta)$, then writing $z^i(t) = z^i(y(t))$, we have

$$\int_a^b T_\alpha \frac{dz^\alpha}{dt}\,dt = \int_a^b T'_\beta \frac{dy^\beta}{dt}\,dt. \tag{13}$$

To see this observe that at every point of the underlying space

$$T'_\beta\,dy^\beta = T_\alpha \frac{\partial z^\alpha}{\partial y^\beta}\,dy^\beta = T_\alpha\,dz^\alpha,$$

i.e. $T'_\beta\,\dot{y}^\beta = T_\alpha \dot{z}^\alpha$, so that the two integrands (and therefore their integrals) in (13) are the same.

To summarize our progress thus far: We have defined integrals (of each of the two kinds) of skew-symmetric tensors of rank n over regions of n-space, and of covector fields (i.e. skew-symmetric tensors of rank 1) along arbitrary arcs. We now proceed to the general rank k case $(1 \le k \le n)$.

It turns out that, as in the cases $k = 1, n$, it is appropriate to define the integral of a skew-symmetric tensor of rank k over (a region of) a k-dimensional surface in the ambient n-dimensional Cartesian space. Thus let

$$x^i = x^i(z^1, \ldots, z^k), \qquad i = 1, \ldots, n, \tag{14}$$

be parametric equations of a k-dimensional surface in the space with co-ordinates $x^1, \ldots, x^n$, and let U be a region of that surface, i.e. a region of the k-dimensional space with co-ordinates $z^1, \ldots, z^k$. How then do we define the integral of a rank k skew-symmetric tensor T defined on an n-space with co-ordinates $x^1, \ldots, x^n$, over a region U of a k-space with co-ordinates $z^1, \ldots, z^k$, given an embedding (i.e. a surface) $x^i = x^i(z^1, \ldots, z^k)$? Before giving the definition we remark first that we shall for convenience' sake use the differential form notation for skew-symmetric tensors, and second that by restricting T to the surface (14) (see §22.1) we may regard it as a tensor defined on the surface, i.e. as a rank k tensor defined on a k-dimensional space (i.e. the surface). Recall that by Theorem 22.1.2 this restriction is given by

$$\sum_{i_1 < \ldots < i_k} J^{i_1 \ldots i_k}_{1 \ldots k} T_{i_1 \ldots i_k}\,dz^1 \wedge \cdots \wedge dz^k = \sum_{i_1 < \ldots < i_k} T_{i_1 \ldots i_k}\,dx^{i_1} \wedge \cdots \wedge dx^{i_k}, \tag{15}$$

where the right-hand side is the original differential form of degree k (i.e. tensor of rank k) considered at the points of the surface.

26.1.2. Definition. *The integral of the second kind* of a skew-symmetric tensor $T = (T_{i_1 \ldots i_k})$ of rank k over a region U of a k-dimensional surface $x^i = x^i(z^1, \ldots, z^k)$, $i = 1, \ldots, n$, is defined to be the ordinary k-fold multiple integral (cf. (15))

$$\int \cdots \int_U T = \int \cdots \int_U \Bigg(\sum_{i_1 < \ldots < i_k} T_{i_1 \ldots i_k} J^{i_1 \ldots i_k}_{1 \ldots k} \Bigg)\,dz^1 \wedge \cdots \wedge dz^k. \tag{16}$$

This integral has the following two invariance properties (noted earlier for $k = 1$):

(i) *The integral is invariant under changes of co-ordinates on the surface.* This follows from the fact that by 18.2.3 under such a change, say $z^q = z^q(z')$, $q = 1, \ldots, k$, the restricted tensor

$$S_{1\ldots k}\, dz^1 \wedge \cdots \wedge dz^k = \left(\sum_{i_1 < \ldots < i_k} T_{i_1 \ldots i_k} J^{i_1 \ldots i_k}_{1 \ldots k} \right) dz^1 \wedge \cdots \wedge dz^k, \tag{17}$$

whose rank k is the same as that of the space on which it is defined, becomes, in terms of the new co-ordinates z',

$$S_{1\ldots k}\, J\, dz'^1 \wedge \cdots \wedge dz'^k, \tag{18}$$

where $J = \det(\partial z^i/\partial z'^j)$. However by (3) the multiple integrals over U of the expressions (17) and (18) are equal.

(ii) *The integral remains unchanged under changes of the co-ordinates of the n-dimensional space.* To see this, let $x^i = x^i(x')$, $i = 1, \ldots, n$, be such a co-ordinate change. Then since $dx^i = (\partial x^i/\partial x'^j)\, dx'^j$, and since by the rule for transforming the components of a tensor,

$$T_{i_1 \ldots i_k} = T'_{j_1 \ldots j_k} \frac{\partial x'^{j_1}}{\partial x^{i_1}} \cdots \frac{\partial x'^{j_k}}{\partial x^{i_k}},$$

it follows that

$$\sum_{i_1 < \ldots < i_k} T_{i_1 \ldots i_k}\, dx^{i_1} \wedge \cdots \wedge dx^{i_k} \equiv \sum_{j_1 < \ldots < j_k} T'_{j_1 \ldots j_k}\, dx'^{j_1} \wedge \cdots \wedge dx'^{j_k}.$$

The invariance of the integral then follows from (15).

We now turn to integrals of the first kind. We shall assume that the n-dimensional space is Euclidean with Euclidean co-ordinates $x^1, \ldots, x^n$. The metric induced on the surface $x^i = x^i(z^1, \ldots z^k)$, $i = 1, \ldots, n$, will be denoted by $g_{ij}\, dz^i\, dz^j$. Thus $g_{ij}\, dz^i\, dz^j = \sum_{q=1}^{n} (dx^q)^2$, where $dx^q = (\partial x^q/\partial z^\alpha)\, dz^\alpha$ (see §7.3). Recall also that $g_{ij} = g_{ji}$ and $dz^i\, dz^j = dz^j\, dz^i$. The element of volume on the surface is given as usual by

$$d\sigma = \sqrt{|g|}\, dz^1 \wedge \cdots \wedge dz^k; \qquad g = \det(g_{ij}). \tag{19}$$

26.1.3. Definition. *The integral* (*of the first kind*) of a function $f(z^1, \ldots, z^k)$ on the surface $x^i = x^i(z^1, \ldots, z^k)$, over a region U of that surface, is defined to be the integral of the function with respect to the volume element $d\sigma$; i.e. the following multiple integral:

$$\int \cdots \int_U f(z^1, \ldots, z^k) \sqrt{|g|}\, dz^1 \wedge \cdots \wedge dz^k. \tag{20}$$

It is important to note that while the integral of the second kind is not connected with any Riemannian geometry which may be present on the surface, or in the ambient space, on the other hand the integral of the first kind is connected with the geometry via the volume element

$$\sqrt{|g|}\, dz^1 \wedge \cdots \wedge dz^k.$$

This last is a skew-symmetric tensor of rank k (under co-ordinate changes with positive Jacobian), which is determined by the Riemannian metric g_{ij} on the surface, which metric is in turn determined (i.e. induced) by the Euclidean metric assumed on the space in which the surface is embedded.

26.2. Examples of Integrals of Differential Forms

(a) Our first example is the trivial one of the integral of a tensor of rank 0 (i.e. of a scalar $f(z)$) over a surface of dimension 0 (i.e. a single point P). By definition this integral is just the value of f at the point P, i.e. $f(P)$. We have an ulterior motive for mentioning this "trivial" example—it will be useful when we come to discuss the general Stokes formula.

(b) We have already discussed (in the preceding subsection) the integral of a covector field (i.e. of a differential form of degree 1) $(T_\alpha) = T_\alpha\, dx^\alpha$ along an arc $x^i = x^i(t)$, $a \le t \le b$. Thus:

$$\text{the integral of the second kind along the arc} = \int_a^b T_\alpha \frac{dx^\alpha}{dt}\, dt. \tag{21}$$

(c) The integral of a tensor field $(T_{ij}) = \sum_{i<j} T_{ij}\, dx^i \wedge dx^j$ (i.e. of a differential form of degree 2) over a region U of a 2-dimensional surface $x^i = x^i(z^1, z^2)$, $i = 1, \ldots, n$, is given by:

$$\begin{aligned} \text{the integral (of the second kind) over the surface} &= \iint_U \sum_{i<j} T_{ij}\, dx^i \wedge dx^j \\ &= \iint_U \left[\sum_{i<j} T_{ij} J^{ij}_{12}\right] dz^1 \wedge dz^2, \end{aligned} \tag{22}$$

where $T_{ij} = T_{ij}(x(z))$, $dx^i = (\partial x^i/\partial z^j)\, dz^j$, $dz^i \wedge dz^j = -\, dz^j \wedge dz^i$.

Before proceeding to our third example we examine (b) and (c) in detail in the case where $n = 3$, and the co-ordinates x^1, x^2, x^3 are Euclidean.

(b′) In this (Euclidean) context the integral of a covector field along an arc is usually written as

$$\int_a^b T_\alpha \frac{dx^\alpha}{dt}\, dt = \int_P^Q T\xi\, dt,$$

where $\xi = \dot{x}$, $T = (T_\alpha) = (T^\alpha)$, $P = (x^1(a), x^2(a), x^3(a))$, $Q = (x^1(b), x^2(b), x^3(b))$, $a \le t \le b$, and $T\xi = \langle T, \xi\rangle$, the Euclidean scalar product. (Recall that, provided we restrict ourselves to co-ordinate changes which are Euclidean isometries, the distinction between lower and upper indices disappears.)

(c′) By §7.2 the vectors

$$\xi = (\xi^i) = \left(\frac{\partial x^i}{\partial z^1}\right) = \frac{\partial x^i}{\partial z^1} e_i,$$

$$\eta = (\eta^i) = \left(\frac{\partial x^i}{\partial z^2}\right) = \frac{\partial x^i}{\partial z^2} e_i,$$

form a basis for the tangent plane at each (non-singular) point of the surface $x^i = x^i(z^1, z^2)$, $i = 1, 2, 3$. If the co-ordinates x^1, x^2, x^3 are Euclidean, then the vector product $[\xi, \eta]$ will be normal to the surface at each point. The vector product $[\xi, \eta]$ is really the tensor $S^{ij} = (\xi^i\eta^j - \xi^j\eta^i)$, with which we customarily associate in a one-to-one fashion the vector S^i defined by

$$S^1 = S^{23}, \qquad S^2 = -S^{13}, \qquad S^3 = S^{12}.$$

It is clear that $S^{ij} = J^{ij}_{12}$, the indicated minor of the Jacobian matrix $(\partial x^\alpha/\partial z^\beta)$. It is also clear that the length of the vector $[\xi, \eta]$ is $\sqrt{g}$, where $g = \det(g_{ij})$, the determinant of the metric g_{ij} induced on the surface by the Euclidean metric (see §7.3):

$$g_{ij}\, dz^i\, dz^j = \sum_{i=1}^{3} (dx^i)^2; \qquad dx^i = \frac{\partial x^i}{\partial z^j} dz^j.$$

It follows from all this and (22) that when the co-ordinates x^1, x^2, x^3 are Euclidean, the integral of a differential form $T_{ij}\, dx^i\, dx^j$ of degree 2, over a region U of a surface $x^i = x^i(z^1, z^2)$ (sometimes called the "flux" of T through the region U), is given by

$$\iint_U T_{ij}\, dx^i \wedge dx^j = \iint_U \left(\sum_{i<j} T_{ij} J^{ij}_{12}\right) dz^1 \wedge dz^2$$

$$= \iint_U \langle T, [\xi, \eta]\rangle\, dz^1 \wedge dz^2 = \iint_U \langle T, n\rangle \sqrt{|g|}\, dz^1 \wedge dz^2,$$

where n is the unit normal vector:

$$n = \frac{[\xi, \eta]}{|[\xi, \eta]|} = \frac{[\xi, \eta]}{\sqrt{|g|}}.$$

We formulate this result as a theorem.

26.2.1. Theorem. *In Euclidean 3-space the integral (of the second kind) of a form T of degree 2 over a region U of a surface $x^i = x^i(z^1, z^2)$, reduces to the following integral of the first kind:*

$$\iint_U T_{ij}\, dx^i \wedge dx^j = \iint_U \langle T, n\rangle \sqrt{|g|}\, dz^1 \wedge dz^2,$$

where x^1, x^2, x^3 are Euclidean co-ordinates, n is the unit normal to the surface, T is the vector associated in the usual way with the tensor T_{ij}, and $g_{ij}\, dz^i\, dz^j = \sum_{i=1}^{3} (dx^i)^2$, $dx^i = (\partial x^i/\partial z^j)\, dz^j$.

Remark. It can be shown that, in contrast with this result, if the space is 4-dimensional then in general the evaluation of the integral of a form of degree 2 over a region of a 2-dimensional surface does not reduce to performing operations on vectors only, even in the Euclidean case.

In the case when (as at present) the space is 3-dimensional Euclidean with Euclidean co-ordinates x^1, x^2, x^3, so that, as already observed, rank 2 skew-symmetric forms reduce in a certain specific sense to vectors, and the distinction between vectors and covectors disappears, there are two integrals of a vector field $(T_\alpha) = (T^\alpha)$ of special importance. These are:

(i) the "line" integral of (T_α) around a closed curve Γ: $\oint_\Gamma T_\alpha\, dx^\alpha$;
(ii) the "surface" integral of (T_α) over a closed surface U:

$$\iint_U \langle T, n\rangle \sqrt{|g|}\, dz^1 \wedge dz^2$$

(with the same symbolism as in the above theorem).

The integral (i) around the closed curve Γ: $x^i = x^i(t)$, $a \le t \le b$, $x^i(a) = x^i(b)$, $i = 1, 2, 3$, which, as we have seen may be written as

$$\oint_\Gamma T_\alpha \xi^\alpha\, dt,$$

is called in analysis the *circulation* of the field around the closed curve Γ.

In the case of the integral (ii) by "closed surface" we mean the boundary of a bounded region $f(x^1, x^2, x^3) \le 0$. The surface integral of (T_α) over the closed surface U, which by Theorem 26.2.1 is just

$$\iint_U T_{ij}\, dx^i \wedge dx^j,$$

where, as usual, $T_1 = T_{23}$, $-T_2 = T_{13}$, $T_3 = T_{12}$, is called the *total* or *net flux* through the surface U of the tensor field $(T_{ij}) = -(T_{ji})$, or (in our present, Euclidean, context) of the vector field $T = (T^1, T^2, T^3)$.

In connexion with the latter integral, recall that if a non-singular surface U is given by an equation $f(x^1, x^2, x^3) = 0$, then there may not exist a set of parametric equations $x^i = x^i(z^1, z^2)$, $i = 1, 2, 3$, for the whole of U. However in a neighbourhood of each (non-singular) point of U there does exist such a parametrization (see §7.1). Thus since the integral is independent of the co-ordinates on the surface, in order to evaluate it over the whole of U (i.e. to extend the definition of integral to surfaces which are not globally parametrizable), we may resort simply to subdividing U into non-overlapping pieces (or "patches") on each of which a parametrization exists, evaluating the integral over each piece using local co-ordinates defined on that piece, and then summing these integrals over the pieces. For example if U is the sphere $(x^1)^2 + (x^2)^2 + (x^3)^2 = 1$, then as the pieces we may take the upper and lower hemispheres.

We now turn to our fourth and final example.

(d) Suppose that in Euclidean n-space with Euclidean co-ordinates $x^1, \ldots, x^n$ we are given a hypersurface M^{n-1} defined by

$$F(x^1, \ldots, x^n) = 0, \qquad \text{grad } F \neq 0, \tag{23}$$

or, locally, by $x^\alpha = x^\alpha(y^1, \ldots, y^{n-1})$. Consider the form

$$K\,d\sigma = K\sqrt{|g|}\,dy^1 \wedge \cdots \wedge dy^{n-1}, \tag{24}$$

on the surface (called the "curvature form" of the surface). Here if $n = 2$, K is the curvature of the curve (23), while if $n = 3$, K is the Gaussian curvature of the surface (23); we shall not give the definition of K for $n > 3$, since our main interest lies with the cases $n = 2, 3$.

Let S^{n-1} denote, as usual, the unit $(n - 1)$-dimensional sphere $\sum_1^n (x^\alpha)^2 = 1$. We denote by Ω_{n-1} the $(n - 1)$-dimensional volume element on the sphere S^{n-1}, invariant under orthogonal co-ordinate transformations, given by:

$$\begin{aligned} \Omega_{n-1} &= d\varphi, \quad \text{for } n = 2; \\ \Omega_{n-1} &= \sin\theta\, d\theta\, d\varphi, \quad \text{for } n = 3. \end{aligned} \tag{25}$$

We define the *Gauss map* (or "normal spherical map") ψ of the surface M^{n-1} to the sphere S^{n-1} to be the map sending each point P of the surface to the tip of the unit normal n_P to the surface at P after the vector n_P has been transported so that its tail is at the origin of co-ordinates. Since the tip of the unit vector n_P is indeed a point of the unit sphere S^{n-1} when its tail is at the origin, this does define a map (the Gauss map)

$$\psi : M^{n-1} \to S^{n-1}. \tag{26}$$

The following theorem links the curvature form (24) and the volume element (25) in a natural way via the Gauss map (26). Specifically, it tells us that the restriction (using the restriction (or "pullback") operation ψ^*—see §22.1) of the form (volume element) Ω_{n-1} on S^{n-1}, is just the curvature form $K\,d\sigma$ on M^{n-1}. (Although we have not supplied the appropriate definitions for $n > 3$ we nonetheless give the general statement. Alternatively, we may regard formula (27) of the theorem as defining K for hypersurfaces M^{n-1}, $n > 3$.)

26.2.2. Theorem. *If $x^\alpha = x^\alpha(y^1, \ldots, y^{n-1})$ is a hypersurface in Euclidean n-space with Euclidean co-ordinates $x^1, \ldots, x^n$, then for all $n \geq 2$*

$$K\,d\sigma = \psi^*(\Omega_{n-1}). \tag{27}$$

Thus in the cases $n = 2, 3$, we have:

for $n = 2$, $K\,dl = \psi^*(d\varphi)$, where K is the curvature of the curve; for $n = 3$, $K\sqrt{g}\,dy^1 \wedge dy^2 = \psi^*(\Omega)$, where K is the Gaussian curvature of the surface, and $\Omega = \sin\theta\,d\theta\,d\varphi$.

PROOF. We give the proof for the case $n = 3$ only. (The proof is similar in the case $n = 2$, and, once given the definition of K, for $n > 3$ also.)

Thus let P be a point on the surface $M^{n-1} = M^2$, and take $x^1 = x$, $x^2 = y$, $x^3 = z$ to be Euclidean co-ordinates with $P = (1, 0, 0)$ and the x-axis normal to the surface at P (so that the y- and z-axes are tangent to M^2 at P). In terms of these co-ordinates the surface will in a neighbourhood of P be given by an equation $x = f(y, z)$, with $f_y = f_z = 0$ at P. By Definition 8.1.1 the Gaussian curvature of the surface in that neighbourhood is

$$K = \det\begin{pmatrix} f_{yy} & f_{yz} \\ f_{zy} & f_{zz} \end{pmatrix}. \tag{28}$$

As before, we take $S^{n-1} = S^2$ to be the sphere $x^2 + y^2 + z^2 = 1$. Thus $\psi(P) = P$, and in a neighbourhood of P the sphere has equation

$$x = \sqrt{1 - y^2 - z^2},$$

so that y, z may also be regarded as local co-ordinates about P on the sphere S^2. In terms of spherical co-ordinates these local co-ordinates on S^2 are given by $y = \sin\theta\sin\varphi$, $z = \cos\theta$, whence

$$\begin{aligned} dy \wedge dz &= (\cos\varphi\sin\theta\,d\varphi + \sin\varphi\cos\theta\,d\theta) \wedge (-\sin\theta\,d\theta) \\ &= \cos\varphi\sin^2\theta\,d\theta \wedge d\varphi. \end{aligned}$$

Hence at P, where $\varphi = 0$, $\theta = \pi/2$, we have

$$dy \wedge dz = d\theta \wedge d\varphi = \sin\theta\,d\theta\,d\varphi = \Omega. \tag{29}$$

Now let P' be a point near P on the surface M^2. By §8.3(23) the normal to M^2 at P' is

$$n_{P'} = \left(\frac{1}{\sqrt{1 + f_y^2 + f_z^2}}, \frac{-f_y}{\sqrt{1 + f_y^2 + f_z^2}}, \frac{-f_z}{\sqrt{1 + f_y^2 + f_z^2}}\right),$$

so that under the map ψ, P' goes to the point

$$(\hat{y}, \hat{z}) = \left(\frac{-f_y}{\sqrt{1 + f_y^2 + f_z^2}}, \frac{-f_z}{\sqrt{1 + f_y^2 + f_z^2}}\right) \tag{30}$$

on the sphere (in the co-ordinates y, z on the sphere). Now by (29) and the definition of the tensor operation ψ^* (see §22.1), we have

$$\psi^*(\Omega)\bigg|_P = \left(\frac{\partial\hat{y}}{\partial y}\frac{\partial\hat{z}}{\partial z} - \frac{\partial\hat{y}}{\partial z}\frac{\partial\hat{z}}{\partial y}\right)\bigg|_P dy \wedge dz.$$

It follows from this and (28) by explicitly calculating the expression $(\partial\hat{y}/\partial y)(\partial\hat{z}/\partial z) - (\partial\hat{y}/\partial z)(\partial\hat{z}/\partial y)$, and then setting $f_x = f_y = 0$ (since we wish to evaluate it at P), that $\psi^*(\Omega)|_P = K\,dy \wedge dz$, where K is the Gaussian curvature of M^2 at P. Now by §7.3(20), the induced metric g_{ij} on M^2 is just δ_{ij} at P, so that $g = 1$ at P. We thus finally arrive at the desired formula

$$K\sqrt{g}\,dy \wedge dz = \psi^*(\Omega_2),$$

completing the proof (for $n = 3$ at least). (Note that when $n = 2$, we have $\Omega = d\varphi$, and, since the surface is in this case a curve $x^1 = x^1(y)$, $x^2 = x^2(y)$, $K\sqrt{g}\,dy$ becomes $K\,dl$ where dl is the element of length.) □

26.3. The General Stokes Formula. Examples

As remarked in the preceding subsection (at the end of Example (c′)) we can extend the definition (26.1.2) of the integral of a form of degree k over a k-dimensional surface in an n-dimensional space, to surfaces U which are not parametrizable as wholes in the form $x^i = x^i(z^1, \ldots, z^k)$. Since the integral as defined is invariant under changes of the spatial co-ordinates and also of the co-ordinates on the surface, and since the integral over a union of non-overlapping regions is the sum of the integrals over the individual regions (i.e. the integral is "additive"), we may extend the definition appropriately to such surfaces U by subdividing U into several patches each of which admits a parametrization, evaluating the integral on each patch, and adding the results.

Before broaching the theme of this subsection we make the further parenthetical remark that one often wishes to integrate over surfaces which, though they may indeed admit global systems of co-ordinates, suffer from the defect that such systems have singular points (see Definition 1.2.4). For instance, as we saw in §1.2, the polar co-ordinate system r, φ on the plane has the singular point $r = 0$, while in 3-space co-ordinatized by cylindrical co-ordinates r, φ, z, all points on the z-axis ($r = 0$) are singular, as is also the case if spherical co-ordinates r, φ, θ are used (the z-axis being characterized by $r = 0$ (the origin), and $\theta = 0, \pi$ (points on the z-axis other than the origin)). On a sphere centred at the origin the system of co-ordinates θ, φ has the north and south poles ($\theta = 0, \pi$) as singular points. (The sphere is the simplest surface which does not admit a global parametrization without singular points.) However in all of these examples the set of singular points is "small" (in technical terms, "of measure zero"), making no contribution to the integral, so that we can ignore them.

We now turn to "Stokes' formula". From a second course in calculus the reader will recall various formulae equating integrals over k-dimensional surfaces (or regions of them) in n-dimensional space ($n = 2, 3$) to related integrals over the boundaries of the surfaces (or regions). (The results in question go under the following names: for $n = 2$, $k = 2$, "Green's formula"; for $n = 3$, $k = 3$, the "Gauss–Ostrogradskiĭ formula"; and for $n = 3$, $k = 2$, "Stokes' formula".) We shall now see how these various formulae are subsumed in a single "general Stokes formula", couched in the language of differential forms.

As mentioned before, in view of the additivity of the multiple integral, we may in our treatment of integrals restrict the discussion to suitable components of the surface over which we are integrating. Set U be a region of a k-dimensional space with co-ordinates $z^1, \ldots, z^k$, defined by an inequality of the form $f(z^1, \ldots, z^k) \leq C$, let Γ denote its boundary, defined by the equation $f(z^1, \ldots, z^k) = C$, and finally, suppose we have an embedding

$$x^i = x^i(z^1, \ldots, z^k), \qquad i = 1, \ldots, n,$$

of the region U with its boundary into an n-dimensional Cartesian space with co-ordinates $x^1, \ldots, x^n$. This embedding turns U into a parametrically defined surface of dimension k, and its boundary Γ into a surface of dimension $(k - 1)$, situated in n-dimensional space.

Stokes' formula equates the integral over the surface U of the differential of a form defined on n-space, with the integral of the form itself over the boundary Γ of U. Before stating the general formula we consider the very simplest case of it, namely $k = 1$. In that case $x^i = x^i(t)$ (putting $z^1 = t$) is a curve, of which U is a segment defined by $a \leq t \leq b$, with boundary Γ consisting of the two end-points $t = a, t = b$. In order for the Stokes' formula to have the correct interpretation in this case, we need to define a 0-dimensional surface to be a finite set of points to each of which a positive or negative sign is attached in a purely formal way. The integral of a function $f(x)$ over such a 0-dimensional surface is then defined to be simply the sum of the values of the function at the points, taken with the corresponding signs (cf. Example 26.2(a)); in other words the integral is $\sum_i \pm f(P_i)$, where $+$ or $-$ is taken according as $+$ or $-$ is assigned to P_i. Thus as an 0-dimensional surface the boundary Γ of U consists of the two points a, b where we assign $-$ to a and $+$ to b, and the integral of f over Γ is just $f(Q) - f(P)$, where $P = (x^i(a))$, $Q = (x^i(b))$. Now the "Fundamental Theorem of Calculus" of Newton and Leibniz tells us that

$$\int_\Gamma f = f(Q) - f(P) = \int_U df = \int_a^b \frac{\partial f}{\partial x^i}\frac{dx^i}{dt}\,dt. \tag{31}$$

Formula (31) is indeed the simplest case of Stokes' formula, equating as it does the integral over the boundary Γ, with an integral over the region U. The general Stokes formula can be regarded simply as a direct generalization of the formula (31) of Newton and Leibniz to several dimensions, and, moreover, in essence reduces to the formula (31).

We now return to the general situation of a region U of a k-dimensional surface $x^i = x^i(z^1, \ldots, z^k)$, where U is defined by $f(z^1, \ldots, z^k) \leq C$, and has boundary Γ given by $f(z^1, \ldots, z^k) = C$. Suppose that we are given a form of degree $(k - 1)$ (i.e. a skew-symmetric tensor of type $(0, k - 1)$) defined on our n-dimensional space with co-ordinates $x^1, \ldots, x^n$, which is to be integrated over the $(k - 1)$-dimensional boundary Γ of U.

26.3.1. Theorem (The General Stokes Formula). *For any differential form*

$$T = \sum_{i_1 < \dots < i_{k-1}} T_{i_1 \dots i_{k-1}} \, dx^{i_1} \wedge \cdots \wedge dx^{i_{k-1}}$$

with smooth coefficients $T_{i_1 \dots i_{k-1}}(x)$, *and for any bounded region* U *of a smooth surface* $x^i = x^i(z^1, \dots, z^k)$, *where the boundary* Γ *of* U *is piecewise smooth and simple (i.e. has no self-intersections), the following formula holds:*

$$\pm \int_\Gamma T = \int_U dT. \tag{32}$$

(*The simplest case of this formula, namely* $k = 1$, *is given in* (31) *above.*) *Here the form* dT *(of degree* k*) is the gradient or differential of the form* T *(of degree* $k - 1$*).*

The various 2- and 3-dimensional cases (i.e. $n = 2, 3$) of this formula (named after Green, Gauss–Ostrogradskiĭ, and Stokes, as we noted earlier) are usually given separate proofs in analysis courses. We shall now examine these important special cases in detail.

(i) *The planar case* ($n = 2$). Here Γ is a simple, closed, piecewise smooth curve $x^i = x^i(t)$, $x^i(a) = x^i(b)$, bounding the region U of the plane. If (T_α) is a smooth covector field (i.e. form of degree 1) defined on U (and its boundary Γ) then the integral of T_α around Γ is defined, and by Stokes' formula (32)

$$\pm \int_\Gamma T_\alpha \frac{dx^\alpha}{dt} \, dt = \iint\limits_U \left(\frac{\partial T_2}{\partial x^1} - \frac{\partial T_1}{\partial x^2} \right) dx^1 \wedge dx^2,$$

or, putting $x^1 = x$, $x^2 = y$,

$$\pm \int_\Gamma T_1 \, dx + T_2 \, dy = \iint\limits_U \left(\frac{\partial T_2}{\partial x} - \frac{\partial T_1}{\partial y} \right) dx \wedge dy. \tag{33}$$

(It can be shown that the appropriate sign is $+$ provided the curve Γ is traversed by t in the counter-clockwise direction.) This particular case of the general Stokes formula is known as "Green's formula".

We now digress a little in order to show how from Green's formula (33) one can obtain the well-known "Residue Theorem" of complex function theory. For this purpose we regard our plane with co-ordinates x, y, as the complex plane co-ordinatized by $z = x + iy$. Let $f(z) = f(x, y) = u(x, y) + i\,v(x, y)$ be a complex-valued function of the complex variable z. The integral of f around Γ may be defined by

$$\oint_\Gamma f(z) \, dz = \oint (u + iv)(dx + i\,dy) = \oint_\Gamma (u\,dx - v\,dy) + i \oint_\Gamma (v\,dx + u\,dy).$$

Applying Green's formula (33), we obtain (assuming f smooth on U)

$$\oint_\Gamma f(z) \, dz = \iint\limits_U \left(\frac{\partial u}{\partial y} + \frac{\partial v}{\partial x} \right) dx \wedge dy + i \iint\limits_U \left(\frac{\partial v}{\partial y} - \frac{\partial u}{\partial x} \right) dx \wedge dy.$$

From this we conclude that if the functions u, v are smooth on U (including the boundary Γ), i.e. if f is smooth on U, and if they satisfy the Cauchy–Riemann equations (see §12.1(15))

$$\frac{\partial u}{\partial y} = -\frac{\partial v}{\partial x}, \qquad \frac{\partial v}{\partial y} = \frac{\partial u}{\partial x},$$

then the form $f(z)\,dz$ has zero integral around Γ (i.e. satisfies the conclusion of Cauchy's Integral Theorem). It follows in particular that for any non-negative integer n, and any simple closed path Γ, we have $\oint_\Gamma z^n\,dz = 0$; for negative integers n the same is true provided only that 0 does not lie in the interior of Γ, while in the contrary case the integral is independent of Γ provided that it is specified in which direction (counter-clockwise, say) Γ is to be traversed in evaluating the integral. Thus to calculate the integral around such closed paths Γ we may choose at our convenience any particular one. Taking therefore the unit circle $z(t) = e^{it}$, $0 \le t \le 2\pi$, as our simple closed path, after a simple calculation we obtain

$$\oint_\Gamma z^n\,dz = \begin{cases} 0, & \text{if } n \neq -1, \\ 2\pi i, & \text{if } n = -1, \end{cases} \tag{34}$$

for any simple, closed path Γ enclosing the origin. From this we can now easily deduce the important "Residue Theorem". Suppose $f(z)$ is represented by a power series on its region of convergence.

$$f(z) = \sum_{n=-\infty}^{\infty} c_n(z-a)^n.$$

Then if Γ is any simple closed path enclosing the point a and lying in the interior of the region of convergence of the series (where the series converges uniformly and so can be integrated term-by-term), it follows from (34) that

$$\oint_\Gamma f(z)\,dz = 2\pi\, i c_{-1},$$

$$\oint_\Gamma (z-a)^{-k} f(z)\,dz = 2\pi\, i c_{k-1}. \tag{35}$$

These formulae for the "residues" of the left-hand side integrals are useful for computing the coefficients of the Taylor series (the case $c_n = 0$ for all $n < 0$), or Laurent series of an analytic function $f(z)$, and, conversely, knowledge of the coefficients in the series allows calculation of the integrals.

We now turn to the two 3-dimensional cases ($n = 3$). As usual we let x^1, x^2, x^3 be co-ordinates on the underlying space.

(ii) *The three-dimensional case with* $k = 3$. Here U is a bounded region of 3-space and Γ its boundary. By the general Stokes formula (32) we have

$$\pm \iint_\Gamma \sum_{i<j} T_{ij}\, dx^i \wedge dx^j = \iiint_U \left(\frac{\partial T_{12}}{\partial x^3} + \frac{\partial T_{23}}{\partial x^1} - \frac{\partial T_{13}}{\partial x^2} \right) dx^1 \wedge dx^2 \wedge dx^3.$$

We shall now assume that x^1, x^2, x^3 are Euclidean co-ordinates. Then by Theorem 26.2.1

$$\iint_\Gamma \sum_{i<j} T_{ij}\, dx^i \wedge dx^j = \iint_\Gamma \langle T, n \rangle \sqrt{|g|}\, dz^1 \wedge dz^2, \tag{36}$$

where T is the vector defined by $T^1 = T_{23}$, $T^2 = -T_{13}$, $T^3 = T_{12}$, n is the unit normal to the surface Γ at each point, z^1, z^2 are co-ordinates on Γ, and $g = \det(g_{ij})$ where g_{ij} is the induced metric on Γ. Since

$$\frac{\partial T_{12}}{\partial x^3} - \frac{\partial T_{13}}{\partial x^2} + \frac{\partial T_{23}}{\partial x^1} = \frac{\partial T^i}{\partial x^i} = \operatorname{div} T,$$

we finally obtain (putting $d\sigma = \sqrt{|g|}\, dz^1 \wedge dz^2$, the element of area on Γ)

$$\iint_\Gamma \langle T, n \rangle \sqrt{|g|}\, dz^1 \wedge dz^2 = \iint_\Gamma \langle T, n \rangle\, d\sigma = \iiint_U (\operatorname{div} T)\, dx^1 \wedge dx^2 \wedge dx^3, \tag{37}$$

which is the well-known "Gauss–Ostrogradskiĭ formula" in 3-dimensional Euclidean space.

(iii) *The three-dimensional case with* $k = 2$. Here U is a region on the surface $x^i = x^i(z^1, z^2)$, $i = 1, 2, 3$, and Γ is its boundary. Thus in this case the general Stokes formula (32) becomes

$$\pm \oint_\Gamma T_\alpha\, dx^\alpha = \iint_U \left[\left(\frac{\partial T_2}{\partial x^1} - \frac{\partial T_1}{\partial x^2} \right) dx^1 \wedge dx^2 + \left(\frac{\partial T_3}{\partial x^1} - \frac{\partial T_1}{\partial x^3} \right) dx^1 \wedge dx^3 + \left(\frac{\partial T_3}{\partial x^2} - \frac{\partial T_2}{\partial x^3} \right) dx^2 \wedge dx^3 \right]. \tag{38}$$

When the underlying space is Euclidean with Euclidean co-ordinates x^1, x^2, x^3, then (as already noted on more than one occasion) the distinction between covectors and vectors disappears (provided we restrict the co-ordinate changes to being Euclidean isometries), and there is a one-to-one correspondence between tensors (T_{ij}) and vectors $T = (T^i)$ (see e.g. (ii) above). Taking this (and 26.2.1) into account in (38) we arrive at the following Euclidean version of that formula ("Stokes' formula"):

$$\oint_\Gamma T_\alpha \, dx^\alpha = \iint_U \langle \text{curl } T, n \rangle \sqrt{|g|} \, dz^1 \wedge dz^2. \tag{39}$$

Here curl T is, as usual, the vector corresponding to the skew-symmetric tensor $S_{\alpha\beta} = \partial T_\beta / \partial x^\alpha - \partial T_\alpha / \partial x^\beta$, i.e.

$$\text{curl } T = \left(\frac{\partial T_3}{\partial x^2} - \frac{\partial T_2}{\partial x^3}, \frac{\partial T_1}{\partial x^3} - \frac{\partial T_3}{\partial x^1}, \frac{\partial T_2}{\partial x^1} - \frac{\partial T_1}{\partial x^2} \right).$$

We have thus seen that in the three cases (i), (ii), (iii) the general Stokes formula (32) reduces (with the aid of Theorem 26.2.1) to staple formulae from a second course in calculus. (We may therefore consider the general Stokes formula as proved for $n = 2, 3$, by simply appealing to those known formulae.)

In conclusion we mention that for the general Stokes formula to hold it is not essential that the boundary Γ consist of a single connected component. If Γ consists of several connected pieces (as for instance in Figure 31), then the integral over the whole of Γ is the sum of the integrals over the separate components, each such integral being taken with the correct sign ($\pm$) prefixed. In connexion with choosing the correct signs for these integrals, we mention only that in the case of the integral $\iint_\Gamma \langle T, n \rangle \sqrt{|g|} \, dz^1 \wedge dz^2$, the signs to be prefixed are determined by the direction of the normals n to the components of Γ (or, as they say, the "orientation" of the boundary components is taken to be that induced by the "orientation" of U—see §26.4 for an explanation of this).

We turn now to an important application of the general Stokes formula to electromagnetic field theory. As usual, the underlying space will be 4-

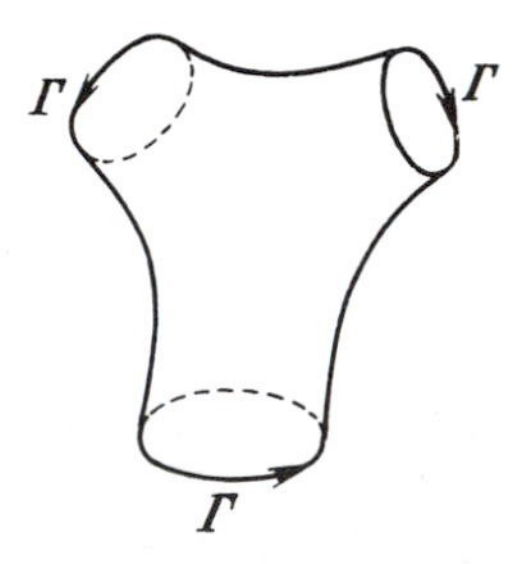

Figure 31

dimensional Minkowski space with co-ordinates $x^0 = ct, x^1, x^2, x^3$ (where c is the speed of light and t the time), with metric

$$(g_{ij}) = \begin{pmatrix} +1 & & & 0 \\ & -1 & & \\ & & -1 & \\ 0 & & & -1 \end{pmatrix},$$

and we shall denote the electromagnetic field tensor by $F_{ik} = -F_{ki}$, $i, k = 0, 1, 2, 3$. We wish now to examine the behaviour of this tensor under co-ordinate transformations which do not change the time co-ordinate $x^0 = ct$, i.e. under under transformations of the form

$$x'^0 = x^0, \qquad x'^i = x'^i(x^1, x^2, x^3).$$

In §21.1 we defined, in terms of the tensor F_{ik}, the electric field covector $E_\alpha = F_{0\alpha}$, $\alpha = 1, 2, 3$, and the magnetic field tensor $H_{\alpha\beta} = -H_{\beta\alpha} = F_{\alpha\beta}$, $\alpha, \beta, = 1, 2, 3$. Since the co-ordinates x^1, x^2, x^3 for fixed x^0 are (essentially) Euclidean, the magnetic field tensor may be regarded (as it was in §21.1) as a vector field H^α defined (in the usual way) by

$$H^1 = H_{23}, \qquad H^2 = -H_{13}, \qquad H^3 = H_{12}.$$

Recall from Example 25.2.5(d) that in the differential form notation the first pair of Maxwell's equations may be written as

$$d(F_{ij}\, dx^i \wedge dx^j) = 0,$$

which in classical notation, in terms of the electric and magnetic fields, comes down to the following pair of equations (see (19) and (20) of §25.2):

$$\operatorname{div} \mathbf{H} = \frac{\partial H^i}{\partial x^i} = 0, \qquad \operatorname{curl} \mathbf{E} + \frac{1}{c}\frac{\partial \mathbf{H}}{\partial t} = 0. \tag{40}$$

From the first of these equations and the Gauss–Ostrogradskiĭ formula (37) we obtain

$$\iiint_U \operatorname{div} \mathbf{H}\, dx^1 \wedge dx^2 \wedge dx^3 = \iint_\Gamma \langle \mathbf{H}, n \rangle\, d\sigma = 0,$$

where Γ is the boundary of a region U in 3-space. This result is usually stated in words as follows: *The flux of a magnetic field through a closed surface is always zero.*

From the second equation in (40) and Stokes' formula (39), which with $T = \mathbf{E}$ becomes

$$\iint_U \langle \operatorname{curl} \mathbf{E}, n \rangle\, d\sigma = \oint_\Gamma E_\alpha\, dx^\alpha,$$

where Γ is the boundary of a region U of a (2-dimensional) surface, we obtain

$$-\iint_U \left\langle \frac{1}{c}\frac{\partial \mathbf{H}}{\partial t}, n \right\rangle d\sigma = \oint_\Gamma E_\alpha \, dx^\alpha.$$

This classical formula can be stated in words as follows: *The time rate of change of the flux of a magnetic field through a surface is equal to the circulation of the electric field around the boundary of the surface.*

The second pair of Maxwell's equations has the form (cf. §25.2(21))

$$\sum_{k=1}^{3} \frac{\partial F_{ik}}{\partial x^k} - \frac{1}{c}\frac{\partial F_{0i}}{\partial t} = \frac{4\pi}{c} j_{(4)i}, \tag{41}$$

where $j_{(4)} = (c\rho, j_1, j_2, j_3)$, ρ being the charge density, and $\mathbf{j} = (j_1, j_2, j_3)$ the 3-dimensional current vector. In 3-dimensional form (41) becomes the pair of equations (see (22), (23) of §25.2)

$$\operatorname{div} \mathbf{E} = 4\pi\rho, \qquad \operatorname{curl} \mathbf{H} + \frac{1}{c}\frac{\partial \mathbf{E}}{\partial t} = \frac{4\pi}{c}\mathbf{j}. \tag{42}$$

The first of these and the Gauss–Ostrogradskiĭ formula (37) give

$$\iiint_U 4\pi\rho \, dx^1 \wedge dx^2 \wedge dx^3 = \iint_\Gamma \langle \mathbf{E}, n \rangle \, d\sigma,$$

or, in words: *The flux of an electric field through the boundary of a region of space is equal to* 4π *times the total charge contained in the region.*

From the second of the equations (42) and Stokes' formula (39) we obtain

$$-\frac{1}{c}\iint_U \left\langle \frac{\partial \mathbf{E}}{\partial t}, n \right\rangle d\sigma + \iint_U \frac{4\pi}{c} \langle \mathbf{j}, n \rangle \, d\sigma = \oint_\Gamma H_\alpha \, dx^\alpha,$$

which can be read as follows: *The net current through a surface less the time rate of change of the flux of the electric field through the surface equals the circulation of the magnetic field around the boundary of the surface.*

Thus by means of the general Stokes formula, the geometric content of Maxwell's equations is made evident.

We end this section by repeating the significant fact that in the first pair of Maxwell's equations the metric is not involved in any way, while the second pair cannot be formulated without it. (This is most easily perceived when Maxwell's equations are written in the notation of differential forms—see (18) and (21) of §25.2.)

26.4. Proof of the General Stokes Formula for the Cube

In the preceding subsection we showed that the classical integral formulae of Green, Gauss–Ostrogradskiĭ, and Stokes, are special cases of the general Stokes formula (32). In the present subsection we shall prove the general Stokes formula for arbitrary k and n ($k \leq n$) indeed, but assuming U to be rather a special surface, namely a k-dimensional cube. (Though special, this case is nonetheless germane to the proof of the general formula.)

A *singular cube* σ in $\mathbb{R}^n$ is defined to be a smooth map $\sigma: I^k \to \mathbb{R}^n$, where I^k is the k-dimensional Cartesian cube

$$I^k = \{(x^1, \ldots, x^k) \mid 0 \leq x^\alpha \leq 1\}.$$

For each α, the equations $x^\alpha = 0$, $x^\alpha = 1$ define two $(k-1)$-dimensional faces of I^k, denoted by I_α^-, I_α^+ respectively (see Figure 32). We denote by ∂I^k the boundary of the cube:

$$\partial I^k = \bigcup_\alpha (I_\alpha^+ \cup I_\alpha^-).$$

Before we state the general Stokes formula for the cube in $\mathbb{R}^n$, we indicate briefly how the sign $(\pm)$ is determined in (32), at least in the present simple situation. We say that two (ordered) bases for a finite-dimensional vector space V define the same *orientation* of V if the determinant of the matrix transforming one to the other is positive. An *orientation of the cube* I^k is then understood to be defined, if at each (non-singular) point x of the cube an orientation μ_x of the tangent space to I^k at x is given in such a way that the (representative) frame μ_x varies continuously as x varies over I^k. (For I^k we can take the standard basis $\{e_1, \ldots, e_k\}$ as defining μ_x for all x.) The *induced orientation on the boundary* ∂I^k is then defined as follows. Let $n(x)$ denote the unit outward normal to the cube at $x \in \partial I^k$. Then the tangent space to ∂I^k at x is given that orientation $\{v_1, \ldots, v_{k-1}\}$ such that

$$\{n(x), v_1, \ldots, v_{k-1}\}$$

is μ_x, the given orientation of I^k, i.e. such that the matrix of the change from $\{e_1, \ldots, e_k\}$ to $\{n(x), v_1, \ldots, v_{k-1}\}$ has positive determinant. (Thus in the 2-dimensional case depicted in Figure 32, the induced orientation of the boundary is the counter-clockwise one, so that for calculating the integral the boundary is to be parametrized so that it is traversed in the counter-clockwise direction as the parameter increases. The reader may like to ponder

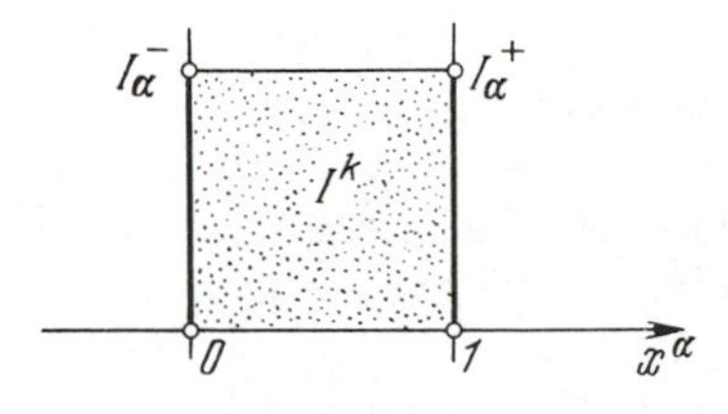

Figure 32

the question of how in higher dimensions the appropriate parametrization of the boundary is determined by the induced orientation.)

Having adumbrated the concept of orientation, we now state the theorem.

26.4.1. Theorem. *Let φ^{k-1} be a $(k-1)$-form in $\mathbb{R}^n$, and let $\sigma: I^k \to \mathbb{R}^n$ be a singular cube in $\mathbb{R}^n$. Then*

$$\int_{\sigma(\partial I^k)} \varphi^{k-1} = \int_{\sigma(I^k)} d\varphi^{k-1},$$

where $d\varphi^{k-1}$ is the differential of φ^{k-1}, and where the orientation on the boundary ∂I^k is taken to be that induced by the standard orientation of the cube I^k.

PROOF. We first remind the reader that of course the left-hand side integral (over $\sigma(\partial I^k)$) is just the sum of the integrals over the faces of the cube.

Write $\omega = \sigma^*(\varphi)$, the restriction of $\varphi = \varphi^{k-1}$ to I^k (see §22.1 for the definition of the restriction (or pullback) operation). One can show, much as in paragraph (i) immediately following Definition 26.1.2, that

$$\int_{\sigma(I^k)} d\varphi = \int_{I^k} \sigma^*(d\varphi), \qquad \int_{\sigma(\partial I^k)} \varphi = \int_{\partial I^k} \sigma^*(\varphi).$$

From this and the fact that $d\omega = d(\sigma^*(\varphi)) = \sigma^*(d\varphi)$ (which can be verified directly from the definitions of the operations d and σ^*), we see that it suffices to show that

$$\int_{\partial I^k} \omega = \int_{I^k} d\omega.$$

We now write the $(k-1)$-form ω defined on I^k explicitly as

$$\omega = a_\alpha(x^1, \ldots, x^k)\, dx^1 \wedge \cdots \wedge \widehat{dx^\alpha} \wedge \cdots \wedge dx^k,$$

where the hat over the symbol dx_α indicates that it does not appear, and where the functions $a_\alpha(x^1, \ldots, x^k)$ are smooth. (As always we assume tacitly enough continuity and differentiability to make the proof work.) Then by §25.2(7)

$$d\omega = \sum_\alpha \frac{\partial a_\alpha}{\partial x^\alpha} dx^\alpha \wedge dx^1 \wedge \cdots \wedge \widehat{dx^\alpha} \wedge \cdots \wedge dx^k = \sum_\alpha (-1)^{\alpha-1} \frac{\partial a_\alpha}{\partial x^\alpha} d^k x,$$

where $d^k x = dx^1 \wedge \cdots \wedge dx^k$. Now there is a well-known theorem of analysis to the effect that any smooth function $a(x^1, \ldots, x^k)$ on a closed and bounded region of $\mathbb{R}^k$ can be uniformly approximated by linear combinations of products $\prod_{q=1}^k b^q(x^q)$ of smooth functions of one variable. Taking this result as given, it is easy to see that we may without loss of generality assume that the functions $a_\alpha(x^1, \ldots, x^k)$ have the form

$$a_\alpha(x^1, \ldots, x^k) = \prod_{q=1}^{k} b_\alpha^q(x^q).$$

This assumed, we now calculate the integral $\int_{I^k} d\omega$ explicitly (using the well known equivalence of multiple and iterated integrals):

$$\int_{I^k} d\omega = \int_{I^k} \left(\sum_\alpha (-1)^{\alpha-1} \frac{\partial a_\alpha}{\partial x^\alpha} \right) d^k x$$

$$= \int_{I^k} \left(\sum_\alpha (-1)^{\alpha-1} \frac{\partial}{\partial x^\alpha} \left(\prod_{q=1}^{k} b_\alpha^q(x^q) \right) \right) d^k x$$

$$= \int_{I^k} \left(\sum_\alpha (-1)^{\alpha-1} b_\alpha^1(x^1) \cdots \frac{\partial b_\alpha^\alpha(x^\alpha)}{\partial x^\alpha} \cdots b_\alpha^k(x^k) \right) d^k x$$

$$= \sum_\alpha \int_{(x^1)} \cdots \int_{(\hat{x}^\alpha)} \cdots \int_{(x^k)} ((-1)^{\alpha-1} b_\alpha(x^1) \cdots b_\alpha^{\alpha-1}(x^{\alpha-1}) b_\alpha^{\alpha+1}(x^{\alpha+1})$$

$$\ldots b_\alpha^k(x^k) \left[(-1)^{\alpha-1} \int_{(x^\alpha)} \frac{\partial b_\alpha^\alpha(x^\alpha)}{\partial x^\alpha} dx^\alpha \right] dx^1 \wedge \cdots \wedge \widehat{dx^\alpha} \wedge \cdots \wedge dx^k$$

$$= \sum_\alpha \int_{(x^1)} \cdots \int_{(\hat{x}^\alpha)} \cdots \int_{(x^k)} (b_\alpha^1(x^1) \cdots \hat{b}_\alpha^\alpha(x^\alpha) \cdots b_\alpha^k(x^k)$$

$$\times [b_\alpha^\alpha(1) - b_\alpha^\alpha(0)]) \, dx^1 \wedge \cdots \wedge \widehat{dx^\alpha} \wedge \cdots \wedge dx^k$$

$$= \sum_\alpha \int_{(x^1)} \cdots \int_{(\hat{x}^\alpha)} \cdots \int_{(x^k)} (b_\alpha^1(x^1) \cdots b_\alpha^\alpha(1) \cdots b_\alpha^k(x^k)$$

$$- b_\alpha^1(x^1) \cdots b_\alpha^\alpha(0) \cdots b_\alpha^k(x^k)) \, dx^1 \wedge \cdots \wedge \widehat{dx^\alpha} \wedge \cdots \wedge dx^k$$

$$= \sum_\alpha \int_{(x^1)} \cdots \int_{(\hat{x}^\alpha)} \cdots \int_{(x^k)} (a_\alpha(x^1, \ldots, x^k) \,|_{x^\alpha = 1}$$

$$- a_\alpha(x^1, \ldots, x^k) \,|_{x^\alpha = 0} \, dx^1 \wedge \cdots \wedge \widehat{dx^\alpha} \wedge \cdots \wedge dx^k = \int_{\partial I^k} \omega.$$

This completes the proof of the theorem. □

26.5. Exercises

1. Calculate the volumes of the groups $SO(3, \mathbb{R})$ and $SU(2)$ endowed with a Killing metric as in §24.4.

2. Let X^i be a vector field in Minkowski space $\mathbb{R}_1^4$. We define the integral of this vector field over a 3-dimensional hypersurface (i.e. the "flux" of the vector field through the hypersurface) as the integral of the 3-form $X^i \, dS_i$, where

$$dS_i = \tfrac{1}{6}\sqrt{|g|} \, \varepsilon_{jkli} \, dx^j \wedge dx^k \wedge dx^l$$

(ε_{jkli} being as defined in §18.2). From the general Stokes formula deduce the following equality:

$$\int_{\partial V} X^i \, dS_i = \int_V \frac{\partial X^i}{\partial x^i} \, dV.$$

3. Let U be a bounded region with smooth boundary in an n-dimensional space with metric g_{ij}, and let Ω_U^p be the space of all smooth p-forms on that space, which vanish outside U. Define a scalar product on the space Ω_U^p by

$$\langle \omega_1, \omega_2 \rangle = \int_U \omega_1 \wedge * \omega_2$$

for every pair of p-forms ω_1, ω_2 in Ω_U^p.

(i) Show that the space Ω_U^p with this scalar product is Euclidean (cf. Exercise 2 in §19).

(ii) The operator $*$ is orthogonal:

$$\langle * \omega_1, * \omega_2 \rangle = \langle \omega_1, \omega_2 \rangle.$$

(iii) The operators $\delta (= (-1)^{np+n+1} * d *)$ and d are "conjugate", i.e.

$$\langle d\omega_1, \omega_2 \rangle = \langle \omega_1, \delta\omega_2 \rangle.$$

(iv) The square of the operator δ is zero: $\delta\delta = 0$.

(v) Write $\Delta = d\delta + \delta d$. Show that the operator Δ on the space Ω_U^p is self-conjugate: $\langle \Delta\omega_1, \omega_2 \rangle = \langle \omega_1, \Delta\omega_2 \rangle$. Verify the following commutativity relations:

$$\Delta d = d\Delta, \qquad \Delta\delta = \delta\Delta, \qquad \Delta * = * \Delta.$$

§27. Differential Forms on Complex Spaces

27.1. The Operators d' and d''

Let D be a region of the complex space $\mathbb{C}^n$ with complex co-ordinates $z^1, \ldots z^n$. We shall denote by $D^{\mathbb{R}}$ the "realization" of the region D, with co-ordinates $x^1, \ldots, x^n, y^1, \ldots, y^n$, where

$$z^k = x^k + iy^k, \qquad k = 1, \ldots, n. \tag{1}$$

The tangent space at each point of $D^{\mathbb{R}}$ may be identified (see §18.1) with the space of operators (on functions $D^{\mathbb{R}} \to \mathbb{R}$) with basis

$$\frac{\partial}{\partial x^1}, \ldots, \frac{\partial}{\partial x^n}, \qquad \frac{\partial}{\partial y^1}, \ldots, \frac{\partial}{\partial y^n}.$$

We shall consider complex tangent vectors, i.e. linear combinations

$$\sum_{k=1}^n a^k \frac{\partial}{\partial x^k} + \sum_{k=1}^n b^k \frac{\partial}{\partial y^k}, \tag{2}$$

where the a^k, b^k are allowed to be complex. It will be convenient to introduce a complex basis $\partial/\partial z^k, \partial/\partial \bar{z}^k$ for this space, defined by (cf. §12.1(6)):

$$\frac{\partial}{\partial z^k} = \frac{1}{2}\left(\frac{\partial}{\partial x^k} - i\frac{\partial}{\partial y^k}\right), \qquad \frac{\partial}{\partial \bar{z}^k} = \frac{1}{2}\left(\frac{\partial}{\partial x^k} + i\frac{\partial}{\partial y^k}\right), \qquad k = 1, \ldots, n. \tag{3}$$

For maximum clarity in the formulae of this section, we shall use the following notational convention. The index (on a tensor) corresponding to a "barred" co-ordinate $\bar{z}^l$ say, will itself be written with a bar over it, i.e. as $\bar{l}$, unless the bar already appears in the symbol involving l (as it does in $\bar{z}^l$). Thus in an expression like $a_{k\bar{l}}z^k\bar{z}^l$ (as in many of the formulae to follow) it is understood that summation takes place over both indices.

Thus in terms of the basis (3) an arbitrary complex vector ξ in the tangent space at a point of D can be written as

$$\xi = \xi^k \frac{\partial}{\partial z^k} + \xi^{\bar{k}} \frac{\partial}{\partial \bar{z}^k}, \qquad \xi = (\xi^k, \xi^{\bar{k}}). \tag{4}$$

It is straightforward to check that ξ is real (i.e. ξ is a linear combination of the operators $\partial/\partial x^k$, $\partial/\partial y^k$ with real coefficients) precisely if $\overline{\xi^k} = \xi^{\bar{k}}$. The 1-forms (or covectors)

$$dz^k = dx^k + i dy^k, \qquad d\bar{z}^k = dx^k - i dy^k, \qquad k = 1, \ldots, n, \tag{5}$$

form the dual basis to the basis (3). (To see this observe that for each $k = 1, \ldots, n$ the values of the 1-forms dz^k, $d\bar{z}^k$ on a vector ξ at a point of D are just ξ^k, $\xi^{\bar{k}}$ respectively; thus dz^k and $d\bar{z}^k$ may be regarded as functionals projecting each vector ξ onto the corresponding components, i.e. they are elements of the dual basis.)

Any complex k-form ω, $1 \leq k \leq 2n$, may be written as a linear combination of the forms $dz^{i_1} \wedge \cdots \wedge dz^{i_p} \wedge d\bar{z}^{j_1} \wedge \cdots \wedge d\bar{z}^{j_q} (p + q = k)$ over the complex numbers. Thus

$$\omega = \omega_{k,0} + \omega_{k-1,1} + \cdots + \omega_{0,k}, \tag{6}$$

where (cf. end of §18.2, where $\omega_{p,q}$ is written $T^{(p,q)}$)

$$\omega_{p,q} = \frac{1}{p!q!} \sum_{\substack{i_1, \ldots, i_p \\ \bar{j}_1, \ldots, \bar{j}_q}} T_{i_1 \ldots i_p \bar{j}_1 \ldots \bar{j}_q} dz^{i_1} \wedge \cdots \wedge dz^{i_p} \wedge d\bar{z}^{j_1} \wedge \cdots \wedge d\bar{z}^{j_q}. \tag{7}$$

Here the components $T_{i_1 \ldots i_p \bar{j}_1 \ldots \bar{j}_q}$ are skew-symmetric in the indices $i_1, \ldots, i_p$ and $\bar{j}_1, \ldots, \bar{j}_q$ separately. We call the form $\omega_{p,q}$ a *form of type* (p, q). The decomposition (6) of ω is independent of the complex co-ordinates on D; i.e. in terms of new co-ordinates $w^1, \ldots, w^n$, which are complex analytic functions of $z^1, \ldots, z^n$, the same tensors $\omega_{p,q}$ will occur as summands in (6) (though expressed in terms of the new co-ordinates). This is clear from the fact the dw^i are linear combinations of the forms $dz^1, \ldots, dz^n$, while the $d\bar{w}^i$ are linear combinations of the forms $d\bar{z}^1, \ldots, d\bar{z}^n$.

27.1.1. Lemma. *The differential operator d on complex forms ω can be uniquely written as*

$$d = d' + d'', \tag{8}$$

where $d'\omega_{p,q}$ has type $(p+1, q)$ and $d''\omega_{p,q}$ has type $(p, q+1)$. The operators d', d'' are invariant with respect to complex analytic co-ordinate changes.

PROOF. For the form $\omega_{p,q}$ as written in (7), we have

$$d\omega_{p,q} = \frac{1}{p!q!} \sum_{\substack{i, i_1, \ldots, i_p \\ \bar{j}_1, \ldots, \bar{j}_q}} \frac{\partial T_{i_1 \ldots i_p \bar{j}_1 \ldots \bar{j}_q}}{\partial z^i} dz^i \wedge dz^{i_1} \wedge \cdots \wedge dz^{i_p} \wedge d\bar{z}^{j_1} \wedge \cdots \wedge d\bar{z}^{j_q}$$

$$+ \frac{(-1)^p}{p!q!} \sum_{\substack{i_1, \ldots, i_p \\ \bar{j}, \bar{j}_1, \ldots, \bar{j}_q}} \frac{\partial T_{i_1 \ldots i_p \bar{j}_1 \ldots \bar{j}_q}}{\partial \bar{z}^j} dz^{i_1} \wedge \cdots \wedge dz^{i_p} \wedge d\bar{z}^j \wedge d\bar{z}^{j_1} \wedge \cdots \wedge d\bar{z}^{j_q}. \tag{9}$$

Denote the first summand in (9) by $d'\omega_{p,q}$ and the second by $d''\omega_{p,q}$. The uniqueness of the decomposition (6) (as applied to both $\omega_{p,q}$ and ω) implies that the operators d' and d'' are well-defined and that the decomposition (8) is unique. By the invariance of the summands in (6) (again as applied to both $\omega_{p,q}$ and ω) the definition of d' and d'' is consistent with respect to complex analytic co-ordinate changes. □

27.1.2. Corollary. *The squares of the operators d', d'' are zero*:

$$(d')^2 = (d'')^2 = 0; \tag{10}$$

and they satisfy the "anti-commutativity" relation:

$$d''d' = -d'd''. \tag{11}$$

PROOF. From $d^2 = 0$ (Theorem 25.2.2) and $d = d' + d''$ it follows that

$$0 = d^2\omega_{p,q} = (d')^2\omega_{p,q} + (d'')^2\omega_{p,q} + (d'd''\omega_{p,q} + d''d'\omega_{p,q}).$$

Since the first term $(d')^2\omega_{p,q}$ has type $(p+2, q)$, the second $(d'')^2\omega_{p,q}$ has type $(p, q+2)$, and the third has type $(p+1, q+1)$, it follows that all three forms are zero, whence (10) and (11). □

27.1.3. Definition. A form ω of type $(p, 0)$ is called *holomorphic* if

$$d''\omega = 0. \tag{12}$$

27.1.4. Example. A form ω of type $(0, 0)$ is a complex-valued function $f(z^1, \ldots, z^n, \bar{z}^1, \ldots, \bar{z}^n)$. In this case the condition $d''f = 0$ is just the condition that f be complex analytic:

$$\frac{\partial f}{\partial \bar{z}^i} \equiv 0, \qquad i = 1, \ldots, n.$$

It is clear that in the general case, a form ω of type $(p, 0)$ is holomorphic precisely if its coefficients are complex analytic functions.

27.2. Kählerian Metrics. The Curvature Form

An *Hermitian metric* on a region D of complex n-space with co-ordinates $z^1, \ldots, z^n$, is given (in terms of these co-ordinates) by a family of functions $g_{j\bar{k}}$ with the following three properties (cf. the definition in §11.2):

(i) $g_{j\bar{k}} = \overline{g_{k\bar{j}}}$;
(ii) under co-ordinate changes $z^j = z^j(z^{1'}, \ldots, z^{n'})$ which are complex analytic (i.e. $\partial z^j/\partial \bar{z}^{k'} \equiv 0$) we have

$$g_{j'\bar{k}'} = g_{j\bar{k}} \frac{\partial z^j}{\partial z^{j'}} \overline{\left(\frac{\partial z^k}{\partial z^{k'}}\right)};$$

(iii) the form $g_{j\bar{k}} \xi^j \bar{\xi}^k$ is positive definite. (Note that by (i) it is real-valued.)

The complex scalar product of a pair of complex vectors (ξ^k), (η^k) is defined in terms of the Hermitian metric $g_{j\bar{k}}$ by:

$$\langle \xi, \eta \rangle_{\mathbb{C}} = g_{j\bar{k}} \xi^j \eta^{\bar{k}}. \tag{13}$$

By virtue of (ii) it follows that this inner product is independent of the particular co-ordinate system in which it is calculated, i.e. is the same for all co-ordinate systems (obtainable from $z^1, \ldots, z^n$ by means of analytic co-ordinate changes). As in the particular case considered in §11.2, so also in the present general case does the complex scalar product (13) define a Riemannian metric $\langle\ ,\ \rangle_{\mathbb{R}}$ on the realized region $D^{\mathbb{R}}$, given by

$$\langle \xi, \eta \rangle_{\mathbb{R}} = \operatorname{Re}\langle \xi, \eta \rangle_{\mathbb{C}}. \tag{14}$$

Clearly the scalar product (13) has the Hermitian property

$$\langle \xi, \eta \rangle_{\mathbb{C}} = \overline{\langle \eta, \xi \rangle}_{\mathbb{C}},$$

so that the expression

$$\{\xi, \eta\} = -\tfrac{1}{2} \operatorname{Im}\langle \xi, \eta \rangle_{\mathbb{C}} = \frac{i}{2}\left[\frac{\langle \xi, \eta \rangle_{\mathbb{C}} - \overline{\langle \xi, \eta \rangle}_{\mathbb{C}}}{2}\right] = \frac{i}{2} g_{j\bar{k}} \left(\frac{\xi^j \bar{\eta}^k - \bar{\xi}^k \eta^j}{2}\right)$$

is real and skew-symmetric in ξ and η, and therefore by the co-ordinate-independence of the scalar product, yields a (real) differential form Ω of degree 2 on $D^{\mathbb{R}}$, namely

$$\Omega = \frac{i}{2} g_{j\bar{k}}\, dz^j \wedge d\bar{z}^k. \tag{15}$$

(Note that if we write $T_{j\bar{k}} = (i/2)\, g_{j\bar{k}}$, then $T_{j\bar{k}} = -\ \overline{T}_{k\bar{j}}$, so that in the co-ordinates x^k, y^k, the form Ω is real.) Clearly Ω is of type (1,1) (see the preceding subsection).

27.2.1. Definition. An Hermitian metric $g_{j\bar{k}}$ on a region D of $\mathbb{C}^n$ is said to be *Kählerian* if the form Ω defined by (15) is "closed", i.e. $d\Omega = 0$.

We shall elucidate the geometrical meaning of this condition on a Hermitian metric in §29.4.

27.2.2. Example. Suppose that D is a region of a (real) 2-dimensional surface endowed with a metric, and let x, y be conformal co-ordinates on the surface (see Theorem 13.1.1), so that in terms of these co-ordinates the metric has the form

$$ds^2 = g\,dz\,d\bar{z}.$$

In this case the form Ω is given by

$$\Omega = \frac{i}{2}\sqrt{g}\,(dx + i\,dy) \wedge (dx - i\,dy) = \sqrt{g}\,dx \wedge dy,$$

i.e. Ω is the element of area on the surface. The metric is certainly Kählerian since the degree of Ω is the same as the dimension of the space on which it is defined (namely 2).

27.2.3. Lemma. *Let $g_{j\bar{k}}$ be a Hermitian metric on a region D of $\mathbb{C}^n$. Define*

$$\omega = \frac{1}{i}\,d''\,d'\ln g.$$

where d', d'' are as in 27.1.1, *and* $g = \det(g_{j\bar{k}})$. *Then ω is a form of type* (1, 1).

Proof. By $d'f(z^1, \ldots, z^n, \bar{z}^1, \ldots, \bar{z}^n)$ we mean $(\partial f/\partial z^i)\,dz^i$, and by $d''f$ we mean $(\partial f/\partial \bar{z}^i)\,d\bar{z}^i$. The only thing to verify is that the definition of ω is consistent, i.e. is independent of the choice of complex co-ordinates. Thus let

$$z^j = z^j(w^1, \ldots, w^n), \qquad \frac{\partial z^j}{\partial \bar{w}^k} \equiv 0, \tag{16}$$

be a complex analytic change of co-ordinates, and write $J = \det(\partial z^j/\partial \omega^k)$. Let $\tilde{g}$ denote the determinant of the Hermitian metric in the new co-ordinates $w^1, \ldots, w^n$. From (ii) above it follows easily that

$$\tilde{g} = J\bar{J}g.$$

Hence

$$d'\ln \tilde{g} = d'\ln J + d'\ln \bar{J} + d'\ln g = d'\ln J + d'\ln g,$$

since $d'\ln \bar{J} \equiv 0$ by the analyticity of the function J. Applying d'' we then obtain

$$d''\,d'\ln \tilde{g} = d''\,d'\ln J + d''\,d'\ln g = -\,d'\,d''\ln J + d''\,d'\ln g = d''\,d'\ln g,$$

where we have used $d'\,d'' = -d''\,d'$ (see Corollary 27.1.2), and the fact that $d''J \equiv 0$, which again is immediate from the analyticity of J. This completes the proof. □

In the context of the above example (27.2.2) of a 2-dimensional real surface, the form ω is related to the Gaussian curvature as follows.

27.2.4. Theorem. *On a region of a real 2-dimensional surface with a metric given in terms of conformal co-ordinates by* $ds^2 = g\, dz\, d\bar{z}$, *the form* $\omega = (1/i)\, d''\, d' \ln g$ *is given by*

$$\omega = -K\Omega, \tag{17}$$

where K is the Gaussian curvature of the surface, and $\Omega = (i/2)\, g\, dz \wedge d\bar{z}$ *is* the element of area.

PROOF. It is immediate that

$$d'\, d'' \ln g = \frac{\partial^2}{\partial z\, \partial \bar{z}} \ln g\, dz \wedge d\bar{z}.$$

By Theorem 13.1.3 the Gaussian curvature is given in terms of conformal co-ordinates by $K = -(2/g)\, [(\partial^2 \ln g)/\partial z\, \partial \bar{z}]$, whence the desired equality (17). □

§28. Covariant Differentiation

28.1. Euclidean Connexions

In §25 we introduced the operation d of taking the differential or gradient of a skew-symmetric tensor (field) $T = (T_{i_1 \dots i_k})$:

$$(dT)_{i_1 \dots i_{k+1}} = \sum_{q=1}^{k+1} (-1)^{q+1} \frac{\partial T_{i_1 \dots \hat{i}_q \dots i_{k+1}}}{\partial x^{i_q}}, \tag{1}$$

where the hat indicates an omitted symbol. In particular for $k = 1$ this becomes

$$(dT)_{ij} = \frac{\partial T_j}{\partial x^i} - \frac{\partial T_i}{\partial x^j}. \tag{2}$$

We indicated there that dT is indeed a skew-symmetric tensor, of rank one greater than that of the original tensor T. (This we proved rigorously for $k = 0, 1$ only.) It was also made apparent that the operation d is the only differential operation independent of whatever geometrical structure the underlying space may possess, in the sense that all other known such tensor operations involving differentiation may be defined in terms of d and the purely algebraic tensor operations (permutation of indices, addition, multiplication, taking the trace, etc.). We shall now introduce another sort of differentiation depending (as we shall ultimately see) on some additional structure on the space. Suppose to begin with that we attempt a more direct generalization of the concept of the gradient of a function to arbitrary tensors $T^{i_1 \ldots i_p}_{j_1 \ldots j_q}$, by defining

$$T^{i_1 \ldots i_p}_{j_1 \ldots j_q; k} = \frac{\partial T^{i_1 \ldots i_p}_{j_1 \ldots j_q}}{\partial x^k}. \tag{3}$$

The result turns out *not* to behave like a tensor under arbitrary co-ordinate transformations. However since this operation is encountered rather frequently, we shall delineate a class of co-ordinate changes under which the components (3) do transform like the components of a tensor.

28.1.1. Theorem. *If $T^{i_1 \ldots i_p}_{j_1 \ldots j_q}$ is a tensor field on a space with co-ordinates $x^1, \ldots, x^n$, then the quantities*

$$T^{i_1 \ldots i_p}_{j_1 \ldots j_q; k} = \frac{\partial T^{i_1 \ldots i_p}_{j_1 \ldots j_q}}{\partial x^k}$$

transform like the components of a tensor of type $(p, q + 1)$ under all linear co-ordinate changes

$$\begin{aligned} x^i &= a^i_j z^j, \qquad i, j = 1, \ldots, n, \\ a^i_j &= \text{const.}, \qquad z^i = b^i_j x^j, \qquad b^i_j a^j_k = \delta^i_k. \end{aligned} \tag{4}$$

Proof. From (4) we have

$$\frac{\partial x^i}{\partial z^j} = a^i_j = \text{const.}, \qquad \frac{\partial^2 x^i}{\partial z^j \partial z^k} = 0,$$

$$\frac{\partial z^i}{\partial x^j} = b^i_j = \text{const.}, \qquad \frac{\partial^2 z^i}{\partial x^j \partial x^k} = 0.$$

From the transformation rule for tensors we have

$$\tilde{T}^{k_1 \ldots k_p}_{l_1 \ldots l_q} = T^{i_1 \ldots i_p}_{j_1 \ldots j_q} \frac{\partial z^{k_1}}{\partial x^{i_1}} \cdots \frac{\partial z^{k_p}}{\partial x^{i_p}} \frac{\partial x^{j_1}}{\partial z^{l_1}} \cdots \frac{\partial x^{j_q}}{\partial z^{l_q}} = T^{(i)}_{(j)} b^{(k)}_{(i)} a^{(j)}_{(l)}, \tag{5}$$

where $(i) = (i_1 \cdots i_p)$, $(k) = (k_1 \cdots k_p)$, $(j) = (j_1 \cdots j_q)$, $(l) = (l_1 \cdots l_q)$. Since the a^i_j and b^j_k are constants, on differentiating (5) with respect to z^r we obtain

$$\tilde{T}^{(k)}_{(l); r} = \frac{\partial \tilde{T}^{(k)}_{(l)}}{\partial z^r} = \frac{\partial T^{(i)}_{(j)}}{\partial z^r} b^{(k)}_{(i)} a^{(j)}_{(l)} = \frac{\partial T^{(i)}_{(j)}}{\partial x^s} \frac{\partial x^s}{\partial z^r} b^{(k)}_{(i)} a^{(j)}_{(l)} = T^{(i)}_{(j); s} a^s_r a^{(j)}_{(l)} b^{(k)}_{(i)}. \tag{6}$$

Since this is the transformation rule for such co-ordinate transformations (compare (6) with (5)), the theorem is proved. □

Note that in this proof the fact that $\partial^2 x^i / \partial z^j \, \partial z^k = 0$ was crucial. Let us consider for instance tensors of type (0, 1) or (1, 0):

$$\frac{\partial T_i}{\partial x^k} = T_{i; k}, \qquad \frac{\partial T^i}{\partial x^k} = T^i{}_{; k}.$$

We know from the above theorem that these quantities transform like the components of tensors under linear co-ordinate changes. How do they transform under more general co-ordinate changes $x^i = x^i(z^1, \ldots, z^n)$, $i = 1, \ldots, n$, where $\partial^2 x^i/\partial z^k\, \partial z^j$ may not be identically zero? The answer is obtained by calculating:

$$\tilde{T}_{j;q} = \frac{\partial \tilde{T}_j}{\partial z^q} = \frac{\partial}{\partial z^q}\left(T_i \frac{\partial x^i}{\partial z^j}\right) = \frac{\partial T_i}{\partial z^q}\frac{\partial x^i}{\partial z^j} + T_i \frac{\partial^2 x^i}{\partial z^q\, \partial z^j}$$

$$= \frac{\partial T_i}{\partial x^p}\frac{\partial x^p}{\partial z^q}\frac{\partial x^i}{\partial z^j} + T_i \frac{\partial^2 x^i}{\partial z^q\, \partial z^j} = T_{i;p}\frac{\partial x^p}{\partial z^q}\frac{\partial x^i}{\partial z^j} + T_i \frac{\partial^2 x^i}{\partial z^q\, \partial z^j}.$$

Here the symbols $\tilde{T}_i$ denote the components of the tensor T in the co-ordinates $z^1, \ldots, z^n$, while the T_i are the components in the co-ordinates $x^1, \ldots, x^n$. Thus the general transformation rule for the quantities $T_{i;p}$ is

$$\frac{\partial \tilde{T}_j}{\partial z^q} = \tilde{T}_{j;q} = T_{i;p}\frac{\partial x^p}{\partial z^q}\frac{\partial x^i}{\partial z^j} + T_i \frac{\partial^2 x^i}{\partial z^q\, \partial z^j}. \tag{7}$$

The term $T_i(\partial^2 x^i/\partial z^q\, \partial z^j)$ is not a tensor. The skew-symmetric part of (7) is, however, a tensor since it is just the differential (or gradient) of T (see §25.1):

$$(d\tilde{T})_{jq} = \tilde{T}_{q;j} - \tilde{T}_{j;q} = (T_{p;i} - T_{i;p})\frac{\partial x^p}{\partial z^q}\frac{\partial x^i}{\partial z^j} = (dT)_{ip}\frac{\partial x^p}{\partial z^q}\frac{\partial x^i}{\partial z^j}.$$

On the other hand the symmetric part of $T_{i;p}$ has as its transformation rule

$$\tilde{T}_{j;q} + \tilde{T}_{q;j} = (T_{i;p} + T_{p;i})\frac{\partial x^p}{\partial z^q}\frac{\partial x^i}{\partial z^j} + 2T_i \frac{\partial^2 x^i}{\partial z^q\, \partial z^j},$$

and therefore clearly does not behave like a tensor with respect to arbitrary co-ordinate changes.

Similarly, for a tensor T^i of type (1, 0) we have:

$$\tilde{T}^j_{;l} = \frac{\partial \tilde{T}^j}{\partial z^l} = \frac{\partial}{\partial z^l}\left(T^i \frac{\partial z^j}{\partial x^i}\right) = \frac{\partial T^i}{\partial z^l}\frac{\partial z^j}{\partial x^i} + T^i \frac{\partial}{\partial z^l}\left(\frac{\partial z^j}{\partial x^i}\right)$$

$$= \frac{\partial T^i}{\partial x^p}\frac{\partial x^p}{\partial z^l}\frac{\partial z^j}{\partial x^i} + T^i \frac{\partial^2 z^j}{\partial x^i\, \partial x^q}\frac{\partial x^q}{\partial z^l}. \tag{8}$$

From this we see that the $T^i_{;p}$ are in general not the components of a tensor (on account of the term $T^i(\partial^2 z^j/\partial x^i\, \partial x^q)(\partial x^q/\partial z^l)$ in (8)). Putting $l = j$ in (8) (and summing) we obtain

$$\tilde{T}^j_{;j} = \frac{\partial \tilde{T}^j}{\partial z^j} = \frac{\partial T^i}{\partial x^p}\frac{\partial x^p}{\partial z^l}\frac{\partial z^l}{\partial x^i} + T^i \frac{\partial^2 z^j}{\partial x^i\, \partial x^q}\frac{\partial x^q}{\partial z^j}$$

$$= T^i_{;p}\delta^p_i + T^i \frac{\partial^2 z^j}{\partial x^i\, \partial x^q}\frac{\partial x^q}{\partial z^j} = T^i_{;i} + T^i \frac{\partial x^q}{\partial z^j}\frac{\partial^2 z^j}{\partial x^i\, \partial x^q}. \tag{9}$$

Remark. As we have mentioned before (see the second remark in §25.1), in contexts where the co-ordinates $x^1, \ldots, x^n$ are Euclidean, the quantity $\partial T^i/\partial x^i = T^i_{;i}$ is called the *divergence of the vector field* (T^i). As is evident from (9), $T^i_{;i}$ is not a scalar (i.e a tensor of type (0, 0)) under non-linear co-ordinate changes (but by Theorem 28.1.1 does behave like a scalar under linear transformations of the co-ordinates). The significance of the divergence may be explained as follows: Under a small displacement of the points of the underlying space:

$$x^i \to x^i + T^i(x^1, \ldots, x^n) = \bar{x}^i,$$

the Euclidean volume element $dx^1 \wedge \cdots \wedge dx^n$ is (up to second-order quantities in the partial derivatives of the T^i) altered by the amount $T^i_{;i}\, dx^1 \wedge \cdots \wedge dx^n$ (cf. §22.2). This follows from Theorem 18.2.3.

We now return to our "generalized gradient"

$$T^{i_1 \ldots i_p}_{j_1 \ldots j_q; k} = \frac{\partial T^{i_1 \ldots i_p}_{j_1 \ldots j_q}}{\partial x^k}. \tag{10}$$

In view of Theorem 28.1.1 and the discussion following it, it is reasonable to agree to henceforth apply this operation only when the co-ordinates $x^1, \ldots, x^n$ are either Euclidean or obtainable from Euclidean co-ordinates $z^1, \ldots, z^n$ by means of a linear transformation

$$x^i = a^i_j z^j, \qquad a^i_j = \text{const}.$$

We saw above (in the case of tensors of types (0, 1) and (1, 0)) that if we apply the operation as defined by (10) in a co-ordinate system which is not linearly related to the system $x^1, \ldots, x^n$, we obtain quantities $\tilde{T}^{k_1 \ldots k_p}_{l_1 \ldots l_q; r}$ related to the $T^{(i)}_{(j); s}$ by a transformation law which is not tensorial.

Let us now take a different view of the situation. Why should we expect that the operation of taking our new "generalized gradient" be defined by a single formula which works uniformly in all co-ordinate systems? We might take a different tack, and admit instead the following possibilities:

(i) the operation is intimately linked to Euclidean geometry;
(ii) it assumes the simple form (10) only in Euclidean co-ordinate systems (and their linear transforms);
(iii) the operation is nonetheless a tensor operation (necessarily defined differently from (10) in co-ordinates not linearly related to Euclidean co-ordinates).

What are the consequences of these assumptions? What formula is to be used in applying the operation in terms of co-ordinates not linearly related to Euclidean co-ordinates? To answer these questions we must first calculate, in terms of Euclidean co-ordinates $x^1, \ldots, x^n$, the effect of our operation on a tensor T, and then apply the transformation rule for tensors to the resulting expression to see what form our hypothetical tensor should take in a different system of co-ordinates $z^1, \ldots, z^n$ (where $x^i = x^i(z^1, \ldots, z^n)$). This we shall now do. We rewrite (10) briefly as

$$T^{(i)}_{(j);s} = \frac{\partial T^{(i)}_{(j)}}{\partial x^s}, \qquad (i) = (i_1, \ldots, i_p), \quad (j) = (j_1, \ldots, j_q).$$

Since we are now regarding $T^{(i)}_{(j);s}$ as a fully-fledged tensor, it follows that

$$\tilde{T}^{(k)}_{(l);r} = T^{(i)}_{(j);s} \frac{\partial x^{(j)}}{\partial z^{(l)}} \frac{\partial z^{(k)}}{\partial x^{(i)}} \frac{\partial x^s}{\partial z^r}, \tag{11}$$

where

$$\frac{\partial x^{(j)}}{\partial z^{(l)}} = \frac{\partial x^{j_1}}{\partial z^{l_1}} \cdots \frac{\partial x^{j_q}}{\partial z^{l_q}}, \qquad \frac{\partial z^{(k)}}{\partial x^{(i)}} = \frac{\partial z^{k_1}}{\partial x^{i_1}} \cdots \frac{\partial z^{k_p}}{\partial x^{i_p}}.$$

We now seek the operation which yields the components $\tilde{T}^{(k)}_{(l);q}$ from the components $\tilde{T}^{(k)}_{(l)}$; i.e. the appropriate analogue of (10) in the new co-ordinates $z^1, \ldots, z^n$.

Again for simplicity we consider first the cases of a vector field (T^i) and a covector field (T_i). From (11) we have

$$\tilde{T}^k_{;q} = T^i_{;s} \frac{\partial z^k}{\partial x^i} \frac{\partial x^s}{\partial z^q}, \qquad \tilde{T}_{l;r} = T_{j;s} \frac{\partial x^j}{\partial z^l} \frac{\partial x^s}{\partial z^r}. \tag{12}$$

Since $T^i_{;s} = \partial T^i / \partial x^s$, it follows from (12) that

$$\tilde{T}^k_{;q} = \frac{\partial T^i}{\partial x^s} \frac{\partial x^s}{\partial z^q} \frac{\partial z^k}{\partial x^i} = \frac{\partial T^i}{\partial z^q} \frac{\partial z^k}{\partial x^i}. \tag{13}$$

Since (T^i) is a vector, we have that $\tilde{T}^k = T^i(\partial z^k / \partial x^i)$. Hence from (13) we obtain

$$\tilde{T}^k_{;q} = \frac{\partial T^i}{\partial z^q} \frac{\partial z^k}{\partial x^i} = \frac{\partial}{\partial z^q} (\tilde{T}^k) - T^i \frac{\partial}{\partial z^q} \left(\frac{\partial z^k}{\partial x^i} \right).$$

Using $T^i = \tilde{T}^s(\partial x^i / \partial z^s)$, we obtain finally

$$\tilde{T}^k_{;q} = \frac{\partial \tilde{T}^k}{\partial z^q} - \tilde{T}^s \frac{\partial x^i}{\partial z^s} \frac{\partial^2 z^k}{\partial x^i \, \partial x^m} \frac{\partial x^m}{\partial z^q}. \tag{14}$$

If we write

$$\Gamma^k_{sq} = - \frac{\partial x^i}{\partial z^s} \frac{\partial x^m}{\partial z^q} \frac{\partial^2 z^k}{\partial x^i \partial x^m}, \tag{15}$$

then (14) becomes

$$\tilde{T}^k_{;q} = \frac{\partial \tilde{T}^k}{\partial z^q} + \Gamma^k_{sq} \tilde{T}^s, \tag{16}$$

We have thus proved the following

28.1.2. Theorem. *Let (T^i) be a vector field, and let $(T^i_{;k})$ be a tensor (with respect to arbitrary co-ordinate changes) given in terms of Euclidean co-ordinates $x^1, \ldots, x^n$ by the formula*

$$T^i_{;k} = \frac{\partial T^i}{\partial x^k},$$

Then in terms of arbitrary co-ordinates $z^1, \ldots, z^n$ the transformed components $\tilde{T}^k_{;r}$ are given by the formula

$$\tilde{T}^k_{;r} = \frac{\partial \tilde{T}^k}{\partial z^r} + \Gamma^k_{sr} \tilde{T}^s,$$

where the coefficients Γ^k_{sr} are defined as in (15).

We now calculate in the same manner the formula for the components $\tilde{T}_{l;r}$ in terms of a given covector field $(\tilde{T}_i)$, on the assumption that $(T_{l;r})$ transforms tensorially under arbitrary co-ordinate changes:

$$\begin{aligned}
\tilde{T}_{l;r} &= T_{j;s} \frac{\partial x^j}{\partial z^l} \frac{\partial x^s}{\partial z^r} = \frac{\partial T_j}{\partial x^s} \frac{\partial x^s}{\partial z^r} \frac{\partial x^j}{\partial z^l} = \frac{\partial T_j}{\partial z^r} \frac{\partial x^j}{\partial z^l} \\
&= \frac{\partial}{\partial z^r}\left(\tilde{T}_k \frac{\partial z^k}{\partial x^j}\right) \frac{\partial x^j}{\partial z^l} = \frac{\partial \tilde{T}_k}{\partial z^r}\left(\frac{\partial z^k}{\partial x^j} \frac{\partial x^j}{\partial z^l}\right) + \tilde{T}_k \frac{\partial x^j}{\partial z^l} \frac{\partial}{\partial z^r}\left(\frac{\partial z^k}{\partial x^j}\right) \\
&= \frac{\partial \tilde{T}_k}{\partial z^r} \delta^k_l + \tilde{T}_k \frac{\partial x^j}{\partial z^l} \frac{\partial}{\partial z^r}\left(\frac{\partial z^k}{\partial x^j}\right) \\
&= \frac{\partial \tilde{T}_l}{\partial z^r} + \tilde{T}_k\left(\frac{\partial x^j}{\partial z^l} \frac{\partial x^s}{\partial z^r} \frac{\partial^2 z^k}{\partial x^j \partial x^s}\right) = \frac{\partial \tilde{T}_l}{\partial z^r} - \Gamma^k_{lr} \tilde{T}_k,
\end{aligned}$$

where, as before, $\Gamma^k_{lr} = -(\partial x^j/\partial z^l)(\partial x^s/\partial z^r)(\partial^2 z^k/\partial x^j \partial x^s)$. We again formulate this result as a theorem.

28.1.3. Theorem. *Let (T_i) be a covector field, and let $(T_{i;k})$ be a tensor (with respect to arbitrary co-ordinate changes) given in terms of Euclidean co-ordinates $x^1, \ldots, x^n$ by the formula*

$$T_{i;k} = \frac{\partial T_i}{\partial x^k}.$$

Then in terms of arbitrary co-ordinates $z^1, \ldots, z^n$ the transformed components $\tilde{T}_{i;k}$ are given by

$$\tilde{T}_{i;k} = \frac{\partial \tilde{T}_i}{\partial z^k} - \Gamma^r_{ik} \tilde{T}_r, \tag{17}$$

where the coefficients Γ^r_{ik} are the same as in the preceding theorem, i.e. are as in (15).

Thus, to summarize, in insisting that our "generalized gradient" transform tensorially under arbitrary co-ordinate changes $x = x(z)$, we are led to the following two (different) formulae for (the "generalized gradients" of) covectors and vectors respectively:

$$\tilde{T}_{i;k} = \frac{\partial \tilde{T}_i}{\partial x^k} - \Gamma^r_{ik}\tilde{T}_r;$$

$$\tilde{T}^i_{;k} = \frac{\partial \tilde{T}^i}{\partial x^k} + \Gamma^i_{rk}\tilde{T}^r,$$

where the coefficients Γ^j_{qk} are the same in both formulae.

We shall forego the calculation of the analogue of the formulae (16) and (17) for the general case of a tensor of type (p, q), and simply state the result without proof.

28.1.4. Theorem. *Let $T^{(i)}_{(j)}$ be the components of a tensor of type (p, q), and let $(T^{(i)}_{(j);k})$ be a tensor (with respect to arbitrary co-ordinate changes) given in terms of Euclidean co-ordinates $x^1, \dots, x^n$ by the formula*

$$T^{(i)}_{(j);k} = \frac{\partial T^{(i)}_{(j)}}{\partial x^k}.$$

Then in terms of arbitrary co-ordinates $z^1, \dots, z^n$ the transformed components $\tilde{T}^{(k)}_{(l);r}$ are given by the formula

$$\tilde{T}^{(k)}_{(l);r} = \frac{\partial \tilde{T}^{(k)}_{(l)}}{\partial x^r} + \sum_{s=1}^{p} \tilde{T}^{k_1 \dots (k_s \to i) \dots k_p}_{l_1 \dots l_q}\Gamma^{k_s}_{ir} - \sum_{s=1}^{q} \tilde{T}^{k_1 \dots k_p}_{l_1 \dots (l_s \to i) \dots l_q}\Gamma^i_{l_s r}, \tag{18}$$

where the family of functions Γ^p_{kq} are defined by (15) (*and where the notation $k_1 \cdots (k_s \to i) \cdots k_p$ indicates that in the p-tuple $k_1 \cdots k_p$ the symbol k_s is to be replaced by i.*

For second-rank tensors the formula (18) becomes:

$$\tilde{T}^i_{j;k} = \frac{\partial \tilde{T}^i_j}{\partial x^k} + \tilde{T}^p_j\Gamma^i_{pk} - \tilde{T}^i_p\Gamma^p_{jk},$$

$$\tilde{T}_{ij;k} = \frac{\partial \tilde{T}_{ij}}{\partial x^k} - \tilde{T}_{pj}\Gamma^p_{ik} - \tilde{T}_{ip}\Gamma^p_{jk},$$

$$\tilde{T}^{ij}_{;k} = \frac{\partial \tilde{T}^{ij}}{\partial x^k} + \tilde{T}^{pj}\Gamma^i_{pk} + \tilde{T}^{ip}\Gamma^j_{pk}.$$

Note that the notation $T^{(i)}_{(j);k}$, $(i) = (i_1 \cdots i_p)$, $(j) = (j_1 \cdots j_q)$, will henceforth be used to denote the components of the tensor of Theorem 28.1.4 (whatever the co-ordinates may be), rather than the expression $\partial T^{(i)}_{(j)}/\partial x^k$.

We emphasize the important fact that our new tensor operation is linked in an essential way with the Euclidean metric with which we assume the underlying space to be endowed. This is clear from the second of the following two defining conditions of the operation:

(i) the result of applying the operation should be a tensor;
(ii) in Euclidean co-ordinates the components are given by

$$T^{(i)}_{(j);k} = \frac{\partial T^{(i)}_{(j)}}{\partial x^k}. \tag{19}$$

(We might turn condition (ii) on its head, and define co-ordinates $x^1, \ldots, x^n$ to be Euclidean (to within a linear co-ordinate change), if in terms of those co-ordinates the components of our "generalized gradient" tensor are given by (19).)

It remains (before giving our culminating definition) to see how the quantities $\Gamma^k_{ij} = \Gamma^k_{ij}(z)$ transform under an arbitrary co-ordinate change $z^i = z^i(z')$, $i = 1, \ldots, n$. As before we let $x^1, \ldots, x^n$ denote Euclidean co-ordinates, with $x^i = x^i(z) = x^i(z(z'))$. By (15) we have

$$\Gamma^k_{pq}(z) = \Gamma^k_{pq} = -\frac{\partial x^i}{\partial z^p}\frac{\partial x^j}{\partial z^q}\frac{\partial^2 z^k}{\partial x^i\,\partial x^j} = \frac{\partial^2 x^m}{\partial z^p\,\partial z^q}\frac{\partial z^k}{\partial x^m}, \tag{20}$$

and in the system z',

$$\Gamma^{k'}_{p'q'} = -\frac{\partial x^i}{\partial z^{p'}}\frac{\partial x^j}{\partial z^{q'}}\frac{\partial^2 z^{k'}}{\partial x^i\,\partial x^j} = \frac{\partial^2 x^m}{\partial z^{p'}\,\partial z^{q'}}\frac{\partial z^{k'}}{\partial x^m}. \tag{21}$$

(The final equalities in (20) and (21) come from

$$0 = \frac{\partial^2 z^k(x(z))}{\partial z^p\,\partial z^q} = \frac{\partial}{\partial z^p}\left(\frac{\partial z^k}{\partial x^i}\frac{\partial x^i}{\partial z^q}\right) = \frac{\partial^2 x^i}{\partial z^p\,\partial z^q}\frac{\partial z^k}{\partial x^i} + \frac{\partial x^j}{\partial z^p}\frac{\partial x^i}{\partial z^q}\frac{\partial^2 z^k}{\partial x^i\,\partial x^j}.\Bigg)$$

From (20) and (21) we obtain

$$\begin{aligned}\Gamma^k_{pq}\frac{\partial z^p}{\partial z^{p'}}\frac{\partial z^q}{\partial z^{q'}} &= -\frac{\partial^2 z^k}{\partial x^i\,\partial x^j}\frac{\partial x^i}{\partial z^p}\frac{\partial x^j}{\partial z^q}\frac{\partial z^p}{\partial z^{p'}}\frac{\partial z^q}{\partial z^{q'}}\\ &= -\frac{\partial^2 z^k}{\partial x^i\,\partial x^j}\frac{\partial x^i}{\partial z^{p'}}\frac{\partial x^j}{\partial z^{q'}} = -\frac{\partial^2 z^k}{\partial z^{p'}\,\partial z^{q'}} + \frac{\partial z^k}{\partial x^j}\frac{\partial^2 x^j}{\partial z^{p'}\,\partial z^{q'}},\end{aligned}$$

where we have used

$$\frac{\partial^2[z^k(x(z'))]}{\partial z^{p'}\,\partial z^{q'}} = \frac{\partial}{\partial z^{p'}}\left(\frac{\partial z^k}{\partial x^i}\frac{\partial x^i}{\partial z^{q'}}\right) = \frac{\partial^2 x^i}{\partial z^{p'}\,\partial z^{q'}}\frac{\partial z^k}{\partial x^i} + \frac{\partial x^i}{\partial z^{p'}}\frac{\partial^2 z^k}{\partial x^i\,\partial x^j}\frac{\partial x^j}{\partial z^{q'}}.$$

Hence

$$\Gamma^k_{pq}\frac{\partial z^p}{\partial z^{p'}}\frac{\partial z^q}{\partial z^{q'}} + \frac{\partial^2 z^k}{\partial z^{p'}\,\partial z^{q'}} = \frac{\partial z^k}{\partial x^j}\frac{\partial^2 x^j}{\partial z^{p'}\partial z^{q'}}.$$

Multiplying both sides of this equation by $\partial z^{k'}/\partial z^k$ and summing, we obtain

$$\frac{\partial z^{k'}}{\partial z^k}\left(\Gamma^k_{pq}\frac{\partial z^p}{\partial z^{p'}}\frac{\partial z^q}{\partial z^{q'}}+\frac{\partial^2 z^k}{\partial z^{p'}\,\partial z^{q'}}\right)=\frac{\partial z^{k'}}{\partial z^k}\frac{\partial z^k}{\partial x^j}\frac{\partial^2 x^j}{\partial z^{p'}\partial z^{q'}}=\frac{\partial^2 x^j}{\partial z^{p'}\,\partial z^{q'}}\frac{\partial z^{k'}}{\partial x^j}=\Gamma^{k'}_{p'q'},$$

whence we obtain finally the desired transformation rule:

$$\Gamma^{k'}_{p'q'}=\frac{\partial z^{k'}}{\partial z^k}\left(\Gamma^k_{pq}\frac{\partial z^p}{\partial z^{p'}}\frac{\partial z^q}{\partial z^{q'}}+\frac{\partial^2 z^k}{\partial z^{p'}\,\partial z^{q'}}\right). \tag{22}$$

The foregoing discussion (culminating in equation (22)) serves as motivation for the following general definition, independent of any initially given Euclidean co-ordinates, of the concept of "covariant differentiation" of tensors.

28.1.5. Definition. An operation of *covariant differentiation* of tensors (of arbitrary type) is said to be defined if we are given, in terms of any system of co-ordinates $z^1, \ldots, z^n$, a family of functions $\Gamma^k_{pq}(z)$ which transform under arbitrary co-ordinate changes $z = z(z')$ according to the formula (22). The quantities Γ^k_{pq} are called *Christoffel symbols.*

(Thus "covariant derivative" is now the official name of our "generalized gradient" tensor. We reemphasize, however, that this definition is independent of any metric which may be present.)

Hence for vectors and covectors the covariant derivatives (relative to given Christoffel symbols) are given by (cf. (16), (17)):

$$T^i_{;k}=\frac{\partial T^i}{\partial x^k}+\Gamma^i_{jk}T^j;$$

$$T_{i;k}=\frac{\partial T_i}{\partial x^k}-\Gamma^j_{ik}T_j,$$

or more generally for tensors of arbitrary type (p, q) by the formula (18). It turns out that the transformation rule (22) for the quantities Γ^k_{ij} is determined by the requirement that the covariant derivative of a tensor be again a tensor, i.e. that covariant differentiation be a tensor operation. (Note however that in general the Γ^k_{ij} are not the components of a tensor.) We shall prove this in the next subsection.

Remarks. 1. An operation of covariant differentiation is often called a *differential-geometric connexion*, or simply a *connexion*.

2. A connexion is said to be *Euclidean* (or *affine*) if there exist coordinates $x^1, \ldots, x^n$ in terms of which $\Gamma^k_{ij} \equiv 0$, i.e. such that, in those co-ordinates,

$$T^{(i)}_{(j);k}=\frac{\partial T^{(i)}_{(j)}}{\partial x^k}.$$

Such co-ordinates are for this reason often called "Euclidean" (stretching our former definition of Euclidean co-ordinates somewhat; the perhaps more appropriate term "affine co-ordinates" is also used).

3. Covariant differentiation is often denoted by the symbol ∇:

$$\nabla_k T^{(i)}_{(j)} = T^{(i)}_{(j);k}.$$

28.2. Covariant Differentiation of Tensors of Arbitrary Rank

In the preceding subsection (see 28.1.5) we defined covariant differentiation of (in particular) vector and covector fields, as the tensor operation given, in any system of co-ordinates $x^1, \ldots, x^n$, respectively by

$$T^i_{;j} = \frac{\partial T^i}{\partial x^j} + \Gamma^i_{kj} T^k, \tag{23}$$

$$T_{i;j} = \frac{\partial T_i}{\partial x^j} - \Gamma^k_{ij} T_k, \tag{24}$$

where the Γ^i_{kj} are previously given functions ("Christoffel symbols") on the points $(x^1, \ldots, x^n)$, which transform according to the rule (22).

Now suppose that we are given a family of functions $\Gamma^i_{kj}(x)$ (no longer presupposed to satisfy (22)) in terms of an arbitrary system of co-ordinates $x^1, \ldots, x^n$. We shall show that for the quantities $T^i_{;j}$, $T_{i;j}$ defined in terms of these $\Gamma^i_{kj}(x)$ by (23) and (24) to transform tensorially under arbitrary co-ordinate changes, i.e. for the operation of covariant differentiation determined by the given $\Gamma^i_{kj}(x)$ to be a tensor operation, the functions Γ^i_{kj} must necessarily transform according to the rule (22).

28.2.1. Theorem. *Let $\Gamma^i_{kj}(x)$ be a family of functions. In order for the operation of covariant differentiation defined in terms of this family to be a tensor operation, it is necessary that under arbitrary co-ordinate changes $x^i = x^i(x^{1'}, \ldots, x^{n'})$ the Γ^i_{kj} transform according to the rule*

$$\Gamma^{i'}_{k'j'} = \Gamma^i_{kj} \frac{\partial x^{i'}}{\partial x^i} \frac{\partial x^k}{\partial x^{k'}} \frac{\partial x^j}{\partial x^{j'}} + \frac{\partial x^{i'}}{\partial x^i} \frac{\partial^2 x^i}{\partial x^{k'} \partial x^{j'}}. \tag{25}$$

PROOF. Since our hypothesis is that the quantities

$$\frac{\partial T^i}{\partial x^j} + \Gamma^i_{kj} T^k = T^i_{;j}$$

are the components of a tensor, it follows (using also $T^i = T^{i'}(\partial x^i/\partial x^{i'})$, that

$$
\begin{aligned}
T^{i'}_{;j'} &= \frac{\partial T^{i'}}{\partial x^{j'}} + \Gamma^{i'}_{k'j'} T^{k'} \\
&= \left(\frac{\partial T^i}{\partial x^j} + \Gamma^i_{kj} T^k\right) \frac{\partial x^{i'}}{\partial x^i} \frac{\partial x^j}{\partial x^{j'}} = \frac{\partial x^{i'}}{\partial x^i} \frac{\partial T^i}{\partial x^{j'}} + \Gamma^i_{kj} T^k \frac{\partial x^{i'}}{\partial x^i} \frac{\partial x^j}{\partial x^{j'}} \\
&= \frac{\partial x^{i'}}{\partial x^i} \frac{\partial}{\partial x^{j'}} \left(T^{k'} \frac{\partial x^i}{\partial x^{k'}}\right) + \Gamma^i_{kj} T^{k'} \frac{\partial x^k}{\partial x^{k'}} \frac{\partial x^{i'}}{\partial x^i} \frac{\partial x^j}{\partial x^{j'}} \\
&= \frac{\partial x^{i'}}{\partial x^i} \frac{\partial x^i}{\partial x^{k'}} \frac{\partial T^{k'}}{\partial x^{j'}} + T^{k'} \frac{\partial x^{i'}}{\partial x^i} \frac{\partial^2 x^i}{\partial x^{k'} \partial x^{j'}} + \Gamma^i_{kj} T^{k'} \frac{\partial x^k}{\partial x^{k'}} \frac{\partial x^{i'}}{\partial x^i} \frac{\partial x^j}{\partial x^{j'}} \\
&= \frac{\partial T^{i'}}{\partial x^{j'}} + \left(\Gamma^i_{kj} \frac{\partial x^k}{\partial x^{k'}} \frac{\partial x^j}{\partial x^{j'}} \frac{\partial x^{i'}}{\partial x^i} + \frac{\partial x^{i'}}{\partial x^i} \frac{\partial^2 x^i}{\partial x^{k'} \partial x^{j'}}\right) T^{k'},
\end{aligned}
$$

whence the desired conclusion. □

28.2.2. Corollary. *Functions Γ^i_{kj} determining a tensorial covariant differentiation, themselves transform like tensors only under co-ordinate changes which are combinations of linear transformations and translations; i.e. affine co-ordinate changes.*

PROOF. If a co-ordinate change $x^i = x^i(x^{1'}, \ldots, x^{n'})$ is such that $\partial^2 x^i/\partial x^{k'} \partial x^{j'} \equiv 0$ for all i, k', j', then it is easy to see that it must be affine.

28.2.3. Corollary. *With the Γ^i_{kj} as in the preceding corollary, the alternating (i.e. skew-symmetric in i and j) expression*

$$
T^i_{kj} = \Gamma^i_{kj} - \Gamma^i_{jk} = \Gamma^i_{[kj]} \tag{26}
$$

is a tensor (the "torsion tensor").

PROOF. The transformation rule (25) differs from that of a tensor of type (1, 2) only by the presence of the summand $(\partial x^{i'}/\partial x^i)(\partial^2 x^i/\partial x^{k'} \partial x^{j'})$. Since this summand is (under our usual smoothness assumptions) symmetric in the indices k' and j', it follows that it will cancel in the difference $\Gamma^{i'}_{k'j'} - \Gamma^{i'}_{j'k'} = T^{i'}_{k'j'}$, so that the latter is a tensor, as required. □

28.2.4. Definition. A connexion Γ^i_{kj} (which we now define to be a family of functions transforming according to (22)) is said to be *symmetric* or *torsion-free* if the torsion tensor $T^i_{kj} = \Gamma^i_{[kj]}$ is identically zero, i.e. if $\Gamma^i_{kj} \equiv \Gamma^i_{jk}$.

28.2.5. Example. If there exist Euclidean co-ordinates $x^1, \ldots, x^n$ (by which we mean that in terms of these co-ordinates $\Gamma^i_{kj} \equiv 0$), then the torsion tensor is identically zero. Thus a Euclidean connexion is symmetric. (In other co-ordinates $x^{1'}, \ldots, x^{n'}$, where $x^i = x^i(x')$, the symbols $\Gamma^{i'}_{k'j'}$ will have the form

$$
\Gamma^{i'}_{k'j'} = \frac{\partial x^{i'}}{\partial x^i} \frac{\partial^2 x^i}{\partial x^{k'} \partial x^{j'}}.
$$

(This follows as in an earlier argument.) We see that these quantities are (as indeed they must be) symmetric in k' and j'.)

Turning now to the consideration of tensors of arbitrary type, we list four conditions which determine uniquely the tensor operation of covariant differentiation corresponding to a given connexion Γ^i_{kj}:

(i) the operation is linear, and commutes with the operation of contraction (i.e. of taking traces);
(ii) the covariant derivative of a zero-rank tensor (i.e. of a function) is the usual gradient of the function:

$$\nabla_k f = \frac{\partial f}{\partial x^k}; \tag{27}$$

(iii) the covariant derivatives of vector and covector fields are given by the formulae (23) and (24) respectively. (It is here that the connexion (Γ^i_{kj}) enters the picture.)
(iv) the covariant derivative of a product $T^{(i)(j)}_{(p)(q)} = R^{(i)}_{(p)} S^{(j)}_{(q)}$ of tensors is calculated using the usual Leibniz product rule:

$$\nabla_k T^{(i)(j)}_{(p)(q)} = (\nabla_k R^{(i)}_{(p)}) S^{(j)}_{(q)} + R^{(i)}_{(p)} (\nabla_k S^{(j)}_{(q)}). \tag{28}$$

We formulate as a theorem the fact that these four properties fully characterize the covariant derivative, singling out for special attention the important case of rank-two tensors.

28.2.6. Theorem. *Let Γ^i_{kj} be a connexion (i.e. a family of functions transforming according to* (22)). *If a tensor operation satisfies conditions* (i), (ii), (iii), (iv) *above, then it is the operation of covariant differentiation determined by the Γ^i_{kj}. In other words it coincides with the operation ∇_k, given for rank-two tensors in particular, by the formulae*

$$\nabla_k T^{ij} = \frac{\partial T^{ij}}{\partial x^k} + \Gamma^i_{lk} T^{lj} + \Gamma^j_{lk} T^{il}, \tag{29}$$

$$\nabla_k T^i_j = \frac{\partial T^i_j}{\partial x^k} + \Gamma^i_{lk} T^l_j - \Gamma^l_{jk} T^i_l, \tag{30}$$

$$\nabla_k T_{ij} = \frac{\partial T_{ij}}{\partial x^k} - \Gamma^l_{ik} T_{lj} - \Gamma^l_{jk} T_{il}, \tag{31}$$

and more generally for tensors $T^{(i)}_{(j)}$, $(i) = (i_1 \cdots i_p)$, $(j) = (j_1 \cdots j_q)$, of arbitrary type (p, q), by the formula (18), *i.e. by*

$$\nabla_k T^{(i)}_{(j)} = \frac{\partial T^{(i)}_{(j)}}{\partial x^k} - T^{(i)}_{jj_2 \ldots j_q} \Gamma^j_{j_1 k} - \cdots - T^{(i)}_{j_1 \ldots j} \Gamma^j_{j_q k} + T^{ii_2 \ldots i_p}_{(j)} \Gamma^{i_1}_{ik} + \cdots + T^{i_1 \ldots i}_{(j)} \Gamma^{i_p}_{ik}. \tag{32}$$

PROOF. We give the proof for tensors T_{ij} of type (0, 2) only; the proof in the general case is completely analogous.

Thus let $e_1, \ldots, e_n$ be the standard basis vector fields, and $e^1, \ldots, e^n$ the dual basis of covector fields; the components of these tensors are given by

$$(e_i)^j = \delta_i^j = (e^j)_i.$$

By the formulae (23), (24) (i.e. by condition (iii)) we have

$$\nabla_k e_i = \Gamma^j_{ik} e_j, \tag{33}$$

$$\nabla_k e^i = -\Gamma^i_{jk} e^j. \tag{34}$$

(From a slightly different point of view, one might regard these equations as defining the Γ^i_{jk}.)

Now every tensor T with components $T^{i_1 \ldots i_p}_{j_1 \ldots j_q}$ has the form (see §17.1(3))

$$T = T^{i_1 \ldots i_p}_{j_1 \ldots j_q} e_{i_1} \otimes \cdots \otimes e_{i_p} \otimes e^{j_1} \otimes \cdots \otimes e^{j_q}.$$

In particular if T is of type (0, 2) then $T = T_{ij} e^i \otimes e^j$. From conditions (i), (ii) and (iv) together with (33) and (34), we obtain

$$\begin{aligned}
\nabla_k(T) &= \nabla_k(T_{ij}) e^i \otimes e^j + T_{ij} \nabla_k e^i \otimes e^j + T_{ij} e^i \otimes \nabla_k e^j \\
&= \frac{\partial T^{ij}}{\partial x^k} e^i \otimes e^j - T_{ij} \Gamma^i_{lk} e^l \otimes e^j - T_{ij} e^i \otimes \Gamma^j_{lk} e^l \\
&= \left(\frac{\partial T_{ij}}{\partial x^k} - \Gamma^l_{ik} T_{lj} - \Gamma_{jk} T_{il}\right) e^i \otimes e^j.
\end{aligned}$$

Hence the components of the tensor $\nabla_k T$ have the form

$$\frac{\partial T_{ij}}{\partial x^k} - \Gamma^l_{ik} T_{lj} - \Gamma^l_{jk} T_{il},$$

as claimed in the theorem. □

Remark. If (T^i) is a vector field and (T_j) a covector field then (see §17.2) the trace $T^i T_i$ of the tensor product of (T^i) with (T_j) is a scalar. Hence by the conditions (i) through (iv) above we have

$$\begin{aligned}
\frac{\partial}{\partial x^k}(T^i T_i) &= \nabla_k(T^i T_i) = (\nabla_k T^i) T_i + T^i (\nabla_k T_i) \\
&= \left(\frac{\partial T^i}{\partial x^k} + \Gamma^i_{jk} T^j\right) T_i + \left(\frac{\partial T_i}{\partial x^k} - \Gamma^j_{ik} T_j\right) T^i \\
&= \frac{\partial}{\partial x^k}(T^i T_i) + \Gamma^i_{jk} T^j T_i - \Gamma^j_{ik} T_j T^i.
\end{aligned}$$

Since the last two terms cancel (after re-indexing) this provides confirmation of the compatibility of conditions (i) through (iv).

EXERCISES

Let ξ denote the velocity vector of the curve $\gamma = \gamma(t)$, i.e. $\xi = \dot{\gamma}$, and let η be any vector field. Define the "covariant derivative of the field η along the curve γ" by $\nabla_\xi \eta = \xi^i \nabla_i \eta$. Show that the field $\nabla_\xi \eta$ depends only on the values of the vector η on the curve γ.

§29. Covariant Differentiation and the Metric

29.1. Parallel Transport of Vector Fields

Let ξ be a vector at an arbitrary point P, and let $T = (T^{(i)}_{(j)})$ be a tensor of type (p, q). The *directional derivative* of T at P relative to (or along) the vector ξ, is defined by

$$\nabla_\xi T^{(i)}_{(j)} = \xi^k \nabla_k T^{(i)}_{(j)}, \tag{1}$$

and is thus a tensor of the same type (p, q). In the case of a rank-zero tensor (i.e. scalar) f, the directional derivative as defined by (1) becomes

$$\nabla_\xi f = \xi^k \frac{\partial f}{\partial x^k} \equiv \partial_\xi f, \tag{2}$$

and so coincides with our former concept of the directional derivative of a function at a given point in the direction ξ. As we saw in Example 23.2.2(a), if we are in motion along some curve $x^i = x^i(t)$, $i = 1, \ldots, n$, in space, and if the directional derivative of a function f in the direction of the velocity vector of that curve is zero at all points of the curve, then the function is constant on the curve; i.e. if $\xi^k(\partial f/\partial x^k) = (d/dt) f(x^1(t), \ldots, x^n(t)) \equiv 0$, where $\xi^k = dx^k/dt$ is the velocity vector (or tangent vector to the curve), then

$$f(x^1(t), \ldots, x^n(t)) = \text{const.}$$

along the curve.

Can we draw a similar conclusion for vector or, more generally, tensor fields? As it stands this question has no clear meaning, since a vector (or more generally a tensor of rank >0) has different components in different co-ordinates; in particular the components of a vector might well be constants in one system of co-ordinates, and yet variable in another. Thus we require some additional geometric structure on the underlying space, which will allow us to compare two vectors (or more generally tensors) attached to different points of the space. It turns out that a prescribed connexion, i.e. covariant differentiation, is just the "additional structure" on the space appropriate for this purpose.

29.1.1. Definition. Suppose we have a connexion Γ^i_{kj} defined on a space co-ordinatized by $x^1, \ldots, x^n$, and let $x^i(t)$, $a \le t \le b$, be a segment of an arbitrary curve. We shall say that a vector (or more generally, tensor) field T

is *covariantly constant* or *parallel* along the given curve segment if the directional covariant derivative of T in the direction of the tangent vector ξ to the curve is zero at all points of the curve segment:

$$\nabla_{\xi} T = \xi^k \nabla_k T = 0, \qquad a \leq t \leq b, \quad \xi^k = \frac{dx^k}{dt}. \tag{3}$$

In particular for vector fields equation (3) takes the form

$$\nabla_{\xi} T^i = \xi^k \nabla_k T^i = \xi^k \left(\frac{\partial T^i}{\partial x^k} + \Gamma^i_{jk} T^j \right) = 0. \tag{4}$$

It is clear from this definition that the concept of parallelism in general depends on the particular given curve segment. However the situation where the co-ordinates $x^1, \ldots, x^n$ are Euclidean, is in this respect exceptional; in terms of such co-ordinates we might define a parallel vector field T as one whose components T^i are constant (in the Euclidean co-ordinates), since the field will then be parallel along any curve whatever. Since covariant differentiation is a tensor operation (whence the co-ordinate independence of the concept of parallelism) the vector field T will then be parallel with respect to any co-ordinates $z^1, \ldots, z^n$, though its components will not in general be constants in terms of such co-ordinates, but may vary with the points.

We see from the definition that the concept of parallelism of two vectors attached to distinct points depends on the covariant differentiation (i.e. on the differential-geometric connexion), and on the path joining the two points. In order to link up our present geometrical ideas with the most basic concepts from school geometry, we recall the famous "fifth postulate" of Euclid: "Given a line through a point P, and a point Q not on the line, there is exactly one line through Q parallel to the given line". For our present purposes it will be convenient to use this postulate in the following (not completely formal) interpretation of it: If in Euclidean geometry we are given at the point P a non-zero vector $(T^i)_P$, then at any point Q there is (up to scalar multiples) one, and only one, parallel vector $(T^i_{\parallel})_Q$.

It is pertinent at this point to ask: What exactly do we mean when we say that two vectors, one at each of two distinct points, are parallel? (As always, of course, a vector (or more generally (a value of) a tensor) is understood as being attached to a given point.) The answer is furnished by the important concept of "parallel transport" of a vector (T^i) from a point P to a point Q along an arc joining P and Q.

29.1.2. Definition. Let $(T^i)_P$ be a vector at a point $P(x_0^1, \ldots, x_0^n)$ and let $x^i = x^i(t), 0 \leq t \leq 1$, be a curve segment joining P to a point $Q = (x_1^1, \ldots, x_1^n)$. The unique (see below) vector field (T^i) defined at all points of the given curve segment, taking the value $(T^i)_P$ at $P(t = 0)$, and parallel along that segment (i.e. satisfying $(dx^k/dt)\nabla_k T^i = 0$ for all $0 \leq t \leq 1$) is said to result from *parallel transport* of the vector $(T^i)_P$ along the given curve to $Q(t = 1)$. The value of the field (T^i) at Q, i.e. when $t = 1$, is denoted by $(T^i)_Q$, and is called the *result of parallel transport* of $(T^i)_P$ along the given curve $x^i = x^i(t)$ from P to Q, relative to the given connexion.

The vector field (T^i) is determined by the equation of parallel transport

$$\frac{dx^k}{dt}\nabla_k T^i = \frac{\partial T^i}{\partial x^k}\frac{dx^k}{dt} + T^j\Gamma^i_{jk}\frac{dx^k}{dt} = \frac{dT^i}{dt} + \left(\frac{dx^k}{dt}\Gamma^i_{jk}\right)T^j = 0, \tag{5}$$

and the initial conditions

$$T^i(0) = T^i, \qquad i = 1, \ldots, n. \tag{6}$$

Since (5) is a system of n linear ordinary differential equations we can appeal to the theory of the existence and uniqueness of solutions of such systems (with the initial conditions (6)) and to the continuability of their solutions, to obtain the following result.

29.1.3. Theorem. *The result of parallel transport along any smooth curve segment exists, is uniquely determined by the curve, the assumed connexion, and the initial vector* $(T^i)_P$, *and depends linearly on* $(T^i)_P$.

In the particular case when the connexion is Euclidean, i.e. $\Gamma^i_{jk} \equiv 0$ in Euclidean co-ordinates, the component form (5) of the equation of parallel transport simplifies to $dT^i/dt = 0$.

29.1.4. Corollary. *If the co-ordinates* $x^1, \ldots, x^n$ *are Euclidean, then vectors attached to different points are parallel (with respect to parallel transport along no matter what arc joining the points) precisely if they have the same components. In terms of arbitrary co-ordinates (for Euclidean n-space) the result of parallel transport of a vector from one point to another is independent of the path.*

Here the intuitive difference between Euclidean geometry and the "curved" non-Euclidean spaces is emerging: If we parallel transport a single vector from P to Q along different curves, then in the presence of non-zero curvature (i.e. if the space is not Euclidean) we may end up with different results.

We address the question of calculating the "curvature" of a space in the next subsection

29.2. Geodesics

We now turn to the following interesting (and fundamental) question: Given an arbitrary connexion, which curves play (with respect to this given connexion) the role played by straight lines when the connexion is Euclidean? The curves in question are called "geodesics".

29.2.1. Definition. A curve $x^i = x^i(t)$ is called a *geodesic* (with respect to a given connexion Γ^i_{jk}) if the vector field defined by its tangent vector $T^i = dx^i/dt$, is parallel along the curve itself, i.e. if the curve parallel transports its own tangent vector:

$$\nabla_T(T) = 0. \tag{7}$$

Rewriting (7) in component form we obtain

$$0 = \nabla_T(T)^j = \frac{dx^i}{dt}\nabla_i\left(\frac{dx^j}{dt}\right) = \frac{dx^i}{dt}\left[\frac{\partial}{\partial x^i}\left(\frac{dx^j}{dt}\right) + \Gamma^j_{ki}\frac{dx^k}{dt}\right]$$
$$= \frac{d^2x^j}{dt^2} + \Gamma^j_{ki}\frac{dx^k}{dt}\frac{dx^i}{dt},$$

whence the equations for the geodesics:

$$\frac{d^2x^j}{dt^2} + \Gamma^j_{ki}\frac{dx^k}{dt}\frac{dx^i}{dt} = 0, \qquad j = 1, \ldots, n. \tag{8}$$

Note that if $\Gamma^j_{ki} \equiv 0$, then the solutions of (8) are just the straight lines, as they should be when the geometry is Euclidean. For an arbitrary connexion Γ^i_{jk} the equations (8) form a (non-linear) system of second-order ordinary differential equations; the theory of the existence and uniqueness of the solutions of such equations tells us that this system in particular has a unique solution satisfying the initial conditions

$$x^j\bigg|_{t=0} = x_0^j, \qquad \frac{dx^j}{dt}\bigg|_{t=0} = \dot{x}_0^j, \qquad j = 1, \ldots, n, \tag{9}$$

for each choice of x_0^j, $\dot{x}_0^j$. We make this a little more precise in the following theorem.

29.2.2. Theorem. *Let Γ^i_{jk} be a connexion defined on a space. Then for any point P and any vector $(T^i)_P$ attached to the point, in some neighbourhood of P there exists a unique geodesic (of the connexion Γ^i_{jk}) starting from P and with initial tangent vector $(T^i)_P$.*

Remark. It is clear from (8) that the geodesics of a given connexion depend only on the "symmetric part" $\Gamma^i_{(jk)} = \Gamma^i_{jk} + \Gamma^i_{kj}$ of the connexion.

29.3. Connexions Compatible with the Metric

By definition, co-ordinates $x^1, \ldots, x^n$ are said to be Euclidean if in terms of these co-ordinates the metric g_{ij} is given by

$$g_{ij} = \delta_{ij}. \tag{10}$$

On the other hand, as we have seen above, the idea of a connexion naturally leads one to an alternative criterion for co-ordinates to be Euclidean (with respect to a given connexion), namely that in terms of those co-ordinates the given connexion be identically 0:

$$\Gamma^k_{ij} \equiv 0, \qquad i, j, k = 1, \ldots, n. \tag{11}$$

(Note that condition (11) is less restrictive than (10) in the sense that if (11) holds in one frame, then it holds in all affine transforms of that frame. Note also that we might alternatively (and even more generously) take the criterion $\Gamma^k_{ij} = -\Gamma^k_{ji}$ as signalling the Euclideanness of the co-ordinates.)

Our aim in the present section is to find a natural relationship linking the two conditions (10) and (11). It behoves us to reiterate that the concepts of a connexion and of a Riemannian metric are, *a priori*, in no way related; neither concept was used to define the other (not to speak of motivations); they are two separate and independent structures on the region of the space under scrutiny. Hence conditions (10) and (11) are logically quite unrelated. However there is a natural way of splicing the two concepts so that (11) becomes (with accuracy up to application of affine co-ordinate transformations) equivalent to (10) as a criterion for co-ordinates to be Euclidean.

29.3.1. Definition. A connexion Γ^k_{ij} is said to be *compatible with a metric* g_{ij} if the covariant derivative of the metric tensor (g_{ij}) is identically zero:

$$\nabla_k g_{ij} \equiv 0, \qquad k, i, j = 1, \ldots, n. \tag{12}$$

We mention two properties enjoyed by a connexion compatible with a metric.

(i) *If a given connexion is compatible with the metric on a space then the corresponding operation of covariant differentiation commutes with the operation of lowering any index of a tensor.*

PROOF. It follows from the condition (iv) in §28.2 (the Leibniz product formula) and the defining condition (12) of compatibility, that

$$\nabla_k(g_{lm} T^{(i)}_{(j)}) = g_{lm}(\nabla_k T^{(i)}_{(j)})$$

for any tensor $T^{(i)}_{(j)}$ of type (p, q). The desired conclusion now follows from the linearity of covariant differentiation (condition (i) in §28.2) and the definition of the operation of lowering an index of a tensor (see §19.1(3)). □

(ii) *If vector fields $T^i(t)$ and $S^i(t)$ are both parallel along a curve $x^i = x^i(t)$, then their scalar product is constant along the curve (provided the connexion is compatible with the metric).*

PROOF. We have

$$\frac{d}{dt}\langle T, S\rangle = \frac{d}{dt}(g_{ij}T^iS^j) = \frac{dx^k}{dt}\nabla_k(g_{ij}T^iS^j) = \frac{dx^k}{dt}g_{ij}\nabla_k(T^iS^j)$$

$$= g_{ij}\left(\frac{dx^k}{dt}\nabla_k T^i\right)S^j + g_{ij}T^i\left(\frac{dx^k}{dt}\nabla_k S^j\right) = 0,$$

whence the result. □

We may paraphrase property (ii) as follows: If the connexion is compatible with the metric, then parallel transport of vectors from a point P to a point Q along a given curve, defines an orthogonal (since inner product preserving) transformation from the tangent space at P to the tangent space at Q.

If a symmetric connexion is compatible with a metric then it is uniquely determined by the metric, and there is a simple formula for the connexion in terms of the metric; this is the substance of the following

29.3.2. Theorem. *If the metric g_{ij} is non-singular (i.e. if $g = \det(g_{ij}) \neq 0$) on the region of n-space under consideration, then there is a unique symmetric connexion which is compatible with the metric. This unique connexion is given in any system of co-ordinates $x^1, \ldots, x^n$, by "Christoffel's formula":*

$$\Gamma^k_{ij} = \tfrac{1}{2}g^{kl}\left(\frac{\partial g_{lj}}{\partial x^i} + \frac{\partial g_{il}}{\partial x^j} - \frac{\partial g_{ij}}{\partial x^l}\right). \tag{13}$$

PROOF. By hypothesis we have $\Gamma^k_{ij} = \Gamma^k_{ji}$. From §28.2(31) and the compatibility we obtain

$$\nabla_k g_{ij} = \frac{\partial g_{ij}}{\partial x^k} - \Gamma^l_{ik}g_{lj} - \Gamma^l_{jk}g_{il} = 0. \tag{14}$$

Our aim is to solve these equations for the Γ^k_{ij}. By definition of the operation of lowering indices we have

$$\Gamma_{k,ij} = g_{kl}\Gamma^l_{ij},$$

so that (14) may be rewritten as

$$\Gamma_{i,jk} + \Gamma_{j,ik} = \frac{\partial g_{ij}}{\partial x^k},$$

where by the assumed symmetry $\Gamma_{i,jk} = \Gamma_{i,kj}$, $\Gamma_{j,ik} = \Gamma_{j,ki}$. By permuting the indices i, j, k we obtain

$$\Gamma_{i,jk} + \Gamma_{j,ik} = \frac{\partial g_{ij}}{\partial x^k},$$

$$\Gamma_{j,ki} + \Gamma_{k,ji} = \frac{\partial g_{kj}}{\partial x^i},$$

$$\Gamma_{i,kj} + \Gamma_{k,ij} = \frac{\partial g_{ik}}{\partial x^j}.$$

If we subtract the first of these three equations from the sum of the last two, we obtain, after taking the assumed symmetry into account,

$$2\Gamma_{k,ij} = \frac{\partial g_{kj}}{\partial x^i} + \frac{\partial g_{ik}}{\partial x^j} - \frac{\partial g_{ij}}{\partial x^k} = 2\, g_{kl}\Gamma^l_{ij}.$$

The desired formula (13) now follows by dividing by 2 and raising the index k. □

29.3.3. Corollary. *Let Γ^k_{ij} be a symmetric connexion compatible with the metric. If the co-ordinates $x^1, \ldots, x^n$ are such that at a given point all first-order partial derivatives of the components g_{ij} of the metric are zero, then at that point the Christoffel symbols Γ^k_{ij} are zero.*

29.3.4. Examples. (a) Consider a surface

$$x^1 = x^1(z^1, z^2), \qquad x^2 = x^2(z^1, z^2), \qquad x^3 = x^3(z^1, z^2),$$

in Euclidean 3-space with Euclidean co-ordinates $x^1 = x$, $x^2 = y$, $x^3 = z$. We shall suppose (as we did in §8) that our Euclidean co-ordinate system has been chosen so that the z-axis is perpendicular to the surface at P and the x-axis and y-axis tangent to the surface at P. Then in some neighbourhood of P the surface is given by an equation of the form (see §§7, 8)

$$z = f(x, y),$$

so that we may take as parameters (i.e. local co-ordinates) $z^1 = x$, $z^2 = y$; moreover the perpendicularity of the z-axis to the surface implies that $\partial f/\partial x|_{(0,0)} = \partial f/\partial y|_{(0,0)} = 0$, i.e. the gradient of f, grad f, is zero at $P = (0, 0)$. By §7.3(20) the induced metric on the surface is given by

$$g_{ij} = \delta_{ij} + \frac{\partial f}{\partial z^i}\frac{\partial f}{\partial z^j}, \qquad z^1 = x, \quad z^2 = y.$$

Thus at the point P, where $\partial f/\partial z^i = 0$, we have $g_{ij} = \delta_{ij}$, and

$$\frac{\partial g_{ij}}{\partial z^k} = \frac{\partial}{\partial z^k}\left(\frac{\partial f}{\partial z^i}\frac{\partial f}{\partial z^j}\right) = \frac{\partial^2 f}{\partial z^i\,\partial z^k}\frac{\partial f}{\partial z^j} + \frac{\partial^2 f}{\partial z^j\,\partial z^k}\frac{\partial f}{\partial z^i} = 0.$$

Hence by the above corollary, in terms of the co-ordinates x, y for the neighbourhood of P on the surface, the (symmetric and compatible) connexion Γ^k_{ij} ($i, j, k = 1, 2$) will be zero at P.

(b) In the remark following (9) in §28.1 we used the term "divergence" of a vector field (T^i) for the scalar

$$\operatorname{div} T = \nabla_i T^i = T^i_{;i} = \frac{\partial T^i}{\partial x^i} + \Gamma^i_{ki} T^k. \tag{15}$$

Now if the connexion Γ^i_{jk} is symmetric, and compatible with a (pseudo-) Riemannian metric g_{ij} assumed on the underlying space, then

$$\Gamma^i_{ki} = \tfrac{1}{2}g^{il}\left(\frac{\partial g_{lk}}{\partial x^i} + \frac{\partial g_{li}}{\partial x^k} - \frac{\partial g_{ki}}{\partial x^l}\right) = \tfrac{1}{2}g^{il}\frac{\partial g_{il}}{\partial x^k} = \frac{1}{2g}\frac{\partial g}{\partial x^k} = \frac{\partial}{\partial x^k}\ln(\sqrt{|g|}),$$

where $g = \det(g_{ij})$. Hence from (15)

$$\nabla_i T^i = \frac{\partial T^i}{\partial x^i} + \frac{1}{2g}\frac{\partial g}{\partial x^k}T^k = \frac{1}{\sqrt{|g|}}\frac{\partial}{\partial x^i}(\sqrt{|g|}\,T^i). \tag{16}$$

We conclude from this that: *The divergence $\nabla_i T^i$ of a vector field (T^i) is given by the simple formula $\nabla_i T^i = \partial T^i/\partial x^i$ precisely in those co-ordinate systems where the volume element $\sqrt{|g|}\,dx^1 \wedge \cdots \wedge dx^n$ has (to within a constant scalar factor) the same form as the Euclidean volume element, i.e. $\sqrt{|g|} = 1$, where $g = \det(g_{ij})$.*

We conclude this subsection with the following summarizing remark: We have now linked together the hitherto separate structures of a connexion (i.e. a covariant differentiation) and a Riemannian metric; the Riemannian geometry gives rise to a uniquely defined symmetric connexion (i.e. covariant differentiation of tensors) relative to which the metric itself is constant (see property (ii) above).

29.4. Connexions Compatible with a Complex Structure (Hermitian Metric)

Let D denote a region of complex n-space with complex co-ordinates $z^1, \ldots, z^n$, $z^k = x^k + iy^k$; as before we write $D^{\mathbb{R}}$ for the realization of D, co-ordinatized by the real co-ordinates $x^1, \ldots, x^n, y^1, \ldots, y^n$. Suppose that there is defined on D a Hermitian metric (see §27.2)

$$ds^2 = h_{i\bar{k}}\,dz^i\,d\bar{z}^k, \qquad h_{k\bar{i}} = \bar{h}_{i\bar{k}}. \tag{17}$$

(Recall that, as in §27, when necessary we place a bar over an index i to indicate that it corresponds to $d\bar{z}^i$.) We saw earlier (in §11.2) that the Hermitian metric gives rise to a Riemannian metric on $D^{\mathbb{R}}$ defined by

$$ds^2_{\mathbb{R}} = \mathrm{Re}(h_{i\bar{k}})[dx^i\,dx^k + dy^i\,dy^k]. \tag{18}$$

We know from the preceding subsection that there is a unique symmetric connexion on $D^{\mathbb{R}}$ compatible with the metric (18). Generally speaking this connexion will not however be compatible with the original complex structure (i.e. Hermitian metric) on D given by (17), i.e. parallel transport of a vector along an arc in D will not in general be a unitary transformation (cf. the remark following (ii) in the preceding subsection). The precise situation is as follows.

29.4.1. Theorem. *The symmetric connexion on* $D^{\mathbb{R}}$ *compatible with the metric* $ds^2_{\mathbb{R}}$, *is compatible with the Hermitian metric* ds^2 *if and only if the latter metric is Kählerian.*

PROOF. Recall that by Definition 27.2.1 the Hermitan metric ds^2 is called Kählerian if the form Ω defined by

$$\Omega = \frac{i}{2} h_{j\bar{k}} dz^j \wedge d\bar{z}^k, \tag{19}$$

is closed, i.e.

$$d\Omega = 0 \Leftrightarrow \frac{\partial h_{j\bar{k}}}{\partial z^i} - \frac{\partial h_{i\bar{k}}}{\partial z^j} = 0, \qquad \frac{\partial h_{j\bar{k}}}{\partial \bar{z}^i} - \frac{\partial h_{j\bar{i}}}{\partial \bar{z}^k} = 0. \tag{20}$$

(For the right-hand side of this double implication, see §25.2(7) or §27.1(9).) We may take the operators

$$\frac{\partial}{\partial z^k} = \frac{1}{2}\left(\frac{\partial}{\partial x^k} - i\frac{\partial}{\partial y^k}\right), \qquad \frac{\partial}{\partial \bar{z}^k} = \frac{1}{2}\left(\frac{\partial}{\partial x^k} + i\frac{\partial}{\partial y^k}\right), \qquad k = 1, \ldots, n,$$

as representing a (complex) basis for the tangent space to the region $D^{\mathbb{R}}$ (see the beginning of §27.1), so that an arbitrary tangent vector ξ can be represented as (cf. §27.1(4))

$$\xi = \xi^k \frac{\partial}{\partial z^k} + \xi^{\bar{k}} \frac{\partial}{\partial \bar{z}^k},$$

and ξ is real (i.e. is a real linear combination of the operators $\partial/\partial x^k$, $\partial/\partial y^k$) precisely if $\xi^{\bar{k}} = \overline{\xi^k}$. It is easy to see (by rewriting (18) in terms of dz^k, $d\bar{z}^k$) that the scalar product of tangent vectors defined by $ds^2_{\mathbb{R}}$ is given, in terms of the basis $\partial/\partial z^k$, $\partial/\partial \bar{z}^k$, by $g_{\alpha\beta}$, $\alpha, \beta = 1, \ldots, n, \bar{1}, \ldots, \bar{n}$, where

$$g_{ij} = g_{\bar{i}\bar{j}} = 0; \qquad g_{i\bar{j}} = h_{i\bar{j}}, \quad g_{\bar{i}j} = \bar{h}_{i\bar{j}}, \tag{21}$$

so that the matrix $G = (g_{\alpha\beta})$ has the form

$$G = \begin{pmatrix} 0 & H \\ \bar{H} & 0 \end{pmatrix}, \qquad H = (h_{i\bar{j}}). \tag{22}$$

The inverse matrix G^{-1} is given by

$$G^{-1} = \begin{pmatrix} 0 & \bar{H}^{-1} \\ H^{-1} & 0 \end{pmatrix}. \tag{23}$$

We shall now compute the components $\Gamma^{\alpha}_{\beta\gamma}$ of the symmetric connexion compatible with the metric $g_{\alpha\beta}$. From Christoffel's formula (13) and (21) we obtain:

$$\Gamma^i_{jk} = \tfrac{1}{2} g^{i\bar{m}} \left(\frac{\partial g_{j\bar{m}}}{\partial z^k} + \frac{\partial g_{\bar{m}k}}{\partial z^j}\right),$$

$$\Gamma^{\bar{i}}_{\bar{j}\bar{k}} = \tfrac{1}{2} g^{m\bar{i}} \left(\frac{\partial g_{\bar{k}m}}{\partial \bar{z}^j} + \frac{\partial g_{m\bar{j}}}{\partial \bar{z}^k}\right) = \overline{\Gamma^i_{jk}}.$$

We also have (using also (23))

$$\Gamma^i_{j\bar{k}} = \tfrac{1}{2} g^{i\bar{m}} \left(\frac{\partial g_{\bar{m}\bar{k}}}{\partial \bar{z}^j} + \frac{\partial g_{j\bar{m}}}{\partial \bar{z}^k} - \frac{\partial g_{j\bar{k}}}{\partial \bar{z}^m} \right) = 0,$$

and, for similar reasons, $\Gamma^{\bar{i}}_{jk} = 0$. Finally, bringing in the hypothesis in the form given by (20), we have

$$\Gamma^i_{j\bar{k}} = \tfrac{1}{2} g^{i\bar{m}} \left(\frac{\partial g_{j\bar{m}}}{\partial \bar{z}^k} - \frac{\partial g_{j\bar{k}}}{\partial \bar{z}^m} \right) = \tfrac{1}{2} g^{i\bar{m}} \left(\frac{\partial h_{j\bar{m}}}{\partial \bar{z}^k} - \frac{\partial h_{j\bar{k}}}{\partial \bar{z}^m} \right) = 0.$$

From the hypothesis that the metric ds^2 be Kählerian it follows in a similar fashion that the components $\Gamma^i_{\bar{j}k}, \Gamma^{\bar{i}}_{j\bar{k}}, \Gamma^{\bar{i}}_{\bar{j}k}$ also vanish. Thus the upshot is that the only possible non-zero components among the $\Gamma^\alpha_{\beta\gamma}$ are the Γ^i_{jk} and the $\Gamma^{\bar{i}}_{\bar{j}\bar{k}}$, and that these are complex conjugates of one another.

Consider now the change effected in a real vector $\xi = (\xi^i, \xi^{\bar{i}})$ by parallelly transporting it through the infinitesimal real interval $(\delta z^j, \delta \bar{z}^j)$, $\delta \bar{z}^j = \overline{\delta z^j}$. It follows from §29.1(5) and what we have just proved about the components $\Gamma^\alpha_{\beta\gamma}$, that (to within second-order expressions in the infinitesimals)

$$\begin{aligned} \xi^i &\to \xi^i - \xi^k \Gamma^i_{kj} \delta z^j, \\ \xi^{\bar{i}} &\to \xi^{\bar{i}} - \xi^{\bar{k}} \Gamma^{\bar{i}}_{\bar{j}\bar{k}} \delta \bar{z}^j. \end{aligned} \tag{24}$$

Hence if we write $A = (a^i_k)$ for the matrix with entries

$$a^i_k = \Gamma^i_{kj} \delta z^j,$$

then the Jacobian of the transformation (24) is

$$\begin{pmatrix} 1 - A & 0 \\ 0 & 1 - \bar{A} \end{pmatrix}. \tag{25}$$

Since by (ii) in §29.3 the realization of the transformation defined by the parallel transport is orthogonal, it follows that the transformation we are approximating in (24) is actually linear (over $\mathbb{R}$), and in terms of the co-ordinates $z^i, \bar{z}^i$ has the form (25). Now it is easy to see that matrices of the form $\begin{pmatrix} B & 0 \\ 0 & \bar{B} \end{pmatrix}$ define transformations of the co-ordinates $z^i, \bar{z}^i$ which are linear over $\mathbb{C}$ (cf. §12.2). This and the orthogonality then together imply that parallel transport along an arc in $D^{\mathbb{R}}$ defines a unitary transformation. This completes the proof of the theorem. (We leave the easy converse to the reader.) □

29.5. Exercises

1. Prove that a connexion is compatible with the metric if and only if for arbitrary vector fields η, ξ_1, ξ_2

$$\partial_\eta \langle \xi_1, \xi_2 \rangle = \langle \nabla_\eta \xi_1, \xi_2 \rangle + \langle \xi_1, \nabla_\eta \xi_2 \rangle.$$

2. Show that if a vector (ξ^i) is parallelly transported through a small interval (δx^k), then its components change as follows:

$$\xi^i \to \xi^i - \xi^j \Gamma^i_{jk} \delta x^k + o(|\delta x|).$$

3. Express the stain tensor (defined in §22.3(23)) in terms of covariant derivatives.

4. Let U be a region of a space on which there is defined a connexion. Denote by $T = T_P$ the tangent space to U at an arbitrary fixed point P of U. Define a map $E: T \to U$ as follows: associate with each vector ξ from T the geodesic $\gamma_\xi(t)$ originating at P and with initial velocity vector ξ; then define $E(\xi) = \gamma_\xi(1)$.

   (i) Show that the map E is defined in some neighbourhood of the origin of the space T, and is (locally) a diffeomorphism.

   (ii) Show that in terms of the co-ordinates on U defined by the map E, the components Γ^k_{ij} of the given connexion all vanish at the point P.

5. The equation of motion of a point electric charge in the field of a magnetic pole has the form

$$\ddot{r} = a\frac{[r, \dot{r}]}{|r|^3}, \qquad a = \text{const.}$$

   Prove that the trajectory of a point charge is a geodesic on a circular cone.

6. Find all the geodesics of the Lobachevskian plane.

7. Show that the geodesics of the sphere are just the great circles.

8. Show using geodesics that an isometry of a space (endowed with a metric) which leaves fixed a point and a reference frame at the point, is necessarily the identity map.

9. Prove that the level curves of the function

$$z(u, v) = \int \frac{du}{\sqrt{f(u) - a}} \pm \int \frac{dv}{\sqrt{g(v) + a}}$$

   are geodesics with respect to the metric

$$dl^2 = (f(u) + g(v))(du^2 + dv^2), \qquad f > 0, \quad g > 0.$$

10. Prove that the inner automorphisms $X \mapsto AXA^{-1}$, $X, A \in SO(3, \mathbb{R})$, are exactly the motions of the Killing metric on $SO(3, \mathbb{R})$ which leave the identity element fixed.

11. Given that Γ^i_{jk} is a symmetric connexion compatible with the metric, establish the following identities:

    (i) $g^{kl}\Gamma^i_{kl} = -\dfrac{1}{\sqrt{|g|}}\dfrac{\partial}{\partial x^k}(\sqrt{|g|}\, g^{ik});$

    (ii) $\Gamma^i_{ki} = \dfrac{1}{2g}\dfrac{\partial g}{\partial x^k}.$

12. Prove that given two points in a Riemannian space sufficiently close to one another there is a geodesic joining them which is locally unique.

13. Given a metric in the form $dl^2 = g_{rr}\,dr^2 + r^2\,d\varphi^2$, show that the line $\varphi = \varphi_0$ through the point $r = 0$ (the origin) is a geodesic.

14. Given a metric (g_{ij}) on an n-dimensional space, establish the following formula (cf. Exercise 2 of §26.5):

$$\oint_{\partial V} X^i\,dS_i = \int_V \nabla_i X^i \sqrt{|g|}\,dx^1 \wedge \cdots \wedge dx^n,$$

where

$$dS_i = \frac{1}{(n-1)!}\sqrt{|g|}\,\varepsilon_{i_1 \dots i_{n-1} i}\,dx^{i_1} \wedge \cdots \wedge dx^{i_{n-1}}.$$

15. Let M be a surface in Euclidean n-space $\mathbb{R}^n$, let π be the linear operator which projects $\mathbb{R}^n$ orthogonally onto the tangent space to M (at an arbitrary fixed point of M), and let X, Y be vector fields in $\mathbb{R}^n$ tangent to the surface M. Show that the connexion on M compatible with the induced metric on M, satisfies

$$\nabla_X Y = \pi\left(X^k \frac{\partial Y}{\partial x^k}\right).$$

§30. The Curvature Tensor

30.1. The General Curvature Tensor

As noted in §29.1, in a non-Euclidean space the result of parallel transport of a vector is in general path-dependent. We saw in that subsection that to obtain the result of parallelly transporting a vector T along an arc of a curve $x(t)$, we need to solve the differential equation

$$\frac{dT^i}{dt} + \Gamma^i_{jk} T^k \frac{dx^j}{dt} = 0. \tag{1}$$

Clearly it would be more convenient if instead of solving this equation, we could find a simpler way of measuring the departure of the given connexion Γ^i_{jk} from being Euclidean. What measure of this kind is available? What simple criterion is there for the existence or otherwise of co-ordinates in terms of which the Γ^i_{jk} vanish? (Of course if we have on our hands a connexion which is not symmetric then the answer is easy, since then it cannot be symmetric in any co-ordinate system, and so, in particular, we cannot have $\Gamma^i_{jk} \equiv 0$. However in this case it is appropriate to define Euclidean co-ordinates as those in terms of which the symmetric part $\Gamma^i_{jk} + \Gamma^i_{kj}$ of the connexion vanishes identically.)

The key to the solution of this problem turns out to be the familiar "equality of mixed partials": if f is a sufficiently smooth function then

$$\frac{\partial^2 f}{\partial x^j\,\partial x^i} = \frac{\partial}{\partial x^j}\frac{\partial f}{\partial x^i} = \frac{\partial}{\partial x^i}\frac{\partial f}{\partial x^j} = \frac{\partial^2 f}{\partial x^i\,\partial x^j}.$$

If there do exist Euclidean co-ordinates relative to a given (symmetric) connexion, then we know that in terms of those co-ordinates covariant differentiation is just ordinary partial differentiation:

$$\nabla_k T^{(i)}_{(j)} = \frac{\partial T^{(i)}_{(j)}}{\partial x^k},$$

so that by the "equality of mixed partials"

$$(\nabla_k \nabla_l - \nabla_l \nabla_k) T^{(i)}_{(j)} = 0,$$

or, equivalently,

$$T^{(i)}_{(j);k;l} = T^{(i)}_{(j);l;k}. \tag{2}$$

Since the $T^{(i)}_{(j);k;l}$ are the components of a tensor, equation (2) will then hold in all co-ordinate systems.

If on the other hand our connexion is arbitrary (and so not necessarily Euclidean), then for a vector field (T^i) we have

$$\begin{aligned}\nabla_k \nabla_l T^i &= \nabla_k \left(\frac{\partial T^i}{\partial x^l} + \Gamma^i_{ql} T^q\right) = \frac{\partial}{\partial x^k}\left(\frac{\partial T^i}{\partial x^l} + \Gamma^i_{ql} T^q\right) + \Gamma^i_{pk}\left(\frac{\partial T^p}{\partial x^l} + \Gamma^p_{ql} T^q\right) \\ &\quad - \Gamma^p_{lk}\left(\frac{\partial T^i}{\partial x^p} + \Gamma^i_{qp} T^q\right) \\ &= \frac{\partial^2 T^i}{\partial x^k \partial x^l} + \frac{\partial T^q}{\partial x^k}\Gamma^i_{ql} + \Gamma^i_{pk}\frac{\partial T^p}{\partial x^l} - \Gamma^p_{lk}\frac{\partial T^i}{\partial x^p} + T^q \frac{\partial \Gamma^i_{ql}}{\partial x^k} + \Gamma^i_{pk}\Gamma^p_{ql} T^q \\ &\quad - \Gamma^p_{lk}\Gamma^i_{qp} T^q.\end{aligned}$$

From this one easily obtains

$$\begin{aligned}(\nabla_k \nabla_l - \nabla_l \nabla_k) T^i = &\left(\frac{\partial \Gamma^i_{ql}}{\partial x^k} - \frac{\partial \Gamma^i_{qk}}{\partial x^l}\right) T^q + (\Gamma^i_{pk}\Gamma^p_{ql} - \Gamma^i_{pl}\Gamma^p_{qk}) T^q \\ &- (\Gamma^p_{lk} - \Gamma^p_{kl}) \frac{\partial T^i}{\partial x^p}.\end{aligned}$$

If we introduce the notation

$$-R^i_{qkl} = \frac{\partial \Gamma^i_{ql}}{\partial x^k} - \frac{\partial \Gamma^i_{qk}}{\partial x^l} + \Gamma^i_{pk}\Gamma^p_{ql} - \Gamma^i_{pl}\Gamma^p_{qk}, \tag{3}$$

then the above becomes

$$(\nabla_k \nabla_l - \nabla_l \nabla_k) T^i = -R^i_{qkl} T^q + T^p_{kl} \frac{\partial T^i}{\partial x^p} \tag{4}$$

(where we have also written T^p_{kl} for the torsion tensor $\Gamma^p_{kl} - \Gamma^p_{lk}$ (see 28.2.3)). It is not difficult to see that the R^i_{qkl} are the components of a tensor, called the *Riemann curvature tensor*. In the case we were considering the given connexion is symmetric, so that $T^p_{kl} \equiv 0$.

30.1.1. Theorem. *For any symmetric connexion* Γ^i_{jk} *and any vector field T, we have*

$$(\nabla_k \nabla_l - \nabla_l \nabla_k) T^i = - R^i_{qkl} T^q,$$

where R^i_{qkl} *is a tensor (the Riemann curvature tensor) given by*

$$- R^i_{qkl} = \frac{\partial \Gamma^i_{ql}}{\partial x^k} - \frac{\partial \Gamma^i_{qk}}{\partial x^l} + \Gamma^i_{pk} \Gamma^p_{ql} - \Gamma^i_{pl} \Gamma^p_{qk}.$$

It follows that if the connexion is Euclidean then $R^i_{qkl} \equiv 0$. *At those points where* $\Gamma^i_{pq} = 0$, *we have*

$$-R^i_{qkl} = \frac{\partial \Gamma^i_{ql}}{\partial x^k} - \frac{\partial \Gamma^i_{qk}}{\partial x^l}.$$

From this theorem and the fact that the R^i_{qkl} are the components of a tensor, we obtain immediately the following significant result.

30.1.2. Corollary. *If the Riemann curvature tensor determined by a given symmetric connexion is not identically zero then the connexion is not Euclidean, i.e. there do not exist co-ordinates in terms of which the components* Γ^k_{ij} *of the connexion vanish identically.*

Remark. As noted, this result follows immediately from the fact that the R^i_{qkl} are the components of a tensor, and that the components of a zero tensor are zero in all co-ordinate systems. Alternatively it may be proved directly without recourse to the tensorial nature of the R^i_{qkl} as follows. We are looking for a co-ordinate change $x = x(x')$ to co-ordinates x' in terms of which $\Gamma^{k'}_{i'j'} \equiv 0$. Now by the transformation rule for the Γ^k_{ij} (see §28.1(22)) we have

$$\Gamma^{k'}_{i'j'} = \frac{\partial x^{k'}}{\partial x^k} \left(\Gamma^k_{ij} \frac{\partial x^i}{\partial x^{i'}} \frac{\partial x^j}{\partial x^{j'}} + \frac{\partial^2 x^k}{\partial x^{i'} \partial x^{j'}} \right).$$

This and the hypothesis $\Gamma^{k'}_{i'j'} \equiv 0$ yield the following equation for the $x^{i'}$:

$$\frac{\partial^2 x^k}{\partial x^{i'} \partial x^{j'}} = -\Gamma^k_{ij} \frac{\partial x^i}{\partial x^{i'}} \frac{\partial x^j}{\partial x^{j'}}, \qquad \Gamma^k_{ij} = \Gamma^k_{ij}(x). \tag{5}$$

Assuming the co-ordinate change to have continuous third-order partial derivatives we have

$$\frac{\partial}{\partial x^{k'}} \left(\frac{\partial^2 x^k}{\partial x^{i'} \partial x^{j'}} \right) = \frac{\partial}{\partial x^{i'}} \left(\frac{\partial^2 x^k}{\partial x^{k'} \partial x^{j'}} \right).$$

Exploiting this as a condition on the right-hand side of (5) and bringing in the assumed symmetry of the Γ^k_{ij}, we obtain finally (after some calculation involving in particular elimination of the primed co-ordinates)

$$R^i_{qkl} \equiv 0.$$

We shall now derive co-ordinate-free formulae for the curvature and torsion tensors. For arbitrary vector fields ξ, η, ζ we set

$$[T(\xi, \eta)]^i = T^i_{kl}\xi^k\eta^l, \tag{6}$$

$$[R(\xi, \eta)\zeta]^i = R^i_{jkl}\xi^k\eta^l\zeta^j. \tag{7}$$

In the following lemma the vector fields $T(\xi, \eta)$ and $R(\xi, \eta)\zeta$ are expressed in terms of the fields ξ, η, ζ.

30.1.3. Lemma. *For arbitrary vector fields ξ, η, ζ, the following equations hold:*

$$T(\xi, \eta) = \nabla_\xi\eta - \nabla_\eta\xi - [\xi, \eta], \tag{8}$$

$$R(\xi, \eta)\zeta = \nabla_\eta\nabla_\xi\zeta - \nabla_\xi\nabla_\eta\zeta + \nabla_{[\xi,\eta]}\zeta, \tag{9}$$

where $[\xi, \eta]$ is the commutator of the vector fields ξ, η.

PROOF. We first show that the right-hand sides in (8) and (9) are linear in ξ, η, ζ. Thus if in the right-hand side of (8) we replace ξ by $f\xi$ where f is a smooth function, we obtain

$$\begin{aligned}\nabla_{f\xi}\eta - \nabla_\eta(f\xi) - [f\xi, \eta] &= f[\nabla_\xi\eta - \nabla_\eta\xi - [\xi, \eta]] - (\partial_\eta f)\xi + (\partial_\eta f)\xi \\ &= f[\nabla_\xi\eta - \nabla_\eta\xi - [\xi, \eta]],\end{aligned}$$

where $\partial_\eta f$ is the directional derivative of f along the field η.

If we replace ξ by $f\xi$ in the right-hand side of (9), we obtain

$$\begin{aligned}\nabla_\eta\nabla_{f\xi} - \nabla_{f\xi}\nabla_\eta + \nabla_{[f\xi,\eta]} &= f\nabla_\eta\nabla_\xi + (\partial_\eta f)\nabla_\xi - f\nabla_\xi\nabla_\eta + \nabla_{f[\xi,\eta]} - (\partial_\eta f)\nabla_\xi \\ &= f[\nabla_\eta\nabla_\xi - \nabla_\xi\nabla_\eta + \nabla_{[\xi,\eta]}].\end{aligned}$$

The linearity in η follows similarly. Finally, replacing ζ by $f\zeta$ we have:

$$\begin{aligned}&\nabla_\eta[(\partial_\xi f)\zeta + f\nabla_\xi\zeta] - \nabla_\xi[(\partial_\eta f)\zeta + f\nabla_\eta\zeta] + (\partial_{[\xi,\eta]}f)\zeta + f\nabla_{[\xi,\eta]}\zeta \\ &\quad = (\partial_\eta\partial_\xi f)\zeta + (\partial_\xi f)\nabla_\eta\zeta + (\partial_\eta f)\nabla_\xi\zeta + f\nabla_\eta\nabla_\xi\zeta - (\partial_\xi\partial_\eta f)\zeta \\ &\qquad -(\partial_\eta f)\nabla_\xi\zeta - (\partial_\xi f)\nabla_\eta\zeta - f\nabla_\xi\nabla_\eta\zeta + (\partial_\xi\partial_\eta f - \partial_\eta\partial_\xi f)\zeta + f\nabla_{[\xi,\eta]}\zeta \\ &\quad = f[\nabla_\eta\nabla_\xi\zeta - \nabla_\xi\nabla_\eta\zeta + \nabla_{[\xi,\eta]}\zeta].\end{aligned}$$

Having shown that the right-hand side expressions of equations (8) and (9) are linear in ξ, η, ζ (that they respect sums being immediate), it suffices to verify (8) and (9) for the basic vector fields $\xi = e_k$, $\eta = e_l$, $\zeta = e_j$ (in which case $\xi^i = \delta^i_k$, $\eta^i = \delta^i_l$, $\zeta^i = \delta^i_j$). However for these fields (8) and (9) follow immediately from the definitions of T^i_{kl} and R^i_{jkl}. □

30.1.4. Application (Tetrad Formalism). Suppose that a region of an n-dimensional space is endowed with a metric g_{ij}. At each point of the region the metric defines a quadratic form $g_{ij}\xi^i\xi^j$ on tangent vectors ξ at the point. From linear algebra we know that at each point there is a basis $\xi_1, \ldots, \xi_n$ for the tangent space in terms of which the matrix of this quadratic form is diagonal. It is not difficult to see that if the g_{ij} are smooth functions on the points of the region, then the preferential frame $\xi_1, \ldots, \xi_n$ can be chosen to also depend smoothly on the points. Hence (locally) there exist n linearly independent smooth vector fields $\xi_1, \ldots, \xi_n$ such that

$$\langle \xi_i, \xi_j \rangle = \varepsilon_i \delta_{ij}, \qquad \varepsilon_i = \pm 1. \tag{10}$$

(For example in the general theory of relativity it is convenient for technical reasons to take at each event an orthonormal "tetrad" of vectors $\xi_0, \xi_1, \xi_2, \xi_3$, where ξ_0 is time-like, and ξ_1, ξ_2, ξ_3 are space-like vectors.)

Consider the pairwise commutators $[\xi_i, \xi_j]$ of these vector fields (see 23.2.3). We may express each of these commutators in terms of the basis $\xi_1, \ldots, \xi_n$ obtaining, say,

$$[\xi_i, \xi_j] = c_{ij}^k \xi_k. \tag{11}$$

Note that the "structural constants" c_{ij}^k are determined by the definition of commutation of vector fields and by the basis $\xi_1, \ldots, \xi_n$ (cf. §24.5). The symmetric connexion compatible with the metric is determined by the quantities $\hat{\Gamma}_{jk}^i$ defined by

$$\nabla_{\xi_k} \xi_j = \hat{\Gamma}_{jk}^i \xi_i. \tag{12}$$

30.1.5. Theorem. *In terms of the structural constants defined in* (11) *the quantities* $\hat{\Gamma}_{jk}^i$ *(defined in* (12)*) determining the symmetric connexion compatible with the metric are given by*

$$\hat{\Gamma}_{jk}^i = \tfrac{1}{2}(c_{kj}^i + \varepsilon_i \varepsilon_j c_{ik}^j + \varepsilon_i \varepsilon_k c_{ij}^k) \tag{13}$$

(where here there is no summation over repeated indices).

Proof. The symmetry of the compatible connexion implies that $T(\xi_i, \xi_j) = 0$. From this, (8) and (12) it then follows that

$$\hat{\Gamma}_{ji}^k - \hat{\Gamma}_{ij}^k = c_{ij}^k. \tag{14}$$

From (10), the compatibility condition, and (12), we obtain

$$0 = \nabla_{\xi_k} \langle \xi_i, \xi_j \rangle = \nabla_{\xi_k}(g_{rs} \xi_i^r \xi_j^s) = \varepsilon_j \hat{\Gamma}_{ik}^j + \varepsilon_i \hat{\Gamma}_{jk}^i$$

(where in the last expression we do not sum over the indices i, j). Permuting the indices i, j, k cyclically we obtain from this the following three equations:

$$\varepsilon_i \hat{\Gamma}_{kj}^i + \varepsilon_k \hat{\Gamma}_{ij}^k = 0,$$
$$\varepsilon_k \hat{\Gamma}_{ji}^k + \varepsilon_j \hat{\Gamma}_{ki}^j = 0,$$
$$\varepsilon_j \hat{\Gamma}_{ik}^j + \varepsilon_i \hat{\Gamma}_{jk}^i = 0.$$

On solving these equations supplemented by (14), we arrive at the desired formula (13), completing the proof. □

We note finally that by using the formula (9) one can express the curvature tensor (in the form $\langle R(\xi_i, \xi_j)\xi_k, \xi_l\rangle$) in terms of the functions c_{ij}^k and their derivatives.

30.2. The Symmetries of the Curvature Tensor. The Curvature Tensor Defined by the Metric

What symmetries do curvature tensors have?

30.2.1. Theorem. (i) *We always have* $R^i_{qkl} = -R^i_{qlk}$.
(ii) *If the connexion is symmetric then*

$$R^i_{qkl} + R^i_{klq} + R^i_{lqk} = 0. \tag{15}$$

(iii) *If the connexion is compatible with the metric, and we define* $R_{iqkl} = g_{ip}R^p_{qkl}$, *then the tensor* R_{iqkl} *is skew-symmetric in the indices* i *and* q:

$$R_{iqkl} = -R_{qikl}. \tag{16}$$

(iv) *If the connexion is both symmetric and compatible with the metric* g_{ik}, *then*

$$R_{iqkl} = R_{kliq}. \tag{17}$$

PROOF. (i) is obvious.

(ii) Let e_i, $i = 1, \ldots, n$, denote the standard basis vectors (at every point) and write in the usual way $[\nabla_k, \nabla_l] = \nabla_k\nabla_l - \nabla_l\nabla_k$ (cf. §24.1). By formula (4) and the symmetry of the given connexion, we have that

$$-R^i_{qkl}e_i = [\nabla_k, \nabla_l]e_q.$$

Hence (15) will follow once we have shown that

$$[\nabla_k, \nabla_l]e_q + [\nabla_l, \nabla_q]e_k + [\nabla_q, \nabla_k]e_l = 0. \tag{18}$$

Expanding the left-hand side of (18), we obtain

$$\nabla_k(\nabla_l e_q - \nabla_q e_l) + \nabla_l(\nabla_q e_k - \nabla_k e_q) + \nabla_q(\nabla_k e_l - \nabla_l e_k). \tag{19}$$

Now

$$(\nabla_l e_q)^i = \frac{\partial e_q^i}{\partial x^l} + \Gamma^i_{jl}e_q^j = 0 + \Gamma^i_{jl}\delta^j_q = \Gamma^i_{ql}.$$

From this and the symmetry of Γ^i_{ql} it follows that $(\nabla_l e_q - \nabla_q e_l) = 0$, and similarly for the two other bracketed expressions in (19).

(iii) It follows from (4) that for each k, l and any vector ξ,

$$\langle [\nabla_k, \nabla_l]\xi, \xi \rangle = -g_{ij} R^i_{qkl} \xi^q \xi^j + g_{ij} T^p_{kl} \frac{\partial \xi^i}{\partial x^p} \xi^j$$

$$= -R_{jqkl} \xi^q \xi^j + g_{ij} T^p_{kl} \frac{\partial \xi^i}{\partial x^p} \xi^j.$$

Thus if we can show that $\langle [\nabla_k, \nabla_l]\xi, \xi \rangle = 0$ for all ξ, then the desired equality (16) will follow (by letting ξ be some suitable specific vector). Now from Leibniz' rule for covariant differentiation (see §28.2(28)), and the assumed compatibility of the connexion with the metric g_{ij}, we have

$$\frac{\partial^2}{\partial x^k \partial x^l} \langle \xi, \xi \rangle = \nabla_k \nabla_l \langle \xi, \xi \rangle = \nabla_k \nabla_l (g_{ij} \xi^i \xi^j)$$

$$= 2(\langle \nabla_k \nabla_l \xi, \xi \rangle + \langle \nabla_l \xi, \nabla_k \xi \rangle),$$

and similarly,

$$\frac{\partial^2}{\partial x^l \partial x^k} \langle \xi, \xi \rangle = 2(\langle \nabla_l \nabla_k \xi, \xi \rangle + \langle \nabla_k \xi, \nabla_l \xi \rangle).$$

On subtracting these two equations and using $\partial^2/\partial x^l \partial x^k = \partial^2/\partial x^k \partial x^l$, we obtain $\langle [\nabla_k, \nabla_l]\xi, \xi \rangle = 0$, as required.

(iv) Referring to the octahedron shown in Figure 33, we see that by virtue of Parts (i), (ii), (iii), the sum of the quantities standing at the vertices of each shaded face is zero. The desired equation (17) follows by adding these vanishing three-term sums for the faces labelled q and i and subtracting from this the sums for the faces k and l. □

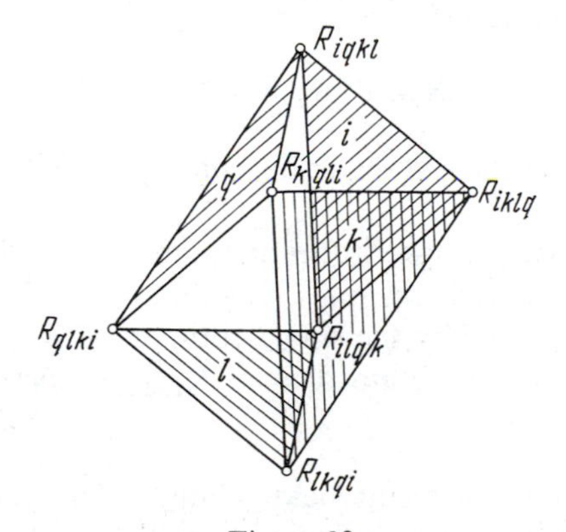

Figure 33

30.3. Examples: The Curvature Tensor in Spaces of Dimensions 2 and 3; the Curvature Tensor Defined by a Killing Metric

The curvature tensor R^i_{qkl} is a tensor of rank 4, skew-symmetric in the indices k and l. We obtained it in a natural way as an operator on vector fields:

$$-R^i_{qkl}T^q = R^i_{qlk}T^q = (\nabla_k\nabla_l - \nabla_l\nabla_k)T^i - T^p_{kl}\frac{\partial T^i}{\partial x^p},$$

where $T^p_{kl} = \Gamma^p_{kl} - \Gamma^p_{lk}$ is the torsion tensor.

If the connexion is symmetric then $T^p_{kl} \equiv 0$. If it is both symmetric and compatible with the given metric g_{ij}, then, as we have seen, the components Γ^k_{ij} and R^i_{qkl} are expressible in terms of the g_{ij} and their derivatives, and satisfy the following symmetry conditions:

$$\begin{aligned} R^i_{qkl} &= -R^i_{qlk}; \\ R_{iqkl} &= g_{im}R^m_{qkl} = -R_{qikl}; \\ R_{iqkl} &= R_{kliq}; \\ R^i_{qkl} &+ R^i_{lqk} + R^i_{klq} = 0. \end{aligned} \tag{20}$$

In this subsection we consider the Riemann curvature tensor (determined by the symmetric connexion compatible with the metric) in low-dimensional spaces, and also when the metric is a Killing metric on a matrix group. In particular we shall be concerned with the number of independent components the Riemann tensor has in these various spaces.

The following definitions apply in all dimensions.

30.3.1. Definition. The trace (or contraction) $R_{ql} = R^i_{qil}$ of the Riemann curvature tensor is called the *Ricci tensor*.

30.3.2. Definition. The scalar

$$R = g^{lq}R_{ql} = g^{lq}R^i_{qil} \tag{21}$$

is called the *scalar curvature* of the underlying space with metric g_{ij}.

(i) *The two-dimensional case.* From the symmetry conditions (20) and the assumption that the underlying space has dimension 2, it follows that the tensor (R_{iqkl}) is determined by the single component R_{1212}; the other components are either zero or obtained from R_{1212} by permuting the indices and using (20).

The following important result yields Gauss' "Theorema Egregium" (so we now fulfil the promise given in §8.1 to prove that celebrated theorem).

30.3.3. Theorem. *For a 2-dimensional surface in a 3-dimensional space endowed with a Riemannian metric, the scalar curvature R is twice the Gaussian curvature. It follows that the Gaussian curvature of a surface (in contrast with the mean curvature) is expressible in terms of the induced metric on the surface alone, and is therefore an intrinsic invariant of the surface.*

PROOF. Let P be any non-singular point of the surface. As on several earlier occasions we choose co-ordinates x, y, z such that P is the origin, the z-axis is normal to the surface at P, and the x- and y-axes are tangent to the surface at P. Then in a neighbourhood of P the surface is given by an equation of the form $z = f(x, y)$ where $\operatorname{grad} f|_P = 0$. By §7.3(20) the induced metric on the surface is given in that neighbourhood of P by

$$g_{ij} = \delta_{ij} + \frac{\partial f}{\partial z^i}\frac{\partial f}{\partial z^j}, \qquad z^1 = x, \quad z^2 = y. \tag{22}$$

It follows from this and $\operatorname{grad} f|_P = 0$ that at $P = (0, 0)$ we have $\partial g_{ij}/\partial z^k = 0$, $i, j, k = 1, 2$. Hence by Theorem 29.3.2 the (unique) symmetric connexion Γ^k_{ij} on the surface compatible with the induced metric, is zero at P, so that by (3) above at the point P we shall have

$$-R^i_{qkl} = \frac{\partial \Gamma^i_{ql}}{\partial z^k} - \frac{\partial \Gamma^i_{qk}}{\partial z^l}.$$

From this and again from Theorem 29.3.2, we obtain

$$R_{iqkl} = \frac{1}{2}\left(\frac{\partial^2 g_{il}}{\partial z^q\,\partial z^k} + \frac{\partial^2 g_{qk}}{\partial z^i\,\partial z^l} - \frac{\partial^2 g_{ik}}{\partial z^q\,\partial z^l} - \frac{\partial^2 g_{ql}}{\partial z^i\,\partial z^k}\right),$$

which specializes to

$$R_{1212} = \frac{1}{2}\left(\frac{\partial^2 g_{12}}{\partial x\,\partial y} + \frac{\partial^2 g_{12}}{\partial x\,\partial y} - \frac{\partial^2 g_{11}}{\partial y^2} - \frac{\partial^2 g_{22}}{\partial x^2}\right). \tag{23}$$

Now from (22) and $\operatorname{grad} f|_P = 0$ we infer

$$\left.\frac{\partial^2 g_{11}}{\partial y^2}\right|_P = 2(f_{xy})^2, \qquad \left.\frac{\partial^2 g_{22}}{\partial x^2}\right|_P = 2(f_{xy})^2, \qquad \left.\frac{\partial^2 g_{12}}{\partial x\,\partial y}\right|_P = f_{xx}f_{yy} + f^2_{xy}.$$

This and (23) then yield, at the point P,

$$R_{1212} = f_{xx}f_{yy} - f^2_{xy} = \det\begin{pmatrix} f_{xx} & f_{xy} \\ f_{yx} & f_{yy} \end{pmatrix} = K,$$

where K is the Gaussian curvature of the surface at the point P (see Definition 8.1.1). Now the scalar curvature R is defined by $R = g^{ql}R^i_{qil}$; hence, using $R^i_{qkl} = g^{is}R_{sqkl}$, we have

$$R = 2\det(g^{ql})R_{1212} = \frac{2}{\det(g_{ij})}R_{1212} = \frac{2}{g}R_{1212}.$$

Since in terms of the co-ordinates x, y we have at the point P that $g = 1$ and $R_{1212} = K$, it follows that $R = 2K$. Since R and K are scalars, and so co-ordinate-independent, we conclude that $R = 2K$ at every point. □

Remark. It is worth noting that in the course of the above proof we obtained a formula for the components of the Riemann tensor on the surface:

$$R_{1212} = \frac{R}{2}(g_{11}g_{22} - g_{12}^2) = \frac{Rg}{2} = Kg, \qquad g = \det(g_{ij}). \tag{24}$$

30.3.4. Examples. (a) In the Euclidean plane we have

$$dl^2 = dx^2 + dy^2, \qquad R^i_{qkl} = 0, \qquad K = \frac{R}{2} \equiv 0.$$

(b) The metric on the sphere in Euclidean 3-space is given by (see §9(3))

$$dl^2 = d\rho^2 + \sin^2\frac{\rho}{R_0}\,d\varphi^2,$$

where R_0 is the radius of the sphere and $\rho = R_0\theta$. Here we have constant positive curvature: $K = R/2 = 1/R_0^2 > 0$ (cf. §9).

(c) The metric on the Lobachevskian plane has the form

$$dl^2 = dr^2 + \sinh^2\frac{r}{R_0}\,d\varphi^2,$$

where $r = R_0\chi$ (see §10.1(11)). Here we have constant negative curvature: $K = R/2 = -1/R_0^2 < 0$.

(The geometrical significance of the sign of the Gaussian curvature was noted in §8.3.)

(ii) *The three-dimensional case.* Here the situation is somewhat more complicated. At each point of the underlying 3-dimensional space the Riemann tensor

$$R_{iqkl} = -R_{qikl} = -R_{iqlk} = R_{kliq}$$

may by virtue of these symmetry relations be regarded as a quadratic form on the linear space of all skew-symmetric rank-2 tensors at the point. To see this denote the symbol $[i, q] = -[q, i]$ by A and the symbol $[k, l] = -[l, k]$ by B; then the symmetry relations imply

$$R_{[iq][kl]} = R_{AB} = R_{BA}.$$

Since $R_{iqkl} = 0$ when either $k = l$ or $i = q$, it follows that the Riemann tensor is determined by the six components R_{AB} where $A, B = [1, 2], [1, 3], [2, 3]$.

The Ricci tensor $R^i_{qil} = R_{ql} = R_{lq}$, which is symmetric of rank 2, also requires six components for its determination, namely the R_{ql} with $q > l$. The scalar curvature $R = \mathrm{tr}(R_{ql}) = g^{ql}R_{ql} = g^{ql}R^i_{qil}$, in contrast with the 2-dimensional case, does not determine the curvature tensor R^i_{qkl}. However the Ricci tensor *does* determine the Riemann tensor, as is shown by the following formula:

$$R_{\alpha\beta\gamma\delta} = R_{\alpha\gamma}g_{\beta\delta} - R_{\alpha\delta}g_{\beta\gamma} + R_{\beta\delta}g_{\alpha\gamma} - R_{\beta\gamma}g_{\alpha\delta} + \frac{R}{2}(g_{\alpha\delta}g_{\beta\gamma} - g_{\alpha\gamma}g_{\beta\delta}). \quad (25)$$

We leave the verification of this to the reader. (Try raising the indices α and β on both sides.)

Note also the invariants λ_1, λ_2, λ_3, the eigenvalues of the Ricci tensor, defined as usual by

$$\det(R_{ql} - \lambda g_{ql}) = 0.$$

Since R is the trace of the Ricci tensor, we have $\lambda_1 + \lambda_2 + \lambda_3 = R$.

To conclude we remark that the standard phrase "a space of positive curvature" refers to a 3-dimensional space whose Riemann tensor R_{AB} defines a positive definite quadratic form on skew-symmetric rank-2 tensors. (On the other hand a space of dimension $n > 3$ is said to have positive (resp. negative) curvature if $R_{\alpha\beta\gamma\delta}\xi^\alpha\eta^\beta\xi^\gamma\eta^\delta > 0$ (resp. < 0).)

(iii) *The four-dimensional case.* In four dimensions the Ricci tensor in general no longer determines the curvature tensor; however it continues to be of great importance, for example in the general theory of relativity. In that theory the underlying space is of course taken to be 4-dimensional space-time, and the gravitational field to be the metric g_{ij}, $i, j = 0, 1, 2, 3$. The properties of matter are combined in the "energy-momentum tensor" T_{ij} (see §21.2(25)). Einstein was led by various criteria to the following equations for the metric g_{ij} of space-time:

$$R_{ij} - \tfrac{1}{2}Rg_{ij} = \lambda T_{ij}, \qquad \nabla_j T^j_i = 0.$$

In the absence of matter these become

$$R_{ij} - \tfrac{1}{2}Rg_{ij} = 0, \qquad \text{or} \quad R_{ij} = 0.$$

The metric has also to satisfy the requirement that its diagonal form have one positive and three negative entries. (For details and explanations see §37.4.)

(iv) *The curvature tensor defined by a Killing metric.* Let G be one of the "classical" groups of matrix transformations and let $\mathfrak{g}$ be its Lie algebra, which we assume to be endowed with a Killing metric (or form) $\langle\ ,\ \rangle_0$ (see §24.4). As in §24.3 for each $X, Y \in \mathfrak{g}$ we denote by L_X, L_Y the left-invariant vector fields on G defined at each point A of G by $L_X(A) = AX$, $L_Y A = AY$. We introduce a connexion on G by defining

$$\nabla_{L_X} L_Y = \tfrac{1}{2}L_{[X,Y]} = \tfrac{1}{2}[L_X, L_Y]. \quad (26)$$

Since by Lemma 24.3.10 for each $A \in G$ the tangent space to G at A is comprised of the matrices of the form $L_X(A)$, $X \in \mathfrak{g}$, it follows that formula (26) does indeed fully determine a connexion on G. (To see that (26) defines a covariant differentiation we need to verify conditions (i) to (iv) of §28.2; we shall, however, omit the details.)

30.3.5. Lemma. *The connexion defined by* (26) *is symmetric and compatible with the Killing metric on G (determined by the given Killing form on* $\mathfrak{g}$).

Proof. For the symmetry we need to show that $T(L_X, L_Y) = 0$ for all X, $Y \in \mathfrak{g}$ (see (6)). Now by (8)

$$T(L_X, L_Y) = \nabla_{L_X} L_Y - \nabla_{L_Y} L_X - [L_X, L_Y] = \tfrac{1}{2} L_{[X,Y]} - \tfrac{1}{2} L_{[Y,X]} - L_{[X,Y]} = 0.$$

We now prove the compatibility of the connexion. It is easy to see that the compatibility of a connexion with a metric determining at each point a scalar product $\langle\ ,\ \rangle$ of tangent vectors, is equivalent to the condition that for all vectors ξ, η, ζ (at each point) we have

$$\partial_\xi \langle \eta, \zeta \rangle = \langle \nabla_\xi \eta, \zeta \rangle + \langle \eta, \nabla_\xi \zeta \rangle. \tag{27}$$

Recall that the Killing metric on G is given at each $A \in G$ by

$$\langle L_Y(A), L_Z(A) \rangle = \langle X, Y \rangle_0,$$

$L_Y(A)$, $L_Z(A)$ being typical tangent vectors to G at A. We see from this that $\langle L_Y(A), L_Z(A) \rangle$ is independent of A, so that $\partial_{L_X} \langle L_Y, L_Z \rangle \equiv 0$, i.e. in the present context the left-hand side of (27) is identically zero. The right-hand side is

$$\begin{aligned} \langle \nabla_{L_X} L_Y, L_Z \rangle + \langle L_Y, \nabla_{L_X} L_Z \rangle &= \tfrac{1}{2}\{\langle L_{[X,Y]}, L_Z \rangle + \langle L_Y, L_{[X,Z]} \rangle\} \\ &= \tfrac{1}{2}\{\langle [X, Y], Z \rangle_0 + \langle Y, [X, Z] \rangle_0\} = 0, \end{aligned}$$

where in the last line we have used the defining property of a Killing form, namely that with respect to such a form the linear operator ad X is skew-symmetric (see 24.4.1). This completes the proof of the lemma. □

The following result is an easy consequence of this lemma, formula (9), and Jacobi's identity.

30.3.6. Corollary. *The curvature of the symmetric connexion compatible with a Killing metric* $\langle\ ,\ \rangle$ *on a matrix group G is given by the formula*

$$R(L_X, L_Y) L_Z = \tfrac{1}{4} L_{[[X,Y],Z]}.$$

It follows from the definition of a Killing metric that

$$\langle R(L_X, L_Y) L_Z, L_W \rangle = \tfrac{1}{4} \langle [X, Y], [Z, W] \rangle. \tag{28}$$

Finally we characterize the geodesics corresponding to the symmetric connexion compatible with a Killing metric on the group G. Since translations (i.e. multiplications of G by fixed matrices from G) are motions of the Killing metric on G, it suffices to characterize the geodesics passing through the identity of G.

30.3.7. Theorem. *The geodesics with respect to a Killing metric on a matrix group G, which pass through the identity of G, are precisely the one-parameter subgroups of G.*

PROOF. The velocity vector of a one-parameter subgroup $A(t) = \exp(tX)$ where $X \in \mathfrak{g}$ and t is a real parameter, is the left-invariant vector field L_X (or rather its restriction to the curve $A(t)$). Hence

$$\nabla_{\dot{A}} \dot{A} = \nabla_{L_X} L_X = \tfrac{1}{2} L_{[X, X]} \equiv 0, \tag{29}$$

which proves that the one-parameter subgroups are geodesics.

Conversely, since for any matrix $X \in \mathfrak{g}$ there always exists a one-parameter subgroup of G with initial velocity vector X, it follows from the uniqueness of the geodesic with specified initial conditions, that all geodesics are one-parameter subgroups. □

30.4. The Peterson–Codazzi Equations. Surfaces of Constant Negative Curvature, and the "sine–Gordon" Equation

Let $r = r(x^1, x^2)$ be a surface in Euclidean space $\mathbb{R}^3$, with induced metric g_{ij}. By §8.1(3), the second fundamental form $b_{ij} dx^i\, dx^j$ on the surface is defined by

$$b_{ij} = \left\langle \frac{\partial}{\partial x^j}\left(\frac{\partial r}{\partial x^i}\right), n \right\rangle, \tag{30}$$

where n denotes the unit normal to the surface and $\langle\ ,\ \rangle$ denotes the Euclidean scalar product in $\mathbb{R}^3$.

The following result provides means for calculating the components of the symmetric connexion compatible with the induced metric g_{ij}.

30.4.1. Proposition. *The components of the symmetric connexion compatible with the metric induced on the surface $r = r(x^1, x^2)$ by the Euclidean metric in $\mathbb{R}^3$, are given by the following formulae:*

$$\Gamma^k_{ij} = \left\langle \frac{\partial^2 r}{\partial x^i\, \partial x^j}, \frac{\partial r}{\partial x^l} \right\rangle g^{lk} \qquad (i, j, k = 1, 2), \tag{31}$$

or, equivalently,

$$\frac{\partial^2 r}{\partial x^i\, \partial x^j} = b_{ij} n + \Gamma^k_{ij} \frac{\partial r}{\partial x^k}. \tag{32}$$

PROOF. By §7.3(19), $g_{ij} = \langle r_{x^i}, r_{x^j} \rangle$, whence

$$\frac{\partial g_{ij}}{\partial x^s} = \left\langle \frac{\partial^2 r}{\partial x^i \, \partial x^s}, \frac{\partial r}{\partial x^j} \right\rangle + \left\langle \frac{\partial r}{\partial x^i}, \frac{\partial^2 r}{\partial x^j \, \partial x^s} \right\rangle.$$

The equation (31) follows easily from this, Christoffel's formula (§29.3(13)), the symmetry relation $\Gamma^k_{ij} = \Gamma^k_{ji}$, and the fact that there are only two variables x^1, x^2.

To see (32) note first that n is perpendicular to $\partial r/\partial x^1$, $\partial r/\partial x^2$. It follows from (30) that the component of $\partial^2 r/\partial x^i \, \partial x^j$ in the direction of n is b_{ij}. If we then write

$$\frac{\partial^2 r}{\partial x^i \, \partial x^j} = b_{ij} n + x^k_{ij} \frac{\partial r}{\partial x^k},$$

we obtain

$$\left\langle \frac{\partial^2 r}{\partial x^i \, \partial x^j}, \frac{\partial r}{\partial x^l} \right\rangle = x^k_{ij} \left\langle \frac{\partial r}{\partial x^k}, \frac{\partial r}{\partial x^l} \right\rangle = x^k_{ij} g_{kl}.$$

It is then immediate from (31) that $x^k_{ij} = \Gamma^k_{ij}$, as required. This completes the proof of the proposition. □

Writing briefly $\partial r/\partial x^i = r_i$, we have (at each point of the surface) a reference frame (r_1, r_2, n), with n perpendicular to r_1 and r_2, which depends smoothly on the points of the surface. Since n is a unit vector in Euclidean 3-space, we have by 5.1.1 that $\partial n/\partial x^i$ is perpendicular to n. Differentiating the equation $\langle n, r_j \rangle = 0$, we obtain (using (30))

$$0 = \frac{\partial}{\partial x^i} \langle n, r_j \rangle = \left\langle \frac{\partial n}{\partial x^i}, r_j \right\rangle + \left\langle n, \frac{\partial r_j}{\partial x^i} \right\rangle = \left\langle \frac{\partial n}{\partial x^i}, r_j \right\rangle + b_{ij},$$

whence (taking into account that $n \perp \partial n/\partial x^i$)

$$\frac{\partial n}{\partial x^i} = -\, b^j_i r_j, \qquad b^j_i = g^{jl} b_{il}. \tag{33}$$

From this and (32) (and the equality of the mixed partials) we obtain

$$\frac{\partial^2 r_i}{\partial x^k \, \partial x^j} = \frac{\partial b_{ij}}{\partial x^k} n - b_{ij} b^l_k r_l + \frac{\partial \Gamma^l_{ij}}{\partial x^k} r_l + \Gamma^s_{ij} [b_{ks} n + \Gamma^l_{ks} r_l]$$

$$= \frac{\partial b_{ik}}{\partial x^j} n - b_{ik} b^l_j r_l + \frac{\partial \Gamma^l_{ik}}{\partial x^j} r_l + \Gamma^s_{ik} [b_{js} n + \Gamma^l_{js} r_l] = \frac{\partial^2 r_i}{\partial x^j \, \partial x^k},$$

whence it follows (using the independence of r_1, r_2) that

$$\frac{\partial \Gamma^l_{ij}}{\partial x^k} - \frac{\partial \Gamma^l_{ik}}{\partial x^j} + \Gamma^s_{ij}\Gamma^l_{ks} - \Gamma^s_{ik}\Gamma^l_{js} = b_{ij}b^l_k - b_{ik}b^l_j, \tag{34}$$

$$\frac{\partial b_{ij}}{\partial x^k} - \frac{\partial b_{ik}}{\partial x^j} = \Gamma^s_{ik}b_{js} - \Gamma^s_{ij}b_{ks}. \tag{35}$$

The equations (34) are called the "Gauss equations". Note that the left-hand side of (34) is just the general component of the curvature tensor R^l_{ijk}; it is not difficult to see that the equations (34) are simply a reformulation of the equality of the scalar curvature with the double of the Gaussian curvature (Theorem 30.3.3). (Verify this !)

The equations (35) are the "Peterson–Codazzi equations". In deriving these equations, we have shown that they provide necessary conditions for a form $b_{ij}(x^1, x^2)$ to be the second fundamental form of a surface in Euclidean $\mathbb{R}^3$ with induced metric g_{ij} (the Γ^k_{ij} occurring in (35) being given by Christoffel's formula). It can be shown that in fact they are also sufficient conditions.

Finally, suppose our surface has negative Gaussian curvature: $K < 0$. By §8.3(26) this implies that at each point of the surface we have $b^2_{12} - b_{11}b_{22} > 0$, so that the discriminant of the quadratic form $b_{ij}dx^i\,dx^j$ is positive; it follows that this quadratic form factors as a product of two linear forms. Hence there exist (locally) co-ordinates p, q on the surface in terms of which the quadratic form is

$$b_{ij}dx^i\,dx^j = 2b_{pq}\,dp\,dq, \tag{36}$$

i.e. the terms in dp^2 and dq^2 vanish. If in addition to being negative, K is constant, then by differentiating the equation $K(g_{pp}g_{qq} - g^2_{pq}) = -b^2_{pq}$, and using the Peterson–Codazzi equations and Christoffel's formula (§29.3(13)), one obtains (we leave the details to the reader)

$$g_{pq}\frac{\partial g_{pp}}{\partial q} = 0 = g_{pq}\frac{\partial g_{qq}}{\partial p}. \tag{37}$$

Now it is easy to see that there is enough latitude in the choice of local co-ordinates p, q satisfying (36) for them to be adjusted so that, locally, $g_{pq} \neq 0$. Then from (37) it follows that

$$\frac{\partial g_{pp}}{\partial q} = 0, \qquad \frac{\partial g_{qq}}{\partial p} = 0,$$

so that g_{pp} is a function of p only, and g_{qq} is a function of q only. This allows us to define new local co-ordinates x, y on the surface, by

$$x = \int_{p_0}^{p} \sqrt{g_{pp}}\, dp, \qquad y = \int_{q_0}^{q} \sqrt{g_{qq}}\, dq. \tag{38}$$

In terms of these new co-ordinates the first and second fundamental forms are:

$$ds^2 = dx^2 + 2g_{xy}\, dx\, dy + dy^2, \qquad b_{ij}\, dx^i\, dx^j = 2b_{xy}\, dx\, dy. \tag{39}$$

We now put $g_{xy} = \cos \omega$. (Here ω is the angle between the tangent vectors (at the point) parallel to the co-ordinate axes.) From the Gauss equations (34) (or from (24)) with $K = -1$, it follows that $R_{1212} = (b_{xy})^2 = -\sin^2 \omega$. By the definition of the curvature tensor (see the left-hand side of (34))

$$R_{1212} = g_{1l}\left(\frac{\partial \Gamma^l_{21}}{\partial x^2} - \frac{\partial \Gamma^l_{22}}{\partial x^1} + \Gamma^s_{21}\Gamma^l_{2s} - \Gamma^s_{22}\Gamma^l_{1s}\right).$$

If one uses Christoffel's formula to express the right-hand side of this equation in terms of the derivatives of g_{12} and g, one eventually ends up with (we again leave the details to the reader!)

$$\omega_{xy} = \sin \omega. \tag{40}$$

This equation is known in the physics literature as the "sine–Gordon" equation. The change of variables $x = (\tau + \xi)/\sqrt{2}$, $y = (\tau - \xi)/\sqrt{2}$, brings it into the form

$$\frac{\partial^2 \omega}{\partial \tau^2} - \frac{\partial^2 \omega}{\partial \xi^2} = \sin \omega. \tag{41}$$

30.5. Exercises

1. Show that the solutions of equation (41) above which are independent of ξ and decrease as $\tau \to +\infty$, correspond to the surface of revolution of curvature $K = -1$, given by

$$z = -\sqrt{1 - x^2 - y^2} + \ln \frac{1 + \sqrt{1 - x^2 - y^2}}{\sqrt{x^2 + y^2}}.$$

(This is sometimes called "Beltrami's pseudosphere"; cf. beginning of §13.3.)

2. Prove formula (25) of §30.3.

3. Let $x^i(t)$, $i = 1, 2$, be a piecewise smooth, closed curve bounding a region U of a surface in Euclidean 3-space. Show that $\Delta\varphi = \iint_U K\sqrt{g}\, dx^1 \wedge dx^2$ (where K is the Gaussian curvature) is the angle through which a vector rotates in being parallel-transported around the curve $x^i(t)$.

4. If in the preceding exercise the curve consists of three geodesic arcs, and if K is constant, show that the sum of the angles of this geodesic triangle is $\pi + K\sigma$ where σ is the area of the triangle. Consider the cases of the sphere and Lobachevskian plane.

5. Let $\xi_1, \ldots, \xi_n$ be vector fields defined on an n-dimensional Riemannian (or pseudo-Riemannian) space, such that $\langle \xi_i, \xi_j \rangle = g_{ij}$. Write $[\xi_i, \xi_j] = c_{ij}^k \xi_k$. Compute the components $\hat{\Gamma}_{ij}^k$ determining the symmetric connexion compatible with the metric. (Here $\nabla_{\xi_j} \xi_i = \hat{\Gamma}_{ij}^k \xi_k$, as in Theorem 30.1.5.)

6. Denote by $\tilde{\xi}(\varepsilon)$ the result of parallel-transporting a vector $\xi = (\xi^k)$ around the boundary of a square of side ε with its sides parallel to the co-ordinate x^i- and x^j-axes. Prove that

$$\lim_{\varepsilon \to 0} \frac{\tilde{\xi}^k(\varepsilon) - \xi^k}{\varepsilon^2} = -R_{lij}^k \xi^l.$$

7. Establish "Bianchi's identities" for the curvature tensor of the symmetric connexion compatible with a given metric:

$$\nabla_m R_{ikl}^n + \nabla_l R_{imk}^n + \nabla_k R_{ilm}^n = 0.$$

8. From the preceding exercise deduce the following equation for the divergence of the Ricci tensor:

$$\nabla_l R_m^l = \frac{1}{2} \frac{\partial R}{\partial x^m}.$$

9. Let $X_1, \ldots, X_n$ be orthonormal vector fields in an n-dimensional Riemannian space, and denote by $\omega_1, \ldots, \omega_n$ the dual basis of 1-forms: $\omega_i(X_j) = \delta_{ij}$. Define 1-forms ω_{ij} and 2-forms Ω_{ij} by

$$\omega_{ij} = \Gamma_{jk}^i \omega_k; \qquad \Omega_{ij} = \tfrac{1}{2} R_{ijkl} \omega_k \wedge \omega_l.$$

(Here $\nabla_{X_k} X_j = \Gamma_{jk}^i X_i$, $\langle R(X_k, X_l) X_j, X_i \rangle = R_{ijkl}$, and summation is understood to take place over repeated indices.)

(i) Prove that $\omega_{ij} = -\omega_{ji}$.
(ii) Deduce the following "Cartan structure equations":

$$d\omega_i = -\omega_j \wedge \omega_{ij},$$

$$d\omega_{ij} = \omega_{il} \wedge \omega_{lj} - \Omega_{ij},$$

$$d\Omega_{ij} = -\Omega_{il} \wedge \omega_{lj} + \omega_{il} \wedge \Omega_{lj}.$$

10. In the notation of the preceding exercise, define forms $\Omega_{(k)}$ and (in even dimensions) Ω by:

$$\Omega_{(k)} = \Omega_{i_1 i_2} \wedge \Omega_{i_2 i_3} \wedge \cdots \wedge \Omega_{i_{k-1} i_k} \wedge \Omega_{i_k i_1};$$

$$\Omega = \varepsilon^{i_1 \ldots i_n} \Omega_{i_1 i_2} \wedge \cdots \wedge \Omega_{i_{n-1} i_n} \qquad (n = 2m).$$

(i) Prove that the forms $\Omega_{(k)}$, Ω are independent of the original orthonormal basis $X_1, \ldots, X_n$.

(ii) The forms $\Omega_{(k)}$, Ω are closed (i.e. $d\Omega_{(k)}$, $d\Omega = 0$).

(iii) Find expressions for these forms in terms of co-ordinates.

(iv) When $n = 2$, the form Ω becomes $\Omega = K\sqrt{g}\, dx^1 \wedge dx^2$, where K is the Gaussian curvature.

(v) Find formulae analogous to those of this and the preceding exercise when the metric is pseudo-Riemannian.

CHAPTER 5

The Elements of the Calculus of Variations

§31. One-Dimensional Variational Problems

31.1. The Euler–Lagrange Equations

In §29.2 we defined the geodesics $x^i = x^i(t)$ relative to a given connexion Γ^i_{jk} by means of the equation

$$\nabla_T(T) = 0,$$

where $T^i = dx^i/dt$ is the velocity vector of the curve; in other words the geodesics are the solutions of the equations

$$\frac{d^2x^i}{dt^2} + \Gamma^i_{jk}\frac{dx^j}{dt}\frac{dx^k}{dt} = 0. \tag{1}$$

If the underlying space is endowed with a metric g_{ij}, and if the connexion Γ^i_{jk} is symmetric, and compatible with that metric, then as we have seen the Γ^i_{jk} are expressible in terms of the g_{ij}, so that the geodesics are determined by the metric. Apart from their defining property (that parallel transport of the tangent vector to a geodesic along the geodesic results again in the tangent vector), what other geometrical properties are possessed by geodesics? Everyone is familiar with the common-sense idea of a geodesic arc joining two points as having the shortest length of all arcs joining the points (at least locally, i.e. for points sufficiently close together). We shall now clarify this idea.

In examining this expected "extremal" property of geodesics we shall find it useful to take a more general stance. Thus suppose $L(x, \xi)$ is a function defined on pairs (x, ξ) where x is any point $x = (x^1, \ldots, x^n)$ of the underlying space and ξ is any tangent vector at x. Let $P = (x_1^1, \ldots, x_1^n)$ and $Q = (x_2^1, \ldots, x_2^n)$ be any two (fixed) points, and consider the set of all smooth arcs $\gamma\colon x^i = x^i(t), a \leq t \leq b$ (where a and b are arranged to be the same for all arcs), joining these two points: $x^i(a) = x_1^i$, $x^i(b) = x_2^i$. For each arc γ define

$$S[\gamma] = \int_P^Q L(x(t), \dot{x}(t))\, dt. \tag{2}$$

Thus S is a functional; it associates with each arc γ from P to Q, a number $S[\gamma]$, called the *action*. The question of interest to us is: For which γ is the action $S[\gamma]$ least?

31.1.1. Examples. (a) Take $L(x, \xi)$ to be $g_{ij}\xi^i\xi^j$. Then $S[\gamma] = \int_P^Q L(x, \dot{x})\, dt = \int_P^Q g_{ij}\dot{x}^i\dot{x}^j\, dt = \int_P^Q |\dot{x}|^2\, dt$. For which arc(s) $\gamma{:}x = x(t)$ does the functional $S[\gamma]$ take its least value?

(b) Take $L(x, \xi) = \sqrt{g_{ij}\xi^i\xi^j} = \sqrt{\langle \xi, \xi \rangle} = |\xi|$. Thus here $L(x, \xi)$ is just the length of the vector ξ, and $S[\gamma] = \int_P^Q \sqrt{g_{ij}\dot{x}^i\dot{x}^j}\, dt$ is the length of the arc γ. Hence in this case the above question becomes that of finding the arc(s) from P to Q of shortest length.

(c) Suppose the metric is Euclidean, i.e. $g_{ij} = \delta_{ij}$. Set $L = (m/2)\delta_{ij}\xi^i\xi^j - U(x)$, where m is a constant and $U(x)$ is a function of the points of the space; then

$$S[\gamma] = \int_P^Q \left[\sum_i \frac{m}{2}(\dot{x}^i)^2 - U(x)\right] dt.$$

The arc γ along which $S[\gamma]$ is least in this case is just the trajectory of a point-particle of mass m in a force field $f_i = -\partial U/\partial x^i$.

The following classical theorem gives necessary conditions (in the form of partial differential equations—the "Euler–Lagrange equations") for an arc γ to minimize $S[\gamma]$.

31.1.2. Theorem. *If the functional* $S[\gamma] = \int_P^Q L(x, \dot{x})\, dt$ *attains its minimum over all smooth arcs* γ *at the smooth arc* $\gamma_0\colon x^i = x^i(t)$, *i.e. if* $S[\gamma_0] \leq S[\gamma]$ *for all smooth arcs* γ *from* P *to* Q, *then along the arc* γ_0 *the following equations hold*:

$$\frac{d}{dt}\left(\frac{\partial L}{\partial \dot{x}^i}\right) - \frac{\partial L}{\partial x^i} = 0, \qquad i = 1, \ldots, n, \tag{3}$$

where

$$\frac{\partial L}{\partial \dot{x}^i} = \left.\frac{\partial L(x, \xi)}{\partial \xi^i}\right|_{\xi = \dot{x}}, \qquad \frac{d}{dt}\left(\frac{\partial L}{\partial \dot{x}^i}\right) = \left.\frac{\partial^2 L}{\partial \xi^i\, \partial \xi^j}\ddot{x}^j + \frac{\partial^2 L}{\partial \xi^i\, \partial x^j}\dot{x}^j\right|_{\xi = \dot{x}}.$$

(Here in $L = L(x^1, \ldots, x^n, \xi^1, \ldots, \xi^n)$, the variables x and ξ are regarded as independent; only after completing the calculation of the derivatives do we put $\xi^i = dx^i/dt$, i.e. do we take ξ to be the tangent vector to γ_0.)

Proof. Let $\eta^i = \eta^i(t)$, $a \le t \le b$, be any n smooth functions satisfying $\eta^i(a) = 0 = \eta^i(b)$. Consider the expression

$$\lim_{\varepsilon \to 0} \frac{S[\gamma_0 + \varepsilon\eta] - S[\gamma_0]}{\varepsilon} = \frac{d}{d\varepsilon} S[\gamma_0 + \varepsilon\eta]|_{\varepsilon=0};$$

here $\gamma_0 + \varepsilon\eta$ is the arc $x^i = x^i(t) + \varepsilon\eta^i(t)$ which also joins P to Q and which is close to γ_0 for small ε. Since as a real-valued function of the real variable ε, $S[\gamma_0 + \varepsilon\eta]$ takes on a minimum value at $\varepsilon = 0$, elementary differential calculus tells us that

$$\frac{d}{d\varepsilon} S[\gamma_0 + \varepsilon\eta]|_{\varepsilon=0} = 0.$$

On the other hand, by differentiating the expression for $S[\gamma_0 + \varepsilon\eta]$ under the integral sign we obtain

$$0 = \frac{d}{d\varepsilon} S[\gamma_0 + \varepsilon\eta]|_{\varepsilon=0} = \int_a^b \left\{\frac{\partial L}{\partial x^i}\eta^i(t) + \frac{\partial L}{\partial \xi^i}\dot{\eta}^i\right\} dt, \tag{4}$$

where the integral is taken along the arc γ_0, i.e. $x^i = x^i(t)$, $\xi^i = \dot{x}^i(t)$.

Integrating by parts we find that

$$\int_a^b \frac{\partial L}{\partial \xi^i}\dot{\eta}^i\, dt = \left(\frac{\partial L}{\partial \xi^i}\eta^i\right)_{t=b} - \left(\frac{\partial L}{\partial \xi^i}\eta^i\right)_{t=a} - \int_a^b \eta^i \frac{d}{dt}\left(\frac{\partial L}{\partial \xi^i}\right) dt,$$

whence, recalling that $\eta^i(a) = \eta^i(b) = 0$, we obtain

$$\int_a^b \frac{\partial L}{\partial \xi^i}\dot{\eta}^i\, dt = -\int_a^b \frac{d}{dt}\left(\frac{\partial L}{\partial \xi^i}\right)\eta^i\, dt.$$

Substitution from this into (4) yields the following equation, valid for any smooth vector function $\eta(t)$ vanishing at the ends of the time interval $[a, b]$, under our hypothesis that the curve γ_0: $x^i = x^i(t)$ minimizes the functional $S[\gamma]$ on the set of all smooth arcs joining P to Q:

$$\frac{d}{d\varepsilon} S[\gamma_0 + \varepsilon\eta]|_{\varepsilon=0} = \int_a^b \left[\frac{\partial L}{\partial x^i} - \frac{d}{dt}\frac{\partial L}{\partial \dot{x}^i}\right]\eta^i\, dt = 0. \tag{5}$$

It follows almost immediately that

$$\psi^i(t) = \frac{\partial L}{\partial x^i} - \frac{d}{dt}\frac{\partial L}{\partial \dot{x}^i} = 0, \qquad i = 1, \ldots, n,$$

since if this were not so, i.e. if we had $\psi^i(t) \neq 0$ for some i and some $t = t_0$, $a \le t_0 \le b$, then (assuming as usual sufficient smoothness) we should have $\psi^i(t) \neq 0$ on an interval of values of t, and then by choosing $\eta^i(t)$ suitably (for instance by taking $\eta^i = \psi^i(t)f(t)$ where $f(a) = f(b) = 0$ but $f(t) > 0$ for $a < t < b$, so that the integrand in (5) is positive on a subinterval of $[a, b]$) we could force the integral in (5) to be non-zero. This completes the proof of the theorem. □

The solutions of the equations $(d/dt)(\partial L/\partial \dot{x}^i) = \partial L/\partial x^i$ are called the *extremal arcs*, or *extremals* of the functional S.

We conclude this subsection by introducing some conventional terminology.†

(i) The function

$$L = L(x, \xi) = L(x, \dot{x}) \tag{6}$$

whose integral we seek to minimize is called the *Lagrangian*.

(ii) The *energy* E is defined by

$$E = E(x, \dot{x}) = E(x, \xi) = \xi^i \frac{\partial L}{\partial \xi^i} - L = \dot{x}^i \frac{\partial L}{\partial \dot{x}^i} - L. \tag{7}$$

(iii) The *momentum* is the covector defined by

$$p_i = \frac{\partial L}{\partial \dot{x}^i} = \frac{\partial L}{\partial \xi^i}. \tag{8}$$

(To see that this is a covector use the transformation rule $\zeta^j = (\partial z^j/\partial x^i)\xi^i$ for vectors.)

(iv) The *force*, also a covector, is defined by

$$f_i = \frac{\partial L}{\partial x^i}. \tag{9}$$

(v) The *Euler–Lagrange equation* is the (covector) equation (3) of Theorem 31.1.2, i.e. the equation for the extremal curves:

$$\frac{d}{dt}\left(\frac{\partial L}{\partial \dot{x}^i}\right) = \frac{\partial L}{\partial x^i} \quad \text{or} \quad \dot{p}_i = f_i.$$

(vi) The expression

$$\frac{\delta S}{\delta x^i} = \frac{\partial L}{\partial x^i} - \frac{d}{dt}\frac{\partial L}{\partial \dot{x}^i} \tag{10}$$

is called the *variational derivative* of the functional $S[\gamma]$. From (5) we see that the variational derivative may alternatively be defined as that quantity $\delta S/\delta x^i$ satisfying, for all smooth η vanishing at a and b, the following equation:

$$\frac{d}{d\varepsilon} S[\gamma + \varepsilon\eta]|_{\varepsilon=0} = \int_a^b \frac{\delta S}{\delta x^i} \eta^i \, dt. \tag{11}$$

† Note that the extremals corresponding to a Lagrangian L will also be the extremals corresponding to the Lagrangian $\hat{L}(t, x, \dot{x}) = L + (d/dt)f(x, t)$, where f is any suitably well-behaved function. The corresponding energy and momentum are then given respectively by $\hat{E} = E - (d/dt)f(x, t)$ and $\hat{p} = p + \partial f/\partial x$.

31.2. Basic Examples of Functionals

(a) If $L = [(m/2)\sum_i (\xi^i)^2] - U(x)$, where U is a function of the points x of the underlying space, then $f_i = -\partial U/\partial x^i$, $p_i = m\dot{x}^i$, and

$$\dot{p}_i = f_i, \quad \text{or} \quad m\ddot{x}^i = -\frac{\partial U}{\partial x^i}. \tag{12}$$

(b) If $L = \frac{1}{2}g_{ij}\xi^i\xi^j$, then

$$p_i = g_{ij}\xi^j, \qquad f_k = \frac{1}{2}\frac{\partial g^{ij}}{\partial x^k}\xi^i\xi^j,$$

and the Euler–Lagrange equations become

$$\frac{dp_k}{dt} = \frac{1}{2}\frac{\partial g_{ij}}{\partial x^k}\dot{x}^i\dot{x}^j, \quad \text{where } p_k = g_{kj}\dot{x}^j, \tag{13}$$

that is,

$$\frac{dp_k}{dt} = \ddot{x}^j g_{jk} + \dot{x}^j\frac{\partial g_{jk}}{\partial x^i}\dot{x}^i = \frac{1}{2}\frac{\partial g_{ij}}{\partial x^k}\dot{x}^i\dot{x}^j,$$

whence, using $g^{km}g_{jk} = \delta_j^m$, we obtain

$$\ddot{x}^m + g^{km}\left(\frac{\partial g_{jk}}{\partial x^i} + \frac{1}{2}\frac{\partial g_{ij}}{\partial x^k}\right)\dot{x}^i\dot{x}^j = 0. \tag{14}$$

Now it is easily verified that

$$\frac{\partial g_{jk}}{\partial x^i}\dot{x}^i\dot{x}^j = \tfrac{1}{2}\dot{x}^i\dot{x}^j\left(\frac{\partial g_{ik}}{\partial x^j} + \frac{\partial g_{jk}}{\partial x^i}\right).$$

Substituting from this in (14) we obtain finally

$$\ddot{x}^m + \Gamma_{ij}^m\dot{x}^i\dot{x}^j = 0, \tag{15}$$

where

$$\Gamma_{ij}^m = \tfrac{1}{2}g^{km}\left(\frac{\partial g_{ik}}{\partial x^j} + \frac{\partial g_{jk}}{\partial x^i} - \frac{\partial g_{ij}}{\partial x^k}\right), \tag{16}$$

whence we see that the Γ_{ij}^m are just the components of the symmetric connexion compatible with the metric g_{ij}! We frame this result as a

31.2.1. Theorem. *If $L = g_{ij}\dot{x}^i\dot{x}^j = \langle\dot{x},\dot{x}\rangle = |\dot{x}|^2$, $S[\gamma] = \int_P^Q |\dot{x}|^2\,dt$, then the Euler–Lagrange equation for the extremal (in particular minimal) arcs is equivalent to the equation for the geodesics relative to the metric g_{ij}.*

(c) If $L = \sqrt{g_{ij}\dot{x}^i\dot{x}^j} = |\dot{x}|$, then as we know the expression $S = \int_P^Q |\dot{x}|\,dt$, the length of the arc of the curve $x = x(t)$ joining P and Q, is independent of the parameter t. The Euler–Lagrange equations $(d/dt)(\partial L/\partial \dot{x}^k) = \partial L/\partial x^k$, take on in this case the form

$$\frac{d}{dt}\left(\frac{g_{kj}\dot{x}^j}{\sqrt{g_{ij}\dot{x}^i\dot{x}^j}}\right) = \frac{\dfrac{\partial g_{ij}}{\partial x^k}\dot{x}^i\dot{x}^j}{2\sqrt{g_{ij}\dot{x}^i\dot{x}^j}}. \tag{17}$$

If we parametrize our curve $x = x(t)$ by a parameter proportional to the natural parameter l, i.e. if $t = \text{const.} \times l$, then $\sqrt{g_{ij}\dot{x}^i\dot{x}^j} = |\dot{x}| = \text{const.}$, and (17) becomes

$$\frac{d}{dt}(g_{kj}\dot{x}^j) = \frac{1}{2}\frac{\partial g_{ij}}{\partial x^k}\dot{x}^i\dot{x}^j,$$

which has the same form as equation (13) in example (b) above, but in the present case holds only under the assumption that the curve $x = x(t)$ is parametrized by a parameter proportional to the natural one. Henceforth in this chapter we shall give the term "natural parameter" a wider meaning than hitherto, allowing it to embrace also parameters which are merely proportional to the natural parameter. Note that in the present example we may indeed restrict the parameter to being natural since arc length is independent of the parametrization of the arc.

We summarize the above in the following

31.2.2. Theorem. *The Euler–Lagrange equation for the extremal (in particular minimal) arcs of the functional corresponding to the Lagrangian* $L = \sqrt{g_{ij}\xi^i\xi^j}$, *is equivalent to the equation for the geodesics relative to the metric* g_{ij}, *provided the curves are parametrized by natural parameters. Thus a smooth arc of smallest length among all smooth arcs joining* P *and* Q, *will, provided it is parametrized by a natural parameter, satisfy the equation defining the geodesics.*

We now consider two properties of the energy and momentum corresponding to an arbitrary Lagrangian.

The first property is the celebrated "law of conservation of energy": *Along an extremal the energy* E *is constant*:

$$\begin{aligned}\frac{dE}{dt} &= \frac{d}{dt}\left(\dot{x}^i\frac{\partial L}{\partial \dot{x}^i} - L\right) = \ddot{x}^i\frac{\partial L}{\partial \dot{x}^i} + \dot{x}^i\frac{d}{dt}\left(\frac{\partial L}{\partial \dot{x}^i}\right) - \frac{\partial L}{\partial x^i}\dot{x}^i - \frac{\partial L}{\partial \dot{x}^i}\ddot{x}^i\\ &= \dot{x}^i\left(\frac{d}{dt}\frac{\partial L}{\partial \dot{x}^i} - \frac{\partial L}{\partial x^i}\right) \equiv 0.\end{aligned}$$

(Recall there that $L = L(x, \xi)$ depends only implicitly on t.)

The second property is the "law of conservation of momentum":

If the co-ordinates $x^1, \ldots, x^n$ *are chosen so that* $\partial L/\partial x^i \equiv 0$, *then along an extremal we shall have* $\dot{p}_i = (d/dt)(\partial L/\partial \dot{x}^i) \equiv 0$. This is immediate from the Euler–Lagrange equations. Co-ordinates x^i for which $\partial L/\partial x^i \equiv 0$, are called *cyclic co-ordinates.*

We now give two examples illustrating these "laws".

(α) (Cf. example (b).) If $L = \frac{1}{2}g_{ij}\dot{x}^i\dot{x}^j$, then $E = L = \frac{1}{2}|\dot{x}|^2$. The law of conservation of energy tells us that along every extremal of the functional $S = \int L\, dt$, we shall have $dE/dt \equiv 0$. Hence the speed $|\dot{x}|$ with which extremals (which are all geodesics—see (b)) are traversed, is constant, so that the parameter t is natural.

Remark. If the Lagrangian $L(x, \dot{x})$ is homogeneous in $\xi = \dot{x}$, of degree of homogeneity 1, i.e. if $L(x, \lambda\xi) = \lambda L(x, \xi)$ (for instance if $L = \sqrt{\langle \xi, \xi \rangle}$), then the energy E is identically zero along all curves, and there arise no restrictions on the parametrization of the extremals.

(β) If a surface in Euclidean 3-space is given in cylindrical coordinates by an equation of the form $f(z, r) = 0$ (so that it is a surface of revolution), then we may take as local co-ordinates on the surface the angle φ together with either r or z. As we know, the Euclidean metric is given in terms of cylindrical co-ordinates by

$$dl^2 = dz^2 + dr^2 + r^2\, d\varphi^2.$$

If the surface is given locally by an equation $r = r(z)$, then the induced metric is

$$dl^2 = g_{zz}(dz)^2 + r^2(z)(d\varphi)^2, \qquad g_{ij} = \begin{pmatrix} g_{zz} & 0 \\ 0 & r^2 \end{pmatrix}.$$

(Here $g_{zz} = 1 + (dr/dz)^2$.) The Lagrangian for the geodesics is then

$$L = \tfrac{1}{2}(g_{zz}\dot{z}^2 + r^2(z)\dot{\varphi}^2)$$

(or $L = \frac{1}{2}(g_{rr}\dot{r}^2 + r^2\dot{\varphi}^2)$ in terms of the local co-ordinates r, φ on the surface $z = z(r)$). From the conservation laws we know that the energy $E = L$, and the components $p_\varphi = \partial L/\partial\dot{\varphi} = r^2\dot{\varphi}$, $p_z = \partial L/\partial \dot{z} = g_{zz}\dot{z}$, are constant along each geodesic. Let e_z, e_φ be the standard basis vectors at the point (z, φ) of the surface; the various scalar products of these vectors at that point are as follows:

$$\langle e_z, e_z \rangle = g_{zz}, \qquad \langle e_z, e_\varphi \rangle = 0, \qquad \langle e_\varphi, e_\varphi \rangle = r^2(z).$$

Let $z = z(t)$, $\varphi = \varphi(t)$ be the parametric equations of a geodesic (i.e. solutions of the equations defining geodesics on the surface). The angle ψ between the velocity (or tangent) vector $v = (\dot{z}, \dot{\varphi})$ to this geodesic, and the vector e_φ, is given by

$$\cos \psi = \frac{\langle v, e_\varphi \rangle}{\sqrt{\langle v, v \rangle \langle e_\varphi, e_\varphi \rangle}} = \frac{r^2 \dot{\varphi}}{\sqrt{E} r} = \frac{p_\varphi}{\sqrt{E} r}.$$

Hence $r \cos \psi = p_\varphi / \sqrt{E}$, and so $r \cos \psi$ is constant along the geodesic. We have by these means thus proved the following

31.2.3. Theorem (Clairaut). *The quantity $r \cos \psi$ is constant along each geodesic of any surface of revolution in Euclidean $\mathbb{R}^3$.*

We note finally that since $p_\varphi = r^2 \dot{\varphi}$ and $2E = g_{zz} \dot{z}^2 + r^2 \dot{\varphi}^2 = g_{zz} \dot{z}^2 + p_\varphi^2 / r^2$ are constant along each geodesic on the surface $r = r(z)$, it follows that a geodesic $z = z(t)$, $\varphi = \varphi(t)$ corresponding to prescribed values of E and p_φ, will be a solution of the following "completely integrable" pair of differential equations:

$$d\varphi = \frac{p_\varphi}{r^2(z)} dt, \qquad dt = \frac{dz}{\sqrt{\dfrac{2E}{g_{zz}} - \dfrac{p_\varphi^2}{g_{zz} r^2(z)}}}. \tag{18}$$

Exercises

Suppose that higher derivatives enter into the Lagrangian; for instance let

$$L = L(t, x, \dot{x}, \ddot{x}, \ldots, x^{(k)});$$

$$S[\gamma] = \int_\gamma L(t, x, \ldots, x^{(k)})\, dt.$$

Prove that the extremals of the functional $S[\gamma]$ satisfy the "Euler–Lagrange" equation

$$\frac{\delta S}{\delta x} = \frac{\partial L}{\partial x} - \frac{d}{dt} \frac{\partial L}{\partial \dot{x}} + \frac{d^2}{dt^2} \frac{\partial L}{\partial \ddot{x}} - \cdots + (-1)^k \frac{d^k}{dt^k} \frac{\partial L}{\partial x^{(k)}} = 0.$$

§32. Conservation Laws

32.1. Groups of Transformations Preserving a Given Variational Problem

The law of conservation of momentum, which was derived in the preceding section, can be given a more convenient invariant form by utilizing the concept of a one-parameter group of transformations (see §23).

Suppose that in the space $\mathbb{R}^n$ we are given a local one-parameter group of local transformations S_τ, $-\infty < \tau < \infty$; recall from §23 that such a group has the following properties:

(i) corresponding to each point P of $\mathbb{R}^n$ there is a real number $\tau_0 > 0$ and a neighbourhood U of P on which for $|\tau| < \tau_0$ the transformations $S_\tau: U \to \mathbb{R}^n$ are defined (and smooth);
(ii) $S_0 = 1$ (the identity transformation) and

$$S_{\tau_1+\tau_2} = S_{\tau_1} \cdot S_{\tau_2}, \qquad S_{-\tau} = S_\tau^{-1}, \tag{1}$$

wherever these transformations are simultaneously defined.

Recall from §23 also that there is associated with each local one-parameter group of transformations a vector field (X^i) tangential to the trajectories $S_\tau(x)$:

$$(X^i) = X(x) = \frac{d}{d\tau} S_\tau(x)|_{\tau=0}, \tag{2}$$

and that, conversely, from the vector field (X^i) the group of transformations S_τ can be recaptured via a standard theorem on the existence and uniqueness of solutions of ordinary differential equations. In fact for each point $x \in \mathbb{R}^n$ where (X^i) is non-zero it is at least intuitively clear that there is a neighbourhood U of the point, on which co-ordinates $y^1, \ldots, y^n$ can be introduced with the property that for all small enough τ the transformation S_τ has the form

$$S_\tau(y^1, \ldots, y^n) = (y^1 + \tau, y^2, \ldots, y^n). \tag{3}$$

This version of the existence theorem for solutions of the appropriate ordinary differential equations will be useful to us in what follows.

32.1.1. Definition. We shall say that a one-parameter group of transformations S_τ *preserves a Lagrangian* $L(x, \xi)$ where ξ is attached to (i.e. tangent at) the point x, if

$$\frac{d}{d\tau} L(S_\tau(x), S_{\tau*}\xi) = 0, \tag{4}$$

where $(S_\tau)_*$ is the map between tangent spaces defined as in §22.2(6).

If we write $x(\tau) = S_\tau(x)$ (so that $x = x(0)$), and $\xi(\tau) = (S_\tau)_*\xi$ (so that $\xi = \xi(0)$), then the left-hand side of (4) becomes

$$\frac{dL}{d\tau} = \dot{x}^i(\tau)\frac{\partial L}{\partial x^i(\tau)} + \frac{\partial L}{\partial \xi^i(\tau)}\frac{d\xi^i(\tau)}{d\tau}.$$

Now by §22.2(6), we have $\xi^i(\tau) = \xi^j(\partial x^i(\tau)/\partial x^j)$ (regarding $x^i(\tau)$ as also a function of the $x^j(0)$, $j = 1, \ldots, n$), whence

$$\frac{dL}{d\tau} = \dot{x}^i(\tau)\frac{\partial L}{\partial x^i(\tau)} + \frac{\partial L}{\partial \xi^i(\tau)}\xi^j\frac{\partial}{\partial x^j}\left(\frac{dx^i(\tau)}{d\tau}\right).$$

Putting $\tau = 0$ we obtain as a consequence of (4) that

$$\left.\frac{dL}{d\tau}\right|_{\tau=0} = X^i \frac{\partial L}{\partial x^i} + \frac{\partial X^i}{\partial x^j} \xi^j \frac{\partial L}{\partial \xi^i} = 0, \tag{5}$$

which is thus a necessary condition for the Lagrangian to be preserved by the group of transformations S_τ. (Note that the left-hand expression in (5) might naturally be called the Lie derivative of $L(x, \xi)$ along the field (X^i)—cf. 23.2.1.)

32.1.2. Theorem. *If a Lagrangian L is preserved by a one-parameter group of transformations S_τ then on any extremal of L the "component" of the momentum along the associated vector field is conserved:*

$$\frac{d}{dt}\left(X^i \frac{\partial L}{\partial \dot{x}^i}\right) = \frac{d}{dt}(X^i p_i) = 0, \tag{6}$$

where (X^i) is the vector field defined (as above) by $X(x) = (d/d\tau)S_\tau(x)|_{\tau=0}$.

PROOF. Let $x \in \mathbb{R}^n$ be such that $X(x) \neq 0$; then as noted above (see (3)) there is a neighbourhood U of x on which there exist co-ordinates $y^1, \ldots, y^n$ in terms of which the action of S_τ takes the simple form $S_\tau(y^1, \ldots, y^n) = (y^1 + \tau, y^2, \ldots, y^n)$. Since by hypothesis L is preserved by the transformation S_τ, it follows that changes in y^1 have no effect on L, i.e.

$$\frac{\partial L(y, \dot{y})}{\partial y^1} \equiv 0.$$

It follows from the Euler–Lagrange equations that on an extremal $y = y(t)$ of L, we shall have

$$\frac{d}{dt}\left(\frac{\partial L}{\partial \dot{y}^1}\right) \equiv 0.$$

Now in terms of the co-ordinates $y^1, \ldots, y^n$ the vector field X calculates out as $(1, 0, \ldots, 0)$, i.e. as the unit vector in the direction of the y^1-axis. Hence $\partial L/\partial \dot{y}^1$ is the "partial" directional derivative (with respect to the tangent vector variable $\dot{y}$) of L along the vector field X. Since in terms of the original co-ordinates $x^1, \ldots, x^n$ this directional derivative is given by $X^i(\partial L/\partial \dot{x}^i)$, the desired conclusion (6) follows. □

32.2. Examples. Applications of the Conservation Laws

(a) According to the general physical principle known as the "Principle of Least Action," the world-line of a relativistic, free particle of positive mass m should be a time-like curve in Minkowski space $\mathbb{R}^4_{1,3}$ with co-ordinates

$x^0 = ct, x^1, x^2, x^3$, extremizing one or the other of the following two actions:

$$S_1 = \frac{mc}{2}\int\left\langle\frac{dx}{d\tau}, \frac{dx}{d\tau}\right\rangle d\tau, \qquad \left\langle\frac{dx}{d\tau}, \frac{dx}{d\tau}\right\rangle = \left(\frac{dx^0}{d\tau}\right)^2 - \sum_{\alpha=1}^{3}\left(\frac{dx^\alpha}{d\tau}\right)^2, \tag{7}$$

$$S_2 = -mcl = -mc\int\sqrt{\left\langle\frac{dx}{d\tau}, \frac{dx}{d\tau}\right\rangle}\, d\tau = -mc\int dl. \tag{8}$$

It is easy to verify (see §31.2) that the extremals of the two functionals in (7) and (8) coincide. It turns out that for purposes of comparison with classical mechanics it is more convenient to work with the action S_2 given by (8).

Although in (8) any parameter τ may be used, the fact that the world-line is time-like leads one to the obvious choice $\tau = t = x^0/c$. With this parameter, (8) becomes

$$S_2 = -mcl = -mc^2\int\sqrt{1 - \left(\frac{w}{c}\right)^2}\, dt, \qquad w^2 = (\dot{x}^1)^2 + (\dot{x}^2)^2 + (\dot{x}^3)^2. \tag{9}$$

(Thus w is the 3-dimensional speed of the particle. The quantity l/c is called the *proper time*.) Following the usual convention we write

$$L = -mc^2\sqrt{1 - \frac{w^2}{c^2}}, \tag{10}$$

so that $S_2 = \int L\, dt$; note that the Lagrangian L is in (10) expressed in terms of 3-dimensional entities. The energy and momentum corresponding to this Lagrangian are given by (see §31.1)

$$E = \dot{x}^\alpha\frac{\partial L}{\partial \dot{x}^\alpha} - L = \frac{mc^2}{\sqrt{1 - \frac{w^2}{c^2}}}, \tag{11}$$

$$p_\alpha = \frac{mw^\alpha}{\sqrt{1 - \frac{w^2}{c^2}}}, \quad \text{where } w^\alpha = \frac{dx^\alpha}{dt}, \quad \alpha = 1, 2, 3. \tag{12}$$

We therefore have

$$E = mc^2\left(1 + \frac{w^2}{2c^2} + \cdots\right),$$

$$p_\alpha = mw^\alpha(1 + \cdots).$$

Thus for small w/c, we obtain to a first approximation the classical formula for the momentum, since then

$$p_\alpha \approx mw^\alpha, \qquad \alpha = 1, 2, 3 \tag{13}$$

(see Example (a) of §31.2), while for the energy we obtain

$$E \approx mc^2 + \frac{mw^2}{2}, \tag{14}$$

which differs from the classical formula by the constant mc^2. Note also the identity

$$E^2 - c^2p^2 = m^2c^4, \qquad E = c\sqrt{p^2 + m^2c^2}, \quad p^2 = p_1^2 + p_2^2 + p_3^2. \quad (15)$$

From this we see that as E runs through the positive reals, the point (E, cp_1, cp_2, cp_3) traces out a 3-dimensional surface, the so-called "mass hyperboloid" of the particle, in Minkowski space $\mathbb{R}_1^4$.

The above treatment (which might be called "3-dimensional formalism") has two shortcomings: the energy and momentum of the particle seem quite disparate concepts, and the time co-ordinate plays a very special role. In order to eliminate these special features we now turn to the action (7) given by

$$S_1 = \frac{mc}{2} \int \left\langle \frac{dx}{d\tau}, \frac{dx}{d\tau} \right\rangle d\tau.$$

It follows from Example (α) of §31.2 that the parameter τ will automatically be natural (in the broader sense of §31.2) for the extremals corresponding to this action. As in the general definition (§31.1(8)) of momentum, we define the 4-*momentum* $\tilde{p}_i = (\tilde{p}_0, \tilde{p}_1, \tilde{p}_2, \tilde{p}_3)$ by

$$\tilde{p}_i = \frac{\partial L}{\partial x'^i}, \qquad i = 0, 1, 2, 3, \quad \text{where } x'^i = \frac{dx^i}{d\tau}.$$

Hence

$$\tilde{p}_0 = \frac{\partial L}{\partial x'^0} = mcx'^0, \qquad \tilde{p}_\alpha = \frac{\partial L}{\partial x'^\alpha} = -mcx'^\alpha, \qquad \alpha = 1, 2, 3.$$

If, using the Minkowski metric, we raise the index of the covector $\tilde{p}_i$, we obtain the vector

$$\tilde{p}^i = mcx'^i, \qquad i = 0, 1, 2, 3.$$

As noted above, we know that along an extremal τ is a constant multiple of l; if we make the (modest) assumption that $\tau = l$, i.e. if we make the (mild) change of parameter ensuring this along a particular extremal, then we obtain

$$dl = c\sqrt{1 - \frac{w^2}{c^2}}\, dt, \qquad x'^i = \frac{dx^i}{dl},$$

$$\tilde{p}^0 = mcx'^0 = \frac{mc}{\sqrt{1 - \frac{w^2}{c^2}}} = \frac{E}{c}, \quad (16)$$

$$\tilde{p}^\alpha = mcx'^\alpha = mc\frac{dx^\alpha}{dt} \cdot \frac{1}{c\sqrt{1 - \frac{w^2}{c^2}}} = \frac{mw^\alpha}{\sqrt{1 - \frac{w^2}{c^2}}} = p^\alpha,$$

where w, E, $p^\alpha = p_\alpha$ are defined as before. We conclude that:

(i) *The (vector) 4-momentum* ($\tilde{p}^i$) *of a particle is linked to its 3-dimensional energy and momentum by the equations*

$$\tilde{p}^0 = E/c, \qquad \tilde{p}^\alpha = p^\alpha; \tag{17}$$

(ii) *Under co-ordinate changes (in particular Lorentz transformations) the energy-momentum vector* (E/c, p^1, p^2, p^3) *transforms like a 4-vector; the possible 4-momenta of a particle of mass* $m > 0$ *lie on the mass hyperboloid*

$$\frac{E^2}{c^2} - p^2 = (\tilde{p}^0)^2 - \sum_{\alpha=1}^{3} (\tilde{p}^\alpha)^2 = m^2c^2, \tag{18}$$

which has induced on it (in the usual way) a Lobachevskian metric.

In particular if we change to a co-ordinate frame x' moving uniformly with speed v in the direction of the x^1-axis, then from §6.2(20) and the transformation rule for vectors we obtain:

$$\frac{E}{c} = \frac{\dfrac{E'}{c} + \dfrac{v}{c}\tilde{p}'^1}{\sqrt{1 - \dfrac{v^2}{c^2}}}, \quad \text{i.e.} \quad E = \frac{E' + p'_1 v}{\sqrt{1 - \dfrac{v^2}{c^2}}};$$

$$\tilde{p}^1 = \frac{\dfrac{v}{c}\tilde{p}'^0 + \tilde{p}'^1}{\sqrt{1 - \dfrac{v^2}{c^2}}}, \quad \text{i.e.} \quad p_1 = \frac{\dfrac{E'v}{c^2} + \tilde{p}'_1}{\sqrt{1 - \dfrac{v^2}{c^2}}};$$

$$p_2 = p'_2, \ p_3 = p'_3. \tag{19}$$

For small v/c these yield (approximately) $E \approx E' + p_1 v$, $p_1 \approx E'v/c^2 + \tilde{p}'_1$ ($\approx mv + \tilde{p}'_1$ if w/c is also small).

(b) The most fundamental of the postulates of the general theory of relativity is the hypothesis ("Einstein's hypothesis") whereby the gravitational field is identified with a metric g_{ij} (of signature $(+ - - -)$ at each point) on 4-dimensional space-time. To put it more precisely, it is assumed that in the absence of other (i.e. non-gravitational) forces, a test particle of arbitrary mass $m > 0$ in a gravitational field will move along a time-like geodesic extremizing the action

$$S_1 = \frac{mc}{2} \int \langle \dot{x}, \dot{x} \rangle \, d\tau, \tag{20}$$

while a particle of zero-mass will move along a null geodesic, i.e. one along which $\langle \dot{x}, \dot{x} \rangle = 0$ (see §6.1).

A *weak gravitational field* is defined to be a metric g_{ab} which can be expressed as a series in powers of $1/c$ of the form

$$g_{ab} = g_{ab}^{(0)} + c^{-2} g_{ab}^{(2)} + O\left(\frac{1}{c^3}\right), \qquad g_{00}^{(2)} = 2\varphi(x^0, x), \tag{21}$$

where $g_{ab}^{(0)}$ is the Minkowski metric $\begin{pmatrix} 1 & & & 0 \\ & -1 & & \\ & & -1 & \\ 0 & & & -1 \end{pmatrix}$ (so that the field is almost "flat"), the terms in $1/c$ are zero, and where the scalar $\varphi(x^0, x)$, $x^0 = ct$, $x = (x^1, x^2, x^3)$, is the classical gravitational potential. (The reason for defining a weak field in this way will appear in Proposition 32.2.2 below. See also §39.2.)

32.2.1. Definition. A time-like geodesic $x^\alpha = x^\alpha(t)$, $t = x^0/c$, of a weak gravitational field is said to be *slow* if $|dx^\alpha/dt| \ll c$, $\alpha = 1, 2, 3$.

Taking the parameter τ to be the proper time l/c, we have

$$d\tau = \frac{dl}{c} = \sqrt{\frac{1}{c^2} g_{ab} \frac{dx^a}{dt} \frac{dx^b}{dt}}\, dt, \qquad t = \frac{x^0}{c},$$

so that in the case of a weak field we shall have

$$d\tau = \left[1 + O\left(\frac{1}{c^2}\right)\right] dt. \tag{22}$$

Hence in the equation for the geodesics

$$\frac{d^2 x^a}{d\tau^2} + \Gamma_{bc}^a \frac{dx^b}{d\tau} \frac{dx^c}{d\tau} = 0, \tag{23}$$

where, as we have seen, the parameter is natural, we may to within an error $O(1/c^2)$ replace $d\tau$ by dt.

32.2.2. Proposition. *In a slowly varying weak field the equation for the slow geodesics has the form*

$$\frac{d^2 x^\alpha}{dt^2} = -\frac{\partial \varphi}{\partial x^\alpha} + O\left(\frac{1}{c}\right), \qquad \alpha = 1, 2, 3, \tag{24}$$

which approximates Newton's equation for the motion of a particle in a classical gravitational field with potential φ.

Proof. By "slowly varying" we mean that the quantities $\partial g_{ab}/\partial t$, $\partial g_{ab}/\partial x^\alpha$, $\alpha = 1, 2, 3$, are small (in absolute value) relative to c. The Christoffel symbols are given by

$$\Gamma^a_{bc} = \tfrac{1}{2} g^{ad}\left(\frac{\partial g_{bd}}{\partial x^c} + \frac{\partial g_{cd}}{\partial x^b} - \frac{\partial g_{bc}}{\partial x^d}\right).$$

Now from (21) and the above condition on $\partial g_{ab}/\partial t, \partial g_{ab}/\partial x^\alpha$, it follows that the partial derivatives $\partial g_{ab}/\partial x^0 = (1/c)(\partial g_{ab}/\partial t)$ are of order $O(1/c^3)$, while the partial derivatives $\partial g_{ab}/\partial x^\alpha$, $\alpha = 1, 2, 3$, are of order $O(1/c^2)$. Since, again by (21), $g_{ab} \approx \begin{pmatrix} 1 & & & 0 \\ & -1 & & \\ & & -1 & \\ 0 & & & -1 \end{pmatrix}$, it follows that in (23) (with t replacing τ) all terms involving a Christoffel symbol are of order at most $O(1/c)$ except $\Gamma^\alpha_{00}\dot{x}^0\dot{x}^0 = \Gamma^\alpha_{00}c^2 + O(1/c)$. (It is here that the slowness condition is exploited.) For Γ^α_{00} we have

$$\Gamma^\alpha_{00} = \tfrac{1}{2} g^{\alpha d}\left(-\frac{\partial g_{00}}{\partial x^d}\right) + O\left(\frac{1}{c^3}\right) = \tfrac{1}{2} g^{\alpha\alpha}\left(-\frac{\partial g_{00}}{\partial x^\alpha}\right) + O\left(\frac{1}{c^3}\right)$$

$$= \frac{1}{c^2}\frac{\partial \varphi}{\partial x^\alpha} + O\left(\frac{1}{c^3}\right), \qquad \alpha = 1, 2, 3.$$

The desired conclusion now follows from (23) (with t in place of τ). □

(c) The behaviour of a classical system of n mutually interacting particles is described by the Lagrangian (in $\mathbb{R}^{3n}$)

$$L = \sum_{i=1}^{n} \frac{m_i \dot{x}_i^2}{2} - U(x_1, \dots, x_n), \qquad x_i = (x_i^1, x_i^2, x_i^3), \tag{25}$$

where $U = \frac{1}{2}\sum_{i \neq j} V(x_i, x_j)$, and by $\dot{x}_i^2$ we mean $(\dot{x}_i^1)^2 + (\dot{x}_i^2)^2 + (\dot{x}_i^3)^2$. We shall assume that the system is translation-invariant, or, in other words, that the Lagrangian L is not altered by transformations of the form

$$x_i \to x_i + \xi, \qquad \xi = (\xi^1, \xi^2, \xi^3).$$

For this it suffices that the function V be a function of the difference in its arguments:

$$V(x_i, x_j) = V(x_i - x_j). \tag{26}$$

It follows in this case from the law of conservation of momentum that:

The total momentum of the system is conserved, i.e.

$$\frac{dP_{\text{total}}}{dt} = 0, \quad \textit{where } P_{\text{total}} = \sum_{i=1}^{n} m_i \dot{x}_i. \tag{27}$$

Proof. Consider the three one-parameter groups S_τ^α, $\alpha = 1, 2, 3$, whose action on $\mathbb{R}^{3n}$ is given by

$$S_\tau^\alpha\colon x_i^\alpha \to x_i^\alpha + \tau, \qquad x_i^\beta \to x_i^\beta \quad \text{for } \beta \neq \alpha, \quad \alpha, \beta = 1, 2, 3. \tag{28}$$

The corresponding vector fields $X^{(\alpha)} = (d/d\tau)S_\tau^\alpha(x)|_{\tau=0}$ are then

$$\begin{aligned} X^{(1)} &= (1, 0, 0, 1, 0, 0, \ldots, 1, 0, 0), \\ X^{(2)} &= (0, 1, 0, 0, 1, 0, \ldots, 0, 1, 0), \\ X^{(3)} &= (0, 0, 1, 0, 0, 1, \ldots, 0, 0, 1). \end{aligned} \tag{29}$$

From Theorem 32.1.2 it then follows that each of the components

$$P_{\text{total}, \alpha} = \sum_{i=1}^n m_i \dot{x}_i^\alpha, \qquad i = 1, 2, 3,$$

is conserved, so that the total momentum is conserved. This completes the proof. □

In the particular case $n = 2$, we have (with V not quite the same function as before)

$$L = m_1(\dot{x}_1)^2 + m_2(\dot{x}_2)^2 - V(x_1 - x_2).$$

In view of the conservation of total momentum we can choose our coordinate frame (moving uniformly relative to the original one) so that $P_{\text{total}} = 0$, i.e. so that

$$m_1\dot{x}_1 + m_2\dot{x}_2 = 0, \tag{30}$$

and then, since in this frame $m_1x_1 + m_2x_2 = \text{const.}$, we can adjust its origin to be at the centre of mass of the system, i.e. so that $m_1x_1 + m_2x_2 = 0$. Hence

$$x_2 = -\frac{m_1x_1}{m_2}, \qquad V(x_1 - x_2) = V\left(x_1 + \frac{m_1}{m_2}x_1\right).$$

Newton's equations applied to the particle of mass m_1 take the form

$$m_1\ddot{x}_1^\alpha = -\frac{\partial V(x_1 - x_2)}{\partial x_1^\alpha}, \qquad \alpha = 1, 2, 3. \tag{31}$$

Writing $m^* = m_1(1 + m_1/m_2)$, $U(x_1) = V(x_1 + (m_1/m_2)x_1)$, these equations become

$$m^*\ddot{x}_1^\alpha = -\frac{\partial U(x_1)}{\partial x_1^\alpha}. \tag{32}$$

We frame our conclusion as a

32.2.3. Theorem. *The problem of describing the motion of a classical two-particle system in a field with a translation-invariant potential* $V(x_1 - x_2)$ *relative to a coordinate frame with its origin at the centre of mass of the system, is equivalent to the problem of describing the motion of a single particle of mass* $m^* = m_1(1 + m_1/m_2)$ *in a field with potential* $U(x_1) = V(x_1 - x_2)$ *where* $m_1 x_1 + m_2 x_2 = 0$.

Thus utilization of the translation group led us to the law of conservation of total momentum, and thence to the reduction of the classical two-particle problem to that of one particle.

(d) We shall now consider Lagrangians preserved by the rotation group $SO(3)$. Let x, y, z be Euclidean co-ordinates for Euclidean 3-space.

32.2.4. Definition. We shall say that a Lagrangian $L(x, y, z, \dot{x}, \dot{y}, \dot{z})$ is *spherically symmetric* if it is preserved by all rotations of $\mathbb{R}^3$.

We single out the following three important subgroups of $SO(3)$:

(i) $S_\varphi^{(x)}$—the rotations about the x-axis through the angles φ;
(ii) $S_\varphi^{(y)}$—the rotations about the y-axis through the angles φ;
(iii) $S_\varphi^{(z)}$—the rotation about the z-axis through the angles φ.

In §24.3 we calculated the associated vector fields L_x, L_y, L_z in $\mathbb{R}_3$ to be given by

$$L_x = (0, -z, y), \qquad L_y = (z, 0, -x), \qquad L_z = (-y, x, 0). \tag{33}$$

By Theorem 32.1.2, for a spherically symmetric Lagrangian $L(x, y, z, \dot{x}, \dot{y}, \dot{z})$, the following quantities (the components of the "angular momentum", corresponding to rotations about the three axes) are conserved along extremals:

$$M_x = L_x^\alpha p_\alpha, \qquad M_y = L_y^\alpha p_\alpha, \qquad M_z = L_z^\alpha p_\alpha,$$

where $p_\alpha = \partial L/\partial \dot{x}^\alpha$. Thus in explicit form we have:

$$M_x = y p_z - z p_y, \qquad M_y = z p_x - x p_z, \qquad M_z = x p_y - y p_x;$$
$$(\dot{M}_x) = (\dot{M}_y) = (\dot{M}_z) = 0. \tag{34}$$

Hence the vector

$$M = (M_x, M_y, M_z) = [\bar{x}, p], \tag{35}$$

where $\bar{x} = (x, y, z)$, is conserved. The vector $[\bar{x}, p]$, the cross product of $\bar{x}$ with p, is called the *angular momentum*.

Consider now a (classical) system of two particles invariant under the full isometry group of Euclidean 3-space. The appropriate Lagrangian will then have the form $L = m_1\dot{x}_1^2/2 + m_2\dot{x}_2^2/2 - V(x_1 - x_2)$ where $V(x_1 - x_2) = V(|x_1 - x_2|)$. Since L is in particular invariant under the translation group we can invoke Theorem 32.2.3 to reduce the problem to that of a single particle of mass $m^* = m$ at the point x_1 in a field $U(r)$, where $r = |x_1|$, $x_1 - x_2 = x_1(1 + m_1/m_2)$. Since the Lagrangian of this one-particle system will be invariant under $SO(3)$, it follows from (35) *et seqq.* that the angular momentum $[x_1, p]$ is conserved. Since U is time-independent it follows from the law of conservation of energy that the energy $E = mv^2/2 + U(r)$ is also conserved.

32.2.5. Lemma. *The motion of the particle is confined to the plane spanned by the vectors x_1 and p.*

PROOF. Since the angular momentum is conserved and $\dot{x}_1 = p/m$, it follows that the direction of $[x_1, \dot{x}_1] = M/m$ is fixed. The lemma now follows from the fact that this vector is perpendicular to the plane determined by x_1 and p. □

In view of this lemma it is convenient to change to cylindrical co-ordinates (z, r, φ) where now the z-axis is in the direction of M. If our particle of mass m is at the point (z, r, φ), then the square of its speed v is

$$v^2 = \left(\frac{d(r\cos\varphi)}{dt}\right)^2 + \left(\frac{d(r\sin\varphi)}{dt}\right)^2 + \left(\frac{dz}{dt}\right)^2 = \dot{r}^2 + r^2\dot{\varphi}^2,$$

since $\dot{z} = 0$. Hence

$$L = \frac{mv^2}{2} - U(r) = \frac{m}{2}(\dot{r}^2 + r^2\dot{\varphi}^2) - U(r),$$

$$p_\varphi = \frac{\partial L}{\partial\dot{\varphi}} = mr^2\dot{\varphi}.$$

Since $z = 0 = p_z$, we have from (34) that

$$|M| = M_z = xp_y - yp_x = p_\varphi = mr^2\dot{\varphi} = \text{const.} \tag{36}$$

Thus $\dot{\varphi} = |M|/mr^2$, and

$$E = \frac{m}{2}\left(\dot{r}^2 + \frac{|M|^2}{m^2r^2}\right) + U(r),$$

or

$$E = \frac{m\dot{r}^2}{2} + U_{\text{eff}}(r), \tag{37}$$

where $U_{\text{eff}}(r) = U(r) + |M|^2/2mr^2$. Thus the problem of describing the motion of the particle has been reduced to a one-dimensional problem (in r) with potential $U_{\text{eff}}(r)$; the solution can be carried out in the following steps:

$$\dot{r}^2 = \frac{2}{m}(E - U_{\text{eff}}); \qquad t - t_0 = \int \frac{dr}{\sqrt{\frac{2}{m}(E - U_{\text{eff}})}}; \qquad \varphi - \varphi_0 = \int \frac{|M|}{mr^2}\,dt. \tag{38}$$

Using the second equation to eliminate t from the third, we end up with a solution $\varphi = \varphi(r)$ or $r = r(\varphi)$ for the trajectory of the particle.

In the two important special cases $U = \alpha/r$, $U = \alpha r^2$, it can be shown that there is a region of the 6-dimensional space co-ordinatized by x, $\dot{x}$, which is completely filled by closed trajectories (orbits) (i.e. for every (x_0, ξ_0) in the region there is a closed orbit $x = x(t)$ such that for some t_0 we have $x_0 = x(t_0)$, $\xi_0 = \dot{x}(t_0)$). These regions are as follows:

$$\text{for } U = \frac{\alpha}{r} \quad \text{with } \alpha < 0, \text{ the region } E < 0;$$

$$\text{for } U = \alpha r^2 \quad \text{with } \alpha > 0, \text{ the region } E \geq 0.$$

(In the case $U = \alpha/r$, the region $E < 0$ is that of the Keplerian ellipses.) Note that for an orbit to be closed the non-trivial condition that

$$r(\varphi + 2\pi n) \equiv r(\varphi) \tag{39}$$

for some positive integer n, needs to be imposed.

It is also known that the potentials of the form $U(r) = \alpha/r, \alpha r^2$, are the only spherically symmetric, analytic potentials for which condition (39) holds throughout some region (of positive measure) of phase space (see the book [47]); in general the subset of phase space determined by the closed orbits has measure zero. For the motion of a particle to determine a bounded region of phase space for $-\infty < t < \infty$ (i.e. for the particle to stay in a bounded region of space and take on only a bounded set of velocities) it suffices that in the solution set of the inequality

$$E - U_{\text{eff}}(r) \leq 0$$

there is a maximal finite interval $0 \leq r_{\min} \leq r \leq r_{\max} < \infty$, containing the initial position r_0.

(e) Let $L(x, \dot{x})$, $x = (x^0, x^1, x^2, x^3)$, be a Lagrangian defined on Minkowski space, preserved by the Lorentz group $O(3, 1)$.

By §24.2(14) and Theorem 24.3.1 (with the matrix acting on the right rather than the left), for each fixed skew-symmetric matrix (A^{ki}) the linear vector field

$$X^i(x) = g^0_{kl} x^l A^{ki} = x_k A^{ki}, \tag{40}$$

where $g^0_{kl} = \begin{pmatrix} 1 & & & 0 \\ & -1 & & \\ & & -1 & \\ 0 & & & -1 \end{pmatrix}$ is the Minkowski metric, determines a one-parameter subgroup of $O(3, 1)$. Hence along extremals we have by Theorem 32.1.2 the following conservation law:

$$\tilde{p}_i X^i = \tilde{p}_i x_k A^{ki} = \text{const.}, \tag{41}$$

where $\tilde{p}_i = \partial L/\partial \dot{x}^i$ is the 4-momentum. In view of the skew-symmetry of the matrix (A^{ki}), the conservation law (41) may be rewritten as

$$\tilde{p}_i x_k A^{ki} = \tfrac{1}{2}(\tilde{p}_i x_k - \tilde{p}_k x_i)A^{ki} = \text{const.}$$

Since the matrix (A^{ki}) is otherwise arbitrary we deduce that the following tensor is conserved along the extremals:

$$M_{ik} = x_i p_k - x_k p_i = \text{const.} \tag{42}$$

This skew-symmetric tensor is called the *moment 4-tensor.*

It is illuminating to consider in the context of Example (a) the corresponding tensor M^{ik} (obtained from M_{ik} by raising the indices relative to the Minkowski metric g^0_{ik}). From (17) it follows that the spatial components $M^{\alpha\beta}$, $\alpha, \beta = 1, 2, 3$, of this tensor are given by

$$M^{\alpha\beta} = x^\alpha p^\beta - x^\beta p^\alpha, \tag{43}$$

where the vector $p = (p^1, p^2, p^3)$ is given by (12). Thus these spatial components coincide with the components of the moment 3-vector (or angular momentum-cf. (35))

$$M = [x, p]. \tag{44}$$

It also follows from (17) that M^{01}, M^{02}, M^{03} are the components of the 3-vector $ctp - (E/c)x$, i.e.

$$(M^{01}, M^{02}, M^{03}) = ctp - \frac{E}{c}x. \tag{45}$$

Consider next a system of n relativistic particles with positions $x_1, \ldots, x_n$, $x_i = (x_i^0, x_i^1, x_i^2, x_i^3)$. Suppose that the Lagrangian $L(x_1, \ldots, x_n, \dot{x}_1, \ldots, \dot{x}_n)$ of this system is preserved by the Poincaré group, i.e. the full group of motions of Minkowski space. Then from the above we obtain the following conservation law:

$$\sum_{i=1}^{n} M_i^{kl} = \text{const.},$$

where M_i^{kl} is the moment tensor of the ith particle, while by arguing much as in Example (c), we also have that the total momentum is conserved:

$$\left(\frac{1}{c}\sum_{i=1}^{n} E_i, \sum_{i=1}^{n} p_i\right) = \text{const.}$$

Hence in particular, using (45), we deduce that

$$\sum c^2 t p_i - \sum E_i x_i = \text{const.} \tag{46}$$

Since by the law of conservation of energy the total energy $\sum E_i$ is also conserved, we deduce from (46) that

$$\frac{\sum E_i x_i}{\sum E_i} = t \cdot \frac{\sum c^2 p_i}{\sum E_i} + \text{const.}$$

Hence the point

$$x = \frac{\sum E_i x_i}{\sum E_i} \tag{47}$$

moves with constant velocity v where

$$v = \frac{\sum c^2 p_i}{\sum E_i}. \tag{48}$$

The point x is the relativistic analogue of the centre of mass; if the particles move with speeds small compared with c, then as we have seen (in Example (a)) $E_i \approx m_i c^2$, so that from (47) we obtain

$$x \approx \frac{\sum m_i x_i}{\sum m_i},$$

which is (approximately) the classical formula for the centre of mass. We remark that the relativistic centre of mass is in general not invariant under coordinate changes.

§33. Hamiltonian Formalism

33.1. Legendre's Transformation

Recall that the energy and momentum were defined for a Lagrangian $L(x, \dot{x})$ by

$$E = \dot{x}^\alpha \frac{\partial L}{\partial \dot{x}^\alpha} - L, \qquad p_\alpha = \frac{\partial L}{\partial \dot{x}^\alpha}. \tag{1}$$

33.1.1. Definition. A Lagrangian L is called *non-singular* on a region of values of $(x, \dot{x})$ if

$$\det\left(\frac{\partial^2 L}{\partial \dot{x}^\alpha\, \partial \dot{x}^\beta}\right) \neq 0 \tag{2}$$

on that region, and *strongly non-singular* if the equations $p_\alpha = \partial L(x, \dot{x})/\partial \dot{x}^\alpha$ determine the $\dot{x}^\alpha$ uniquely as smooth functions $v^\alpha(x, p)$ on the region.

The space with co-ordinates (x, p) is called the *phase space* (corresponding to L). For strongly non-singular Lagrangians, the change from co-ordinates (or variables) $(x, \dot{x})$ to the co-ordinates (x, p) is smoothly invertible; this co-ordinate change is called the *Legendre transformation of co-ordinates.* In terms of these new variables the function L cedes its central role to the function $H(x, p) = H = pv - L$, called the *Hamiltonian.* (Here pv denotes $p_i \dot{x}^i$.)

33.1.2. Theorem. *Let $L(x, \dot{x})$ be a strongly non-singular Lagrangian and let $H(x, p)$ denote the corresponding Hamiltonian, $p = \partial L(x, \dot{x})/\partial \dot{x}$, $\dot{x} = v(x, p)$. Then the Euler–Lagrange equation*

$$\frac{\partial L}{\partial x} - \frac{d}{dt}\left(\frac{\partial L}{\partial \dot{x}}\right) = 0 \tag{3}$$

and the equation $\partial L/\partial \dot{x} = p$, are together equivalent to the following equations in phase space ("Hamilton's equations"), in which x and p are regarded as independent variables:

$$\dot{p} = -\frac{\partial H}{\partial x}, \qquad \dot{x} = \frac{\partial H}{\partial p}, \tag{4}$$

Proof. Writing as before $v = \dot{x}$ we have

$$\left.\frac{\partial H}{\partial p}\right|_{x\,\text{fixed}} = \frac{\partial}{\partial p}(pv - L) = v + p\frac{\partial v}{\partial p} - \frac{\partial L}{\partial v}\frac{\partial v}{\partial p} = v,$$

since $p = \partial L/\partial v$. This establishes the second equation in (4).

The first of Hamilton's equations is obtained as follows:

$$-\frac{\partial H}{\partial x} = -\frac{\partial}{\partial x}(pv - L) = -p\frac{\partial v}{\partial x} + \frac{\partial L}{\partial x} + \frac{\partial L}{\partial v}\frac{\partial v}{\partial x} = \frac{\partial L}{\partial x} = \dot{p},$$

where in the last step we have used the Euler–Lagrange equation (3).

For the proof in the opposite direction note first that $L = p\dot{x} - H$ where $\dot{x}$ is given by the second of Hamilton's equations. Then

$$\frac{\partial L}{\partial x} = \dot{x}\frac{\partial p}{\partial x} - \frac{\partial H}{\partial p}\frac{\partial p}{\partial x} - \frac{\partial H}{\partial x} = -\frac{\partial H}{\partial x} = \dot{p},$$

using the first of Hamilton's equations. Finally

$$\frac{\partial L}{\partial \dot{x}} = p + \dot{x}\frac{\partial p}{\partial \dot{x}} - \frac{\partial H}{\partial p}\frac{\partial p}{\partial \dot{x}} = p,$$

completing the proof. □

The action $S = \int L\,dt$ can now be rewritten as

$$S = \int L\,dt = \int [p\dot{x} - H(x, p)]\,dt, \tag{5}$$

where $\dot{x} = \partial H/\partial p$.

If $x = x(t)$ is any curve and if we define $p(t) = \partial L(x, \dot{x})/\partial \dot{x}$, then the "integrability condition"

$$v(t) = \frac{dx}{dt}, \quad \text{where } v = \frac{\partial H}{\partial p}, \tag{6}$$

holds along the curve; this follows from the fact that the second of Hamilton's equations holds independently of the Euler–Lagrange equation.

On the other hand it is clear that for an arbitrary curve $x = x(t)$, $p = p(t)$ in the phase space, the integrability condition $dx/dt = \partial H/\partial p$ need not hold. We now consider the Lagrangian

$$\tilde{L}(x, p; \dot{x}, \dot{p}) = p\dot{x} - H(x, p) \tag{7}$$

defined on phase space.

33.1.3. Lemma. *The Euler–Lagrange equations for the Lagrangian $\tilde{L}$ defined on the $2n$-dimensional space with co-ordinates x, p, are exactly Hamilton's equations* (4), *and therefore imply the integrability condition* (6).

PROOF. Write $\tilde{L}(x, p; \dot{x}, \dot{p}) = \tilde{L}(y, \dot{y})$ where $y = (x, p)$. The Euler–Lagrange equations for $\tilde{L}$ are as follows:

$$\frac{d}{dt}\left(\frac{\partial \tilde{L}}{\partial \dot{y}^\alpha}\right) = \frac{\partial \tilde{L}}{\partial y^\alpha}.$$

Putting first $y^\alpha = x^i$, these become

$$\frac{d}{dt}\left(\frac{\partial \tilde{L}}{\partial \dot{x}^i}\right) = \frac{\partial \tilde{L}}{\partial x^i}, \quad \text{or} \quad \dot{p} = -\frac{\partial H}{\partial x}.$$

Putting $y^\alpha = p_i$, they become

$$\frac{d}{dt}\left(\frac{\partial \tilde{L}}{\partial \dot{p}_i}\right) = \frac{\partial \tilde{L}}{\partial p_i}, \quad \text{or} \quad 0 = \dot{x}^i - \frac{\partial H}{\partial p_i}.$$

□

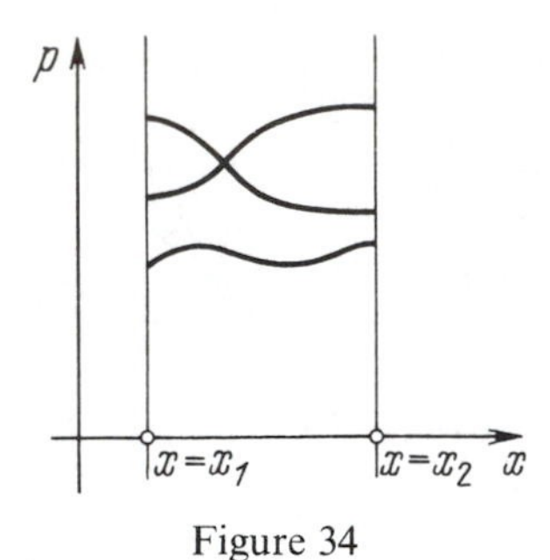

Figure 34

Remark. At the very beginning of this chapter we derived the Euler–Lagrange equations as necessary conditions for an arc to be extremal among all arcs with common end-points P, Q (i.e. in that "variational class"). The corresponding variational class of arcs in the phase space consists of arcs with only the x-co-ordinates of their end-points necessarily in common (see Figure 34).

33.2. Moving Co-ordinate Frames

We now consider the more general situation of a Lagrangian L in $\mathbb{R}^n$ which varies with time: $L = L(x, \dot{x}, t)$. We wish to discover how various entities (in particular energy and momentum) transform under co-ordinate changes of the form $x = x(x', t')$, $t = t'$.

It is clear that as our new Lagrangian we may take any function $\tilde{L}$ of the form (cf. the footnote to §31.1)

$$\tilde{L}(x', \dot{x}', t) = L(x(x', t), \dot{x}, t) + \frac{d}{dt} f(x, t);$$

the extremals (joining a given pair of points) corresponding to this Lagrangian, i.e. extremizing the action $\int \tilde{L}\, dt$, will then be as prior to the change.

We first consider the case that the co-ordinate change does not involve t, i.e. $x = x(x')$. Then, as we have seen several times before, the velocity $\dot{x}(t)$ is, for any curve $x(t)$, a vector; i.e. its components transform according to the rule

$$\dot{x}^i = \frac{\partial x^i}{\partial x^{i'}} \dot{x}^{i'}, \quad \text{or} \quad v^i = \frac{\partial x^i}{\partial x^{i'}} v^{i'}. \tag{8}$$

The transformation rule for the components of the momentum is derived as follows: Taking

$$\tilde{L}(x', \dot{x}', t) = L\left(x(x'), \frac{\partial x}{\partial x'} v', t\right),$$

we have

$$p_{i'} = \frac{\partial \tilde{L}}{\partial v^{i'}} = \frac{\partial L}{\partial v^i} \frac{\partial x^i}{\partial x^{i'}} = p_i \frac{\partial x^i}{\partial x^{i'}}, \tag{9}$$

so that (as already noted in the definition of momentum in §31.1) the momentum is a covector. It is then immediate that the energy $E = p_i v^i - L$ is a scalar, i.e. is unaltered:

$$E = p_i v^i - L = p_{i'} v^{i'} - \tilde{L} = E'.$$

We now turn to the consideration of co-ordinate changes $x = x(x', t)$, $t = t'$ which do involve the time t. (The reader may find the following easier to understand if he visualizes the frame x as fixed, and the primed frame x' as moving.) In this more general situation we have

$$\dot{x}^i = \frac{\partial x^i}{\partial x^{i'}} \dot{x}^{i'} + \frac{\partial x^i}{\partial t} = \frac{\partial x^i}{\partial x^{i'}} \dot{x}^{i'} + a^i(x', t). \tag{10}$$

Since we have already dealt with the case $x = x(x')$ we might as well suppose that at the particular instant $t = t_0$ of interest the frame x coincides with the frame x', i.e. $x = x(x', t_0) = x'$. We shall call this frame *instantaneous*. Thus at the instant $t = t_0$ we have from (10) that

$$v^i = v^{i'} + \frac{\partial x^i}{\partial t} = v^{i'} + a^i(x', t_0) = v^{i'} + a^i(x, t_0), \qquad t = t_0.$$

Hence as a result of changing to the moving frame we have (at the instant of interest $t = t_0$):

$$\begin{aligned} &L \to L' = L, \qquad L(x, v, t) = L(x', v' + a, t); \\ &v \to v' = v - a(x', t); \\ &p \to p' = p, \quad \text{since} \quad \frac{\partial L}{\partial v} = \frac{\partial L'}{\partial v'}; \\ &E \to E' = p'v' - L' = p_i(v^i - a^i) - L = E - p_i a^i(x', t). \end{aligned} \tag{11}$$

Thus the momentum is not changed, but the energy is increased by the amount $-p_i a^i = \langle p, v' - v \rangle$. The Hamiltonian therefore is transformed as follows (at $t = t_0$):

$$E = H(x, p) \to E' = H(x', p') - p'_i a^i(x', t) = H'(x', p', t), \tag{12}$$

where we have used the assumption that at $t = t_0$ we have $x = x'$ (whence also $p = p'$), and where $a = \partial x/\partial t|_{t=t_0}$. Finally we compare Hamilton's equations in the two frames (at $t = t_0$). In the frame x the equations are $\dot{x} = \partial H/\partial p$, $\dot{p} = -\partial H/\partial x$, and in the moving frame x', they are $\dot{x}' = \partial H'/\partial p'$, $\dot{p}' = -\partial H'/\partial x'$. Since at $t = t_0$ we have $x = x'$, $p = p'$, we obtain from (12) that

$$\begin{aligned} \dot{x}' &= \frac{\partial H'}{\partial p'} = \frac{\partial H}{\partial p} - a = \dot{x} - a, \\ \dot{p}' &= -\frac{\partial H'}{\partial x'} = -\frac{\partial H}{\partial x} + p'_i \frac{\partial a^i}{\partial x'} = \dot{p} + p_i \frac{\partial a^i}{\partial x'}. \end{aligned} \tag{13}$$

Thus the relationship between the two pairs of equations is a reasonably simple one.

33.2.1. Theorem. *Under the change to an instantaneous moving co-ordinate frame, the co-ordinates (by definition of such a frame) and the momentum are not altered* ($x' = x$, $p' = p$), *while the energy (i.e. Hamiltonian) is increased by the amount* $-\langle p, a\rangle$:

$$H \to H' = H - \langle p, a\rangle, \qquad a = a(x', t) = \left.\frac{\partial x}{\partial t}\right|_{t=t_0}.$$

Remark. In the special case (of a classical particle) where the change is to a uniformly moving frame, i.e. where $a = \text{const.}$, L' is always taken to be $m(v')^2/2$ $(=L + (d/dt)f(x, t))$, whence $p' = p + ma$ and $H' = H - pa + ma^2/2$. In this as in the more general situation the equations (11) are often regarded as holding only to within a constant (cf. Example (a) below).

We now consider three important examples.

33.2.2. Examples. (a) As our first example we take the Lagrangian to be $L = m\dot{x}^2/2 - U(x)$, and a moving frame which moves "translationally", i.e. one for which $x = x(x', t)$ has the form

$$x(x', t) = x' + b(t),$$

so that $a = a(t) = \partial x/\partial t$ is independent of x'. In this case we have

$$p = mv = m(v' + a) = mv' + ma = p'(+\ \text{const.}),$$

$$H' = H - \langle p', a\rangle\ (+\ \text{const.}).$$

The Euler–Lagrange equation for the extremals (i.e. in this case Newton's equation) takes the form

$$\frac{d}{dt}(mv) = f = m\dot{v}' + m\dot{a},$$

whence

$$m\dot{v}' = f - m\dot{a} = f'.$$

Thus from the point of view of the moving frame a classical particle of mass m will experience an additional (inertial) force equal to $-m\dot{a}$.

(b) Consider a Euclidean frame x' in Euclidean $\mathbb{R}^3$ rotating with constant angular velocity Ω relative to a frame x. Suppose that the x^3-axis and $x^{3'}$-axis coincide with the axis of rotation and that at $t = 0$ the frames coincide; then

$$x = \begin{pmatrix} x^1 \\ x^2 \\ x^3 \end{pmatrix} = \begin{pmatrix} \cos|\Omega|t & -\sin|\Omega|t & 0 \\ \sin|\Omega|t & \cos|\Omega|t & 0 \\ 0 & 0 & 1 \end{pmatrix} \begin{pmatrix} x^{1'} \\ x^{2'} \\ x^{3'} \end{pmatrix}, \tag{14}$$

whence it follows easily that $a = \partial x/\partial t = [\Omega, x]$. From Theorem 33.2.1 we then obtain (at $t = 0$)

$$H' = H - \langle p', a\rangle = H - \langle p', [\Omega, x]\rangle,$$
$$p' = p, \quad x' = x, \quad t = 0,$$

so that at $t = 0$, Hamilton's equations take the form (see (13))

$$\dot{p}' = -\frac{\partial H'}{\partial x'} = -\frac{\partial H}{\partial x} + \frac{\partial}{\partial x}\langle p', [\Omega, x]\rangle = -\frac{\partial H}{\partial x} + [p', \Omega],$$
$$\dot{x}' = \frac{\partial H'}{\partial p'} = \frac{\partial H}{\partial p} - \frac{\partial}{\partial p'}\langle p', [\Omega, x]\rangle = \frac{\partial H}{\partial p} - [\Omega, x] = \dot{x} - a. \qquad (15)$$

If we take the Lagrangian to be $L = m\dot{x}^2/2 - U(x)$ (where as usual $\dot{x}^2$ is brief notation for $|\dot{x}|^2$) then $p = m\dot{x}$. Hence by (10)

$$p = m\left(\frac{\partial x}{\partial x^{i'}}\dot{x}^{i'} + a\right),$$

so that

$$\dot{p} = m\left(\frac{d}{dt}\left(\frac{\partial x}{\partial x^{i'}}\right)\dot{x}^{i'} + \frac{\partial x}{\partial x^{i'}}\ddot{x}^{i'} + \frac{d}{dt}[\Omega, x]\right). \qquad (16)$$

From (14) we have that

$$\frac{\partial x}{\partial x^{1'}} = (\cos|\Omega|t, \sin|\Omega|t, 0), \qquad \frac{\partial x}{\partial x^{2'}} = (-\sin|\Omega|t, \cos|\Omega|t, 0),$$
$$\frac{\partial x}{\partial x^{3'}} = (0, 0, 1).$$

From this and (16) it follows readily that at time $t = 0$

$$\dot{p} = m[\Omega, \dot{x}'] + m\ddot{x}' + m[\Omega, \dot{x}], \qquad t = 0,$$

or, putting $\dot{x} = \dot{x}' + a$ (see (15)),

$$m\ddot{x}' = 2m[\dot{x}', \Omega] + f + m[[\Omega, x], \Omega], \qquad t = 0, \qquad (17)$$

where f is the force relative to the fixed frame. The last term on the right-hand side of (17) represents a small force provided $|\Omega|$ is small and $|x|$ is not large. The force $2m[\dot{x}', \Omega]$ is the well-known "Coriolis force". With these conditions we have

$$m\dot{v}' = 2m[v', \Omega] + f + O(\Omega^2). \qquad (18)$$

Remark. If $L = m\dot{x}^2/2 - U(x)$, then under the change to an (instantaneous) moving co-ordinate frame, we have (see the previous remark)

$$H \to H' = H - p_i a^i (+\text{ const.}), \qquad p_i = mv^i = m(v^{i'} + a^i) = p'_i(+\text{ const.}),$$

where

$$H = p_i v^i - L = \frac{p^2}{2m} + U(x),$$

$$H' = \frac{(p')^2}{2m} + U(x) - p_i a^i = \frac{(p' - ma)^2}{2m} + U(x) - \frac{ma^2}{2}.$$

Thus the change to the instantaneous co-ordinate frame is equivalent to two (classically intuitive) operations:

(i) a change in momentum $p \to p - ma\ (= p' - ma)$;
(ii) a change in potential $U \to U_{\text{eff}} = U(x) - ma^2/2$.

(c) *Inclusion of an electromagnetic field.* Given a Lagrangian $L(x, \dot{x})$, we define a new Lagrangian $\tilde{L}$ by the formula

$$\tilde{L} = L + \frac{e}{c} A_i \dot{x}^i,$$

where A_i is the vector-potential of the electromagnetic field, and e is the amount of charge (for instance on a (relativistic) particle of mass m). For the respective actions we write as usual

$$S = \int L\, dt, \qquad \tilde{S} = \int \tilde{L}\, dt = \int \left(L\, dt + \frac{e}{c} A^i\, dx^i \right).$$

This supplementing of the Lagrangian L by the term $(e/c)A_i\dot{x}^i$, is described as "inclusion of the electromagnetic field".

Writing as usual $H(x, p) = pv - L$ and $\tilde{H}(x, \tilde{p}) = \tilde{p}v - \tilde{L}$, where $p = \partial L/\partial v$, $\tilde{p} = \partial \tilde{L}/\partial v$ (and where, as before, pv, etc. is shorthand for $\langle p, v \rangle = p_i v^i$, etc.), we have

$$\tilde{p}_i = p_i + \frac{e}{c} A_i, \qquad p_i = \tilde{p}_i - \frac{e}{c} A_i,$$
$$\tilde{H}(x, \tilde{p}) = H\left(x, \tilde{p} - \frac{e}{c} A \right). \tag{19}$$

Hence inclusion of the electromagnetic field is equivalent to replacement of p in the original Hamiltonian $H(x, p)$, by $\tilde{p} - (e/c)A$, and so in this respect resembles the change to a moving system of co-ordinates (see (i) in the remark immediately preceding the present example).

The Euler–Lagrange equations for the Hamiltonian $\tilde{L}$ are

$$\dot{\tilde{p}}_i = \frac{\partial \tilde{L}}{\partial x^i} = \frac{\partial L}{\partial x^i} + \frac{e}{c} \frac{\partial A_j}{\partial x^i} \dot{x}^j. \tag{20}$$

From the first equation in (19) we have $\dot{\tilde{p}}_i = \dot{p}_i + (e/c)(\partial A_i/\partial x^j)\dot{x}^j$. Substituting from this in (20), we arrive at the equations

$$\dot{p}_i = f_i = \frac{\partial L}{\partial x^i} \quad \text{(in the absence of the electromagnetic field);}$$

$$\dot{p}_i = f_i + \frac{e}{c} F_{ij}\dot{x}^j \quad \text{(in the presence of the field).}$$

The tensor $F_{ij} = \partial A_j/\partial x^i - \partial A_i/\partial x^j$ is the *electromagnetic field tensor* (see §21.1). In view of the skew-symmetry of this tensor the scalar $F_{ij}\dot{x}^i\dot{x}^j$ is identically zero, whence

$$\dot{p}_i \dot{x}^i = \dot{p}_i v^i = f_i v^i, \tag{21}$$

where f is the force in the absence of the field, i.e. $f = \partial L/\partial x$.

Remark. In 3-dimensional formalism, the component $A_0 = c\varphi$, where φ is the familiar electric potential, and (A_1, A_2, A_3) is the magnetic vector potential. In this notation the term added to the Lagrangian L takes the form $e\varphi + (e/c)A_\alpha \dot{x}^\alpha$, where $x^0 = ct$, $\alpha = 1, 2, 3$.

33.3. The Principles of Maupertuis and Fermat

Before stating Maupertuis' principle we rephrase the law of conservation of energy in terms of the Hamiltonian: Along any trajectory $x(t)$, $p(t)$ in $2n$-dimensional phase space (i.e. any solution of Hamilton's equations) the Hamiltonian $H(x(t), p(t))$ is constant. (This constant is the "energy level" for the extremal or trajectory.)

33.3.1. Theorem (Maupertuis' Principle). *Let $H(x, p)$ be a time-independent Hamiltonian. Then any arc $x(t)$, $p(t)$ extremizing $S = \int L\,dt = \int (p\dot{x} - H)\,dt$, with a particular energy level E, will also extremize, over the class of arcs with the same constant energy level E, the "truncated" action $S_0 = \int p\dot{x}\,dt = \int p\,dx$.*

PROOF. If a curve $(x(t), p(t))$ in phase space, of energy level $H(x(t), p(t)) =$ const. $= E$, extremizes

$$S = \int L\,dt = \int (p\dot{x} - H)\,dt,$$

then it certainly extremizes $\int (p\dot{x} - E)\,dt$ over the narrower class of all arcs with the same constant energy level E, and therefore, since $\int E\,dt$ is constant, also extremizes $S_0 = \int p\dot{x}\,dt$ over that class of arcs. This completes the proof. □

We now consider two important examples.

(a) If $L = m\dot{x}^2/2 - U(x)$ (so that $H = p^2/2m + U(x)$), then $p = m\dot{x}$, whence for a fixed energy level E we have

$$|p| = \sqrt{2m(E - U(x))}.$$

By Maupertuis' principle, any extremal (with energy level E) corresponding to the Lagrangian L, will (among all curves with energy level E) be an extremal also for the action

$$S_0 = \int \langle p, \dot{x} \rangle \, dt = \int |p||\dot{x}| \, dt = \int |\dot{x}|\sqrt{2m(E - U(x))} \, dt = \int \sqrt{g_{ij}\dot{x}^i\dot{x}^j} \, dt,$$

where

$$g_{ij} = 2m(E - U(x))\delta_{ij},$$

(assuming that the co-ordinates x are Euclidean). We have thus proved the following result.

33.3.2. Theorem. *Curves $x(t)$ in Euclidean space which are extremals corresponding to the Hamiltonian $H = p^2/2m + U(x)$, and have fixed energy level E, are geodesics (with non-natural parameter) with respect to the new metric*

$$g_{ij} = 2m(E - U(x))\delta_{ij}. \tag{22}$$

(b) The appropriate Hamiltonian for the trajectories of light rays in a continuous, isotropic medium (i.e. a medium in which the (variable) velocity of light $c(x)$ at each point x is direction-independent) in Euclidean space, turns out to be $H = c(x)|p|$. (This choice of Hamiltonian is justified by its consequences, e.g. by Fermat's principle below.)

If the energy level E is fixed, then by Maupertuis' principle any extremal (with that energy level) will also be an extremal (among all curves with that energy level) for the truncated action $S_0 = \int \langle p, \dot{x} \rangle \, dt$. Now along such an extremal we have $H(x, p) = E$, whence

$$|p| = \frac{E}{c(x)}, \tag{23}$$

and we also have

$$\dot{x} = \frac{\partial H}{\partial p} = c(x)\frac{p}{|p|} \quad \text{(whence } |\dot{x}| = c(x)\text{)}. \tag{24}$$

(Note that we are assuming $E \neq 0$, so that $p \neq 0$.) From (23) and (24) it follows that

$$\langle p, \dot{x} \rangle = |p||\dot{x}| = \frac{E}{c(x)}|\dot{x}|.$$

Hence along an extremal arc γ of energy level E, we have

$$S_0 = \int_\gamma \langle p, \dot{x} \rangle \, dt = \int_\gamma |\dot{x}| \frac{E}{c(x)} \, dt = E \int_\gamma \frac{|\dot{x}|}{c(x)} \, dt = E \int_\gamma \sqrt{g_{ij} \dot{x}^i \dot{x}^j} \, dt,$$

where $g_{ij} = [1/c^2(x)]\delta_{ij}$. Since $\int_\gamma [|\dot{x}|/c(x)] \, dt = \int_\gamma dt$ is just the time it takes for the light to traverse the arc γ, we deduce the following celebrated result (which in its original formulation stated that the path taken by a light ray between two points is always shortest).

33.3.3. Theorem (Fermat's Principle). *In an isotropic medium the paths taken by light rays in passing from a point P to a point Q are extremals corresponding to the traversal-time (as action). Such paths are geodesics with respect to the new metric*

$$g_{ij} = \frac{1}{c^2(x)} \delta_{ij}. \tag{25}$$

Remark. As we have seen (in §31.2), the geodesics determined by a metric g_{ij} are the extremals corresponding to the particular Lagrangian $L = g_{ij} v^i v^j$ (or, equivalently, to the Hamiltonian $g^{ij} p_i p_j$ regarded, as in Lemma 33.1.3, as a Lagrangian defined on the phase space; to see this equivalence note that $p_r = \partial L / \partial v^r = g_{rj} v^j$, whence $g^{rs} p_r p_s = g^{rs}(g_{sj} v^j)(g_{ir} v^i)$, etc.). In fact the extremals (in phase space) corresponding to the Hamiltonian $H'(x, p) = \sqrt{H} = \sqrt{g^{ij} p_i p_j}$, coincide with the extremals with respect to H. This is a consequence of the fact that at a particular (constant) energy level E, the respective vector fields (on the phase space) determined by H and H' are proportional (with the constant proportionality factor $1/2\sqrt{E}$). To see this, observe that the vector field determined by H in phase space is given by Hamilton's equations

$$\dot{p} = -\frac{\partial H}{\partial x}, \qquad \dot{x} = \frac{\partial H}{\partial p},$$

while the corresponding equations for $H' = \sqrt{H}$ are

$$\dot{p} = -\frac{\partial \sqrt{H}}{\partial x} = \frac{-1}{2\sqrt{E}} \frac{\partial H}{\partial x}, \qquad \dot{x} = \frac{1}{2\sqrt{E}} \frac{\partial H}{\partial p}.$$

Applying this argument to the latter example, with $H' = c(x)|p| = \sqrt{c^2(x) p^2}$, we can deduce immediately that the metric $g^{ij} = c^2(x)\delta^{ij}$ (i.e. $g_{ij} = [1/c^2(x)]\delta_{ij}$) will have the property described in Theorem 33.3.3.

We mention in conclusion that in an anisotropic medium the analogous tensor g_{ij} (or g^{ij}), which determines the velocity of light in the medium, will no longer be conformally Euclidean.

33.4. Exercises

1. Consider the differential equation

$$\dot{u} = \frac{\partial}{\partial x}\frac{\delta S}{\delta u},$$

on the space of functions $u(x)$ periodic with period T, and satisfying

$$\int_{x_0}^{x_0+T} u(x)\,dx = 0,$$

where $S = S[u]$ is a functional of the form

$$S[u] = \int_{x_0}^{x_0+T} L(u, u', \ldots)\,dx.$$

Transform the differential equation into standard Hamiltonian form. (Hint. Work with the Fourier coefficients u_n of the function u, determined by $u = \sum_{n=-\infty}^{\infty} u_n e^{(2\pi i n x)/T}$.)

2. Suppose we have a Lagrangian L depending on derivatives higher than the first: $L = L(u, u', \ldots, u^{(n)})$, $S[u] = \int L\,dx$. Then the Euler–Lagrange equation $\delta S/\delta u = 0$ for the extremals of the functional S can be rewritten in Hamiltonian form as

$$q_i' = \frac{\partial H}{\partial p_i}, \qquad p_i' = -\frac{\partial H}{\partial q_i}, \qquad i = 1, \ldots, n,$$

where

$$q_i = u^{(i-1)}, \qquad p_i = \frac{\partial L}{\partial u^{(i)}} + \sum_{s=1}^{n-i}(-1)^s \frac{d^s}{dx^s}\frac{\partial L}{\partial u^{(i+s)}}, \qquad 1 \le i \le n,$$

provided the equation $p_n = \partial L/\partial q_n'$ can be uniquely solved for $q_n' = u^{(n)}$.

§34. The Geometrical Theory of Phase Space

34.1. Gradient Systems

Let $f = f(y^1, \ldots, y^m)$ be a function defined on a space $\mathbb{R}^m$ with co-ordinates $y^1, \ldots, y^m$, endowed with a (no longer necessarily symmetric) metric g_{ij}. We define the *gradient* (or *vector gradient*) ∇f of f relative to the metric g_{ij} by

$$(\nabla f)^i = g^{ij}\frac{\partial f}{\partial y^j}. \tag{1}$$

Thus the vector gradient is obtained from the usual (covector) gradient by means of the tensor operation of raising the index (see §19.1).

As in §23.1 we associate with the vector field ∇f the system of autonomous differential equations

$$\dot{y}^i = (\nabla f)^i, \tag{2}$$

which in this case we call a *gradient system.*

34.1.1. Lemma. *Let $y = y(t)$ be an integral curve of the gradient system (2), and let $h = h(y)$ be any function. Then the derivative of h with respect to t (or "relative to the gradient system") is given by*

$$\dot{h} = \frac{d}{dt} h(y(t)) = \langle \nabla h, \nabla f \rangle = g^{ij} \frac{\partial h}{\partial y^i} \frac{\partial f}{\partial y^j}. \tag{3}$$

PROOF. We have

$$\dot{h} = \frac{\partial h}{\partial y^i} \dot{y}^i = \frac{\partial h}{\partial y^i} (\nabla f)^i = \frac{\partial h}{\partial y^i} g^{ij} \frac{\partial f}{\partial y^j},$$

whence the lemma. □

In what follows we shall be mainly concerned with the case that g_{ij} is a skew-symmetric "metric"; we can then write g_{ij} as a skew form

$$\Omega = g_{ij}\, dy^i \wedge dy^j \qquad (i < j), \tag{4}$$

where $g_{ij} = -g_{ij}$, $\det(g_{ij}) = g \neq 0$. This last condition, that g_{ij} be non-singular, forces on us the requirement that m be even, say $m = 2n$ (since a skew-symmetric matrix of odd degree has zero determinant.) It follows also that $g > 0$.

34.1.2. Lemma. *The following formula holds:*

$$\frac{-1}{n!} \underbrace{\Omega \wedge \cdots \wedge \Omega}_{n \text{ times}} = \sqrt{g}\, dy^1 \wedge \cdots \wedge dy^{2n}. \tag{5}$$

It follows that $\sqrt{g}$ is a polynomial in the g_{ij} (called the "Pfaffian").

PROOF. It suffices to prove that (5) holds at each particular point P (at which (g_{ij}) is non-singular). At a particular point P, the tensor g_{ij} is a skew-symmetric (or "alternating") form on the tangent space at P. Just as a symmetric form is diagonal relative to a suitable basis of the tangent space, so the skew-symmetric form $(g_{ij}(P))$ takes the form $\begin{pmatrix} O_n & I_n \\ -I_n & O_n \end{pmatrix}$ relative to a suitable basis of the tangent space, where I_n is the $n \times n$ identity matrix and O_n the $n \times n$ zero matrix. Hence we can choose new co-ordinates

$$(x^1, \ldots, x^n, p_1, \ldots, p_n) = (z^1, \ldots, z^{2n}), \tag{6}$$

which are linearly related to the old, i.e. $y = Az$, where $A = A_P$, such that at the point P

$$(\tilde{g}_{ij}(P)) = A^{\mathrm{T}}(g_{ij}(P))A = \begin{pmatrix} O_n & I_n \\ -I_n & O_n \end{pmatrix}.$$

Thus in terms of the new co-ordinates (6), we have at the point P

$$\Omega = \sum_i dx^i \wedge dp_i, \tag{7}$$

whence by definition of the exterior product (see §18.3)

$$\Omega \wedge \cdots \wedge \Omega = -n!\, dz^1 \wedge \cdots \wedge dz^{2n},$$

at P. Since $\sqrt{\tilde{g}} = 1$, this establishes (5) at the point P in the system of co-ordinates z. However by Corollary 18.2.4 (which did not depend on the symmetry of g_{ij}) the expression on the right-hand side of (5) transforms like a tensor with respect to transformations with positive Jacobian. Hence if A has positive determinant, then since the left-hand side of (5) is a tensor, it follows that (5) holds also in terms of our initial co-ordinates $y^1, \ldots, y^{2n}$. This completes the proof in the case where $\det A > 0$; in the case $\det A < 0$, a modification of the proof yields the desired conclusion. □

Remark. Note the consequence of this lemma that the non-singularity condition $g \neq 0$ is equivalent to the condition $\Omega^n \neq 0$.

34.1.3. Definition. Let g_{ij} be a skew-symmetric metric defined on a $2n$-dimensional space. If there are co-ordinates x, p for the space in terms of which

$$(g_{ij}) = \begin{pmatrix} O_n & I_n \\ -I_n & O_n \end{pmatrix},$$

then we call the space with its metric g_{ij} an (*abstract*) *phase space*, and the special co-ordinates we term *canonical*. A gradient system (2) defined by such a metric is said to be *Hamiltonian*.

Writing $H(x, p)$ instead of $f(x, p)$ for the initially given function, we see that in terms of the canonical co-ordinates x, p for a skew-symmetric metric on a phase space, the Hamiltonian system (2) takes the form

$$\dot{y}^i = (\nabla H)^i, \qquad i = 1, \ldots, 2n,$$

or, more explicitly,

$$\dot{x}^i = \frac{\partial H}{\partial p_i}, \qquad \dot{p}_i = -\frac{\partial H}{\partial x^i}, \qquad i = 1, \ldots, n, \tag{8}$$

which the reader will recognize as having the same form as Hamilton's equations. This is of course no coincidence! Hamiltonian systems derive their importance from the equivalence of Hamilton's equations in phase space with the Euler–Lagrange equations (at least in the case of a strongly nonsingular Langrangian—see Theorem 33.1.2).

The following result is immediate from Lemma 34.1.1.

34.1.4. Lemma. *The derivative of an arbitrary function $f(x, p, t)$ relative to the Hamiltonian system* (8) *is given by*

$$\dot{f} = \frac{\partial f}{\partial t} + \langle \nabla f, \nabla H \rangle, \tag{9}$$

where

$$\langle \nabla f, \nabla H \rangle = g^{ij} \frac{\partial f}{\partial y^i} \frac{\partial H}{\partial y^j} = \sum_{i=1}^{n} \left(\frac{\partial f}{\partial x^i} \frac{\partial H}{\partial p_i} - \frac{\partial H}{\partial x^i} \frac{\partial f}{\partial p_i} \right). \tag{10}$$

In particular taking $f = H = H(x, p, t)$, we have

$$\dot{E} = \dot{H} = \frac{\partial H}{\partial t},$$

since $\langle \nabla H, \nabla H \rangle = -\langle \nabla H, \nabla H \rangle = 0$.

We now consider functions (Hamiltonians) $H = H(x, p, t)$ possibly depending explicitly on the time t. The corresponding Hamiltonian system is

$$\dot{x} = \frac{\partial H}{\partial p}, \qquad \dot{p} = -\frac{\partial H}{\partial x}.$$

These equations taken in conjunction with the equation $\dot{E} = \partial H / \partial t$ which they imply (see the preceding lemma), and the trivial equation $\dot{t} = 1$, conduce to the introduction of an *extended phase space* with co-ordinates

$$(x^1, p_1, x^2, p_2, \ldots, x^n, p_n, x^{n+1}, p_{n+1}), \qquad x^{n+1} = t, \qquad p_{n+1} = E,$$

where the metric is (naturally, as we shall see) given by

$$\hat{g}_{ij} = \begin{pmatrix} 0 & 1 & & & & & & 0 \\ -1 & 0 & & & & & & \\ & & \ddots & & & & & \\ & & & 0 & 1 & & & \\ & & & -1 & 0 & & & \\ & & & & & & 0 & -1 \\ 0 & & & & & & 1 & 0 \end{pmatrix}, \tag{11}$$

or, expressed as a form, by

$$\hat{\Omega} = \sum_{i=1}^{n} dx^i \wedge dp_i - dt \wedge dE. \tag{12}$$

If we define a new Hamiltonian $\tilde{H} = \tilde{H}(x, p, t, E)$ on this extended phase space by

$$\tilde{H}(x, p, t, E) = H(x, p, t) - E,$$

then the Hamiltonian system corresponding to this new Hamiltonian is

$$\dot{x} = \frac{\partial \tilde{H}}{\partial p} = \frac{\partial H}{\partial p}, \qquad \dot{p} = -\frac{\partial \tilde{H}}{\partial x} = -\frac{\partial H}{\partial x},$$

$$\dot{t} = -\frac{\partial \tilde{H}}{\partial E} = 1, \qquad \dot{E} = \frac{\partial \tilde{H}}{\partial t} = \frac{\partial H}{\partial t}, \tag{13}$$

whence we have the following

34.1.5. Corollary. *Let g_{ij} be a metric on $\mathbb{R}^{2n}$ with canonical co-ordinates x, p, and let $H(x, p, t)$ be any function (Hamiltonian). Then the Hamiltonian system corresponding to the new Hamiltonian $\tilde{H} = \tilde{H}(x, p, t, E) = H(x, p, t) - E$ defined on the extended phase space with co-ordinates x, p, t, E with the metric $\hat{g}_{ij}$ given by* (11), *is equivalent to the original Hamiltonian system supplemented by the equations $\dot{t} = 1$, $\dot{E} = \partial H/\partial t$.*

Note finally that if the original Hamiltonian H does not depend explicitly on the time t, i.e. if $\partial H/\partial t \equiv 0$, then $t = x^{n+1}$ is a cyclic co-ordinate (see §31.2) for the Hamiltonian $\tilde{H}$, and consequently the corresponding component of the momentum, namely $p_{n+1} = E$, is conserved, i.e. we find ourselves back again at the law of conservation of energy along extremals.

34.2. The Poisson Bracket

Let g_{ij} be a skew-symmetric metric on $2n$-dimensional phase space with (canonical) co-ordinates x, p.

34.2.1. Definition. The *Poisson bracket* (or *Poisson commutator*) of two functions $f(x, p)$, $g(x, p)$ defined on $2n$-dimensional phase space with metric $g_{ij} = \begin{pmatrix} O_n & I_n \\ -I_n & O_n \end{pmatrix}$, is defined to be the scalar product of their gradients:

$$\{f, g\} = \langle \nabla f, \nabla g \rangle = g^{ij} \frac{\partial f}{\partial y^i} \frac{\partial g}{\partial y^j} = \sum_{i=1}^{n} \frac{\partial f}{\partial x^i} \frac{\partial g}{\partial p_i} - \frac{\partial g}{\partial x^i} \frac{\partial f}{\partial p_i}, \tag{14}$$

where $(\nabla f)^i = g^{ij}(\partial f/\partial y^j)$, etc.

34.2.2. Theorem. *The Poisson bracket has the following properties:*

(i) $\{f, g\} = -\{g, f\}, \{\lambda f_1 + \mu f_2, g\} = \lambda\{f_1, g\} + \mu\{f_2, g\},$

where λ, μ *are arbitrary constants;*

(ii) $\{f, \{g, h\}\} + \{h, \{f, g\}\} + \{g, \{h, f\}\} = 0$ (*Jacobi's identity*); (15)

(iii) $\{fg, h\} = f\{g, h\} + g\{f, h\};$ (16)

(iv) $\nabla\{f, g\} = -[\nabla f, \nabla g],$

where [,] *is the operation of taking the commutator of vector fields.*

PROOF. Since properties (i) and (iii) are almost immediate, we need to prove only (ii) and (iv).

Consider the map $f \mapsto \nabla f$, from the set of all (smooth) functions on the phase space, to the set of vector fields. Property (iv) and the easy fact that this map is linear, together imply that it is an algebra homomorphism from the algebra of functions f under the Poisson bracket operation, to the Lie algebra of vector fields with the usual bracket operation. It is in fact a monomorphism from the algebra of equivalence classes of functions with the same gradient (and so differing by a constant) to the Lie algebra of vector fields. Since the latter algebra satisfies Jacobi's identity, so does the former. Hence (ii) will hold if (iv) does, so that we need verify only (iv).

To this end we compute the commutator of the vector fields ∇f, ∇g, which in terms of the co-ordinates x, p are given by

$$\nabla f = \left(\frac{\partial f}{\partial p}, -\frac{\partial f}{\partial x}\right), \qquad \nabla g = \left(\frac{\partial g}{\partial p}, -\frac{\partial g}{\partial x}\right).$$

Recall from §23.2 that the commutator of vector fields X and Y is in terms of co-ordinates (y^j) given by the formula

$$[X, Y]^i = X^j \frac{\partial Y^i}{\partial y^j} - Y^j \frac{\partial X^i}{\partial y^j}.$$

Using this formula to calculate the x^1-co-ordinate of $[\nabla f, \nabla g]$, we obtain

$$\begin{aligned}
[\nabla f, \nabla g]^1 &= \frac{\partial f}{\partial p_j}\frac{\partial}{\partial x^j}\left(\frac{\partial g}{\partial p_1}\right) - \frac{\partial f}{\partial x^j}\frac{\partial}{\partial p_j}\left(\frac{\partial g}{\partial p_1}\right) \\
&\quad - \frac{\partial g}{\partial p_j}\frac{\partial}{\partial x^j}\left(\frac{\partial f}{\partial p_1}\right) + \frac{\partial g}{\partial x^j}\frac{\partial}{\partial p_j}\left(\frac{\partial f}{\partial p_1}\right) \\
&= \frac{\partial f}{\partial p_j}\frac{\partial}{\partial p_1}\left(\frac{\partial g}{\partial x^j}\right) + \frac{\partial g}{\partial x^j}\frac{\partial}{\partial p_1}\left(\frac{\partial f}{\partial p_j}\right) - \frac{\partial f}{\partial x^j}\frac{\partial}{\partial p_1}\left(\frac{\partial g}{\partial p_j}\right) - \frac{\partial g}{\partial p_j}\frac{\partial}{\partial p_1}\left(\frac{\partial f}{\partial x^j}\right) \\
&= \frac{\partial}{\partial p_1}\left(\frac{\partial f}{\partial p_j}\frac{\partial g}{\partial x^j} - \frac{\partial f}{\partial x^j}\frac{\partial g}{\partial p_j}\right) = (-\nabla\{f, g\})^1,
\end{aligned}$$

where we have at several points used the interchangeability of the order of differentiation. For the remaining x^i-co-ordinates and the p_i-co-ordinates of $[\nabla f, \nabla g]$ the verification is similar. □

34.2.3. Corollary. *The set of smooth functions $f(x, p)$ defined on phase space forms a Lie algebra with respect to the Poisson bracket.*

(As noted at the beginning of the above proof, this Lie algebra is "essentially" isomorphic to the Lie algebra of gradient vector fields determined by the skew-symmetric metric $g_{ij} = \begin{pmatrix} O_n & I_n \\ -I_n & O_n \end{pmatrix}$.)

It turns out that the converse of Corollary 34.2.3 is true: *If for any non-singular skew-symmetric metric g_{ij} on $\mathbb{R}^{2n}$ we define the Poisson bracket in the same way by*

$$\{f, g\} = g^{ij} \frac{\partial f}{\partial y^i} \frac{\partial g}{\partial y^j}, \tag{17}$$

then the set of smooth functions on $\mathbb{R}^n$ forms a Lie algebra with respect to the commutation operation only if there exist local co-ordinates $x^1, \ldots, x^n$, $p_1, \ldots, p_n$ in terms of which (g_{ij}) takes the form $\begin{pmatrix} O_n & I_n \\ -I_n & O_n \end{pmatrix}$.

We give only a partial proof of this converse. We begin by proving the following result.

34.2.4. Theorem. *The smooth functions on $\mathbb{R}^{2n}$ form a Lie algebra with respect to the commutation operation $\{\ ,\ \}$ defined by (17) if and only if the form $\Omega = g_{ij}\, dy^i \wedge dy^j$ is closed, i.e. $d\Omega = 0$.*

PROOF. Recall from Definition 25.1.1 that

$$(d\Omega)_{ijk} = \frac{\partial g_{jk}}{\partial y^i} - \frac{\partial g_{ik}}{\partial y^j} + \frac{\partial g_{ij}}{\partial y^k}. \tag{18}$$

Thus we wish to show that the Jacobi identity (15) is equivalent to the vanishing of the right-hand side of (18). From (17) we have

$$\{f_1, \{f_2, f_3\}\} = g^{pq} \frac{\partial f_1}{\partial y^p} \left(\frac{\partial g_{ij}}{\partial y^q} \frac{\partial f_2}{\partial y^i} \frac{\partial f_3}{\partial y^j} + g^{ij} \frac{\partial^2 f_2}{\partial y^q\, \partial y^i} \frac{\partial f_3}{\partial y^j} + g^{ij} \frac{\partial f_2}{\partial y^i} \frac{\partial^2 f_3}{\partial y^q\, \partial y^j} \right).$$

Permuting 1, 2, 3 cyclically and adding (and using repeatedly $g^{rs} = -g^{sr}$ and the interchangeability of order of differentiation) we find that Jacobi's identity is equivalent to

$$g^{pq} \frac{\partial g^{ij}}{\partial y^q} \left(\frac{\partial f_1}{\partial y^p} \frac{\partial f_2}{\partial y^i} \frac{\partial f_3}{\partial y^j} + \frac{\partial f_3}{\partial y^p} \frac{\partial f_1}{\partial y^i} \frac{\partial f_2}{\partial y^j} + \frac{\partial f_2}{\partial y^p} \frac{\partial f_3}{\partial y^i} \frac{\partial f_1}{\partial y^j} \right) = 0,$$

which is the same as

$$\frac{\partial f_1}{\partial y^p}\frac{\partial f_2}{\partial y^i}\frac{\partial f_3}{\partial y^j}\left(g^{pq}\frac{\partial g^{ij}}{\partial y^q} + g^{jq}\frac{\partial g^{pi}}{\partial y^q} + g^{iq}\frac{\partial g^{jp}}{\partial y^q}\right) = 0.$$

In view of the arbitrariness of the functions f_1, f_2, f_3, this is equivalent to

$$g^{pq}\frac{\partial g^{ij}}{\partial y^q} + g^{jq}\frac{\partial g^{pi}}{\partial y^q} + g^{iq}\frac{\partial g^{jp}}{\partial y^q} = 0, \qquad i, j, p = 1, \ldots, 2n. \tag{19}$$

If we multiply (19) by $g_{rp}g_{sj}g_{ti}$ (and as usual sum over appropriately placed repeated indices) then after using the equation

$$g_{ri}\frac{\partial g^{ij}}{\partial y^q} = -g^{ij}\frac{\partial g_{ri}}{\partial y^q},$$

and the like, we obtain finally

$$\frac{\partial g_{ts}}{\partial y^r} - \frac{\partial g_{tr}}{\partial y^s} + \frac{\partial g_{sr}}{\partial y^t} = 0, \tag{20}$$

as required. This completes the proof. □

(Note that it can be shown by a similar sort of calculation that the condition (iv) in Theorem 34.2.2 is also equivalent to (20), so that (ii) and (iv) are actually equivalent.)

Our touted converse of Corollary 34.2.3 now follows from the preceding result together with the following theorem, which we state without proof. (A proof may be found beginning on p. 230 of the book [29].)

34.2.5. Theorem (Darboux). *Let Ω be a differential form of rank 2 satisfying $\Omega^n = \Omega \wedge \cdots \wedge \Omega \neq 0$ (cf. 34.1.2). If Ω is closed then there exist local coordinates $x^1, \ldots, x^n, p_1, \ldots, p_n$ in terms of which Ω is given by*

$$\Omega = \sum_i dx^i \wedge dp_i.$$

Returning to the main thread of our development, suppose that in phase space we are given a Hamiltonian $H(x, p)$; the corresponding Hamiltonian system is then

$$\dot{y}^i = (\nabla H)^i, \qquad i = 1, \ldots, 2n, \quad (y^i) = (x, p). \tag{21}$$

By Lemma 34.1.4 the derivative of an arbitrary function $f(x, p)$ relative to this system satisfies

$$\dot{f} = \{f, H\}. \tag{22}$$

(Hence in particular $\dot{x}^i = \{x^i, H\} = \partial H/\partial p_i$, $\dot{p}_i = \{p_i, H\} = -\partial H/\partial x^i$.) We shall say that the function $f(x, p)$ is an *integral function* of the Hamiltonian system (or *integral of motion*) if f is constant along the trajectories of the

Hamiltonian system. From (22) and Theorem 34.2.2 we obtain immediately the following

34.2.6. Corollary. *A function $f(x, p)$ is an integral function of the Hamiltonian system* (21) *precisely if it commutes with the Hamiltonian*: $\{f, H\} = 0$. *The totality of integral functions of a given Hamiltonian system forms a Lie algebra with respect to the Poisson bracket, which Lie algebra is also closed under multiplication of functions.*

Suppose that we are dealing with a Hamiltonian $H(x, p)$ which is not explicitly dependent on the time t, i.e. $\partial H/\partial t \equiv 0$. Then the law of conservation of energy applies, so that every extremal (or "trajectory") lies in some level surface $H(x, p) = E$ ($=$ const.). Now by the above corollary a function $f = f(x, p)$ is an integral function of the corresponding Hamiltonian system if and only if $\{f, H\} = 0$. However then $\{H, f\} = 0$, so that $H(x, p)$ is, *mutatis mutandis*, constant on each integral curve of the vector field ∇f. Hence this vector field is parallel to all level surfaces $H(x, p) = E$, so that by Theorem 24.1.5, the restriction to a particular surface $H(x, p) = E$ of all integral functions of a given Hamiltonian system forms a Lie algebra (the *Lie algebra of integral functions on that surface*).

34.2.7. Examples. (a) Let $L(x, y, z, \dot{x}, \dot{y}, \dot{z})$ be a spherically symmetric Lagrangian in Euclidean $\mathbb{R}^3$ (see Definition 32.2.4). As we saw in Example (d) of §32.2, along extremals the three functions (components of the moment) M_x, M_y, M_z given by

$$M_x = L_x^i \frac{\partial L}{\partial \dot{x}^i} = yp_z - zp_y, \qquad M_y = L_y^i \frac{\partial L}{\partial \dot{x}^i} = zp_x - xp_z,$$

$$M_z = L_z^i \frac{\partial L}{\partial \dot{x}^i} = xp_y - yp_x, \tag{23}$$

are conserved, where here L_x, L_y, L_z are the three linear vector fields corresponding to three particular one-parameter groups (see the above-mentioned example). Thus in our present terminology the functions M_x, M_y, M_z are integral functions of the Hamiltonian system determined by L. In §24.2 we saw that the various commutators of these vector fields calculate out as

$$[L_x, L_y] = L_z, \qquad [L_y, L_z] = L_x, \qquad [L_z, L_x] = L_y. \tag{24}$$

From (23) and (14) one finds by direct computation that

$$\{M_x, M_y\} = M_z, \qquad \{M_y, M_z\} = M_x, \qquad \{M_z, M_x\} = M_y. \tag{25}$$

From this (and Example 24.2.3(a)) we conclude that:

The functions M_x, M_y, M_z on the phase space of a Hamiltonian system with a spherically symmetric Lagrangian $L(x, y, z, \dot{x}, \dot{y}, \dot{z})$, generate with respect to the Poisson bracket a Lie algebra isomorphic to the Lie algebra so(3).

(b) In the Keplerian problem the Hamiltonian (in $\mathbb{R}^6$) has the form

$$H(x, p) = \frac{p^2}{2m} + \frac{\alpha}{|x|}, \qquad \alpha < 0. \tag{26}$$

As in the preceding example spherical symmetry yields the three integral functions M_1, M_2, M_3 ($= M_x, M_y, M_z$ respectively) where

$$M = (M_1, M_2, M_3) = [x, p] = \text{const.}$$

It turns out that there are a further three functions W_1, W_2, W_3 which are integral for this Hamiltonian; they are given by

$$W = (W_1, W_2, W_3) = \left[\frac{p}{m}, M\right] + \frac{\alpha x}{|x|} = \text{const.} \tag{27}$$

(The vector W is the "Laplace–Runge–Lenz vector".) We wish now to compute the Poisson brackets $\{M_i, W_j\}, \{W_i, W_j\}$. To begin with, it is easy to verify from (23) and the definition (14) that

$$\begin{aligned} &\{p_i, M_i\} = 0, && \{p_1, M_2\} = p_3, && \{p_1, M_3\} = -p_2, \\ &\{p_2, M_1\} = -p_3, && \{p_2, M_3\} = p_1, && \{p_3, M_1\} = p_2, \\ &\{p_3, M_2\} = -p_1. \end{aligned} \tag{28}$$

(Verify these!) We know that (writing $x^i = x_i$, $i = 1, 2, 3$)

$$M_1 = x_2 p_3 - x_3 p_2,$$

$$W_1 = \frac{1}{m}(p_2 M_3 - p_3 M_2) + \frac{\alpha x_1}{|x|}, \qquad W_2 = \frac{1}{m}(p_3 M_1 - p_1 M_3) + \frac{\alpha x_2}{|x|}.$$

Using properties (i) and (iii) of Theorem 34.2.2 together with (25) and (28), we obtain:

$$\begin{aligned} \{M_1, W_1\} &= \frac{1}{m}\{M_1, p_2 M_3 - p_3 M_2\} + \alpha\left\{M_1, \frac{x_1}{|x|}\right\} \\ &= \frac{1}{m}[\{M_1, p_2\}M_3 + \{M_1, M_3\}p_2 - \{M_1, p_3\}M_2 \\ &\quad - \{M_1, M_2\}p_3] + \frac{\alpha}{|x|}\{M_1, x_1\} + \alpha x_1\left\{M_1, \frac{1}{|x|}\right\} \\ &= \frac{1}{m}\{p_3 M_3 - M_2 p_2 + p_2 M_2 - p_3 M_3\} = 0; \end{aligned}$$

$$\begin{aligned} \{M_1, W_2\} &= \frac{1}{m}\{M_1, p_3 M_1 - p_1 M_3\} + \alpha\left\{M_1, \frac{x_2}{|x|}\right\} \\ &= \frac{1}{m}[\{M_1, p_3\}M_1 - \{M_1, M_3\}p_1] + \frac{\alpha}{|x|}\{M_1, x_2\} \\ &= \frac{1}{m}[-p_2 M_1 + p_1 M_2] + \frac{\alpha x_3}{|x|} = W_3. \end{aligned}$$

After calculating in a similar fashion the remaining brackets $\{M_i, W_j\}$, one arrives at the following table:

$\{M_i, W_j\}$	W_1	W_2	W_3
M_1	0	W_3	$-W_2$
M_2	$-W_3$	0	W_1
M_3	W_2	$-W_1$	0

(29)

Analogous, though lengthier, calculations lead to the following expressions for the $\{W_i, W_j\}$:

$$\{W_1, W_2\} = -\frac{2E}{m} M_3, \qquad \{W_2, W_3\} = -\frac{2E}{m} M_1,$$

$$\{W_3, W_1\} = -\frac{2E}{m} M_2. \tag{30}$$

It is clear from this that the structure of the Lie algebra generated by the functions W_i, M_j restricted to a surface $H(x, p) = E = \text{const.}$, i.e. at a constant energy level (see the paragraph following Corollary 34.2.6), will depend on that energy level (see Exercise 3 below).

34.3. Canonical Transformations

Suppose we have a Hamiltonian system

$$\dot{x}^i = \frac{\partial H}{\partial p_i}, \qquad \dot{p}_i = -\frac{\partial H}{\partial x^i}, \qquad i = 1, \ldots, n,$$

arising from an arbitrary Hamiltonian $H(x, p)$.

34.3.1. Theorem. *The form $\Omega = \sum_{i=1}^n dx^i \wedge dp_i$ is preserved along the integral curves (extremals) of any Hamiltonian system, i.e. for any Hamiltonian $H = H(x, p)$ the Lie derivative of Ω along the vector field ∇H is zero:*

$$\dot{\Omega} = L_{\nabla H}\Omega = 0.$$

PROOF. In tensor notation the form Ω is just the tensor $(g_{ij}) = \begin{pmatrix} O_n & I_n \\ -I_n & O_n \end{pmatrix}$. Hence by formula (23) of §23.2 we have

$$L_{\nabla H} g_{ij} = (\nabla H)^k \frac{\partial g_{ij}}{\partial x^k} + g_{kj}\frac{\partial (\nabla H)^k}{\partial x^i} + g_{ik}\frac{\partial (\nabla H)^k}{\partial x^j}.$$

Since $g_{ij} = \text{const.}$, and $(\nabla H)^k = g^{ks}(\partial H/\partial y^s)$, it follows that

$$L_{\nabla H} g_{ij} = g_{kj} g^{ks} \frac{\partial^2 H}{\partial y^i \, \partial y^s} + g_{ik} g^{ks} \frac{\partial^2 H}{\partial y^j \, \partial y^s}$$

$$= \frac{\partial^2 H}{\partial y^i \, \partial y^j} - \frac{\partial^2 H}{\partial y^j \, \partial y^i} = 0,$$

where we have used also the skew-symmetry of g_{ij} and the fact that $g_{kj} g^{ks} = \delta_j^s$, etc. □

34.3.2. Corollary (Liouville). *The "volume element" of phase space*

$$\sqrt{g} \, dx^1 \wedge \cdots \wedge dx^n \wedge dp_1 \wedge \cdots \wedge dp_n$$
$$= dx^1 \wedge \cdots \wedge dx^n \wedge dp_1 \wedge \cdots \wedge dp_n$$

with respect to the metric $(g_{ij}) = \begin{pmatrix} O_n & I_n \\ -I_n & O_n \end{pmatrix}$, *is preserved relative to any Hamiltonian system (in the sense that its Lie derivative along the vector field ∇H is zero).*

PROOF. Since $L_{\nabla H}\Omega = 0$ by the preceding theorem, and since for any two forms Ω_1, Ω_2 we have (see Exercise 1 of §23.3)

$$L_{\nabla H}(\Omega_1 \wedge \Omega_2) = (L_{\nabla H}\Omega_1) \wedge \Omega_2 + \Omega_1 \wedge (L_{\nabla H}\Omega_2),$$

it follows that the Lie derivative along ∇H of $\underbrace{\Omega \wedge \cdots \wedge \Omega}_{n \text{ times}}$ is also zero. The corollary is now immediate from Lemma 34.1.2. □

34.3.3. Definition. A map $\Phi: \mathbb{R}^{2n} \to \mathbb{R}^{2n}$ is a *canonical transformation of phase space* (with co-ordinates x, p) if it preserves the form $\Omega = dx^i \wedge dp_i$ (or, equivalently, the metric $(g_{ij}) = \begin{pmatrix} O_n & I_n \\ -I_n & O_n \end{pmatrix}$).

Thus a canonical transformation is an "isometry" of phase space (or "motion" of the skew-symmetric metric). In these terms Theorem 34.3.1 tells us that the self-transformations of phase space which effectively slide it along the trajectories of some Hamiltonian system, form a one-parameter group of canonical transformations. The converse statement is also true; this is the import of the following theorem.

34.3.4. Theorem. *Let $\Phi_t(x, p)$ be a local, one-parameter group of canonical transformations of phase space, and let X denote the corresponding vector field: $X = d\Phi_t/dt|_{t=0}$. Then there exists (locally) a function $H(x, p)$ such that $X = \nabla H$, i.e.*

$$X = \left(\frac{\partial H}{\partial p}, -\frac{\partial H}{\partial x} \right).$$

PROOF. Suppose that in terms of the co-ordinates x, p the components of X are given by $X = (A^i, B_i)$, $i = 1, \ldots, n$. For small time intervals Δt we have

$$\Phi_{\Delta t}: \begin{cases} x^i \to x^i + A^i(x, p)\Delta t + O(\Delta t^2) = x^{i'}, \\ p_i \to p_i + B_i(x, p)\Delta t + O(\Delta t^2) = p_i'. \end{cases}$$

It follows (from the transformation rules for rank-one tensors, if you like) that

$$dx^{i'} = dx^i + (dA^i)\Delta t + O(\Delta t^2), \qquad dp_i' = dp_i + (dB_i)\Delta t + O(\Delta t^2),$$

whence

$$\begin{aligned} dx^{i'} \wedge dp_i' &= (dx^i + (dA^i)\Delta t) \wedge (dp_i + (dB_i)\Delta t) + O(\Delta t^2) \\ &= dx^i \wedge dp_i + \Delta t\left[\frac{\partial A^i}{\partial x^j} dx^j \wedge dp_i + \frac{\partial A^i}{\partial p_j} dp_j \wedge dp_i + \frac{\partial B_i}{\partial p_j} dx^i \wedge dp_j \right. \\ &\quad \left. + \frac{\partial B_i}{\partial x^j} dx^i \wedge dx^j\right] + O(\Delta t^2). \end{aligned}$$

Now we have from the assumption that the transformations Φ_t preserve the metric, that

$$\sum_i dx^{i'} \wedge dp_i' = \sum_i dx^i \wedge dp_i.$$

It follows from this and the expression for $dx^{i'} \wedge dp_i'$ preceding it, that:

$$\frac{\partial A^i}{\partial p_j} dp_i \wedge dp_j = 0, \quad \text{whence} \quad \frac{\partial A^i}{\partial p_j} = \frac{\partial A^j}{\partial p_i};$$

$$\frac{\partial B_i}{\partial x^j} dx^i \wedge dx^j = 0, \quad \text{whence} \quad \frac{\partial B_i}{\partial x^j} = \frac{\partial B_j}{\partial x^i};$$

$$\frac{\partial A^i}{\partial x^j} dx^j \wedge dp_i = -\frac{\partial B_i}{\partial p_j} dx^i \wedge dp_j, \quad \text{whence} \quad \frac{\partial A^i}{\partial x^j} = -\frac{\partial B_i}{\partial p_j}.$$

These are precisely the conditions for the form $-B_i\, dx^i + A^i\, dp_i$ to be closed (see Example 25.1.2(b)). From the general Stokes formula (26.3.1) it follows that throughout some neighbourhood of each point (x_0, p_0) where $X \neq 0$ the integral

$$\int_{t_0}^{t} (-B_i \dot{x}^i + A^i \dot{p}_i)\, dt, \qquad x_0 = x(t_0), \qquad p_0 = p(t_0),$$

is independent of the path $x(t)$, $p(t)$. Hence on that neighbourhood we may define a function $H(x, p)$ unambiguously by

$$H(x, p) = \int_{(x_0, p_0)}^{(x, p)} (-B_i\, dx^i + A^i\, dp_i),$$

whence it follows that $\partial H/\partial x^i = -B_i$, $\partial H/\partial p_i = A^i$. This completes the proof. □

Remark. Although a Hamiltonian $H(x, p, t)$ depending explicitly on the time t yields in the same way a one-parameter family of canonical transformations Φ_t, it can be shown that in general this family need not form a local group, i.e. it can happen that $\Phi_{t_1+t_2} \neq \Phi_{t_1} \circ \Phi_{t_2}$.

Observe that when $n = 1$ the form Ω is just $\Omega = dx \wedge dp$, so that preservation of Ω is equivalent to preservation of area in phase space; in other words the set of canonical transformations of phase space coincides, in this case, with the set of area-preserving transformations. For $n > 1$ however, it turns out that the former set is properly contained in the latter, i.e. there are transformations of phase space preserving $\underbrace{\Omega \wedge \cdots \wedge \Omega}_{n \text{ times}}$, but not preserving Ω.

Those canonical transformations of phase space (x, p) which are linear transformations of $\mathbb{R}^{2n}$ are called *symplectic* (cf. Definition 14.3.5). It is easy to see that in the case $n = 1$ the group of symplectic transformations is just $SL(2, \mathbb{R})$.

Finally we investigate the Lie algebra of the group of symplectic transformations. A typical matrix K of this Lie algebra is of the form $K = (d/dt)A(t)|_{t=0}$, for some smooth family of symplectic transformations $A(t)$ with $A(0) = I_{2n}$. By Theorem 34.3.4 (which, as its proof shows, holds for any smooth family of (local) canonical transformations), there exists a Hamiltonian $H(x, p)$ such that

$$Ky = \left(\frac{\partial H}{\partial p}, -\frac{\partial H}{\partial x}\right), \tag{31}$$

where $y = (x, p)$. It follows that to within an additive constant such a Hamiltonian must have the form

$$H = \frac{1}{2}\sum_{i,j=1}^{n}(a_{ij}x^i x^j + 2b_{ij}x^i p_j + c_{ij}p_i p_j), \tag{32}$$

whose associated (symmetric) matrix is $\frac{1}{2}\begin{pmatrix} A & B \\ B^T & C \end{pmatrix}$, where $A = (a_{ij})$, $B = (b_{ij})$, $C = (c_{ij})$, and the matrices A and C are symmetric. It then follows further from (31) that

$$K = \begin{pmatrix} B^T & C \\ -A & -B \end{pmatrix}. \tag{33}$$

Thus the matrices of the Lie algebra of the group of symplectic transformations are precisely those of the form (33) where A and C are symmetric. A direct computation shows that this Lie algebra is isomorphic to the Lie algebra of quadratic Hamiltonians of the form (32) with the Poisson bracket as commutation. The classical harmonic oscillator furnishes the simplest example of such a quadratic Hamiltonian: $H = p^2/2m + m\omega^2 x^2$, where ω is the frequency. For this reason quadratic Hamiltonians of the form (32) are often called "generalized oscillators".

34.4. Exercises

1. With each vector field X on configuration space (i.e. on the space with co-ordinates x), we associate a function F_X on phase space by defining $F_X = p_i X^i$. Show that $\{F_X, F_Y\} = -F_{[X,Y]}$.

2. Let $f = f(x)$ be a function on phase space independent of the momentum p. Show that $\{f, F_X\} = \partial_X f$.

3. Prove that in the Keplerian problem (Example 34.2.7(b)) the Lie algebras generated by the integral functions W_i, M_j at various fixed energy levels E are isomorphic to the following Lie algebras:

   (i) for $E < 0$, to $so(4)$;
   (ii) for $E = 0$, to the Lie algebra of the full isometry group of Euclidean $\mathbb{R}^3$.
   (iii) for $E > 0$, to $so(1, 3)$

4. Let $\Omega = g_{ij}\, dy^i \wedge dy^j = dx^i \wedge dp_i$, and let $X = (X^k)$ be a vector field. Show that the form $g_{ik} X^k\, dy^i$ is closed if and only if the Lie derivative of the form Ω along the field X is zero: $L_X \Omega = 0$.

5. Let M_x, M_y, M_z be as in Example 34.2.7(a) (the components of the angular momentum), and write $M^2 = M_x^2 + M_y^2 + M_z^2$. Verify that

$$\{M^2, M_x\} = \{M^2, M_y\} = \{M^2, M_z\} = 0.$$

6. With the same notation as in the preceding exercise, show that the Lie algebra with respect to the Poisson bracket, generated by the functions $p_x, p_y, p_z, M_x, M_y, M_z$, is isomorphic to the Lie algebra of the full isometry group of Euclidean 3-space (see Exercise 11 of §24.7).

7. Let (g_{ij}) be a non-singular skew-symmetric matrix. Such a matrix defines in the usual way a "skew" scalar product on the $2n$-dimensional vector space $\mathbb{R}^{2n}$. Prove that any subspace on which the restriction of this skew scalar product is identically zero, has dimension $\leq n$.

§35. Lagrange Surfaces

35.1. Bundles of Trajectories and the Hamilton–Jacobi Equation

For various purposes it is essential to know the properties not just of individual trajectories of a Hamiltonian system, but of whole bundles of such trajectories. In more precise terms the problem is as follows. Consider at time $t = 0$ an n-dimensional surface $\Gamma = \Gamma^n$ in $2n$-dimensional phase space with canonical co-ordinates x, p, whose equations are given in the "graphical" form

$$p_i = f_i(x^1, \ldots, x^n), \qquad i = 1, \ldots, n. \tag{1}$$

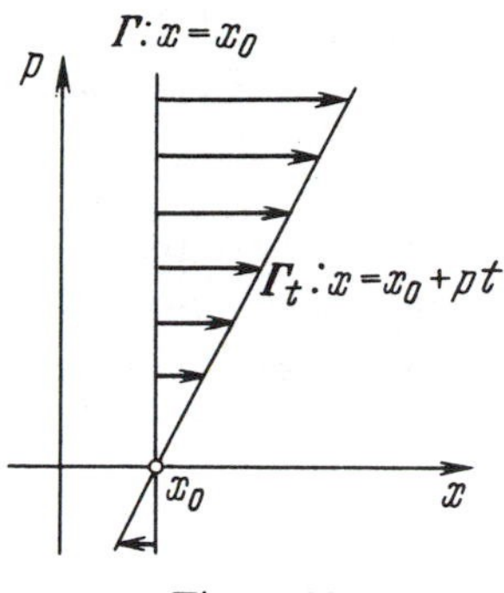

Figure 35

This surface is imagined to move so that each of its points slides along a trajectory of a given Hamiltonian $H(x, p, t)$; at time $t > 0$, the resulting surface is denoted by Γ_t, so that $\Gamma_0 = \Gamma$. It is convenient to regard the family of surfaces Γ_t in phase space as an $(n + 1)$-dimensional surface Γ^{n+1} in the extended phase space (of $2(n + 1)$ dimensions) with co-ordinates x, p, t, E (see Corollary 34.1.5). The surface Γ^{n+1} is comprised of the points $(x(t), p(t), t, E(t))$ where $E = H(x, p, t)$, $p_i(t) = f_i(x(t))$, and $x(t)$, $p(t)$ is a generic trajectory, starting at the surface Γ, of the original Hamiltonian $H(x, p, t)$. (Note that the surface Γ is the intersection of Γ^{n+1} with the hyperplane $t = 0$ in extended phase space.)

However we need to consider a wider class of surfaces $\Gamma = \Gamma^n$ in phase space (and concomitant surfaces Γ^{n+1} in extended phase space) than those which can be given graphically. For instance if we take $\Gamma = \Gamma_0$ to be the surface $x = x_0$, with p arbitrary (see Figure 35), then we obtain as Γ^{n+1} the bundle of all trajectories emanating from the surface $x = x_0$. For $t = 0$ the surface Γ_0 is not of the graphical form $p = f(x)$, yet for $t > 0$, Γ_t may be of that form (at least locally). It turns out that the surfaces in phase space and in extended phase space appropriate for our purposes, are as specified in the following definitions.

35.1.1. Definition. An n-dimensional surface $\Gamma = \Gamma^n$ in the phase space with co-ordinates x, p is said to be a *Lagrange surface* if for each point $Q \in \Gamma^n$, the (truncated) action $S(P) = \int_\gamma p\, dx$ is (locally) independent of the arc γ in Γ, joining Q to P, or in other words if in some neighbourhood of Q we have that $S(P)$ is a (single-valued) function of P, depending only on the end-point of γ.

Now let Γ^{n+1} be the surface in extended phase space defined (analogously to the above) in terms of a given Lagrange surface Γ in phase space and a given Hamiltonian $H(x, p, t)$.

35.1.1′. Definition. The surface Γ^{n+1} is called a *Lagrange surface in extended phase space* if for each point $Q \in \Gamma^{n+1}$, the action $S(P) = \int_\gamma (p\, dx - E\, dt)$

(note that $E = H(x, p, t)$ on the surface Γ^{n+1}) is (locally) independent of the arc γ in Γ^{n+1} joining Q to P, so that $S(P)$ is locally a well-defined function of the points P of Γ^{n+1}.

With regard to these definitions, note that if Γ is given in graphical form as $p = f(x)$, then S is a function of the x-co-ordinates (of P) alone, i.e. $S = S(x)$, and the analogous statement holds for the action in the second definition: S is a function of $x^0(=t), x^1, \ldots, x^n$ alone, i.e. $S = S(x, t)$.

35.1.2. Lemma.

(i) *If Γ is a Lagrange surface in phase space given in graphical form as $p_i = f_i(x^1, \ldots, x^n)$, $i = 1, \ldots, n$. then*

$$\frac{\partial S(x)}{\partial x^i} = p_i, \qquad i = 1, \ldots, n. \tag{2}$$

Conversely if Γ has graphical form and (2) *holds for some function $S(x)$, then Γ is a Lagrange surface.*

(ii) *If Γ^{n+1} is a Lagrange surface in extended phase space given (graphically) by $p_i = f_i(x^0, x^1, \ldots, x^n)$, $i = 0, 1, \ldots, n$ (where $x^0 = t$, $p_0 = E$), then*

$$\frac{\partial S(x, t)}{\partial x^i} = p_i, \quad \text{and} \quad \frac{\partial S(x, t)}{\partial t} = -H(x, p, t). \tag{3}$$

Conversely if Γ^{n+1} has graphical form and (3) *holds for some function $S(x, t)$, then Γ^{n+1} is a Lagrange surface in extended phase space.*

PROOF. We prove (ii) only; the proof of (i) is similar (and simpler).

Suppose $S(x, t) = \int_\gamma (p(dx/d\tau) - H(dt/d\tau)d\tau$ is single-valued, where $\gamma(\tau)$ is any arc in Γ^{n+1} joining Q to $P(x, t)$. Then the Fundamental Theorem of the Calculus tells us that for all such γ we have

$$\frac{dS(x(\tau), t(\tau))}{d\tau} = p(\tau)\frac{dx(\tau)}{d\tau} - H\left(x(\tau), p(\tau), t(\tau)\frac{dt(\tau)}{d\tau}\right),$$

whence $dS(x, t) = p\,dx - H\,dt$, yielding $p = \partial S/\partial x$, $-H = \partial S/\partial t$.

For the converse, suppose $p = \partial\hat{S}/\partial x$, $-H = \partial\hat{S}/\partial t$ for some single-valued function $\hat{S}(x, t)$. Then

$$d\hat{S} = \frac{\partial\hat{S}}{\partial x}dx + \frac{\partial\hat{S}}{\partial t}dt = p\,dx - H\,dt.$$

It is then immediate from the general Stokes formula (26.3.1) that $\int_\gamma (p\,dx - H\,dt)$ is independent of the path γ from Q to P. This completes the proof. □

The function $S(x, t)$ is called the *action of the trajectory bundle*, and the equation

$$\frac{\partial S}{\partial t} + H\left(x, \frac{\partial S}{\partial x}, t\right) = 0, \tag{4}$$

which follows from (3), is the *Hamilton–Jacobi equation.*

Note that if we know the function $S(x, t)$ of part (ii) of the above lemma, then we can retrieve the surface Γ^{n+1} since it is given by the equations (3): $p = \partial S/\partial x$, $E = -\partial S/\partial t$. Then by intersecting this surface with any hyperplane of the form $t = t_0$, we shall have come back to the surface $\Gamma^n_{t_0}$ obtained from $\Gamma^n = \Gamma^n_0$ by sliding the points of the latter along the trajectories of the Hamiltonian system $\dot{p} = -\partial H/\partial x$, $\dot{x} = \partial H/\partial p$, on which they lie.

We give now an alternative definition of a Lagrange surface in phase space co-ordinatized by x, p (which can be easily adapted also to extended phase space with co-ordinates x, p, E, t). This definition has the advantage that it makes evident the fact that under canonical transformations of phase space, Lagrange surfaces go into Lagrange surfaces.

35.1.3. Definition. An n-dimensional surface Γ in phase space is called a *Lagrange surface* if at each of its points P the skew-symmetric scalar product of any pair of vectors tangent at P is zero, i.e. if the restriction to the surface Γ of the form $\Omega = \sum_i dx^i \wedge dp_i$ yields the zero (rank-two) tensor on the space Γ.

The simplest Lagrange surfaces (according to either definition) are as follows:

(i) $\Gamma_{x_0} = \{(x, p) \mid x = x_0 (= \text{const.}), p \text{ arbitrary}\}$;
(ii) $\Gamma_{p_0} = \{(x, p) \mid p = p_0 (= \text{const.}), x \text{ arbitrary}\}$.

Under the transformation $x \to p$, $p \to -x$ we have

$$\Omega = dx^i \wedge dp_i \to -dp_i \wedge dx^i = dx' \wedge dp_i = \Omega,$$

so that this transformation, which interchanges the families of surfaces $\{\Gamma_{x_0}\}$ and $\{\Gamma_{p_0}\}$, is canonical.

35.1.4. Theorem (cf. Lemma 35.1.2). *A surface given in graphical form as $p_i = f_i(x)$ is a Lagrange surface in the sense of the latter definition* (35.1.3) *if and only if there exists a function $S(x)$ such that $f_i = \partial S(x)/\partial x^i$, $i = 1, \ldots, n$ (and the analogous statement holds in extended phase space).*

PROOF. If a surface Γ is given by equations $p_i = f_i(x)$, then the restriction to Γ of the form $\Omega = dx^i \wedge dp_i$ (i.e. the induced skew-symmetric metric) is calculated in terms of the co-ordinates $x^1, \ldots, x^n$ on Γ as follows:

$$\Omega|_\Gamma = \sum_i dx^i \wedge dp_i(x) = \sum_{i,j} dx^i \wedge \frac{\partial f_i}{\partial x^j} dx^j = \sum_{i<j} \left(\frac{\partial f_i}{\partial x^j} - \frac{\partial f_j}{\partial x^i}\right) dx^i \wedge dx^j.$$

It follows from Example 25.1.2(b) that $\Omega|_\Gamma$ is the differential $d\omega$ of the form $\omega = f_i\,dx^i$. Hence the condition that $\Omega|_\Gamma \equiv 0$ is equivalent to the condition $d\omega \equiv 0$, i.e. that the form ω (defined on Γ) be closed. By the general Stokes formula (26.3.1), this implies that the integral $S(x) = \int_{x_0}^{x} f_i(x)\,dx^i$ is independent of the path on Γ joining x_0 to x.

The proof is completed by the observation that the converse follows from the equality of mixed partials. □

35.1.5. Theorem. *Let $H = H(x, p)$ be a Hamiltonian not explicitly dependent on the time t. Then the following hold*:

(i) *The vector $\nabla H = (\partial H/\partial p, -\partial H/\partial x)$ is tangent to the surface $H = E_0 =$* const., *at each point (x, p) on that surface.*
(ii) *At each point of the surface $H = E_0$ the vector ∇H has zero (skew-symmetric) scalar product with every vector tangent to the surface at that point.*
(iii) *If an n-dimensional Lagrange surface Γ has constant energy level, i.e. if $H(x, p) = E_0$ for all $(x, p) \in \Gamma$, then ∇H is tangent to Γ at every point $(x, p) \in \Gamma$. Thus in particular any trajectory of the Hamiltonian system corresponding to H which intersects the surface Γ, lies entirely in Γ.*

PROOF. For a vector ξ to be tangent to the surface $H = E_0$ at a point on it, it is necessary and sufficient that at that point $\xi^i(\partial H/\partial y^i) = 0$, where $y = (x, p)$. From this (i) is immediate. To see (ii) let ξ be any vector tangent to the surface $H = E_0$ at (x, p). Then

$$\langle \xi, \nabla H \rangle = \xi^i g_{ij}(\nabla H^j) = \xi^i g_{ij} g^{jk} \frac{\partial H}{\partial y^k} = \xi^i \frac{\partial H}{\partial y^i} = 0.$$

Finally we prove (iii). Let P be any point of the Lagrange surface Γ (of dimension n) entirely at the energy level $H(x, p) = E_0$, and let $\{\xi_1, \ldots, \xi_n\}$ be a basis for the tangent space to Γ at P. By the (second) definition of a Lagrange surface we have that $\langle \xi_i, \xi_k \rangle = 0$, $i, k = 1, \ldots, n$, and by Part (ii), already proven, we have that at the point P, $\langle \nabla H, \xi_i \rangle = 0$, $i = 1, \ldots, n$. Hence the skew-symmetric scalar product vanishes on the space spanned by the ξ_i together with ∇H. Since by Exercise 7 of §34.5 such a vector space can have dimension at most n, it follows that $\nabla H = \sum_i \lambda_i \xi_i$, which proves the first statement in (iii). The second statement now follows since any trajectory intersecting Γ must then be tangent to Γ at each point of intersection, and consequently must lie in Γ. □

The following corollary (and its notation) pertains to the latter part of the next subsection.

35.1.6. Corollary. *Let S^{n-1} be an arbitrary $(n-1)$-dimensional Lagrange surface (in the sense that the skew-symmetric scalar product vanishes on it), with constant energy level $H(x, p) = E_0$, and consider the surface Γ formed by* *all trajectories of the Hamiltonian system* $\dot{y} = \nabla H$, $y = (x, p)$, *which intersect* S^{n-1}. *If the surface* $\Gamma = \Gamma^n$ *is n-dimensional then it also is a Lagrange surface with constant energy level* E_0. *If in addition the surface* Γ^n *is given (locally) by equations of the form*

$$p_i = f_i(x) = \frac{\partial S_0}{\partial x^i} \tag{5}$$

(*where* $S_0(x)$ *is as in* Theorem 35.1.4), *then the function* $S_0(x)$ *satisfies the "truncated" Hamiltonian–Jacobi equation* (*see* (2))

$$E_0 = H\left(x, \frac{\partial S_0}{\partial x}\right). \tag{6}$$

Finally, if Γ^{n+1} *is the surface in extended phase space defined, as before, in terms of* Γ^n, *and if* $S(x, t)$ *is as in* Lemma 35.1.2 (ii), *i.e.* $S(x, t) = \int_\gamma (p\, dx - E\, dt)$ *where* γ *is any arc in* Γ^{n+1} *joining* $(x_0, 0)$ *to* (x, t), *then we have*

$$S(x, t) = -E_0 t + S_0(x), \qquad -\frac{\partial S}{\partial t} = H\left(x, \frac{\partial S}{\partial x}\right).$$

35.2. Hamiltonians Which Are First-Order Homogeneous with Respect to the Momentum

In this subsection we consider separately the important special case of a Hamiltonian which is first-order homogeneous with respect to the momentum:

$$H(x, \lambda p) = \lambda H(x, p), \qquad \lambda > 0. \tag{7}$$

(For instance for the trajectories of a light ray in an isotropic medium the appropriate Hamiltonian is (as we saw in Example (b) of §33.3) given by $H(x, p) = c(x)|p|$, which is first-order homogeneous in p.)

Such Hamiltonians are generally not considered at $p = 0$ (cf. the case $H(x, p) = c(x)|p|$ just mentioned, which is not smooth at $p = 0$). Note that once we know the trajectories of such a Hamiltonian of any particular energy level $H = E_0$ (e.g. $E_0 = 1$), then those at other energy levels can be obtained by simply applying the appropriate similarity transformation $p \to \lambda p$, $x \to x$, under which H becomes λH.

Recall that while earlier in the present chapter we obtained the geodesics of a metric $g_{ij}(x)$ as the extremals of the Lagrangian $L = g_{ij} v^i v^j$ (with corresponding Hamiltonian $H = g^{ij} p_i p_j$), the same purpose is served by using instead the Hamiltonian $\hat{H} = \sqrt{H} = \sqrt{g^{ij} p_i p_j}$. (See the remark towards the end of §33.3 where in particular it is shown that at a constant energy level, the respective Hamiltonian systems are proportional, with constant proportionality factor.) The relevance of this to our present context resides in the fact that the latter Hamiltonian $\hat{H}$ is first-order homogeneous in p:

$$\hat{H}(x, \lambda p) = \lambda \hat{H}(x, p), \qquad \lambda > 0.$$

35.2.1. Theorem. *If $H(x, p)$ is such that $H(x, \lambda p) = \lambda H(x, p)$ for $\lambda > 0$, then the one-parameter group of transformations Φ_t arising from H (i.e. determined by the vector field $X = (\dot{x}, \dot{p}) = (\partial H/\partial p, \ -\partial H/\partial x)$, preserves the form $p\,dx = p_i\,dx^i$, i.e. the Lie derivative along X of the phase-covector $(p, 0)$ is zero.*

PROOF. By the formula (20) of §23.2, if (T_j) is any covector in $2n$-dimensional phase space, then its Lie derivative along the vector field X is given by

$$(L_X T)_j = X^k \frac{\partial T_j}{\partial y^k} + T_k \frac{\partial X^k}{\partial y^j}, \qquad y = (x, \ p).$$

Taking $T = (p_1, \ldots, p_n, 0, \ldots, 0)$, it follows that for $j = 1, \ldots, n$,

$$\begin{aligned} (L_X T)_j &= X^k \frac{\partial p_j}{\partial y^k} + p_l \frac{\partial X^l}{\partial x^j} \quad \text{(summing over } l \text{ from 1 to } n) \\ &= X^{j+n} + p_l \frac{\partial^2 H}{\partial x^j\,\partial p_l} = -\frac{\partial H}{\partial x^j} + \frac{\partial}{\partial x^j}\left(p_l \frac{\partial H}{\partial p_l}\right). \end{aligned} \tag{8}$$

Now since $H(x, \lambda p) = \lambda H(x, p)$, it follows that

$$p_l \frac{\partial H}{\partial p_l} = \frac{\partial H(x, \lambda p)}{\partial \lambda} = H(x, p), \tag{9}$$

so that the last expression in (8) is zero, whence the first n components of $L_X T$ vanish. For $j = n + 1, \ldots, 2n$, we have

$$(L_X T)_j = p_l \frac{\partial X^l}{\partial p_j} = \frac{\partial}{\partial p_j}\left(p_l \frac{\partial H}{\partial p_l}\right) - \frac{\partial H}{\partial p_j},$$

which again vanishes in view of (9). This completes the proof. □

This result prompts the following definition. (See §22.1 for the precise definition of the restriction operation on tensors.)

35.2.2. Definition. A surface Γ in phase space is said to be a *conical Lagrange surface* if the restriction to Γ of the form $p\,dx$ is identically zero.

With this definition we deduce immediately the following

35.2.3. Corollary *The set of conical Lagrange surfaces is preserved by Hamiltonian systems for which $H(x, \lambda p) = \lambda H(x, p)$, $\lambda > 0$.*

Observe that the Lagrange surface Γ_{x_0} (defined by $x = x_0 = \text{const.}$, p arbitrary) is conical since the restriction of the form $p\,dx$ to it is zero, while on the other hand the surface Γ_{p_0} (defined by $p = p_0 \neq 0$, x arbitrary) is not, since clearly the restriction to it of $p\,dx$ is not identically zero.

Given any Hamiltonian $H(x, p)$ satisfying $H(x, \lambda p) = \lambda H(x, p), \lambda > 0$, the bundle of all trajectories emanating from a particular point x_0, and of a particular energy level E_0, is of obvious importance (e.g. in studying light emitted from a point source). We define the "wave front" of such a trajectory in terms of the surface Γ_{x_0} as follows. Consider the $(n - 1)$-dimensional phase-surface contained in Γ_{x_0} satisfying the additional equation $H = E_0$; we denote this surface by $S_{x_0}^{n-1}$. If we now imagine each point of $S_{x_0}^{n-1}$ slid along the trajectory emanating from it, then after a time interval t we shall have obtained a new surface $S_{x_0}^{n-1}(t)$.

35.2.4. Definition. The *wave front* at an instant $t > 0$ of the trajectory bundle of constant energy level E_0 emanating (at time $t = 0$) from a point x_0, is the projection of the phase-surface $S_{x_0}^{n-1}(t)$ onto the x-space (i.e. onto the surface $p = 0$).

Thus the wave front at time $t > 0$ is a surface of dimension $n - 1$ in ordinary x-space ($p = 0$). (Clearly at time $t = 0$ the projection of the surface $S_{x_0}^{n-1}$ onto the x-space is just the single point x_0.)

If one knows the wave front at any particular time $t_0 > 0$, then one can (obviously) employ what is known as "Huygens' principle" to obtain the wave front at a later time $t_1 > t_0 > 0$. This works as follows: Regarding each point on the known wave front (corresponding to time t_0) as a new centre of emission, one considers the wave front corresponding to the time $t_1 - t_0 > 0$, emanating from that point. The envelope of all such wave fronts (centering on the points of the known front) is then the desired wave front corresponding to the time t_1 (see Figure 36).

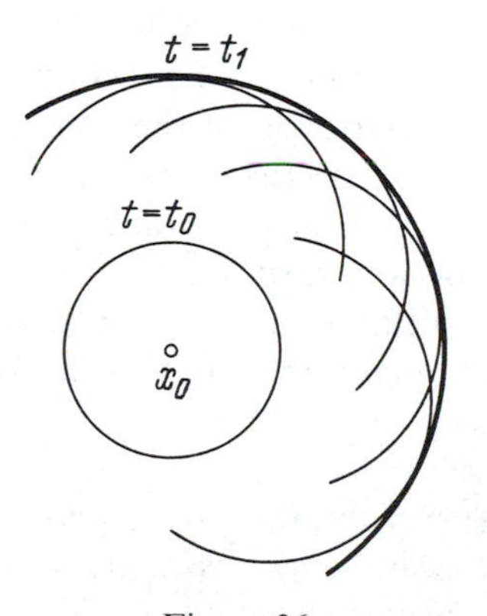

Figure 36

We now consider more general conical Lagrange surfaces.

35.2.5. Lemma. *If Γ^n is an arbitrary (n-dimensional) conical Lagrange surface in phase space, then the projection of Γ^n onto the x-space ($p = 0$) has dimension $\leq n - 1$.*

PROOF. If the projection in question were (even locally) of dimension n, then we could use $x^1, \ldots, x^n$ as local co-ordinates for the surface Γ^n; but then if $\xi^1, \ldots, \xi^n$ were the components, in terms of these local co-ordinates, of any tangent vector to Γ^n at any point, it would follow from the defining condition $p\,dx|_{\Gamma^n} = 0$, that $p_i \xi^i = 0$, whence it would follow in turn that the tangent plane at the point has dimension $<n$. This contradiction completes the proof. □

Hence our conical Lagrange surface Γ^n cannot, even locally, be given in the graphical form $p_i = f_i(x)$, so that we cannot associate with it (in the manner of Theorem 35.1.4) a function $S(x)$. We therefore proceed as for the above special surface Γ_{x_0}. We consider the surface S^{n-1} defined by the requirements that it be contained in Γ^n and satisfy $H(x, p) = E_0$. By Corollary 35.1.6 the surface

$$\tilde{\Gamma}^n = \bigcup_{-\infty < t < \infty} S^{n-1}(t)$$

consisting entirely of the trajectories intersecting S^{n-1}, is a Lagrange surface of constant energy level E_0. It is reasonable to assume that this surface might be given (locally) in the form $p_i = \partial S_0(x)/\partial x^i$, where S_0 is as in Theorem 35.1.4. This assumed, S_0 satisfies the "truncated" Hamilton–Jacobi equation (see (6))

$$E_0 = H\left(x, \frac{\partial S_0(x)}{\partial x}\right). \tag{10}$$

Note that this is all possible provided only that Γ^n is Lagrange and the Hamiltonian is not explicitly time-dependent; we now bring into play the additional assumption that the surface Γ^n is conical.

Since Γ^n is conical the restriction of the form $p\,dx$ to its subspace S^{n-1} is zero. Theorem 35.2.1 the implies that the restriction of $p\,dx$ to every surface $S^{n-1}(t)$ is zero. It follows that the function $S_0(P) = \int_{P_0}^{P} p\,dx$ (which we are assuming exists for the Lagrange surface $\tilde{\Gamma}^n$) is constant on each of the surfaces $S^{n-1}(t)$. Thus the level surfaces $S_0(x) = \text{const.}$ (in ordinary space) must coincide with the projections of the surfaces $S^{n-1}(t)$ onto the x-space. We restate this conclusion as a

35.2.6. Theorem. *The wave fronts (i.e. projections of the surfaces $S^{n-1}(t)$ onto the x-space) are just the level surfaces $S_0(x) = \text{const.}$, where $S_0(x)$ is the truncated action of the trajectory bundle $\tilde{\Gamma}^n$.*

EXERCISES

Suppose that in phase space $\mathbb{R}^{2n}$ we are given n independent functions $f_1(x, p), \ldots, f_n(x, p)$ (i.e. having linearly independent vector-gradients) which pairwise commute: $\{f_i, f_j\} = 0$. Show that the surface in $\mathbb{R}^{2n}$ defined by the n equations $f_1 = 0, \ldots, f_n = 0$, is Lagrange.

§36. The Second Variation for the Equation of the Geodesics

36.1. The Formula for the Second Variation

In §31.1 we showed that the extremals $\gamma: x = x(t)$ of the functional

$$S[\gamma] = \int_\gamma L(x, \dot{x})\, dt, \tag{1}$$

needs must satisfy the Euler–Lagrange equations

$$\frac{\partial S}{\partial x^i} = \frac{\partial L}{\partial x^i} - \frac{d}{dt}\left(\frac{\partial L}{\partial \dot{x}^i}\right) = 0, \qquad i = 1, \ldots, n; \tag{2}$$

i.e. that these equations provide necessary conditions for an arc $\gamma: x = x(t)$ joining a given point P to a given point Q to minimize $S[\gamma]$ (over all arcs from P to Q). We shall now address the problem of finding sufficient conditions, i.e. conditions ensuring that an arc $\gamma: x^i = x^i(t)$ satisfying the Euler–Lagrange equations actually gives $S[\gamma]$ a minimum value (at least among arcs "close" to γ).

As the reader knows from "advanced calculus" courses, a necessary condition for a function $f(x^1, \ldots, x^N)$ of several variables to assume a (local) minimum value at a point P is that

$$\left.\frac{\partial f}{\partial x^i}\right|_P = 0, \qquad i = 1, \ldots, N, \tag{3}$$

while a sufficient condition (under the assumption that P satisfies (3)) is that the quadratic form $(\partial^2 f/\partial x^i\, \partial x^j)\, dx^i\, dx^j$ be positive definite at the point P (i.e. as a form on tangent vectors at P). Thus as a first step in the search for sufficient conditions for $S[\gamma]$ to attain a "local" minimum at γ, where γ satisfies the Euler–Lagrange equations, it is natural to consider the following analogous bilinear form (called the *second variation of the arc* γ):

$$\left[\frac{\partial^2}{\partial\lambda\, \partial\mu} S[\gamma + \lambda\xi + \mu\eta]\right]_{\substack{\lambda=0\\ \mu=0}} = G_\gamma(\xi, \eta) = G_\gamma(\eta, \xi), \tag{4}$$

where η, ξ are vector fields defined along the arc $\gamma(t)$, $a \le t \le b$, which vanish at the end-points $\gamma(a) = P$, $\gamma(b) = Q$. It can be shown without too much difficulty that for $S[\gamma]$ to be a minimum (given that γ satisfies the Euler–Lagrange equations), it is necessary that $G_\gamma(\xi, \xi) \ge 0$ for all ξ, and sufficient that $G_\gamma(\xi, \xi) > 0$ for all nonzero ξ.

36.1.1. Lemma. *If $\gamma: x^i = x^i(t)$ $(i = 1, \dots, n)$ satisfies the Euler–Lagrange equations, then the following formula holds:*

$$G_\gamma(\xi, \eta) = \left.\frac{\partial^2 S[\gamma + \lambda\xi + \mu\eta]}{\partial\lambda\,\partial\mu}\right|_{\substack{\lambda=0\\ \mu=0}} = \int_a^b (J_{ij}\xi^j)\eta^i\,dt, \tag{5}$$

where

$$J_{ij}\xi^j = \frac{d}{dt}\left(\frac{\partial^2 L}{\partial\dot{x}^i\,\partial\dot{x}^j}\dot{\xi}^j + \frac{\partial^2 L}{\partial\dot{x}^i\,\partial x^j}\xi^j\right) - \frac{\partial^2 L}{\partial x^i\,\partial\dot{x}^j}\dot{\xi}^j - \frac{\partial^2 L}{\partial x^i\,\partial x^j}\xi^j. \tag{6}$$

Proof. Using the formula for the first variation (formula (11) in §31.1) we have

$$\left.\frac{\partial^2 S[\gamma + \lambda\xi + \mu\eta]}{\partial\lambda\partial\mu}\right|_{\substack{\lambda=0\\ \mu=0}} = \left[\frac{\partial}{\partial\lambda}\left(\frac{\partial}{\partial\mu}S[\gamma + \lambda\xi + \mu\eta]\right)_{\mu=0}\right]_{\lambda=0}$$

$$= \frac{\partial}{\partial\lambda}\int_a^0 \left(\frac{\partial L}{\partial y^i} - \frac{d}{dt}\frac{\partial L}{\partial\dot{y}^i}\right)\eta^i\,dt\Bigg|_{\lambda=0},$$

where $y(t) = x(t) + \lambda\xi(t)$, and $L = L(y, \dot{y})$. Then, by using the same trick as was used in the derivation of the Euler–Lagrange equations (Theorem 31.1.2), the last expression converts to

$$\int_a^b \frac{\partial}{\partial\lambda}\left(\frac{\partial L}{\partial y^i} - \frac{d}{dt}\frac{\partial L}{\partial\dot{y}^i}\right)_{\lambda=0}\eta^i(t)\,dt = \int_a^b \left(\frac{\partial M_i}{\partial y^k}\xi^k + \frac{\partial M_i}{\partial\dot{y}^k}\dot{\xi}^k + \frac{\partial M_i}{\partial\ddot{y}^k}\ddot{\xi}^k\right)_{\lambda=0}\eta^i\,dt,$$

where

$$M_i = M_i(y, \dot{y}, \ddot{y}) = \frac{\partial L}{\partial y^i} - \frac{d}{dt}\frac{\partial L}{\partial\dot{y}^i} = \frac{\partial L}{\partial y^i} - \frac{\partial^2 L}{\partial y^j\,\partial\dot{y}^i}\dot{y}^j - \frac{\partial^2 L}{\partial\dot{y}^j\,\partial\dot{y}^i}\ddot{y}^j.$$

The latter integral is clearly just

$$\int_a^b \left(\frac{\partial M_i}{\partial x^k}\xi^k + \frac{\partial M_i}{\partial\dot{x}^k}\dot{\xi}^k + \frac{\partial M_i}{\partial\ddot{x}^k}\ddot{\xi}^k\right)\eta^i\,dt, \tag{7}$$

where here $M_i = M_i(x, \dot{x}, \ddot{x})$. Explicit calculation of the integrand in (7) yields finally

$$G_\gamma(\xi, \eta) = \int_a^b \left(\frac{\partial^2 L}{\partial x^i\,\partial x^j}\xi^j + \frac{\partial^2 L}{\partial x^i\,\partial\dot{x}^j}\dot{\xi}^j - \frac{d}{dt}\left(\frac{\partial^2 L}{\partial\dot{x}^i\,\partial x^j}\xi^j + \frac{\partial^2 L}{\partial\dot{x}^i\,\partial\dot{x}^j}\dot{\xi}^j\right)\right)\eta^i\,dt,$$

as required. □

36.1.2. Definition. The linear operator J which acts on vector fields $\xi(t)$ defined along the curve γ, is called the *Jacobi operator* (corresponding to the given Lagrangian L).

We shall now consider (as our only example) the important special case where the extremals coincide with the geodesics of a given metric. For this purpose it will be convenient to employ the Lagrangian $L = \frac{1}{2}g_{ij}\dot{x}^i\dot{x}^j$, i.e. to take as the action

$$S = \int_a^b \tfrac{1}{2}g_{ij}\dot{x}^i\dot{x}^j\,dt \tag{8}$$

(rather than the length functional $l = \int\sqrt{g_{ij}\dot{x}^i\dot{x}^j}\,dt$, although, as we know from §31.2, for both of these actions the extremals coincide with the geodesics of the metric g_{ij}).

36.1.3. Theorem. *In the case when $S = \int_a^b \frac{1}{2}g_{ij}\dot{x}^i\dot{x}^j\,dt$, and $\gamma\colon x^i = x^i(t)$ is any geodesic of the metric g_{ij}, with t a natural parameter, the bilinear form (second variation)*

$$\left.\frac{\partial^2 S[\gamma + \lambda\xi + \mu\eta]}{\partial\lambda\,\partial\mu}\right|_{\substack{\lambda=0\\ \mu=0}} = G_\gamma(\xi, \eta),$$

becomes

$$G_\gamma(\xi, \eta) = -\int_a^b (\nabla_{\dot{x}}^2\xi^i + \dot{x}^j\dot{x}^k\xi^l R^i_{jkl})\eta^m g_{im}\,dt, \tag{9}$$

or

$$G_\gamma(\xi, \eta) = -\int_a^b \langle J\xi, \eta\rangle\,dt, \tag{10}$$

where

$$(J\xi)^i = \nabla_{\dot{x}}^2\xi^i + \dot{x}^j\dot{x}^k\xi^l R^i_{jkl}, \tag{11}$$

R^i_{jkl} denoting the curvature tensor.

Proof. From Example (b) of §31.2 we have, for any arc γ,

$$\frac{\delta S}{\delta x^l} = -(\ddot{x}^k + \Gamma^k_{ij}\dot{x}^i\dot{x}^j)g_{kl} = -\nabla_{\dot{x}}(\dot{x})^k g_{kl},$$

where $\nabla_{\dot{x}}T$ denotes the covariant derivative of the vector T in the direction of the tangent vector $\dot{x}$ (see §§29.1, 29.2). Hence by formula (11) of §31.1

$$\frac{\partial}{\partial\mu}S[\gamma + \mu\eta]_{\mu=0} = -\int_a^b (\ddot{x}^k + \Gamma^k_{ij}\dot{x}^i\dot{x}^j)g_{kl}\eta^l\,dt.$$

Hence

$$\left[\frac{\partial^2}{\partial\lambda\,\partial\mu} S[\gamma + \lambda\xi + \mu\eta]\right]_{\substack{\lambda=0\\ \mu=0}} = \left[-\frac{\partial}{\partial\lambda}\int_a^b (\ddot{y}^k + \Gamma^k_{ij}\dot{y}^i\dot{y}^j)g_{kl}\eta^l\,dt\right]_{\lambda=0},$$

where $y(t) = x(t) + \lambda\xi(t)$, $\Gamma^k_{ij} = \Gamma^k_{ij}(y)$, $g_{kl} = g_{kl}(y)$, $\eta = \eta(t)$. Arguing once again as in the derivation of the Euler–Lagrange equations, the last expression becomes

$$-\int_a^b \frac{\partial}{\partial\lambda}(\nabla_{\dot{y}}(\dot{y})^k g_{kl})_{\lambda=0}\,\eta^l\,dt = -\int_a^b \left(\frac{\partial D_l}{\partial x^r}\xi^r + \frac{\partial D_l}{\partial \dot{x}^r}\dot{\xi}^r + \frac{\partial D_l}{\partial \ddot{x}^r}\ddot{\xi}^r\right)\eta^l\,dt,$$

where

$$D_l = D_l(x, \dot{x}, \ddot{x}) = \nabla_{\dot{x}}(\dot{x})^k g_{kl}(x) = (\ddot{x}^k + \Gamma^k_{ij}\dot{x}^i\dot{x}^j)g_{kl}.$$

Now

$$\frac{\partial D_l}{\partial x^r} = \frac{\partial \Gamma^k_{ij}}{\partial x^r}\dot{x}^i\dot{x}^j g_{kl} + \nabla_{\dot{x}}(\dot{x})^k\frac{\partial g_{kl}}{\partial x^r} = \frac{\partial \Gamma^k_{ij}}{\partial x^r}\dot{x}^i\dot{x}^j g_{kl},$$

since $\nabla_{\dot{x}}(\dot{x}) = 0$ along the geodesic γ. We also have

$$\frac{\partial D_l}{\partial \dot{x}^r} = 2\,\Gamma^k_{rj}\dot{x}^j g_{kl}, \qquad \frac{\partial D_l}{\partial \ddot{x}^r} = g_{rl}.$$

Hence

$$G_\gamma(\xi, \eta) = -\int_a^b \left(\frac{\partial \Gamma^k_{ij}}{\partial x^r}\dot{x}^i\dot{x}^j\xi^r + 2\Gamma^k_{rj}\dot{x}^j\dot{\xi}^r + \ddot{\xi}^k\right)g_{kl}\,\eta^l\,dt. \qquad (12)$$

We now compare this with the right-hand side of (9). By the definition of covariant differentiation (see §29.1) we have

$$\nabla_{\dot{x}}(\xi)^i = \dot{\xi}^i + \Gamma^i_{kl}\dot{x}^k\xi^l,$$

whence

$$\nabla^2_{\dot{x}}(\xi)^i = \ddot{\xi}^i + \frac{d}{dt}(\Gamma^i_{kl}\dot{x}^k\xi^l) + \Gamma^i_{ks}\dot{x}^k\dot{\xi}^s + \Gamma^i_{ks}\Gamma^s_{pq}\dot{x}^k\dot{x}^p\xi^q.$$

By using the formula for R^i_{qkl} given in Theorem 30.1.1, together with the fact that γ is a geodesic, it follows after a little calculation that

$$\nabla^2_{\dot{x}}(\xi)^i + R^i_{qkl}\dot{x}^q\dot{x}^k\xi^l = \ddot{\xi}^i + 2\Gamma^i_{ks}\dot{x}^k\dot{x}^s + \frac{\partial \Gamma^i_{qk}}{\partial x^l}\dot{x}^q\dot{x}^k\xi^l. \qquad (13)$$

This together with (12) now yields the desired result. □

(Note that if ξ (or η) is a constant multiple of $\dot{x}$ then the right-hand side of (13) is that same constant multiple of $(d/dt)(\ddot{x}^i + \Gamma^i_{ks}\dot{x}^k\dot{x}^s) = 0$, whence $G_\gamma(\xi, \eta) = 0$.)

36.1.4. Example. Let $\gamma(t)$, $a \leq t \leq b$, t natural, be a geodesic arc in a 2-dimensional space with metric g_{ij}. In a neighbourhood of the arc we introduce special co-ordinates x, y with the following properties:

(i) the x-axis, i.e. the set of points $(x, 0)$, is just the geodesic arc γ itself, and in addition x is the natural parameter, i.e. $x = t$.

(ii) the curves $x =$ const. are arranged to be orthogonal to the arc γ, with moreover the y-co-ordinates scaled so that in terms of the co-ordinates x, y we have $g_{ij}(x, 0) = \delta_{ij}$.

If now $\xi(t)$, $\eta(t)$ are vector fields normal to γ (and therefore with zero x-components) then in view of the fact that on γ we have $\Gamma^i_{j1} = -\Gamma^j_{i1}$ (which follows from §29.3(14)) we infer that $\nabla^2_{\dot{x}}(\xi) = \ddot{\xi}$. From this and §30.3(i) it is then easy to see that the formula for $G_\gamma(\xi, \eta)$ given in the preceding theorem simplifies to

$$G_\gamma(\xi, \eta) = -\int_a^b \left(\frac{d^2}{dt^2}\xi^i + K(t)\xi^i\right)\eta_i \, dt,$$

where K is the Gaussian curvature. (Note that the summand with $i = 1$ is zero.)

Remark. The formula (10) for the second variation can be generalized so as to apply to "broken" vector fields, i.e. vector fields ξ for which $\nabla_{\dot{x}}\xi$ has jump discontinuities (only).

EXERCISE

Show that if ξ is a broken vector field defined on the geodesic γ, then the following analogue of formula (10) is valid:

$$G_\gamma(\xi, \eta) = -\sum_{P_i} \langle \eta, \Delta_{P_i}(\nabla_{\dot{x}}\xi)\rangle - \int_a^b \langle J\xi, \eta\rangle \, dt, \tag{14}$$

where $\Delta_P(\nabla_{\dot{x}}\xi)$ denotes the size of the jump in the covariant derivative at the point P, and the summation is over all points P_i of discontinuity of $\nabla_{\dot{x}}\xi$ on γ.

36.2. Conjugate Points and the Minimality Condition

We noted above, in connexion with the definition of $G_\gamma(\xi, \eta)$, that for $S[\gamma]$ to be a minimum (given that γ satisfies the Lagrange–Euler equations), it is sufficient that the quadratic form $G_\gamma(\xi, \xi)$ be positive definite, and necessary that $G_\gamma(\xi, \xi)$ be non-negative, for all ξ vanishing at the end-points of the arc γ; we shall in what follows assume this result. We shall also suppose that $L = \frac{1}{2}g_{ij}\dot{x}^i\dot{x}^j$ (as in Theorem 36.1.3), and further that the metric g_{ij} is Riemannian. We wish to examine further the question of when a geodesic arc between two points is actually the shortest arc joining the points. To begin with we establish a criterion for the bilinear form $G_\gamma(\xi, \eta)$ (where γ is now a geodesic arc) to be non-degenerate. (We remind the reader that the bilinear form $G_\gamma(\xi, \eta)$ (on fields ξ, η) vanishing at the end-points P, Q of γ) is said to be *degenerate* if there exists a non-zero field ξ such that $G_\gamma(\xi, \eta) = 0$ for all fields η.) The following definition applies to the general situation of an arbitrary Lagrangian L (as dealt with in Lemma 36.1.1).

36.2.1. Definition. A vector field ξ defined on an extremal arc γ (corresponding to a Lagrangian L) is said to be a *Jacobi field* if it satisfies "Jacobi's equation" $J\xi = 0$, and vanishes at the end-points P, Q of γ.

Thus for the particular Lagrangian $L = \frac{1}{2}\langle \dot{x}, \dot{x} \rangle$ we are considering, Jacobi's equation has the form

$$J\xi = (\nabla_{\dot{x}}^2 \xi^i + \dot{x}^j \dot{x}^k \xi^l R^i_{jkl}) = 0, \qquad i = 1, \dots, n. \tag{15}$$

36.2.2. Definition. Points P, Q of a geodesic γ are called *conjugate points* of γ if there exists a non-zero Jacobi field on the arc of γ joining P to Q.

36.2.3. Lemma. *Let γ be a geodesic arc joining P and Q. The bilinear form $G_\gamma(\xi, \eta)$ is non-degenerate if and only if the endpoints P and Q of γ are not conjugate points of γ.*

PROOF. If ξ is Jacobi, then from Theorem 36.1.3 we have that $G_\gamma(\xi, \eta) = 0$ for all vector fields η (defined on γ and vanishing at P and Q).

For the converse, suppose that for some non-zero field ξ, we have $G_\gamma(\xi, \eta) = 0$ for all fields η. Then in particular if we take $\eta = \alpha(t)J\xi$ where $\alpha(t)$ is any (smooth) function vanishing at $t = a$, $t = b$ (the values of t at P and Q), then once again we obtain from Theorem 36.1.3 that

$$G_\gamma(\xi, \eta) = -\int_a^b \alpha(t)\langle J\xi, J\xi \rangle = 0.$$

Since this holds for all functions $\alpha(t)$ and since the metric is Riemannian, it follows that $J\xi = 0$. This concludes the proof. □

We are now in a position to establish "Jacobi's minimality condition", an important necessary condition for a geodesic arc γ to minimize $S[\gamma]$, i.e. to be "minimal".

36.2.4. Theorem. *If a geodesic arc γ joining points P and Q contains in its interior a pair of conjugate points P', Q', then the arc γ is not minimal.*

PROOF. We shall give the proof only under the additional assumption that the end-points P, Q of γ are not conjugate. From this assumption and Lemma

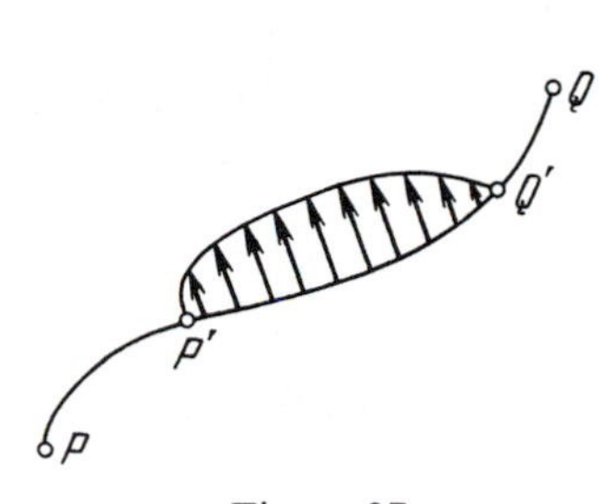

Figure 37

36.2.3, it follows that the bilinear form $G_\gamma(\xi, \eta)$ is nondegenerate. This nondegeneracy and the necessary condition that $G_\gamma(\xi, \xi)$ be non-negative (for γ to be minimal) together imply that $G_\gamma(\xi, \xi)$ is positive definite, i.e. that $G_\gamma(\xi, \xi) > 0$ for all non-zero ξ. We shall show that the assumption of the presence of interior conjugate points P', Q' leads to a contradiction of this positive definiteness. Since P', Q' are conjugate points of γ, there exists a non-zero Jacobi field ξ' defined on γ', the sub-arc of γ joining P' to Q', which by definition vanishes at P' and Q'. We define a "broken" field ξ on the full arc γ by setting $\xi = \xi'$ between P' and Q', and $\xi = 0$ on the remainder of γ (see Figure 37). Then from formula (14) (where the contribution of the two "jumps" is zero since $\xi = 0$ at the corresponding points) we deduce that

$$G_\gamma(\xi, \xi) = 0,$$

contradicting the positive definiteness of $G_\gamma(\xi, \xi)$. □

36.2.5. Theorem. *Any sufficiently short arc γ of a geodesic minimizes the action $S[\gamma]$ over all smooth arcs (with the same end-points as γ). Hence any sufficiently short arc of a geodesic is locally the shortest of all smooth arcs sharing its end-points.*

Proof. As noted at the beginning of this subsection, to establish the minimality of a geodesic arc γ it suffices to show that $G_\gamma(\xi, \xi)$ is positive definite (where ξ ranges over all vector fields on γ which vanish at its end-points). From Theorem 36.1.3 we have

$$\begin{aligned} G_\gamma(\xi, \xi) &= -\int_a^b [\langle \nabla_{\dot{x}}^2 \xi, \xi\rangle + \langle R(\dot{x}, \xi)\dot{x}, \xi\rangle]\, dt \\ &= \int_a^b \langle \nabla_{\dot{x}}\xi, \nabla_{\dot{x}}\xi\rangle\, dt - \int_a^b [\langle R(\dot{x}, \xi)\dot{x}, \xi\rangle + \nabla_{\dot{x}}\langle \nabla_{\dot{x}}\xi, \xi\rangle]\, dt \\ &= \int_a^b \langle \nabla_{\dot{x}}\xi, \nabla_{\dot{x}}\xi\rangle\, \mathrm{dt} - \int_a^b \langle R(\dot{x}, \xi)\dot{x}, \xi\rangle\, dt, \end{aligned} \tag{16}$$

where we have used firstly

$$\nabla_{\dot{x}}\langle \nabla_{\dot{x}}\xi, \xi\rangle = \langle \nabla_{\dot{x}}^2\xi, \xi\rangle + \langle \nabla_{\dot{x}}\xi, \nabla_{\dot{x}}\xi\rangle,$$

which holds by virtue of the compatibility of the metric, and secondly that, since $\langle \nabla_{\dot{x}}\xi, \xi \rangle$ is a scalar, and since $\xi(a) = \xi(b) = 0$,

$$\int_a^b \nabla_{\dot{x}} \langle \nabla_{\dot{x}}\xi, \xi \rangle \, dt = \int_a^b \frac{d}{dt} \langle \nabla_{\dot{x}}\xi, \xi \rangle \, dt = 0.$$

It is not difficult to see that for segments whose length Δl is sufficiently small, we have (see the exercise below)

$$\left| \int_a^b \langle R(\dot{x}, \xi)\dot{x}, \xi \rangle \, dt \right| < c(\Delta l) \int_a^b \langle \nabla_{\dot{x}}\xi, \nabla_{\dot{x}}\xi \rangle \, dt, \tag{17}$$

where $c(\Delta l)$ depends only on the metric g_{ij} and the length Δl, and moreover tends to zero as $\Delta l \to 0$. Since $\nabla_{\dot{x}}\xi$ cannot be identically zero on γ unless ξ is also (this follows from §29.1(5) *et seqq.*, taking into account that $\xi(a) = \xi(b) = 0$), we have

$$\int_a^b \langle \nabla_{\dot{x}}\xi, \nabla_{\dot{x}}\xi \rangle \, dt > 0.$$

The positive definiteness of $G_\gamma(\xi, \xi)$ now follows from this, (16) and (17). □

EXERCISE

Prove the inequality (17). (Hint. Show that on an interval of length Δl, we have $|\xi| <$ const. $\times (\max |\nabla_{\dot{x}}\xi|)\Delta l$.)

CHAPTER 6

The Calculus of Variations in Several Dimensions. Fields and Their Geometric Invariants

§37. The Simplest Higher-Dimensional Variational Problems

37.1. The Euler–Lagrange Equations

Let D denote a region with piecewise smooth boundary ∂D, of the Euclidean space $\mathbb{R}^n$ with Euclidean co-ordinates $x^1, \ldots, x^n$. Consider the linear space F of smooth vector-functions $f(x^1, \ldots, x^n) = (f^1, \ldots, f^k)$ defined on D, i.e. with domain D. Let $L(x^\beta; p^j; q^i_\alpha)$ be a smooth real-valued function of the three arguments x^β, $1 \le \beta \le n$; p^j, $1 \le j \le k$; q^i_α, $1 \le i \le k$, $1 \le \alpha \le n$ (making altogether $n + k + nk$ real arguments); we call such a function a *Lagrangian*, and from a given such Lagrangian we construct a functional $I[f]$ defined on F, as follows:

$$I[f] = \int_D L(x^\beta; f^j(x^\beta); f^i_{x^\alpha}(x^\beta))\, dx^1 \wedge \cdots \wedge dx^n,$$

where the integral is the multiple integral (see §26) over the region D (which we shall later assume to be bounded), and where $f^i_{x^\alpha}(x^\beta) = (\partial/\partial x^\alpha) f^i(x^\beta)$. We shall often write $I[f]$ more briefly as $I[f] = \int_D L(x^\beta; f^j; f^i_{x^\alpha})\, d^n x$.

The simplest case, namely that of 1-dimensional variational problems ($n = 1$), formed the subject of the preceding chapter. There we paid particular attention to the arc-length functional $l(\gamma) = \int_0^1 \sqrt{g_{ij}(y)\dot{y}^i\dot{y}^j}\, dt$, and the "action" functional $S(\gamma) = \int_0^1 g_{ij}(y)\dot{y}^i\dot{y}^j\, dt$, these functionals being defined on the set of all piecewise smooth arcs $\gamma(t) = (y^1(t), \ldots, y^k(t))$, $0 \le t \le 1$, in the k-dimensional Riemannian space with metric g_{ij}.

In the present chapter however, we shall be concerned with the case $n > 1$, i.e. with higher-dimensional variational problems. The simplest instance of such a problem is provided by the area functional, which associates with each 2-dimensional surface in $\mathbb{R}^3$ its area (over D). Thus if

$$f(x, y) = (u^1(x, y), u^2(x, y), u^3(x, y))$$

is a surface in Euclidean $\mathbb{R}^3(u^1, u^2, u^3)$, defined for all $(x, y) \in D \subseteq \mathbb{R}^2$, and if the induced metric on the surface is given by $dl^2 = E\,dx^2 + 2F\,dx\,dy + G\,dy^2$, then the *area functional* is (see §7.4)

$$I[f] = \iint_D \sqrt{EG - F^2}\,dx\,dy.$$

Here the Lagrangian is (see §7.3(18))

$$L(x, y; f; f_x, f_y) = L(f_x, f_y) = \sqrt{EG - F^2} = \sqrt{\langle f_x, f_x\rangle\langle f_y, f_y\rangle - \langle f_x, f_y\rangle^2}.$$

We return now to the general case. What questions of interest arise in the first instance in connexion with a functional $I[f]$? Since $I[f]$ may be regarded as a function defined on the infinite-dimensional space F, it is natural to turn for an answer to this to the analogous though much simpler situation of functions of only finitely many variables. Consider a function $\alpha(u, v)$ of two variables; to a large extent the behaviour of α is determined by the nature of its critical points (u_0, v_0) (i.e. the points where grad $\alpha = 0$). At each of these points the graph of the function α will have a local maximum, or a local minimum, or a ("non-degenerate") saddle point, or of course may exhibit the more complicated behaviour appertaining to a "degenerate" saddle point (see Figure 38).

Hence in studying functionals $I[f]$ it is natural to look for those functions f_0 at which $I[f]$ attains a (local) maximum or minimum value, or has a

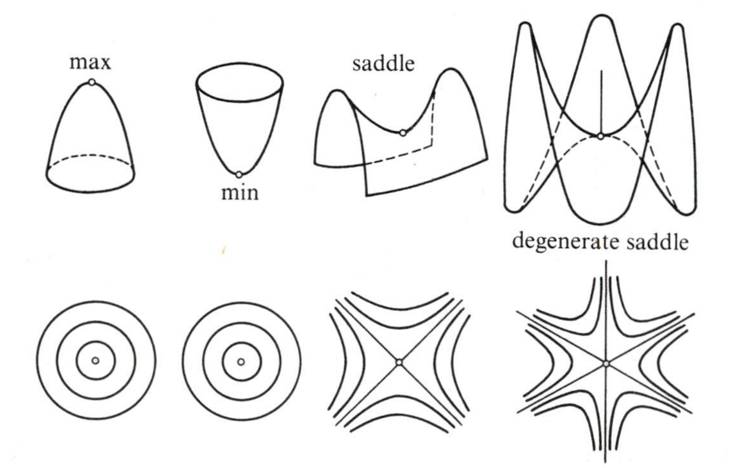

Figure 38

saddle point of one sort or another. Now since in the finite-dimensional case the critical points are obtained as the solutions of the equation grade $\alpha = 0$, we shall first need to find the appropriate analogue of this equation in the case of a functional $I[f]$, i.e. in the infinite-dimensional case.

In the finite-dimensional case the condition grad $\alpha = 0$ is equivalent to the condition that $\partial_a(\alpha) = 0$ for all vectors a at the (critical) point. Here by $\partial_a(\alpha)$ we mean of course the directional derivative

$$\partial_a(\alpha) = a^1 \frac{\partial \alpha}{\partial u} + a^2 \frac{\partial \alpha}{\partial v} = \langle a, \text{grad } \alpha \rangle$$

in the direction a. The reader will recall that $\partial_a(\alpha)$ can be defined alternatively by

$$\partial_a(\alpha) = \lim_{\varepsilon \to 0} \frac{1}{\varepsilon} [\alpha(u + \varepsilon a^1, v + \varepsilon a^2) - \alpha(u, v)].$$

It is on this form of the definition of $\partial_a(\alpha)$ that we shall model the concept of the "directional derivative" of a functional $I[f]$.

Consider a "point" f in F, the domain of the functional I, and let $\eta \in F$ be a function vanishing on ∂D; such functions η will be used as increments or "perturbations" of f. The perturbation $\eta = \delta f$ determines the "direction of change" in going from f to $f + \varepsilon\eta$, where ε is a small (real) parameter (just as above in the latter definition of $\partial_a(\alpha(u_0, v_0))$ the vector a determined the direction of displacement from the point $(u, v) = (u_0, v_0)$). If we now form the expression $(1/\varepsilon)(I[f + \varepsilon\eta] - I[f])$ and proceed to the limit as $\varepsilon \to 0$, we obtain a function (of f and η) which it is natural to call the "derivative of the functional I at the point f in the direction η". Thus this directional derivative is given by

$$\left. \frac{dI[f + \varepsilon\eta]}{d\varepsilon} \right|_{\varepsilon=0} = \lim_{\varepsilon \to 0} \frac{1}{\varepsilon} (I[f + \varepsilon\eta] - I[f]) = \int \frac{\delta I}{\delta f} \delta f \, d^n x, \tag{1}$$

where the final expression has been obtained by taking the limit under the integral sign. In the integrand $(\delta I/\delta f)\, \delta f (=(\delta I/\delta f^i)\delta f^i)$, δf is just η, while an explicit expression for the vector $\delta I/\delta f = (\delta I/\delta f^i)$, called the *variational derivative of the functional* $I[f]$, is given below.

The analogy with the finite-dimensional case prompts the following terminology.

37.1.1. Definition. A function $f_0 \in F$ is said to be *stationary* (or *extremal*, or *critical*) for a functional I, if $\delta I[f_0]/\delta f \equiv 0$ for every perturbation $\delta f = \eta$ identically zero on the boundary of D.

We now derive an explicit expression for $\delta I[f]/\delta f$, yielding a necessary and sufficient condition for a function f_0 to be extremal. Thus

$$\Delta I[f] = I[f + \varepsilon\eta] - I[f] = \int_D [L(x^\beta; f^j + \varepsilon\eta^j; f^i_{x^\alpha} + \varepsilon\eta^i_{x^\alpha})$$

$$- L(x^\beta; f^j; f^i_{x^\alpha})]\, d^n x.$$

Expanding the first term in the integrand in a Taylor series we obtain

$$\Delta I[f] = \int_D \left[\sum_{j=1}^{k} \frac{\partial L}{\partial f^j} \varepsilon\eta^j + \sum_{i=1}^{k} \sum_{\alpha=1}^{n} \frac{\partial L}{\partial f^i_{x^\alpha}} \varepsilon\eta^i_{x^\alpha} + o(\varepsilon) \right] d^n x$$

$$= \varepsilon \int_D \sum_{i=1}^{k} \left[\frac{\partial L}{\partial f^i} \eta^i + \sum_{\alpha=1}^{n} \frac{\partial L}{\partial f^i_{x^\alpha}} \eta^i_{x^\alpha} \right] d^n x + \int_D o(\varepsilon)\, d^n x.$$

Integrating by parts, we obtain from

$$\frac{\partial}{\partial x^\alpha}\left(\frac{\partial L}{\partial f^i_{x^\alpha}} \eta^i \right) = \frac{\partial}{\partial x^\alpha}\left(\frac{\partial L}{\partial f^i_{x^\alpha}} \right) \eta^i + \frac{\partial L}{\partial f^i_{x^\alpha}} \eta^i_{x^\alpha},$$

that

$$\Delta I[f] = \varepsilon \int_D \sum_{\alpha=1}^{n} \frac{\partial}{\partial x^\alpha}\left(\frac{\partial L}{\partial f^i_{x^\alpha}} \eta^i \right) d^n x$$

$$+ \varepsilon \int_D \sum_{i=1}^{k} \left(\frac{\partial L}{\partial f^i} - \sum_{\alpha=1}^{n} \frac{\partial}{\partial x^\alpha}\left(\frac{\partial L}{\partial f^i_{x^\alpha}} \right) \right) \eta^i\, d^n x + \int_D o(\varepsilon)\, d^n x.$$

Since all functions involved are assumed to be piecewise smooth (as also the boundary of D), the multiple integrals are the same as the corresponding iterated integrals, and the order of integration in the latter is immaterial. Hence we may integrate first with respect to any variable x^α, obtaining for a typical term of the first integral above

$$\int_D \frac{\partial}{\partial x^\alpha}\left(\frac{\partial L}{\partial f^i_{x^\alpha}} \eta^i \right) d^n x = \int_{x^1, \ldots, \hat{x}^\alpha, \ldots, x^n} \left[\int_P^Q \frac{\partial}{\partial x^\alpha}\left(\frac{\partial L}{\partial f^i_{x^\alpha}} \eta^i \right) dx^\alpha \right] d^{n-1}x,$$

where as usual the hat over a symbol indicates that that symbol is omitted, where P and Q depend on $x^1, \ldots, \hat{x}^\alpha, \ldots, x^n$, and where $d^{n-1}x = dx^1 \wedge \cdots \wedge d\hat{x}^\alpha \wedge \cdots \wedge dx^n$. Since in the inner integral (from P to Q) the variables $x^1, \ldots, \hat{x}^\alpha, \ldots, x^n$ function merely as independent parameters, it follows that

$$\int_D \frac{\partial}{\partial x^\alpha}\left(\frac{\partial L}{\partial f^i_{x^\alpha}} \eta^i \right) d^n x = \int_{x^1, \ldots, \hat{x}^\alpha, \ldots, x^n} \left[\frac{\partial L}{\partial f^i_{x^\alpha}} \eta^i \bigg|_Q - \frac{\partial L}{\partial f^i_{x^\alpha}} \eta^i \bigg|_P \right] d^{n-1}x \equiv 0,$$

where in deducing that the integral is identically zero, we have used the fact that P and Q are on ∂D where the perturbation η vanishes. Thus

$$\Delta I[f] = \varepsilon \int_D \sum_{i=1}^k \left[\frac{\partial L}{\partial f^i} - \sum_{\alpha=1}^n \frac{\partial}{\partial x^\alpha}\left(\frac{\partial L}{\partial f^i_{x^\alpha}}\right)\right] \eta^i \, d^n x + \int_D o(\varepsilon) \, d^n x.$$

Since

$$\lim_{\varepsilon \to 0} \frac{1}{\varepsilon} \int_D o(\varepsilon) \, d^n x = 0,$$

we obtain finally the desired expression for $\delta I/\delta f$:

$$\frac{\delta I[f]}{\delta f^i} = \frac{\partial L}{\partial f^i} - \sum_{\alpha=1}^n \frac{\partial}{\partial x^\alpha}\left(\frac{\partial L}{\partial f^i_{x^\alpha}}\right). \tag{2}$$

From Definition 37.1.1 we immediately obtain the following theorem.

37.1.2. Theorem. *A function $f_0 \in F$ is extremal for the functional $I[f]$ if and only if it satisfies the system of equations*

$$\frac{\delta I[f]}{\partial f^i} = \frac{\partial L}{\partial f^i_0} - \sum_{\alpha=1}^n \frac{\partial}{\partial x^\alpha}\left(\frac{\partial L}{\partial f^i_{0,x^\alpha}}\right) = 0 \qquad (1 \le i \le k). \tag{3}$$

The equations (3) are called the *Euler–Lagrange equations for the functional I*. If $I[f]$ has a minimum (or maximum) at $f_0 \in F$, then from (1) it is clear that we must have

$$\int_D \frac{\delta I}{\delta f^i_0} \eta^i \, d^n x = 0$$

for all perturbations η ($\eta = 0$ on ∂D). It follows readily that for $I[f]$ to have a minimum or maximum at f_0, the function f_0 must be an extremal, i.e. must satisfy the Euler–Lagrange equations (3). (This was of course the reason for Definition 37.1.1.)

37.2. The Energy-Momentum Tensor

Consider a functional of the form

$$I[f] = \int_D L(f^j, f^i_{x^k}) \, d^n x,$$

where the Lagrangian L does not explicitly depend on the variables x^i, $1 \le i \le n$. The corresponding extremal functions f (determining the "behaviour of the system") are, as we have seen, just the solutions of the Euler–Lagrange system of equations

$$\frac{\partial L}{\partial f^i} - \frac{\partial}{\partial x^k}\left(\frac{\partial L}{\partial f^i_{x^k}}\right) = 0 \tag{4}$$

(where as usual summation takes place over the index k).

Imitating the derivation in mechanics of the law of conservation of energy we substitute from (4) in the equation

$$\frac{\partial L}{\partial x^i} = \frac{\partial L}{\partial f^\alpha}\frac{\partial f^\alpha}{\partial x^i} + \frac{\partial L}{\partial f^\alpha_{x^k}}\frac{\partial (f^\alpha_{x^k})}{\partial x^i},$$

to obtain

$$\frac{\partial L}{\partial x^i} = \frac{\partial}{\partial x^k}\left(\frac{\partial L}{\partial f^\alpha_{x^k}}\right)\frac{\partial f^\alpha}{\partial x^i} + \frac{\partial L}{\partial f^\alpha_{x^k}}\frac{\partial (f^\alpha_{x^k})}{\partial x^i}.$$

Using

$$\frac{\partial}{\partial x^i}(f^\alpha_{x^k}) = \frac{\partial^2 f^\alpha}{\partial x^i\,\partial x^k} = \frac{\partial}{\partial x^k}(f^\alpha_{x^i}),$$

this becomes

$$\frac{\partial L}{\partial x^i} = \frac{\partial}{\partial x^k}\left(\frac{\partial L}{\partial f^\alpha_{x^k}}\right)\frac{\partial f^\alpha}{\partial x^i} + \frac{\partial L}{\partial f^\alpha_{x^k}}\cdot\frac{\partial}{\partial x^k}(f^\alpha_{x^i}) = \frac{\partial}{\partial x^k}\left(\frac{\partial L}{\partial f^\alpha_{x^k}} f^\alpha_{x^i}\right).$$

Since $\partial L/\partial x^i = \delta^k_i(\partial L/\partial x^k)$, the last equation can be rewritten as

$$\frac{\partial L}{\partial x^k}\delta^k_i = \frac{\partial}{\partial x^k}\left(\frac{\partial L}{\partial f^\alpha_{x^k}} f^\alpha_{x^i}\right),$$

or equivalently,

$$\frac{\partial}{\partial x^k}\left(\delta^k_i L - f^\alpha_{x^i}\frac{\partial L}{\partial f^\alpha_{x^k}}\right) = 0. \tag{5}$$

If the co-ordinates $x^1, \ldots, x^n$ are Euclidean then the distinction between upper and lower indices disappears, and if we then write

$$T_{ik} = f^\alpha_{x^i}\frac{\partial L}{\partial f^\alpha_{x^k}} - \delta_{ik}L, \tag{6}$$

equation (5) takes the form

$$\sum_k \frac{\partial T_{ik}}{\partial x^k} = 0 \qquad (T_{ik} = T^k_i), \tag{7}$$

i.e. the divergence of the tensor T_{ik} (and tensor it is) is zero at all points of the region D.

If on the other hand the co-ordinates are pseudo-Euclidean then we define the tensor T_{ik} by

$$T_{ik} = g_{kl} f^\alpha_{x^i}\frac{\partial L}{\partial f^\alpha_{x^l}} - g_{ik}L, \qquad \frac{\partial T^k_i}{\partial x^k} \equiv 0. \tag{8}$$

37.2.1. Definition. The tensor T_{ik} is called the *energy-momentum tensor* of the given system (with Lagrangian $L(f, f_{x^\alpha})$).

Note that equation (7) does not define the energy-momentum tensor uniquely, since if we supplement any tensor T_{ik} satisfying (7) by a summand of the form $\sum_l (\partial/\partial x^l)\psi_{ikl}$, where ψ_{ikl} is any tensor skew-symmetric in the indices k and l, then clearly the new tensor

$$\tilde{T}_{ik} = T_{ik} + \sum_l \frac{\partial}{\partial x^l}\psi_{ikl}, \qquad \tilde{T}_i^k = T_i^k + \frac{\partial}{\partial x^l}\psi_i^{kl},$$

will also satisfy (7), in view of the fact that $\sum_{k,l} (\partial^2\psi_{ikl}/\partial x^k\,\partial x^l) = 0$. The tensor T_{ik} defined by (6) or (8) is in general not symmetric; however if we take instead (7) as defining T_{ik}, then we can often select a symmetric solution by adding to the appropriate one of (6), (8) a suitable tensor of the form $\sum_l (\partial\psi_{ikl}/\partial x^l)$. (For instance a tensor ψ_{ikl} skew-symmetric in k and l, and satisfying

$$\sum_l \frac{\partial}{\partial x^l}(\psi_{ikl}) = \tfrac{1}{2}(T_{ik} - T_{ki}) \tag{9}$$

(where T_{ik} is given by (6) or (8)) would suffice.) The definition of the energy-momentum tensor as a symmetric tensor is important in several physical questions (see below). In what follows our main concern among the applications of the concept of the energy-momentum tensor, will be with variational problems in 4-dimensional pseudo-Riemannian spaces, or more particularly in Minkowski space $\mathbb{R}_1^4$ with the usual co-ordinates x^0, x^1, x^2, x^3, where x^0 is proportional to t (the time), and x^1, x^2, x^3 are spatial co-ordinates.

Assuming therefore that the underlying space is $\mathbb{R}_1^4$, we now define the 4-dimensional "momentum vector" of a system with Lagrangian L, in terms of the energy-momentum tensor. To begin with consider for each $k = 0, 1, 2, 3$, the "standard" 3-form

$$dS_k = \tfrac{1}{6}\varepsilon_{ljmk}\, dx^l \wedge dx^j \wedge dx^m,$$

defined on the hypersurface $x^k = 0$ (cf. Exercise 2 of §26.5). From these forms we construct the 3-form $T^{ik}\, dS_k$ defined on $D \subseteq \mathbb{R}_1^4$, and thence in turn the momentum vector.

37.2.2. Definition. The *momentum 4-vector* of a system with Lagrangian L is the vector $P = (P^0, P^1, P^2, P^3)$ where

$$P^i = \lambda \int_{x^0 = \text{const.}} T^{ik}\, dS_k = \lambda \int_{x^0 = \text{const.}} T^{i0}\, dS_0, \qquad i = 0, 1, 2, 3, \quad \lambda = \text{const.}$$

(Here we have used the fact that the restriction of the form dS_k to any hypersurface $x^0 = \text{const.}$, is zero if $k \neq 0$.)

Remark. By analogy with formula (7) of §31.1 for the energy of a system in the one-dimensional case, we shall call the component $T^{00} = \dot{f}^\alpha(\partial L/\partial \dot{f}^\alpha) - L$ (where the dot indicates differentiation with respect to $x^0 = ct$) the *energy density*, so that $\int_{x^0=\text{const.}} T^{00}\, d^3x$ (where $d^3x = dx^1 \wedge dx^2 \wedge dx^3$) represents the total energy of the system at that particular time. From Definition 37.2.2 we then see that the "time"-component P^0 is just the total energy multiplied by the constant λ, which is usually taken to be $1/c$. (Cf. §32.2(17) where the same conclusion was reached for the time-component of the 4-momentum of a free relativistic particle.)

37.2.3. Proposition. *The condition $\partial T^{ik}/\partial x^k = 0$ is (under certain general assumptions) equivalent to the conservation of the momentum vector P.*

PROOF. We shall prove only half of the equivalence, namely that the conservation of P follows from the condition $\partial T^{ik}/\partial x^k = 0$. For this we shall assume that the components T^{ik} approach zero at least as fast as $1/R^2$ as $R \to \infty$, where $R = \sqrt{(x^1)^2 + (x^2)^2 + (x^3)^2}$.

Let x_1^0 and x_2^0 be any two values of x^0, and denote by C the cylinder in $\mathbb{R}_1^4$ with "lid" D_1 and "base" D_2 in the hyperplanes $x^0 = x_1^0$, $x^0 = x_2^0$ respectively, and with "lateral" surface Π of radius R. From the condition $\partial T^{ik}/\partial x^k = 0$ and the general Stokes formula, it follows that $\int_C T^{ik}\, dS_k = 0$ (see Exercise 2 of §26.5). Thus we have

$$\left(\int_{D_2} - \int_{D_1} + \int_{\Pi}\right) T^{ik}\, dS_k = 0.$$

If we now let $R \to \infty$, the integral over Π tends to 0 by virtue of our assumption, while the integrals over D_1 and D_2 tend to $P^i(x_1^0)$ and $P^i(x_2^0)$ respectively. Hence

$$P^i(x_1^0) = \int_{x^0 = x_1^0} T^{ik}\, dS_k = \int_{x^0 = x_2^0} T^{ik}\, dS_k = P^i(x_2^0),$$

as required. □

37.2.4. Lemma. *The momentum vector of a given system (suitably behaved out towards ∞) remains unaffected if we replace T_{ik} by the symmetrized tensor $\tilde{T}_{ik} = T_{ik} + \sum_l (\partial \psi_{ikl}/\partial x^l)$, where ψ_{ikl} is a tensor skew-symmetric in the indices k and l.*

PROOF. It is immediate from the definition (37.2.2) of the momentum vector that we need to prove that

$$\int_{x^0=\text{const.}} \frac{\partial \psi^{ikl}}{\partial x^l}\, dS_k = 0. \tag{10}$$

Since $\psi^{i00} = 0$ (by the skew-symmetry), and since the restriction of dS_k to the surface $x^0 = \text{const.}$ is zero if $k \neq 0$, it follows that the left-hand side of (10) is equal to

$$\int_S \operatorname{div}(\psi^{i01}, \psi^{i02}, \psi^{i03})\, dx^1 \wedge dx^2 \wedge dx^3,$$

where S is the Euclidean space $x^0 = \text{const.}$ Now from the Gauss–Ostrogradskiĭ formula (see §26.3(37)) we have

$$\int_{|x| \leq R} \operatorname{div}(\psi^{i01}, \psi^{i02}, \psi^{i03})\, dx^1 \wedge dx^2 \wedge dx^3 = \int_{|x| = R} \langle \psi_i, n \rangle\, d\sigma, \quad (11)$$

where $\psi_i = (\psi^{i01}, \psi^{i02}, \psi^{i03})$, n is the unit outward normal to the 2-sphere $(x^1)^2 + (x^2)^2 + (x^3)^2 = R^2$ in S, and $d\sigma$ is the element of area on that sphere. Assuming that the components ψ_{i0j} approach zero sufficiently quickly as the radius $R \to \infty$ (which is what is meant by "suitably behaved out towards ∞"), we see that as $R \to \infty$ the right-hand side of (11), and therefore also the left-hand side, approach zero, i.e. the integral (10) is zero, as required. □

We thus see that (at least in the cases of interest to physics) given a Lagrangian, we may assume that the energy-momentum tensor is symmetric without thereby altering the momentum vector. We shall suppose in what follows, therefore, that T_{ik} is symmetric.

It is a consequence of the symmetry (assumed) of T_{ik} that the "angular momentum" is conserved.

37.2.5. Definition. The *angular momentum* is the tensor

$$M^{ik} = \int (x^i\, dP^k - x^k\, dP^i) = \frac{1}{c} \int_{x^0 = \text{const.}} (x^i T^{kl} - x^k T^{il})\, dS_l.$$

This is the natural generalization of the classical formula for the angular momentum, defined for a system of particles by (cf. §32.2, Example (d))

$$M^{ik} = \sum (P^i x^k - P^k x^i),$$

where the summation is over all particles in the system.

37.2.6. Lemma. *If the energy-momentum tensor T^{ik} is symmetric, then (for suitable Lagrangians) the angular momentum (the tensor M^{ik}) is conserved.*†

† A more appropriate statement would be that the formula that we have taken as defining the angular momentum follows, under the assumption that the energy-momentum tensor is symmetric, from a more general physical definition of the angular momentum which incorporates its conservation.

Proof. If we imitate the proof of Proposition 37.2.3 (with the expression $x^iT^{kl} - x^kT^{il}$ in the role of T^{ik}) to show that

$$M^{ik}(x_1^0) = \frac{1}{c}\int_{D_1} (x^iT^{kl} - x^kT^{il})\, dS_l$$

is equal to

$$M^{ik}(x_2^0) = \frac{1}{c}\int_{D_2} (x^iT^{kl} - x^kT^{il})\, dS_l,$$

where D_1 and D_2 are given respectively by $x^0 = x_1^0$, $x^0 = x_2^0$, then our task reduces to showing that the integrand in the formula for the M^{ik} has zero divergence, i.e.

$$\frac{\partial}{\partial x^l}(x^iT^{kl} - x^kT^{il}) = 0.$$

Calculating out the left-hand side of this, we obtain

$$\frac{\partial}{\partial x^l}(x^iT^{kl} - x^kT^{il}) = T^{ki} - T^{ik} = 0,$$

where we have used, in turn, the identity $\partial T^{kl}/\partial x^l \equiv 0$, and the symmetry of the tensor T^{ik}. □

We now consider some concrete examples of higher-dimensional variational problems.

37.3. The Equations of an Electromagnetic Field

The equations describing electromagnetic fields (i.e. Maxwell's equations) turn out to be just the Euler–Lagrange equations corresponding to the particular action (i.e. functional) $S = S_f + S_m + S_{mf}$, which we shall now elaborate.

We begin with the term S_m. This is that part of the action due to the particles (i.e. charges) considered separately from the field through which they move, i.e. the action of the charges assuming the field absent. The appropriate action, as it is usually defined, is given by

$$S_m = -\sum_i m_i c \int_a^b dl,$$

where the sum is taken over all of the particles in the field, of masses m_i, c is the speed of light, and the integral $\int_a^b dl$ (where l denotes arc length) is taken over the arc of the world-line of the particle in $\mathbb{R}_{1,3}^4$ between the two fixed events corresponding to the positions of the particle at an initial time t_1 and a later time t_2. As we saw in Example (a) of §32.2, the action $-mc\int_a^b dl$ of each particle can be expressed in the 3-dimensional form $\int_{t_1}^{t_2} L\,dt$ where $L = -mc^2\sqrt{1-(v^2/c^2)}$, v being the 3-dimensional velocity of the particle.

The term S_{mf}, representing that part of the action determined by the mutual interaction of the particles and the field, is usually defined by

$$S_{mf} = -\sum_j \frac{e_j}{c} \int A_k^{(j)}\,dx^k,$$

where again the summation is over all particles (indexed by j), where e_j is the charge on the jth particle, where, as for S_m, the integral is taken along an arc of the world-line of the jth particle, and where (A_i) is a given 4-covector defined on $\mathbb{R}_1^4$ (the so-called "4-potential") which characterizes the field. (In the above defining expression for S_{mf} the superscript (j) indicates merely that the integral is to be taken along the world-line of the jth particle, i.e. that for the calculation of the integral only the values of (A_i) on the jth particle's world-line are relevant.) Note that the interaction of the particles with the field is registered in the expression for S_{mf} only by the presence of the single parameter e_j.

Thus the appropriate action insofar as it involves a charged particle in an electromagnetic field is given by

$$S_m + S_{mf} = \int_a^b \left(-mc\,dl - \frac{e}{c} A_k\,dx^k\right).$$

Finally, the term S_f is that part of the action depending on the properties of the field alone, i.e. the action due to the field in the assumed absence of charges. If we are interested only in the motion of the particles in a given electromagnetic field, then the term S_f need not be considered; on the other hand this term is crucial if our interest lies rather in finding equations characterizing the field. By way of preparing for the definition of S_f we introduce some already familiar concepts of electromagnetic-field theory, defining them in terms of the basic 4-potential (A_i).

The three spatial components A^1, A^2, A^3 of the 4-vector (A^i) obtained by raising the index of the tensor (A_i) (for this purpose resorting, of course, to the Minkowski metric), define a 3-vector $\mathbf{A}$ called the *vector-potential* of the field. The remaining component A^0, perhaps more familiarly denoted by φ, is called the *scalar potential* of the field. The *electric field strength* is then the 3-vector

$$\mathbf{E} = \frac{1}{c}\frac{\partial \mathbf{A}}{\partial t} - \operatorname{grad}\varphi,$$

while the *magnetic field strength* is by definition the 3-vector $\mathbf{H} = \operatorname{curl}\mathbf{A}$. An electromagnetic field (defined by a given 4-covector (A_i)) is said to be an *electric field* if $\mathbf{E} \neq 0$, $\mathbf{H} = 0$, and a *magnetic field* if $\mathbf{E} = 0$, $\mathbf{H} \neq 0$. Finally the *electromagnetic field tensor* (F_{ik}) is defined by

$$F_{ik} = \frac{\partial A_k}{\partial x^i} - \frac{\partial A_i}{\partial x^k}.$$

We are now ready for the definition of S_f: we set

$$S_f = a \int 2(E^2 - H^2)\, d^4x,$$

where $H^2 = \langle \mathbf{H}, \mathbf{H} \rangle$, $E^2 = \langle \mathbf{E}, \mathbf{E} \rangle$ are the (Euclidean) scalar squares of the 3-vectors $\mathbf{H}$ and $\mathbf{E}$, where a is a constant (usually taken as $1/16c\pi$), and where with respect to the spatial co-ordinates x^1, x^2, x^3, the integral is taken over the whole of 3-space, while with respect to the variable x^0 (proportional to the time) it is taken over the interval between two fixed instants. Recalling (from §21.1) that $F_{ik}^2 \equiv F_{ik}F^{ik} = 2(H^2 - E^2)$, and substituting for a its customary value, we have

$$S_f = -\frac{1}{16c\pi} \int 2(H^2 - E^2)\, d^4x = -\frac{1}{16c\pi} \int F_{ik}^2\, d^4x.$$

Putting this together with the formulae for S_m and S_{mf} we obtain the formula for the total action S of an electromagnetic field containing charged particles:

$$S = -\sum \int mc\, dl - \sum \int \frac{e}{c} A_k\, dx^k - \frac{1}{16c\pi} \int F_{ik}^2\, d^4x. \tag{12}$$

We have hitherto regarded the charge as a totality of point-charges. It is however sometimes convenient to regard the total charge as being distributed continuously throughout space. In this case the amount of charge contained in the 3-dimensional volume element $dV = dx^1 \wedge dx^2 \wedge dx^3$ is given by $\rho\, dV$, where ρ denotes the point-density of charge (thus ρ depends on x^1, x^2, x^3 and the time t).

We may parametrize the world-line in $\mathbb{R}_1^4$ of a (variable) point-charge by the time: $x^0 = ct$, $x^i = x^i(t)$, $i = 1, 2, 3$. Then (dx^i/dt) is the 4-dimensional velocity vector of the point-charge, and it is natural to call the 4-vector (j^i) defined by $j^i = \rho(dx^i/dt)$, the *current 4-vector*. The three spatial components of this 4-vector define the usual current 3-vector $\mathbf{j} = \rho v$, where v is the charge velocity at the given point, while the component j^0 is just $c\rho$. Direct calculation shows that in terms of the current vector (j^i) the total action (12) takes the form (verify it!)

$$S = -\sum \int mc\, dl - \frac{1}{c^2} \int A_i j^i\, d^4x - \frac{1}{16c\pi} \int F_{ik}^2\, d^4x. \tag{13}$$

(Here in the second and third terms the integrals are over the same region of $\mathbb{R}_1^4$—see above. Note also that the summation in the first term might also be more appropriately replaced by an integral.)

Having defined the action S for an electromagnetic field we are now ready to show that it is an appropriate one, in the sense that Maxwell's equations for the field are just the Euler–Lagrange equations corresponding to S. Since we are interested only in the field, we may take the motion of the charges (i.e. the current) as predetermined, i.e. known in advance. Thus since we are, as it were, given the trajectories (i.e. world-lines) of the charges in advance, we can restrict our attention to the action $S = S_{mf} + S_f$. Our problem is therefore that of finding the conditions (in the form of the Euler–Lagrange equations) which the 4-potential (A_i) must satisfy for S to have an extreme value. Taking into account the assumption that in the term S_{mf} the current (j^i) is not to be regarded as subject to variation, we have from (13) that the corresponding Lagrangian is

$$L = L\left(A_i, \frac{\partial A_i}{\partial x^\alpha}\right) = -\frac{1}{c}\left(\frac{1}{c} j^i A_i + \frac{1}{16\pi} F_{ik}^2\right), \tag{14}$$

where $F_{ik} = \partial A_k/\partial x^i - \partial A_i/\partial x^k$. With L given by (14), the Euler–Lagrange equations (3) become

$$\frac{\partial L}{\partial A_i} - \frac{\partial}{\partial x^k}\left(\frac{\partial L}{\partial A_{i,x^k}}\right) = 0, \qquad i = 0, 1, 2, 3, \tag{15}$$

where $A_{i,x^k} = \partial A_i/\partial x^k$. (Note here that the derivation of the Euler–Lagrange equations (3) carries over to the case where the region of integration D is unbounded, under appropriate restrictions on L and the perturbations.)

From (14) it follows that

$$\frac{\partial L}{\partial A_i} = -\frac{1}{c^2} j^i,$$

and (after a little calculation) that

$$\frac{\partial L}{\partial A_{i,x^k}} = \frac{1}{4c\pi} F^{ik}. \tag{16}$$

Hence (15) may be rewritten as

$$\frac{\partial F^{ik}}{\partial x^k} = -\frac{4\pi}{c} j^i, \qquad i = 0, 1, 2, 3. \tag{17}$$

Thus these are the Euler–Lagrange equations obtained by variation of the 4-potential in the action $S = S_{mf} + S_f$.

If we write these four equations in their 3-dimensional form, they will reveal themselves as Maxwell's equations (in their familiar classical guise). Recall first (from §21.1) that in terms of the co-ordinates $x^0 = ct$, $x^1 = x$, $x^2 = y$, $x^3 = z$, the tensor (F^{ik}) has the form

$$F^{ik} = \begin{pmatrix} 0 & -E_x & -E_y & -E_z \\ E_x & 0 & H_z & -H_y \\ E_y & -H_z & 0 & H_x \\ E_z & H_y & -H_x & 0 \end{pmatrix}.$$

(This can also be calculated easily from the definitions of **E**, **H** and F_{ik} above.)

In terms of these explicit components, the first ($i = 1$) of the equations (17), namely

$$\frac{1}{c}\frac{\partial F^{10}}{\partial t} + \frac{\partial F^{11}}{\partial x} + \frac{\partial F^{12}}{\partial y} + \frac{\partial F^{13}}{\partial z} = -\frac{4\pi}{c} j^1,$$

becomes

$$\frac{1}{c}\frac{\partial E_x}{\partial t} + \frac{\partial H_z}{\partial y} - \frac{\partial H_y}{\partial z} = -\frac{4\pi}{c} j_x.$$

This together with the second ($i = 2$) and third ($i = 3$) of the equations (17) yields

$$\operatorname{curl} \mathbf{H} = -\frac{1}{c}\frac{\partial \mathbf{E}}{\partial t} + \frac{4\pi}{c}\mathbf{j}. \tag{18}$$

The zero-th equation becomes

$$\frac{\partial(E_x)}{\partial x} + \frac{\partial(E_y)}{\partial y} + \frac{\partial(E_z)}{\partial z} = \frac{4\pi}{c} c\rho,$$

that is,

$$\operatorname{div} \mathbf{E} = 4\pi\rho. \tag{19}$$

Equations (18) and (19) comprise the "second pair" of Maxwell's equations. The "first pair", namely

$$\operatorname{curl} \mathbf{E} = -\frac{1}{c}\frac{\partial \mathbf{H}}{\partial t}, \tag{20}$$

$$\operatorname{div} \mathbf{H} = 0, \tag{21}$$

follow from the definitions (above) of **H** and **E**. Thus we have retrieved Maxwell's equations ((18)–(21)) for the electromagnetic field; these are the fundamental equations of electrodynamics.

To conclude the subsection we find an explicit expression for the energy-momentum tensor of an electromagnetic field under the condition that there are no charges present. In this case the action is just S_f which we defined as $-(1/16c\pi)\int F_{ik}^2\, d^4x$, so that the corresponding Lagrangian L takes the simple form

$$L = -\frac{1}{16c\pi} F_{kl}^2 = -\frac{1}{16c\pi}\left(\frac{\partial A_l}{\partial x^k} - \frac{\partial A_k}{\partial x^l}\right)^2.$$

From the defining equation (8) for the energy-momentum tensor T_{ik}, with the above particular Lagrangian L and with (A^i) in place of (f^i), we obtain

$$T_i^k = \frac{\partial A_l}{\partial x^i} \frac{\partial L}{\partial\left(\dfrac{\partial A_l}{\partial x^k}\right)} - \delta_i^k L,$$

The same calculation as yielded (16) gives

$$\frac{\partial L}{\partial\left(\frac{\partial A_l}{\partial x^k}\right)} = -\frac{1}{4c\pi} F^{kl},$$

whence

$$T_i^k = -\frac{1}{4\pi c}\frac{\partial A_l}{\partial x^i} F^{kl} + \frac{1}{16\pi c}\delta_i^k F_{lm} F^{lm}.$$

Raising the index i (by means of the Minkowski metric g^{ik}), we then obtain

$$T^{ik} = -\frac{g^{im}}{4\pi c}\frac{\partial A_l}{\partial x^m} F^{kl} + \frac{1}{16\pi c} g^{ik} F_{lm} F^{lm}. \tag{22}$$

However this tensor is not symmetric. To rectify this we proceed as outlined above (in the paragraph following Definition 37.2.1). Thus to symmetrize the right-hand side of (22) we add to it the term

$$\frac{1}{4\pi c} g^{im} \frac{\partial A_m}{\partial x^l} F^{kl} = \frac{1}{4\pi c}\frac{\partial A^i}{\partial x^l} F^{kl}, \tag{23}$$

which can be expressed in the requisite form $(\partial/\partial x^l)(\psi^{ikl})$. This can be seen as follows:

$$\frac{\partial A^i}{\partial x^l} F^{kl} = \frac{\partial}{\partial x^l}(A^i F^{kl}) - A^i \frac{\partial F^{kl}}{\partial x^l} = \frac{\partial}{\partial x^l}(A^i F^{kl}),$$

where we have used $\partial F^{kl}/\partial x^l = 0$, this being just the form taken by Maxwell's equations (17) under our current assumption of zero charge (whence $(j^i) = 0$). (By Lemma 37.2.4, under suitable conditions the momentum vector is unaffected by such symmetrization of the energy-momentum tensor.) Thus adding the expression (23) to the right-hand side of (22) and using $F_{il} = \partial A_l/\partial x^i - \partial A_i/\partial x^l$, we finally obtain the formula for the symmetric energy-momentum tensor of a charge-free electromagnetic field (cf. §21.2(25)):

$$T^{ik} = \frac{1}{4\pi c}\left(-F^{il}F_l^k + \tfrac{1}{4} g^{ik} F_{lm} F^{lm}\right). \tag{24}$$

Exercises

1. Suppose that the charge density ρ is zero and that the components A_i of the vector-potential are of the form $A_i(x^1 - ct)$, $i = 0, 1, 2, 3$. Prove that if each $A_i(x)$ is a smooth function, bounded for all x, then the field invariants (i.e. the eigenvalues of (T_{ik})) are zero. (This is the case of electromagnetic waves propagated in a single direction (see the conclusion of §21.2).)

2. Sometimes "non-local" invariants of the field are considered. For example for a monochromatic field of (fixed) frequency ω, but whose dependence on the spatial coordinates x, y, z is arbitrary, the following integral is considered:

$$I = \int_{\mathbb{R}^6} \frac{\mathbf{E}(r_1)\mathbf{E}(r_2) + \mathbf{H}(r_1)\mathbf{H}(r_2)}{|r_1 - r_2|} d^3r_1\, d^3r_2.$$

Show that this quantity is invariant under Lorentz transformations. (It, or some constant multiple of it, is called the "photon number" of the field.)

37.4. The Equations of a Gravitational Field

Let g_{ij} be a pseudo-Riemannian metric of the same type (1, 3) as the Minkowski metric, on 4-dimensional space-time $\mathbb{R}^4$, and let Γ^i_{jk} be the connexion compatible with this metric. In Einsteinian general relativity this metric (with corresponding element of length given by $dl^2 = g_{ij}\, dx^i\, dx^j$) is intended to be identified with the gravitational field. With respect to an "inertial" reference frame employing Euclidean spatial co-ordinates $x^1 = x$, $x^2 = y$, $x^3 = z$, and the time $x^0 = ct$, we have $dl^2 = (dx^0)^2 - (dx^1)^2 - (dx^2)^2 - (dx^3)^2$. (We called such co-ordinates "pseudo-Euclidean" or, more particularly, "Minkowski".) A space-time with the Minkowski metric defined globally on it is called *flat*.

We shall however be interested both here and in the sequel in the more general situation of a so-called curved space-time, i.e. one where the pseudo-Riemannian metric varies from point to point. Of course in some neighbourhood of each individual point $x_0 \in \mathbb{R}^4$ co-ordinates can be chosen in terms of which the quadratic form $g_{ij}(x_0)$ becomes Minkowskian.

Since g_{ij} has type (1,3) we have $g = \det(g_{ij}) < 0$. The standard 4-form for the volume-element is $d\Omega = \sqrt{-g}\, d^4x$. In §30.1 we introduced the Riemann curvature tensor R^i_{jkl} defined in terms of the (compatible) affine connexion Γ^i_{jk} (and so ultimately in terms of the metric). (Note that the definition of a connexion compatible with a pseudo-Riemannian metric is the same as that for a Riemannian metric (Definition 29.3.1), and Christoffel's formulae follow as in the latter case.) In §30.1(3) we defined the curvature tensor essentially by the following formula

$$R_{iklm} = \frac{1}{2}\left(\frac{\partial^2 g_{im}}{\partial x^k\, \partial x^l} + \frac{\partial^2 g_{kl}}{\partial x^i\, \partial x^m} - \frac{\partial^2 g_{il}}{\partial x^k\, \partial x^m} - \frac{\partial^2 g_{km}}{\partial x^i\, \partial x^l}\right) + g_{np}(\Gamma^n_{kl}\Gamma^p_{im} - \Gamma^n_{km}\Gamma^p_{il}).$$

From this, one immediately obtains the following formula for the Ricci tensor $R_{ik} = R^q_{iqk} = g^{lm}R_{limk}$ (see 30.3.1):

$$R_{ik} = \frac{\partial \Gamma^l_{ik}}{\partial x^l} - \frac{\partial \Gamma^l_{il}}{\partial x^k} + \Gamma^l_{ik}\Gamma^m_{lm} - \Gamma^m_{il}\Gamma^l_{km}. \tag{25}$$

Recall also that we defined the "scalar curvature" R by $R = g^{ik}R_{ik} = g^{il}g^{km}R_{iklm}$.

We shall obtain Einstein's equations for a gravitational field as the Euler–Lagrange equations corresponding to an action $S = S_g + S_m$. The action S_g contributed by the field in the absence of matter, is taken to be the "Hilbert action" of the field, which is given simply by

$$S_g = \int R \, d\Omega, \qquad d\Omega = \sqrt{|g|} \, d^4x,$$

where the integral is taken over the region of $\mathbb{R}^4$ determined by $x_1^0 \le x^0 \le x_2^0$ where x_1^0, x_2^0 are fixed (thus the spatial co-ordinates take all real values independently of one another).

The Euler-Lagrange equations $\delta S_g/\delta g^{ij} = 0$ corresponding to the action S_g alone, are given by the following

37.4.1. Theorem. *The variational derivative $\delta S_g/\delta g^{ij}$ is given by*

$$\frac{\delta S_g}{\delta g^{ij}} = \frac{\delta \int R\sqrt{|g|} \, d^4x}{\delta g^{ij}} = (R_{ij} - \tfrac{1}{2} R g_{ij})\sqrt{|g|},$$

that is,

$$\delta S_g = \int \left(R_{ik} - \frac{1}{2} R_{ik} \right) \delta g^{ik} \sqrt{|g|} \, d^4x.$$

PROOF. To begin with we show that the contributions to δS_g of all terms involving the second derivatives $\partial^2 g_{ik}/\partial x^p \, \partial x^q$, are zero, whence we obtain a formula for δS_g (see (28) or (31) below) involving explicitly only the components g_{ik} of the metric, and the Christoffel symbols Γ^i_{jk}. From (25) we have

$$R\sqrt{-g} = \sqrt{-g}\, g^{ik} R_{ik} = \sqrt{-g} \left(g^{ik} \frac{\partial \Gamma^l_{ik}}{\partial x^l} - g^{ik} \frac{\partial \Gamma^l_{il}}{\partial x^k} + g^{ik} \Gamma^l_{ik} \Gamma^m_{lm} - g^{ik} \Gamma^m_{il} \Gamma^l_{km} \right). \tag{26}$$

The first two sums in the last expression can be rewritten as

$$\sqrt{-g}\, g^{ik} \frac{\partial \Gamma^l_{ik}}{\partial x^l} = \frac{\partial}{\partial x^l}(\sqrt{-g}\, g^{ik} \Gamma^l_{ik}) - \Gamma^l_{ik} \frac{\partial}{\partial x^l}(\sqrt{-g}\, g^{ik}),$$

$$\sqrt{-g}\, g^{ik} \frac{\partial \Gamma^l_{il}}{\partial x^k} = \frac{\partial}{\partial x^k}(\sqrt{-g}\, g^{ik} \Gamma^l_{il}) - \Gamma^l_{il} \frac{\partial}{\partial x^k}(\sqrt{-g}\, g^{ik}). \tag{27}$$

By applying the general Stokes formula to the integrals of each of the divergences $(\partial/\partial x^l)(\sqrt{-g}\, g^{ik} \Gamma^l_{ik})$ and $(\partial/\partial x^k)(\sqrt{-g}\, g^{ik} \Gamma^l_{il})$ (initially over a bounded subregion ultimately allowed to expand to the full region of integration), and then arguing as in the derivation of the Euler–Lagrange equations (3), it is readily shown that, with appropriate restrictions on the perturbations, the contributions to δS_g from these divergences are zero. From this, (26) and (27) it follows that

$$\delta \int R\, d\Omega = \delta \int G\sqrt{-g}\, d^4x,$$

where

$$G\sqrt{-g} = \Gamma^l_{il}\frac{\partial}{\partial x^k}(\sqrt{-g}\, g^{ik}) - \Gamma^l_{ik}\frac{\partial}{\partial x^l}(\sqrt{-g}\, g^{ik})$$
$$- (\Gamma^m_{il}\Gamma^l_{km} - \Gamma^l_{ik}\Gamma^m_{lm})g^{ik}\sqrt{-g}, \tag{28}$$

which involves explicitly only the g^{ij} and (via the Γ^i_{jk}) their first derivatives.

From Christoffel's formula (Theorem 29.3.2) expressing the compatible connexion Γ^i_{jk} in terms of the metric, it follows that (see Exercise 11(i) of §29.5)

$$g^{kl}\Gamma^i_{kl} = -\frac{1}{\sqrt{-g}}\frac{\partial}{\partial x^k}(\sqrt{-g}\, g^{ik}).$$

Substituting from this in the first term in the right-hand side of (28), and dividing by $\sqrt{-g}$, we obtain

$$G = -\Gamma^l_{il}\Gamma^i_{kp}g^{kp} - \Gamma^l_{ik}\frac{1}{\sqrt{-g}}\frac{\partial}{\partial x^l}(\sqrt{-g}\, g^{ik}) - (\Gamma^m_{il}\Gamma^l_{km} - \Gamma^l_{ik}\Gamma^m_{lm})g^{ik}. \tag{29}$$

Differentiating the product in the second term, we get

$$-\Gamma^l_{ik}\frac{1}{\sqrt{-g}}\frac{\partial}{\partial x^l}(\sqrt{-g}\, g^{ik}) = \Gamma^l_{ik}\frac{1}{\sqrt{-g}}\frac{1}{2\sqrt{-g}}\frac{\partial g}{\partial x^l}g^{ik} - \Gamma^l_{ik}\frac{\partial g^{ik}}{\partial x^l}$$
$$= -\Gamma^l_{ik}g^{ik}\Gamma^p_{lp} - \Gamma^l_{ik}\frac{\partial g^{ik}}{\partial x^l}, \tag{30}$$

where the last equality follows from the fact that $\Gamma^i_{ki} = (1/2g)(\partial g/\partial x^k)$ (see Exercise 11(ii) of §29.5). From the compatibility of the connexion we also have $\partial g^{ik}/\partial x^l = -\Gamma^i_{ml}\, g^{mk} - \Gamma^k_{ml}g^{im}$. Substituting from this in the last expression in (30), we obtain

$$-\Gamma^l_{ik}\Gamma^p_{lp}g^{ik} - \Gamma^l_{ik}\frac{\partial g^{ik}}{\partial x^l} = -\Gamma^l_{ik}\Gamma^p_{lp}g^{ik} + \Gamma^l_{ik}\Gamma^i_{ml}g^{mk} + \Gamma^l_{ik}\Gamma^k_{ml}g^{im}$$

$$= -\Gamma^l_{ik}\Gamma^m_{lm}g^{ik} + \Gamma^l_{ik}\Gamma^i_{ml}g^{mk} + \Gamma^i_{ml}\Gamma^l_{ki}g^{km}$$

$$= -\Gamma^l_{ik}\Gamma^m_{lm}g^{ik} + 2\Gamma^l_{ik}\Gamma^i_{ml}g^{mk},$$

whence (29) becomes

$$\begin{aligned} G &= 2\Gamma^l_{ik}\Gamma^i_{ml}g^{mk} - \Gamma^l_{ik}\Gamma^m_{lm}g^{ik} - \Gamma^m_{im}\Gamma^i_{kl}g^{kl} - g^{ik}(\Gamma^m_{il}\Gamma^l_{km} - \Gamma^l_{ik}\Gamma^m_{lm}) \\ &= g^{ik}(2\Gamma^l_{mk}\Gamma^m_{il} - \Gamma^l_{ik}\Gamma^m_{lm} - \Gamma^m_{lm}\Gamma^l_{ki}) - g^{ik}(\Gamma^m_{il}\Gamma^l_{km} - \Gamma^l_{ik}\Gamma^m_{lm}) \\ &= 2g^{ik}(\Gamma^l_{mk}\Gamma^m_{il} - \Gamma^l_{ik}\Gamma^m_{lm}) - g^{ik}(\Gamma^m_{il}\Gamma^l_{km} - \Gamma^l_{ik}\Gamma^m_{lm}) = g^{ik}(\Gamma^m_{il}\Gamma^l_{mk} - \Gamma^l_{ik}\Gamma^m_{lm}). \end{aligned}$$

Thus the upshot of these calculations is that the variation of our integral has the following simple form:

$$\delta \int R\sqrt{|g|}\, d^4x = \delta \int g^{ik}(\Gamma^m_{il}\Gamma^l_{mk} - \Gamma^l_{ik}\Gamma^m_{lm})\sqrt{|g|}\, d^4x. \tag{31}$$

Since the integrand L say, in the right-hand side of (31) can be expressed as a function of the g^{ij} and $\partial g^{ij}/\partial x^\alpha$, we might now seek to establish the theorem by taking, in the formula (2) for $\delta I/\delta f^i$, the function L as the Lagrangian, and the g^{ij} in the role of the f^i. However since the computations are rather formidable we abandon this line of argument and begin the proof anew in a more geometrical vein.

To begin with, note that since g is a function of the g^{ij} alone, we have by Taylor's theorem that

$$\begin{aligned} \delta \int R\sqrt{-g}\, d^4x &= \delta \int g^{ik}R_{ik}\sqrt{-g}\, d^4x \\ &= \int \left(R_{ik}\sqrt{-g}\, \delta g^{ik} + R_{ik}g^{ik}\frac{\partial\sqrt{-g}}{\partial g^{rs}}\, \delta g^{rs}\right) d^4x \\ &\quad + \int g^{ik}\sqrt{-g}\,(\delta R_{ik})\, d^4x, \end{aligned} \tag{32}$$

where the δR_{ik} denote the changes in the R_{ik} resulting from the changes δg^{rs} in the g^{rs}.

We first find an expression for $\partial\sqrt{-g}/\partial g^{rs}$ in terms of the g_{rs}. Since the (i, j)th cofactor C_{ij} of the matrix (g^{rs}) does not involve the entry g^{ij} (note that for $i \neq j$ we are considering g^{ij} and g^{ji} as independent), and since $g^{ik}C_{il} = \delta^k_l g$, it follows that $\partial g/\partial g^{rs} = C_{rs} = gg_{rs}$ (where in the last equality we are using $g_{ik}g^{il} = \delta^l_k$). Thus

$$\frac{\partial\sqrt{-g}}{\partial g^{rs}} = \frac{1}{2\sqrt{-g}}\frac{\partial g}{\partial g^{rs}} = -\tfrac{1}{2}\sqrt{-g}\, g_{rs},$$

whence (32) becomes:

$$\delta \int R\sqrt{-g}\, d^4x = \int (R_{ik} - \tfrac{1}{2}Rg_{ik})\delta g^{ik}\sqrt{-g}\, d^4x + \int g^{ik}(\delta R_{ik})\sqrt{-g}\, d^4x. \tag{33}$$

Thus the theorem will follow if we can show that the second integral on the right-hand side of (33) is zero. To this end we first show (with a view to applying the general Stokes formula) that the function $g^{ik}\,\delta R_{ik}$ is the divergence of some vector. Let P be any particular point of the underlying space. By Exercise 4 of §29.5, there exist in some neighbourhood of P co-ordinates ("inertial co-ordinates") in terms of which the $(\partial/\partial x^\alpha)(g^{ik})$ and therefore also the Γ^k_{ij} vanish at P. Thus in terms of such co-ordinates we have that at the point P

$$R_{ik} = \left(\frac{\partial \Gamma^l_{ik}}{\partial x^l} - \frac{\partial \Gamma^l_{il}}{\partial x^k}\right),$$

whence

$$g^{ik}\,\delta R_{ik} = g^{ik}\left(\frac{\partial}{\partial x^l}(\delta\Gamma^l_{ik}) - \frac{\partial}{\partial x^k}(\delta\Gamma^l_{il})\right) = g^{ik}\frac{\partial}{\partial x^l}(\delta\Gamma^l_{ik}) - g^{il}\frac{\partial}{\partial x^l}(\delta\Gamma^k_{ik})$$

$$= \frac{\partial}{\partial x^l}(g^{ik}\,\delta\Gamma^l_{ik} - g^{il}\,\delta\Gamma^k_{ik}). \tag{34}$$

Now it is immediate from Theorem 28.2.1 that if Γ^i_{jk}, $\hat{\Gamma}^i_{jk}$ are two symmetric connexions on a space, compatible with metrics g_{ij}, $\hat{g}_{ij}$ respectively, then although neither the Γ^i_{jk} nor the $\hat{\Gamma}^i_{jk}$ transform like the components of a tensor under arbitrary co-ordinate changes, the differences $\Gamma^i_{jk} - \hat{\Gamma}^i_{jk}$ do transform like the components of a tensor; thus by taking $\hat{g}^{ij} = g^{ij} + \delta g^{ij}$, it follows that the $\delta\Gamma^i_{jk}$ are the components of a tensor. Hence $(g^{ik}\delta\Gamma^l_{ik} - g^{il}\delta\Gamma^k_{ik})$ is a tensor (or, more particularly, a vector). If we denote this vector by (W^l) then from (34), we have that at the point P

$$g^{ik}\,\delta R_{ik} = \frac{\partial W^l}{\partial x^l} = \nabla_l W^l,$$

where in the second equality we have used once again the fact that in terms of the inertial co-ordinates defined on a neighbourhood of P, the Γ^k_{ij} vanish at P. Since covariant differentiation is a tensor operation and since P was arbitrary, it follows that

$$g^{ik}\,\delta R_{ik} = \nabla_l W^l \tag{35}$$

holds true at all points and in terms of any co-ordinates. Now by equation (16) of §29.3 we have (in any co-ordinates)

$$\nabla_l W^l = \frac{1}{\sqrt{-g}}\frac{\partial}{\partial x^l}(\sqrt{-g}\,W^l).$$

From this and (35) we obtain

$$\int g^{ik}\,\delta R_{ik}\sqrt{-g}\,d^4x = \int \frac{\partial}{\partial x^l}(\sqrt{-g}\,W^l)\,d^4x,$$

whence, essentially by applying (as before) the general Stokes formula to the right-hand side integral and then arguing as in the derivation of the Euler–Lagrange equations(3), we infer that

$$\int g^{ik}\,\delta R_{ik}\sqrt{-g}\,d^4x = 0,$$

completing the proof of the theorem. □

The action S_g is usually taken with the constant factor $c^3/16\pi G$, where c is the speed of light and G is the "gravitational constant", in which case the statement of Theorem 37.4.1 amends to

$$\delta S_g = \frac{c^3}{16\pi G}\int\left(R_{ik} - \tfrac{1}{2}Rg_{ik}\right)\delta g^{ik}\sqrt{-g}\,d^4x.$$

If matter is present then the total action is $S_{\text{total}} = S_g + S_m$; here the supplementary action S_m is usually taken to have the form

$$S_m = \frac{1}{c}\int \Lambda\sqrt{-g}\,d^4x,$$

where Λ is a function determined by the properties of the matter (i.e. by the fields defining it), and by the metric, and where the integral is taken over the same region (defined by $x_1^0 \le x^0 \le x_2^0$) as was the integral defining S_g. Given Λ, we can find $\delta S_m/\delta g^{ik}$, and obtain thence the equations for the gravitational field in the form

$$\frac{1}{\lambda\sqrt{-g}}\frac{\delta S_g}{\delta g^{ik}} = R_{ik} - \tfrac{1}{2}Rg_{ik} = -\frac{16\pi G}{c^4}\frac{\delta S_m}{\delta g^{ik}}\frac{1}{\sqrt{-g}}. \tag{36}$$

Note that these equations are non-linear, so that in general the sum of two solutions (for the field g^{ij}) need not be a solution.

In relativity theory the Lagrangian is chosen so that the quantities $-(2/c\sqrt{|g|})(\delta S_m/\delta g^{ik})$ coincide with the components T_{ik} of the energy-momentum tensor of the material system. For instance in the case of an electromagnetic field, taking $\Lambda = \lambda F_{ik}F^{ik}$, (cf. the expression for S_f in the preceding subsection) we obtain directly from (2) that

$$\frac{1}{\sqrt{-g}}\frac{\delta S_m}{\delta g^{ik}} = \frac{1}{\sqrt{-g}}\frac{\partial}{\partial g^{ik}}(\sqrt{-g}\,\Lambda) - \frac{\partial}{\partial x^l}\frac{\partial(\sqrt{-g}\,\Lambda)}{\partial\left(\dfrac{\partial g^{ik}}{\partial x^l}\right)},$$

and it can be shown from this that

$$\frac{1}{\sqrt{|g|}}\frac{\delta S_m}{\delta g^{ik}}\bigg|_{g^{ik} = g^{ik}_{\text{Minkowski}}}$$

coincides (to within a constant factor) with the expression (24) for the energy-momentum tensor of an electromagnetic field.

In the case of so-called "empty" space, i.e. in the absence of matter, we have $T_{ik} \equiv 0$, so that the gravitational field equations take the form $R_{ik} - \frac{1}{2}Rg_{ik} = 0$, or more simply $R_{ik} = 0$ (this was noted in §30.3(iii)). To see how the latter equations are arrived at, note first that (36) may be rewritten as

$$R_k^\alpha - \tfrac{1}{2}R\delta_k^\alpha = \frac{8\pi G}{c^4} T_k^\alpha.$$

On contracting the indices α and k this yields $R - 2R = (8\pi G/c^4)T$, where $T = T_\alpha^\alpha$, whence

$$R = -\frac{8\pi G}{c^4} T.$$

Substituting for R from this into the original equations (36) for the general gravitational field we obtain

$$R_{ik} = \frac{8\pi G}{c^4}(T_{ik} - \tfrac{1}{2}g_{ik}T).$$

Thus if $T_{ik} = 0$, then $R_{ik} = 0$, as claimed.

It does not follow from the equations $R_{ik} \equiv 0$ that empty space-time is flat: the vanishing of the Ricci tensor does not in general entail the vanishing of the Riemann curvature tensor. (Recall that on the other hand if our space were 3-dimensional then $R^i_{jkl} \equiv 0$ would follow from $R_{ik} \equiv 0$, since in this case the Riemann curvature tensor is expressible in terms of the Ricci tensor (see §30.3(ii)).

From Theorem 37.4.1 (or, more precisely, its proof) we obtain the following consequence for 2-dimensional spaces.

37.4.2. Theorem. *Let g_{ij}, $i, j = 1, 2$, be a Riemannian metric on a 2-dimensional space, and let K denote as usual the Gaussian curvature of the space. Then the integral $S[g] = \int K\, dS$, where $dS = \sqrt{g}\, dx^1 \wedge dx^2$, is invariant under smooth local perturbations of the metric.*

PROOF. From Theorem 30.3.3 and its proof we obtain: $K = R/2$, $R_{ij} = \frac{1}{2}Rg_{ij}$, whence $R_{ij} - \frac{1}{2}Rg_{ij} \equiv 0$. Hence from Theorem 37.4.1 (whose proof was, in essence, independent of the number of dimensions and the fact that the metric was pseudo-Riemannian), we have

$$\frac{1}{\sqrt{g}}\frac{\delta S}{\delta g^{ij}} = R_{ij} - \tfrac{1}{2}Rg_{ij} \equiv 0,$$

which yields the desired conclusion. □

Since a closed surface in Euclidean $\mathbb{R}^3$ is by definition compact (and without boundary), we deduce immediately the following celebrated result of Gauss and Bonnet.

37.4.3. Corollary (Gauss–Bonnet). *The integral over a closed surface in 3-dimensional Euclidean space of the Gaussian curvature of the surface, is not changed by smooth deformations of the surface*:

$$\int K\, dS = \text{const.}$$

The value of this constant, and its significance, will be discussed in Part II.

37.5. Soap Films

Consider a smooth hypersurface V^{n-1} in Euclidean $\mathbb{R}^n$ with Euclidean co-ordinates $x^1, \ldots, x^n$, which we shall assume to be given in the graphical form $x^n = f(x^1, \ldots, x^{n-1})$, where the domain of definition of the function f is a bounded region D of $\mathbb{R}^{n-1}$. In this subsection our concern will be with the area functional $S[f]$ defined on the space of all functions f with the fixed domain D. Thus

$$S[f] = \int_D \sqrt{\det A}\, d^{n-1} x,$$

where $A = (g_{ij}(x))$, $x \in D$, is the induced Riemannian metric on the surface V^{n-1} defined by f, and $d^{n-1} x = dx^1 \wedge \cdots \wedge dx^{n-1}$.

We shall now express $\sqrt{\det A}$ explicitly in terms of the function f. We might for this purpose appeal directly to formula (31) of §7.4 (or rather to its generalization to surfaces of $(n-1)$ dimensions). Alternatively we may proceed as follows. Denote by $d\tau^{n-1}$ the form representing the $(n-1)$-dimensional volume element of V^{n-1}; thus $S[f] = \int_D d\tau^{n-1}$. For each point P of V^{n-1}, let $\alpha(P)$ denote the angle between $n(P)$, the unit normal to V^{n-1} at P, and $e_n = (0, \ldots, 0, 1)$. Then as Figure 39 suggests (at least in the case $n = 3$), we have

$$S[f] = \int_D d\tau^{n-1} = \int_D \frac{d^{n-1}x}{\cos \alpha(P)}.$$

Now since $n(P)$ is given by (cf. the formula for the unit normal in the case $n = 3$ given in §8.3(23)):

$$n(P) = \frac{1}{\sqrt{1 + \sum_{i=1}^{n-1} (f_{x^i})^2}}(-f_{x^1}, \ldots, -f_{x^{n-1}}, 1),$$

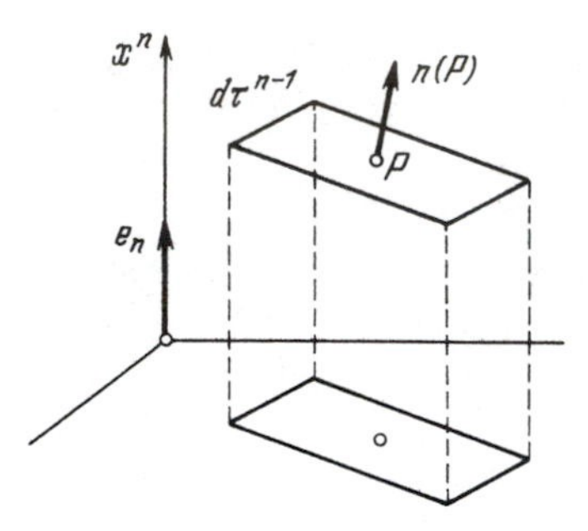

Figure 39

it follows that

$$\cos \alpha(P) = \langle e_n, n(P) \rangle = \frac{1}{\sqrt{1 + \sum_{i=1}^{n-1} (f_{x^i})^2}},$$

whence

$$S[f] = \int_D \sqrt{1 + \sum_{i=1}^{n-1} (f_{x^i})^2}\, dx^1 \wedge \cdots \wedge dx^{n-1}.$$

Hence the Euler–Lagrange equations for the extremal surfaces $x^n = f(x^1, \ldots, x^{n-1})$ over D, take in the present case the form

$$\sum_{i=1}^{n-1} \frac{\partial}{\partial x^i} \left(\frac{f_{x^i}}{\sqrt{1 + \sum_{j=1}^{n-1} (f_{x^j})^2}} \right) = 0. \tag{37}$$

(This follows by substituting in (3); note that here $k = 1$.)

37.5.1. Definition. Surfaces which are extremal with respect to the area functional S, are called *minimal surfaces.*

Remark. In $\mathbb{R}^3$ the minimal surfaces can be modelled by thin soap films (thin in order to render negligible their weight) adhering to closed contours made for instance out of wire.

It is easy to see from (37) that for the two-dimensional minimal surfaces in $\mathbb{R}^3$ (x, y, z) of the form $z = f(x, y)$, the Euler–Lagrange equations have the form (verify it!)

$$(1 + f_x^2) f_{yy} - 2 f_{xy} f_x f_y + (1 + f_y^2) f_{xx} = 0. \tag{38}$$

From the Euler–Lagrange equations (37) for a minimal surface we can obtain a characterization of such a surface in terms of one of its embedding invariants (namely its mean curvature) in $\mathbb{R}^n$.

37.5.2. Theorem. *The mean curvature H of a smooth hypersurface V^{n-1} in n-dimensional Euclidean space is identically zero if and only if in some neighbourhood of each of its points the surface V^{n-1} can be represented in graphical form by a function extremal with respect to the area functional* (*i.e. by a solution of the equations* (37) *for a minimal surface*).

(Thus the condition $H \equiv 0$ is equivalent to (local) minimality of a surface V^{n-1} in $\mathbb{R}^n$.)

This theorem is proved by simply calculating directly the mean curvature H of a surface $x^n = f(x^1, \ldots, x^{n-1})$, and then checking that the equation $H = 0$ is equivalent to the Euler–Lagrange equations (37). (For the calculation of H one uses the definition $H = \operatorname{tr}(A^{-1}Q)$, where A and Q are respectively the matrices of the first and second fundamental forms of the surface; this definition of the mean curvature is the exact n-dimensional analogue of Definition 8.3.1.)

We shall content ourselves with carrying out this programme of proof only in the case $n = 3$, i.e. the case of a 2-dimensional surface V^2 in $\mathbb{R}^3$. Thus let P be any non-singular point of the surface and choose (as several times earlier) Euclidean co-ordinates x, y, z in a neighbourhood of P in $\mathbb{R}^3$ by taking the origin to be P, the z-axis to be perpendicular to V^2 at P, and the x-axis and y-axis tangent to the surface at P. Then in this neighbourhood the surface V^2 is given in terms of these Euclidean co-ordinates by an equation of the form $z = f(x, y)$. By §8.3(25) the entries in the matrix $A = \begin{pmatrix} E & F \\ F & G \end{pmatrix}$ of the first fundamental form are (in this neighbourhood of P) then given in terms of the co-ordinates x, y, z by

$$E = 1 + f_x^2, \qquad F = f_x f_y, \qquad G = 1 + f_y^2, \tag{39}$$

and by §8.3(23) the entries in the matrix $Q = \begin{pmatrix} L & M \\ M & N \end{pmatrix}$ of the second fundamental form are given by

$$L = \frac{f_{xx}}{\sqrt{1 + f_x^2 + f_y^2}}, \qquad M = \frac{f_{xy}}{\sqrt{1 + f_x^2 + f_y^2}}, \qquad N = \frac{f_{yy}}{\sqrt{1 + f_x^2 + f_y^2}}. \tag{40}$$

The mean curvature is

$$H = \operatorname{tr}(A^{-1}Q) = \frac{1}{EG - F^2}(GL - 2FM + EN). \tag{41}$$

Hence the equation $H = 0$ is equivalent to $GL - 2FM + EN = 0$; substituting in the latter from (39) and (40) we obtain the Euler–Lagrange equation (38), as required. □

We wish next to investigate the form taken by the Euler–Lagrange equations for a minimal 2-dimensional surface V^2 in Euclidean $\mathbb{R}^3$, in terms of co-ordinates u, v which are locally conformal. By Theorem 13.1.1 such co-ordinates exist, at least provided that E, F, G are real analytic functions (i.e. are representable as power series in u and v on $D(u, v)$ the region of variation of u and v). Let then $r = r(u, v)$ be (locally) the equation of our surface V^2, where u, v are conformal co-ordinates (and r is the radius-vector at some fixed point). Then (whether u, v are conformal or not)

$$S[r] = \int_{D(u,v)} \sqrt{EG - F^2}\, du\, dv,$$

where by §7.3(19)

$$E = \langle r_u, r_u \rangle, \qquad F = \langle r_u, r_v \rangle, \qquad G = \langle r_v, r_v \rangle, \tag{42}$$

and by §8.1(4) the coefficients of the second fundamental form are given by

$$L = \langle r_{uu}, n \rangle, \qquad M = \langle r_{uv}, n \rangle, \qquad N = \langle r_{vv}, n \rangle,$$

where n is the unit normal vector to the surface. Since our co-ordinates u, v are conformal, we have $F = 0$, $E = G$, whence

$$\begin{aligned} S[r] &= \iint_D \sqrt{\langle r_u, r_u \rangle \langle r_v, r_v \rangle}\, du\, dv \\ &= \iint_D \sqrt{(x_u^2 + y_u^2 + z_u^2)(x_v^2 + y_v^2 + z_v^2)}\, du\, dv, \end{aligned}$$

and also, from (41),

$$H = \frac{1}{E}(L + N) = \frac{1}{E}\langle r_{uu} + r_{vv}, n \rangle = \frac{1}{E}\langle \Delta r, n \rangle, \tag{43}$$

where Δ is the Laplace operator. Now by the case $n = 3$ of Theorem 37.5.2 (which we have just proved) the condition $H = 0$ is equivalent to the Euler–Lagrange equations (37) (for locally defined f). Thus in view of (43) we may regard the equations $\langle \Delta r, n \rangle = 0$ as the Euler–Lagrange equations of a minimal surface in terms of conformal co-ordinates.

We now show that the equations $\langle \Delta r, n \rangle = 0$ are in fact equivalent to $\Delta r = 0$, i.e. to the radius-vector's being as they say "harmonic". It is trivial that $\Delta r = 0$ implies $\langle \Delta r, n \rangle = 0$. For the reverse implication it suffices to show that both $\langle \Delta r, r_u \rangle = 0$ and $\langle \Delta r, r_v \rangle = 0$, since at a non-singular point of V^2 the vectors r_u, r_v, n are linearly independent, so that from these equations together with the additional equation $\langle \Delta r, n \rangle = 0$, it will follow that the inner product of Δr with every vector is zero, whence $\Delta r = 0$. Now the equations $E = G$, $F = 0$ (which hold by virtue of the conformality of the co-ordinates u, v) can by (42) be written more explicitly as $\langle r_u, r_u\rangle = \langle r_v, r_v\rangle$, $\langle r_u, r_v\rangle = 0$. On differentiating with respect to u and v we obtain

$$\begin{aligned} \langle r_{uu}, r_u\rangle &= \langle r_{uv}, r_v\rangle, \\ \langle r_{uv}, r_u\rangle &= \langle r_{vv}, r_v\rangle, \\ \langle r_{uu}, r_v\rangle + \langle r_u, r_{uv}\rangle &= 0, \\ \langle r_{uv}, r_v\rangle + \langle r_u, r_{vv}\rangle &= 0, \end{aligned}$$

from which it follows that

$$\begin{aligned} \langle r_{uu}, r_u\rangle + \langle r_{vv}, r_u\rangle &= 0, \\ \langle r_{uu}, r_v\rangle + \langle r_{vv}, r_v\rangle &= 0, \end{aligned}$$

i.e. that $\Delta r = 0$. We have thus proved

37.5.3. Proposition. *In terms of conformal co-ordinates the radius-vector defining a minimal surface is harmonic.*

Remark. We may speak of the harmonicity of the radius-vector $r(u, v)$ only with respect to a particular co-ordinate system; generally speaking harmonicity is not preserved by co-ordinate transformations.

The actual structure of 2-dimensional minimal surfaces in $\mathbb{R}^3$ can be rather complicated; thus for a given boundary contour $S^1 \subset \mathbb{R}^3$, there are, generally speaking, many "soap films" with that boundary. (In other words there is no uniqueness theorem for the solutions of the differential equation $H = 0$ (or equivalently $\Delta r = 0$).) Figures 40 and 41 depict examples of this.

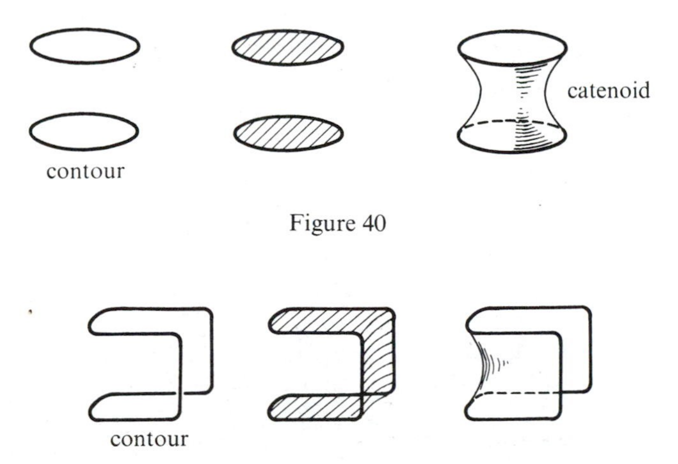

Figure 40

Figure 41

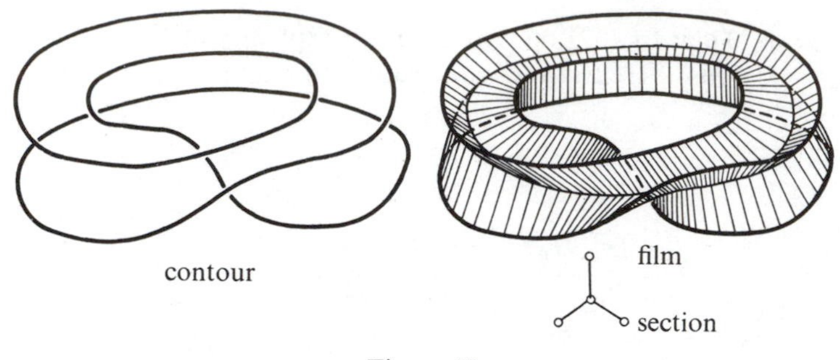

Figure 42

The solutions of the differential equations $H = 0$ (or $\Delta r = 0$) may have singularities. An example (of a "Möbius trefoil") is shown in Figure 42; this particular surface is homeomorphic to the surface with singularities shown in Figure 43.

We now consider a different functional whose Euler–Lagrange equations turn out to be once again equivalent to harmonicity of the radius vector, i.e. for which the extremal surfaces are defined (this time however in terms of arbitrary co-ordinates) by harmonic radius-vectors. Let V^2 be any surface in Euclidean $\mathbb{R}^3$ determined by the 3-dimensional radius-vector $r(u, v)$ (where the (local) co-ordinates u, v are now arbitrary). The *Dirichlet functional* $D[r]$ is defined by

$$D[r] = \int_{D(u,v)} \frac{E + G}{2} \, du \, dv,$$

where as before E, G are the coefficients of du^2 and dv^2 in the first fundamental form of the surface $r(u, v)$. By (42) the Lagrangian is, more explicitly,

$$L(r_u, r_v) = \frac{E + G}{2} = \tfrac{1}{2}(x_u^2 + y_u^2 + z_u^2 + x_v^2 + y_v^2 + z_v^2).$$

It follows by specializing the general formula (3) that the Euler–Lagrange equations corresponding to this Lagrangian are given in vectorial notation by $\Delta r = 0$; thus the extremal surfaces are precisely those with harmonic radius-vector.

Since $(E + G)/2 \geq \sqrt{EG - F^2}$, with equality precisely when both $E = G$ and $F = 0$, it follows that for any piecewise smooth radius-vector $r(u, v)$ we

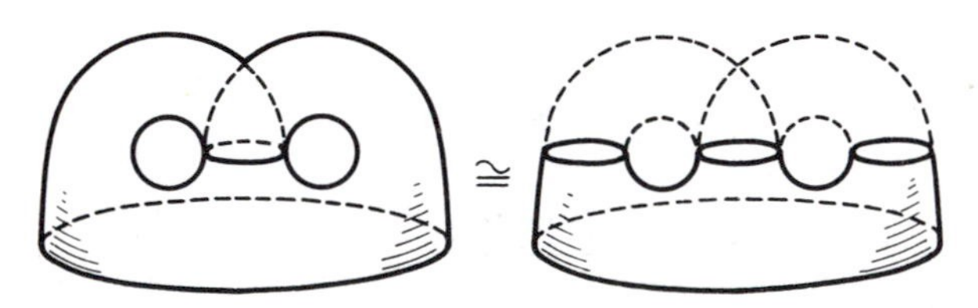

Figure 43

have $D[r] \geq S[r]$, with equality occurring precisely if $E \equiv G$, $F \equiv 0$, i.e. if the co-ordinates u, v are conformal. Thus any extremal surface of the functional $D[r]$ for which the co-ordinates u, v happen to be conformal, will automatically be an extremal surface for the functional $S[r]$. (The converse of this is in general false.) Hence (assuming the existence of conformal co-ordinates) the extremals corresponding to $S[r]$ can all be obtained by selecting from among all harmonic radius-vectors (i.e. extremals corresponding to $D[r]$) those for which the co-ordinates u, v are conformal (and then, if you like, applying any non-singular co-ordinate transformation). (See Proposition 37.5.3 and the discussion preceding it.)

Harmonic radius vectors $r(u, v)$ for which the co-ordinates u, v are not conformal will, by the above, not determine minimal surfaces. As an example consider the radius-vector $r(u, v) = (u, v, \operatorname{Re} f(u + iv))$, the graph of the real part (the imaginary part would serve equally well) of a non-linear complex analytic function $f(u + iv)$.

The relationship between the functionals $D[r]$ and $S[r]$ is in many ways analogous to that between the length functional $l_a^b[\gamma]$ and the action $S_a^b[\gamma] = \int_a^b |\dot{\gamma}|^2 \, dt$, of a path γ (see §31.2). It is clear that

$$(l_a^b[\gamma])^2 \leq (b - a) S_a^b[\gamma],$$

with equality occurring precisely when the parameter t in terms of which the path γ is parametrized is proportional to arc length on γ. (Recall from Theorem 31.2.2 that the extremals of $l_b^a[\gamma]$ satisfy the equation for the geodesics provided the parameter is natural in this sense.) This similarity of relationships is connected with the fact that the functionals $l_a^b[\gamma]$ and $S[r]$ are invariant under arbitrary transformations of the parameters (co-ordinates), while the functionals $S_a^b[\gamma]$ and $D[r]$ are not.

37.6. Equilibrium Equation for a Thin Plate

We now consider a particular example from the theory of the equilibrium of elastic bodies, namely the equilibrium of a thin elastic plate under bending. By "thin" we mean that we assume the plate's thickness to be negligible by comparison with its other two dimensions. We shall assume also that in its undeformed state the plate is planar, and that the deformation (or strain) is small, i.e. that the amounts by which the points of the plate are displaced are negligible by comparison with the thickness of the plate. Our object is to obtain the equations for equilibrium under bending of such a plate by applying our variational method to the functional representing its free (or elastic) energy.

Under bending, those parts of the plate closer to the convex side undergo stretching, while those nearer the concave side undergo compression. The amount of compression (or stretching) decreases with depth (or height) into the plate, until the so-called "neutral surface" is reached, where there

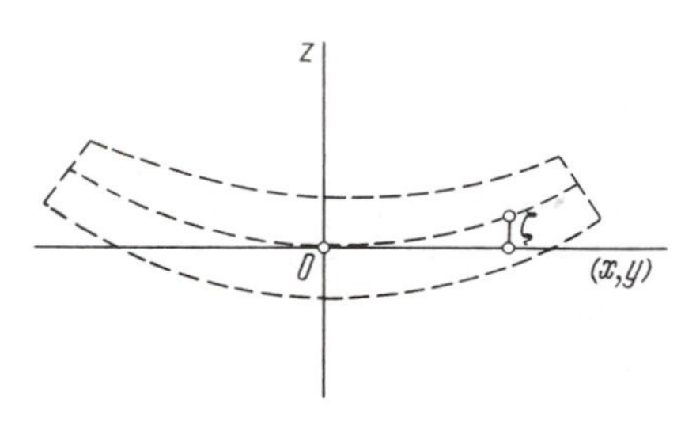

Figure 44

occurs neither compression nor stretching. The neutral surface separates the plate into two halves each of thickness half the original (see Figure 44).

We introduce a rectangular Cartesian (i.e. Euclidean) co-ordinate system, taking the origin 0 to be on the neutral surface, the z-axis to be perpendicular to that surface, and the (x, y)-plane to coincide with the plane of the neutral surface when the plate is in its undeformed state. Denote by $\zeta(x, y)$ the vertical displacement of the points of the neutral surface of the deformed plate. It can be shown that if h denotes the thickness of the plate, then the free energy of the deformed elastic plate is given by

$$F = \frac{Eh^3}{24(1-\sigma^2)} \iint \left\{ \left(\frac{\partial^2 \zeta}{\partial x^2} + \frac{\partial^2 \zeta}{\partial y^2} \right)^2 + 2(1-\sigma) \left[\left(\frac{\partial^2 \zeta}{\partial x\, \partial y} \right)^2 - \frac{\partial^2 \zeta \partial^2 \zeta}{\partial x^2 \partial y^2} \right] \right\} dx\, dy,$$

where the integral is taken over the domain of definition of the displacement function $\zeta = \zeta(x, y)$, E is Young's elastic modulus, and σ is the Poisson coefficient. (The constants E and σ are calculated from

$$E = \frac{9K\mu}{3K + \mu}, \qquad \sigma = \frac{1}{2} \frac{3K - 2\mu}{3K + \mu},$$

where K and μ are respectively the compression and shearing moduli of the material from which the plate is made.)

We now calculate the variation in the functional F due to a perturbation $\delta\zeta$ (which in this case is permitted to be non-zero at the edge of the plate). Note first that in view of the assumption that the deformation is "small", we may in what follows, without undue sacrifice of accuracy, use interchangeably dS, the element of area on the neutral surface, and $dx\, dy$, the element of area on the (x, y)-plane. We begin by rewriting F as a sum of two integrals:

$$F = \frac{Eh^3}{24(1-\sigma^2)} \left\{ \iint (\Delta\zeta)^2\, dS + 2(1-\sigma) \iint \left[\left(\frac{\partial^2 \zeta}{\partial x\, \partial y} \right)^2 - \frac{\partial^2 \zeta}{\partial x^2} \frac{\partial^2 \zeta}{\partial y^2} \right] dS \right\}. \tag{44}$$

(Here Δ denotes, as usual, the Laplace operator.) We shall calculate the variations in each of these two integrals separately. Beginning with the first integral on the right-hand side of (44), we have

$$\delta \frac{1}{2} \iint (\Delta\zeta)^2 \, dS = \iint (\Delta\zeta)(\Delta\delta\zeta) \, dS = \iint (\Delta\zeta)(\operatorname{div} \operatorname{grad} \delta\zeta) \, dS$$

$$= \iint \operatorname{div}(\Delta\zeta \operatorname{grad} \delta\zeta) \, dS - \iint \langle \operatorname{grad} \delta\zeta, \operatorname{grad} \Delta\zeta \rangle \, dS. \quad (45)$$

Denote by γ the (closed) boundary ∂D of the domain of the function $\zeta(x, y)$. (A natural choice for γ would be the closed curve bounding the neutral surface when the plate is in its undeformed state.) Then using Green's formula (§26.3(33)), it is easy to verify that

$$\iint_D \operatorname{div}(\Delta\zeta \operatorname{grad} \delta\zeta) \, dx \, dy = \oint_\gamma \Delta\zeta \langle n, \operatorname{grad} \delta\zeta \rangle \, dl,$$

where dl is the element of arc length along γ, and n is the unit normal vector to γ. Since $\langle n, \operatorname{grad} \delta\zeta \rangle = (\partial/\partial n)(\delta\zeta)$, the directional derivative of $\delta\zeta$ in the direction n, it follows that (to within acceptable limits of accuracy)

$$\iint_D \operatorname{div}(\Delta\zeta \operatorname{grad} \delta\zeta) \, dS = \oint_\gamma \Delta\zeta \frac{\partial}{\partial n}(\delta\zeta) \, dl. \quad (46)$$

Similarly

$$\iint_D \langle \operatorname{grad} \delta\zeta, \operatorname{grad} \Delta\zeta \rangle \, dS = \iint_D \operatorname{div}((\delta\zeta) \operatorname{grad} \Delta\zeta) \, dS - \iint_D (\delta\zeta) \, \Delta^2\zeta \, dS$$

$$= \oint_\gamma \delta\zeta \langle n, \operatorname{grad} \Delta\zeta \rangle \, dl - \iint_D (\delta\zeta) \, \Delta^2\zeta \, dS$$

$$= \oint_\gamma \delta\zeta \frac{\partial(\Delta\zeta)}{\partial n} \, dl - \iint_D (\delta\zeta) \, \Delta^2\zeta \, dS.$$

From this, (46) and (45) we obtain

$$\delta \frac{1}{2} \iint_D (\Delta\zeta)^2 \, dS = \iint_D (\delta\zeta) \Delta^2\zeta \, dS - \oint_\gamma \delta\zeta \frac{\partial(\Delta\zeta)}{\partial n} \, dl + \oint_\gamma \Delta\zeta \frac{\partial(\delta\zeta)}{\partial n} \, dl. \quad (47)$$

Transferring our attention to the second integral on the right-hand side of (44), we have (neglecting, as usual when applying the variational technique, second-order terms in the first and second partial derivatives of $\delta\zeta$)

$$\delta \iint_D \left[\left(\frac{\partial^2 \zeta}{\partial x \, \partial y} \right)^2 - \frac{\partial^2 \zeta}{\partial x^2} \frac{\partial^2 \zeta}{\partial y^2} \right] dS = \iint_D \left[2 \frac{\partial^2 \zeta}{\partial x \, \partial y} \frac{\partial^2 \delta\zeta}{\partial x \, \partial y} - \frac{\partial^2 \zeta}{\partial x^2} \frac{\partial^2 \delta\zeta}{\partial y^2} - \frac{\partial^2 \delta\zeta}{\partial x^2} \frac{\partial^2 \zeta}{\partial y^2} \right] dS.$$

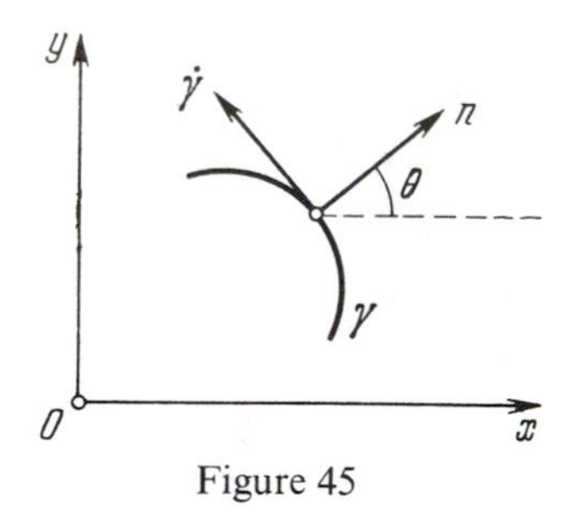

Figure 45

The integrand in the right-hand side integral can be expressed as the divergence of a vector T:

$$\frac{\partial}{\partial x}\left(\frac{\partial \delta\zeta}{\partial y}\frac{\partial^2 \zeta}{\partial x\,\partial y} - \frac{\partial \delta\zeta}{\partial x}\frac{\partial^2 \zeta}{\partial y^2}\right) + \frac{\partial}{\partial y}\left(\frac{\partial \delta\zeta}{\partial x}\frac{\partial^2 \zeta}{\partial x\,\partial y} - \frac{\partial \delta\zeta}{\partial y}\frac{\partial^2 \zeta}{\partial x^2}\right) = \operatorname{div} T,$$

whence we obtain (as before via Green's theorem)

$$\begin{aligned} \delta \iint\limits_D \left[\left(\frac{\partial^2 \zeta}{\partial x\,\partial y}\right)^2 - \frac{\partial^2 \zeta}{\partial x^2}\frac{\partial^2 \zeta}{\partial y^2}\right] dS &= \oint_\gamma \langle n, T\rangle\, dl \\ &= \oint_\gamma \left[\cos\theta\left(\frac{\partial \delta\zeta}{\partial y}\frac{\partial^2 \zeta}{\partial x\,\partial y} - \frac{\partial \delta\zeta}{\partial x}\frac{\partial^2 \zeta}{\partial y^2}\right) + \sin\theta\left(\frac{\partial \delta\zeta}{\partial x}\frac{\partial^2 \zeta}{\partial x\,\partial y} - \frac{\partial \delta\zeta}{\partial y}\frac{\partial^2 \zeta}{\partial x^2}\right)\right] dl, \end{aligned} \tag{48}$$

where θ denotes the angle between the positive x-axis and the unit normal n (see Figure 45).

Denoting by $\partial/\partial n$ the operator ∂_n, and by $\partial/\partial l$ the operation of taking the directional derivative with respect to the unit tangent vector to the curve γ, we have

$$\frac{\partial}{\partial x} = \cos\theta\frac{\partial}{\partial n} - \sin\theta\frac{\partial}{\partial l}, \qquad \frac{\partial}{\partial y} = \sin\theta\frac{\partial}{\partial n} + \cos\theta\frac{\partial}{\partial l},$$

whence (48) can be rewritten as

$$\begin{aligned} &\delta \iint\limits_D \left[\left(\frac{\partial^2 \zeta}{\partial x\,\partial y}\right)^2 - \frac{\partial^2 \zeta}{\partial x^2}\frac{\partial^2 \zeta}{\partial y^2}\right] dS \\ &\quad = \oint_\gamma \frac{\partial \delta\zeta}{\partial n}\left[2\sin\theta\cos\theta\frac{\partial^2 \zeta}{\partial x\,\partial y} - \sin^2\theta\frac{\partial^2 \zeta}{\partial x^2} - \cos^2\theta\frac{\partial^2 \zeta}{\partial y^2}\right] dl \\ &\quad\quad + \oint_\gamma \frac{\partial \delta\zeta}{\partial l}\left[\sin\theta\cos\theta\left(\frac{\partial^2 \zeta}{\partial y^2} - \frac{\partial^2 \zeta}{\partial x^2}\right) + (\cos^2\theta - \sin^2\theta)\frac{\partial^2 \zeta}{\partial x\,\partial y}\right] dl. \end{aligned} \tag{49}$$

We now integrate by parts the second integral in the right-hand side of (49). Its integrand, which we denote briefly by $(\partial\delta\zeta/\partial l)B$ can be written as

$$\frac{\partial \delta\zeta}{\partial l}B = \frac{\partial(\delta\zeta B)}{\partial l} - \delta\zeta\frac{\partial B}{\partial l}.$$

Now from the definition of the line integral (see §26.2) it follows that the line integral of the directional derivative $(\partial/\partial l)(\delta\zeta B)$ over an arc is just the difference of the values of the function $\delta\zeta B$ at the arc's end-points; thus since γ is closed, we have $\oint_\gamma (\partial/\partial l)(\delta\zeta B) = 0$, whence the second integral in (49) is equal to

$$-\oint_\gamma \delta\zeta \frac{\partial}{\partial l}\left[\sin\theta\cos\theta\left(\frac{\partial^2\zeta}{\partial y^2} - \frac{\partial^2\zeta}{\partial x^2}\right) + (\cos^2\theta - \sin^2\theta)\frac{\partial^2\zeta}{\partial x\,\partial y}\right] dl.$$

Putting this and (47) together we obtain finally

$$\delta F = \frac{Eh^3}{12(1-\sigma^2)}\left\{\iint_D (\delta\zeta)\Delta^2\zeta\, dS - \oint_\gamma \delta\zeta\left[\frac{\partial(\Delta\zeta)}{\partial n} + (1-\sigma)\frac{\partial}{\partial l}\left(\sin\theta\cos\theta\left(\frac{\partial^2\zeta}{\partial y^2} - \frac{\partial^2\zeta}{\partial x^2}\right) + (\cos^2\theta - \sin^2\theta)\frac{\partial^2\zeta}{\partial x\,\partial y}\right)\right] dl + \oint_\gamma \frac{\partial(\delta\zeta)}{\partial n} \times \left[\Delta\zeta + (1-\sigma)\left(2\sin\theta\cos\theta\frac{\partial^2\zeta}{\partial x\,\partial y} - \sin^2\theta\frac{\partial^2\zeta}{\partial x^2} - \cos^2\theta\frac{\partial^2\zeta}{\partial y^2}\right)\right] dl\right\}. \tag{50}$$

Now the work done by the external forces acting on the plate in "perturbing" it by an amount $\delta\zeta$ is given by $\iint_D P\,\delta\zeta\, dS$, where $P = P(x, y)$ is the external force on the plate per unit area (of the neutral surface) in the direction perpendicular to the (x, y)-plane. The equilibrum equation for the plate therefore takes the form

$$\delta F - \iint_D P\,\delta\zeta\, dS = 0, \tag{51}$$

where δF is given by (50). Note that this equation involves both surface and contour integrals. Since the equation holds for all perturbations $\delta\zeta$, which therefore may assume any prescribed values on ∂D and yet be non-zero only on arbitrarily small areas of D, it follows that those two parts of the left-hand side of equation (51) comprised on the one hand of the surface integrals and on the other hand of the contour integrals, must vanish separately. Taking the surface integrals first we obtain

$$\iint_D \left(\frac{Eh^3}{12(1-\sigma^2)}\Delta^2\zeta - P\right)\delta\zeta\, dS = 0.$$

Since $\delta\zeta$ is arbitrary it follows that

$$H\Delta^2\zeta - P = 0, \tag{52}$$

where the constant H, which is determined by the material of the plate, is given by $H = Eh^3/12(1 - \sigma^2)$. This constant is called the *rigidity* of the plate under bending (or its *cylindrical rigidity*). Equation (52) is thus the equation for equilibrium of a thin elastic plate deformed by the action of external forces.

From the vanishing of that part of the left-hand side of (51) made up of contour integrals, we obtain supplementary "boundary conditions" for equilibrium. In this connexion it is usual to single out various important special cases.

(a) Suppose that part of the edge $\gamma = \partial D$ of the plate is free, i.e. that no external forces act on it. Then along this portion of the boundary $\delta\zeta$ and $\delta(\partial\zeta/\partial n)$ are arbitrary, so that the corresponding factors in the integrands of the contour integrals in (50) must sum to zero, yielding:

$$-\frac{\partial(\Delta\zeta)}{\partial n} + (1 - \sigma)\frac{\partial}{\partial l}\left(\sin\theta\cos\theta\left(\frac{\partial^2\zeta}{\partial x^2} - \frac{\partial^2\zeta}{\partial y^2}\right) + (\sin^2\theta - \cos^2\theta)\frac{\partial^2\zeta}{\partial x\,\partial y}\right) = 0;$$

$$\Delta\zeta + (1 - \sigma)\left(2\sin\theta\cos\theta\frac{\partial^2\zeta}{\partial x\,\partial y} - \sin^2\theta\frac{\partial^2\zeta}{\partial x^2} - \cos^2\theta\frac{\partial^2\zeta}{\partial y^2}\right) = 0.$$

(b) Suppose that the boundary of the plate is fixed (for instance embedded in some rigid material). In this case the boundary of the plate cannot be displaced vertically, nor can its direction be altered; consequently $\delta\zeta \equiv 0$ and $\delta(\partial\zeta/\partial n) \equiv 0$ (so that certainly the contour integrals in (50) all vanish). In this case the boundary conditions take the form $\zeta = 0$, $\partial\zeta/\partial n = 0$. The import of the first condition is that the points of the plate's boundary do not move vertically, while the second signifies that the edge of the plate remains flat.

37.7. Exercises

1. Consider the set of skew-symmetric tensors F_{ik} defined on Minkowski space $\mathbb{R}^4_1$ which satisfy the condition $d(F_{ik}\,dx^i \wedge dx^k) = 0$. Show that the extremals with respect to this set, of the functional $S(F) = \int F \wedge *F = -\frac{1}{2}\int F_{ik}F^{ik}\,d^4x$, satisfy Maxwell's equations (*in vacuo*).

2. Prove that in the case of curved space-time the covariant divergence of the energy-momentum tensor is zero, i.e. $\nabla_k T^k_i = 0$.

3. Let V be a bounded region of the hypersurface $x^0 = \text{const.}$ in $\mathbb{R}^4_1$. With reference to §37.2, prove the following:

(i)
$$\frac{\partial}{\partial t}\int_V T^{00}\,d^3x = -c\oint_{\partial V} T^{0\alpha}\,d\sigma_\alpha,$$

where $T^{0\alpha}\,d\sigma_\alpha = T^{01}\,dx^2 \wedge dx^3 + T^{02}\,dx^3 \wedge dx^1 + T^{03}\,dx^1 \wedge dx^2$;

(ii) $$\frac{\partial}{\partial t}\int_V \frac{1}{c} T^{\alpha 0}\, d^3x = -\oint_{\partial V} T^{\alpha\beta}\, d\sigma_\beta.$$

(As noted in §37.2, the component T^{00} is called the *energy density* of the system; the 3-vector with components $(1/c)T^{10}$, $(1/c)T^{20}$, $(1/c)T^{30}$, is called the *momentum density* of the system. The 3-dimensional tensor $T^{\alpha\beta}$, α, $\beta = 1, 2, 3$ (which by (ii) represents the density of the flux of momentum) is called the *stress tensor*.)

4. Consider the functional $S[\Gamma] = \int R\sqrt{|g|}\, dV$, where $R = g^{ik}R_{ik}$, $R_{ik} = \partial\Gamma^l_{ik}/\partial x^l - \partial\Gamma^l_{il}/\partial x^k + \Gamma^l_{ik}\Gamma^m_{lm} - \Gamma^m_{il}\Gamma^l_{km}$, and g_{ik} is a fixed metric. Show that the extremals $\Gamma = (\Gamma^k_{ij})$ of this functional satisfy Christoffel's formulae expressing the Γ^k_{ij} in terms of the components of the metric tensor.

5. What form does the energy-momentum tensor of a gravitational field take?

6. Let $F = F_{ik}\, dx^i \wedge dx^k$, $i, k = 0, 1, 2, 3$, denote the electromagnetic field tensor in a 4-dimensional space-time with metric g_{ij}. Show that in this case Maxwell's equations take the form $dF = 0$, $\delta F = 4\pi j/c$, where j is the current, and $\delta = *^{-1} d\, *$.

7. Consider in an even-dimensional Riemannian (or pseudo-Riemannian) space the functional
$$S[g] = \int \Omega,$$
where the form Ω is as defined in Exercise 10 of §30.5. Show that $\delta S/\delta g \equiv 0$.

§38. Examples of Lagrangians

38.1

Let $\varphi(x)$ denote a complex scalar field† in the space $\mathbb{R}^4_1$ with metric
$$(g_{ab}) = \begin{pmatrix} 1 & & & 0 \\ & -1 & & \\ & & -1 & \\ 0 & & & -1 \end{pmatrix},$$
and consider the following action
$$S = \text{const.} \int \left[\hbar^2 \left\langle \frac{\partial\bar{\varphi}}{\partial x}, \frac{\partial\varphi}{\partial x} \right\rangle - m^2c^2\overline{\varphi(x)}\varphi(x)\right] d^4x = \int \Lambda\, d^4x, \tag{1}$$
where the bar denotes complex conjugation, $\hbar$ is Planck's constant, and c is the speed of light. Here $m(\geq 0)$ is the mass of the particle described by the

† It should be noted that the fields in question in this section, as also in §§40, 41 (where non-abelian gauge groups are considered), represent from their inception quantum objects having no direct classical analogues (i.e. are not modelled after antecedent classical fields).

field (or "wave function") φ. Thus the Lagrangian under consideration is $\Lambda = \Lambda(\varphi, \bar{\varphi}, \partial\varphi/\partial x, \partial\bar{\varphi}/\partial x)$, where φ and $\bar{\varphi}$ are formally regarded as independent variables. (This Lagrangian corresponds to the situation of a free relativistic particle.) It is almost immediate from (3) of §37.1 that the Euler–Lagrange equations $\delta S/\delta\varphi = 0$, $\delta S/\delta\bar{\varphi} = 0$, reduce to the single "Klein–Gordon equation"

$$(\hbar^2 \square + m^2c^2)\varphi = 0, \tag{2}$$

where

$$\square = \frac{\partial^2}{(\partial x^0)^2} - \sum_{\alpha=1}^{3} \frac{\partial^2}{(\partial x^\alpha)^2}.$$

It follows easily from this and the definition of the energy-momentum tensor (§37.2(8)) that

$$T^{ba} = T^{ab} = \hbar^2 g^{ac} g^{bd}\left(\frac{\partial\bar{\varphi}}{\partial x^c}\frac{\partial\varphi}{\partial x^d} + \frac{\partial\varphi}{\partial x^c}\frac{\partial\bar{\varphi}}{\partial x^d}\right) - g^{ab}\Lambda, \tag{3}$$

whence the energy density is given by

$$T^{00} = \hbar^2 \sum_c \frac{\partial\bar{\varphi}}{\partial x^c}\frac{\partial\varphi}{\partial x^c} + m^2c^2\bar{\varphi}\varphi. \tag{4}$$

The action (1) is clearly invariant under transformations of the form

$$\varphi \to e^{i\alpha}\varphi, \qquad \bar{\varphi} \to e^{-i\alpha}\bar{\varphi} \qquad (\alpha = \text{const.}), \tag{5}$$

which give rise to a "current"

$$J^\alpha = ig^{ab}\left(\bar{\varphi}\frac{\partial\varphi}{\partial x^b} - \varphi\frac{\partial\bar{\varphi}}{\partial x^b}\right), \tag{6}$$

which is conserved, i.e. $\partial J^a/\partial x^a = 0$. (We leave the deduction of this fact from the Euler–Lagrange equations as an exercise for the reader.) The quantity $Q = \int_{t=\text{const.}} J^0\, d^3x$ is called the "charge" of the field φ.

The presence of an electromagnetic field given by a vector potential $A = A(x)$, is taken into account by replacing the operator $i\hbar(\partial/\partial x)$ by $i\hbar(\partial/\partial x) + (e/c)A$, i.e. by replacing

$$i\frac{\partial}{\partial x^a} \quad \text{by} \quad i\frac{\partial}{\partial x^a} + \frac{e}{c\hbar}A_a(x), \tag{7}$$

where e is the charge on the particle. The full Lagrangian for the case of a particle of mass m in an electromagnetic field has the form (putting $\hbar = 1$ for simplicity):

$$\Lambda(\varphi, \bar{\varphi}, A) = \left\langle \frac{\partial\bar{\varphi}}{\partial x^a} + \frac{ie}{c}A_a\bar{\varphi}, \frac{\partial\varphi}{\partial x^a} - \frac{ie}{c}A_a\varphi \right\rangle - m^2c^2\bar{\varphi}\varphi - \frac{1}{16\pi c}F_{ab}F^{ab}. \tag{8}$$

Exercise

Show that the corresponding action is invariant under the "gauge transformations"

$$\varphi \to e^{i\alpha}\varphi, \qquad \bar{\varphi} \to e^{-i\alpha}\bar{\varphi}; \qquad A_a \to A_a + \frac{\partial \alpha}{\partial x^a}, \qquad \alpha = \alpha(x); \quad S \to S, \tag{9}$$

(assuming for simplicity $\hbar = c = 1$).

If the scalar field is real, or as they say "neutral", i.e. $\varphi \equiv \bar{\varphi}$, then the Lagrangian and the Euler–Lagrange equations take the form

$$\Lambda = \frac{1}{2}\left\langle \frac{\partial \varphi}{\partial x}, \frac{\partial \varphi}{\partial x} \right\rangle - m^2c^2\varphi^2, \qquad (\Box + m^2c^2)\varphi = 0, \tag{10}$$

and the current vector J^a is identically zero. In this case it is not possible to incorporate the electromagnetic field (as above) since the resulting Euler–Lagrange equations then turn out not to have real solutions.

It is easy to verify that the "free" equations (2) and (10) have the solution $\varphi = \text{const.} \times e^{i\langle k, x\rangle}$, where k is a vector independent of x satisfying

$$\langle k, k\rangle = \frac{m^2c^2}{\hbar^2}. \tag{11}$$

Such a solution is taken as describing a free particle of mass m and momentum $p = \hbar k$. Thus the momentum lies on the mass hyperboloid $\langle p, p\rangle = m^2c^2$ (cf. §32.2(18)).

38.2

If the state of the particle (of non-zero mass) is described in terms of a complex vector field $\varphi = (\varphi_0, \varphi_1, \varphi_2, \varphi_3)$ then the appropriate action is $S = \int \Lambda \, d^4x$, where

$$\Lambda = -g^{bd}g^{ac}\frac{\partial \bar{\varphi}_a}{\partial x^b}\frac{\partial \varphi_c}{\partial x^d} + m^2 g^{ab}\bar{\varphi}_a\varphi_b, \tag{12}$$

and where the field is required to satisfy the supplementary condition

$$\frac{\partial \varphi^a}{\partial x^a} = 0, \qquad \frac{\partial \bar{\varphi}^a}{\partial x^a} = 0. \tag{13}$$

By §37.2(8) the energy-momentum tensor is in this case given by

$$T^{ab} = T^{ba} = -g^{ac}g^{bd}g^{kl}\left(\frac{\partial \bar{\varphi}_k}{\partial x^c}\frac{\partial \varphi_l}{\partial x^d} + \frac{\partial \bar{\varphi}_l}{\partial x^d}\frac{\partial \varphi_k}{\partial x^c}\right) - g^{ab}\Lambda. \tag{14}$$

The group of action-preserving transformations $\varphi \to e^{i\alpha}\varphi$, $\bar{\varphi} \to e^{-i\alpha}\bar{\varphi}$ gives rise to a conserved current

$$J^a = -ig^{ab}g^{cd}\left(\frac{\partial \varphi_c}{\partial x^b}\bar{\varphi}_d - \frac{\partial \bar{\varphi}_c}{\partial x^b}\varphi_d\right). \tag{15}$$

Exercises

1. Deduce the conservation of current (i.e. $\partial J^a/\partial x^a = 0$) from the Euler–Lagrange equations $\delta S/\delta\bar{\varphi} = 0$, $\delta S/\delta\varphi = 0$ (which in the present context reduce to $(\Box + m^2)\varphi = 0$).
2. Show that in the absence of the supplementary conditions (13) the energy $\int T^{00}\, d^3x$ need not, in general, be positive.

As in the case of a scalar field, for the solutions (for the components of φ) of the free equation (see Exercise 1 above) of the plane-wave type, namely const. $\times\, e^{i\langle k, x\rangle}$, we must have $\langle k, k\rangle = m^2$ (where we have put $\hbar = c = 1$ for simplicity), and again the inclusion of an electromagnetic field given by a vector-potential A is achieved by replacing the differential operator $i(\partial/\partial x^a)$ as indicated in (7). The total Lagrangian (including an "electromagnetic" term) is again invariant under the gauge transformations (9) (with of course φ now denoting a vector).

The case of a vector field of zero mass ($m = 0$) which we encountered earlier (in the form of an electromagnetic field) is of special interest. Here if the field is real then the Lagrangian has the same form as that of an electromagnetic field (cf. §37.3), namely

$$\Lambda = \text{const.} \times F_{ab}F^{ab}, \quad \text{where } F_{ba} = \frac{\partial \varphi_a}{\partial x^b} - \frac{\partial \varphi_b}{\partial x^a}, \tag{16}$$

which is invariant under the gauge transformations

$$\varphi_a \to \varphi_a - \frac{\partial \alpha(x)}{\partial x^a}. \tag{17}$$

Later in this chapter (in §40) we shall consider yet another kind of field, namely "spinor" fields ψ taking their values in a space of "spinors".

§39. The Simplest Concepts of the General Theory of Relativity

39.1

We first recall (from §6) the basic assumptions of Einstein's special theory of relativity (STR). (It is interesting to note that in the creation of that theory, apart from the physicists Einstein and Lorentz (whose participation is widely known), the foremost geometers of that time, Poincaré and Minkowski, also took part.) According to that theory with each "event" occurring at a particular point of space at a particular instant there is associated a point in the 4-dimensional Minkowski space–time $\mathbb{R}_1^4$, with metric given by $dl^2 = (dx^0)^2 - \sum_{\alpha=1}^3 (dx^\alpha)^2$ in terms of pseudo-Euclidean co-ordinates x^0, x^1, x^2, x^3, where $x^0 = ct$, t being the time and c the speed of light in vacuo ($c \approx 299{,}793$ kilometres per second). A vector ξ is said to be respectively *time-like*, *light* (i.e. *isotropic*) or *space-like*, according as $\langle \xi, \xi \rangle > 0$, $\langle \xi, \xi \rangle = 0$ or $\langle \xi, \xi \rangle < 0$, and then a curve $\gamma(\tau)$ in $\mathbb{R}_1^4$ is time-like, light, or space-like according as its velocity vector $v = d\gamma/d\tau$ is, at all of its points, time-like, light or space-like. The world-line of a particle of mass $m > 0$ is time-like (since its speed should not exceed that of light), while that of a particle of zero mass ($m = 0$) is isotropic. In developing STR as it applies to a free particle of positive mass, either of the following two Lagrangians may be used (see Example (a) of §32.2):

$$S^{(1)} = \int_{\gamma(\tau)} L_{\text{free}}^{(1)}\, d\tau = \alpha \int_{\gamma(\tau)} \langle v, v \rangle \, d\tau;$$

$$S^{(2)} = \int_{\gamma(\tau)} L_{\text{free}}^{(2)}\, d\tau = \beta \int_{\gamma(\tau)} \sqrt{\langle v, v \rangle}\, d\tau.$$

Here $\gamma(\tau)$ is the world-line of the particle in $\mathbb{R}_1^4$, $v = d\gamma(\tau)/d\tau$ is as before its 4-velocity, and α, β are the particular constants $\alpha = \frac{1}{2}mc$, $\beta = -mc$. In the physics literature the Lagrangian one usually finds employed is the second one: $L_{\text{free}}^{(2)} = \beta\sqrt{\langle v, v \rangle}$, whose corresponding functional $S^{(2)}$ is proportional to the 4-dimensional arc length l of the world-line $\gamma(\tau)$. Since this functional is independent of the parameter τ, we may take $\tau = x^0/c = t$ ("ordinary" time), thence obtaining the Lagrangian in terms of "3-dimensional formalism":

$$L_{\text{free}}^{(2)} = \beta \langle v, v \rangle^{1/2} = \beta c \sqrt{1 - \frac{|w|^2}{c^2}}, \tag{1}$$

where $v = d\gamma/dt$, i.e. $v^a = dx^a/[d(x^0/c)] = c(dx^a/dx^0)$, (so that in particular $v^0 = c$), and $w^\alpha = v^\alpha$, $\alpha = 1, 2, 3$, i.e. w is the "ordinary" 3-dimensional velocity vector. This 3-dimensional form of the Lagrangian is useful in that it allows ready comparison with the analogous classical system. Thus if $|w| \ll c$, then writing briefly $w = |w|$, we have

$$L_{\text{free}}^{(2)} = \beta \langle v, v \rangle^{1/2} = \beta c \sqrt{1 - \frac{w^2}{c^2}} = \beta c\left(1 - \frac{w^2}{2c^2} + O\left(\frac{w^4}{c^4}\right)\right). \tag{2}$$

Giving β its value $\beta = -mc$, and letting $|w|/c \to 0$, we arrive (to within a constant (namely mc^2)) at the Lagrangian of a classical free particle (see Example (c) of §31.1).

As we saw in (11) and (12) of §32.2 the energy and (3-dimensional) momentum of our free relativistic particle are given by

$$E = w\frac{\partial L^{(2)}_{\text{free}}}{\partial w} - L^{(2)}_{\text{free}} = \frac{mc^2}{\sqrt{1-\dfrac{w^2}{c^2}}}, \quad p^\alpha = \frac{\partial L^{(2)}_{\text{free}}}{\partial w^\alpha} = \frac{mw^\alpha}{\sqrt{1-\dfrac{w^2}{c^2}}} = \delta^{\alpha\beta}p_\beta. \tag{3}$$

(Recall that in "3-dimensional formalism" indices are raised using the Euclidean metric $+\delta_{\alpha\beta}$.) Setting $E/c = p^0$ we obtain the momentum 4-vector $p = (p^a)$, $a = 0, 1, 2, 3$. It is immediate that the 4-momentum of a free particle of mass m satisfies the equation

$$p^a p^b g_{ab} = \langle p, p\rangle = (p^0)^2 - \sum_{\alpha=1}^{3}(p^\alpha)^2 = m^2c^2, \tag{4}$$

which defines the "mass surface" in the space co-ordinatized by the momenta (p^a), and endowed with the Minkowski metric g_{ab}. This surface (or rather half of it) is isometric to 3-dimensional Lobachevskian space (see §10), and its volume element is easily seen to have the form $d\sigma = \text{const.} \times d^3p/p^0$ (in terms of the co-ordinates p^1, p^2, p^3).

If we use instead the first Lagrangian, namely $L^{(1)}_{\text{free}} = \alpha\langle v, v\rangle$, then the energy (see §31.1) is given by

$$v\frac{\partial L^{(1)}_{\text{free}}}{\partial v} - L^{(1)}_{\text{free}} = L^{(1)}_{\text{free}} = \alpha\langle v, v\rangle.$$

Thus along each extremal the quantity $\langle v, v\rangle$ is conserved, i.e. is constant, so that the extremals must in particular be such that τ is a natural parameter for them, i.e. $dl = \text{const.} \times d\tau$ (and the constant factor may clearly be chosen arbitrarily, i.e. its value does not affect the class of extremals). We choose τ to be the proper time, $\tau = l/c$, for reasons which will soon appear. The components of the 4-momentum as defined in §31.1 are given by

$$p_a = \frac{\partial L^{(1)}_{\text{free}}}{\partial v^a}, \quad \text{that is} \quad p_0 = 2\alpha v^0, \qquad p_\beta = -2\alpha v^\beta, \qquad \beta = 1, 2, 3,$$

where now $v^a = dx^a/d\tau = c(dx^a/dt)(dt/dl)$ (using $\tau = l/c$). Since

$$\frac{dl}{dt} = c\sqrt{1-\frac{w^2}{c^2}},$$

it follows, on putting $\alpha = \frac{1}{2}mc$ and raising the index of p_a using the Minkowski metric, that this 4-momentum coincides with the previous one defined in terms of the Lagrangian $L^{(2)}_{\text{free}}$ (see (3) above). Thus it follows in particular that the 4-momentum (p^a) is indeed a 4-vector under co-ordinate transformations of $\mathbb{R}^4_1$. The vector $v = (v^a)$, where $v^a = dx^a/d\tau$, $\tau = l/c$, is usually called the *invariant 4-velocity* to distinguish it from the 3-velocity $w = (w^1, w^2, w^3)$. It is easy to see that the two velocities are related by $w^\beta = c(v^\beta/v^0)$, $\beta = 1, 2, 3$.

The effect on the particle of an external electromagnetic field given by a vector-potential $A_a(x)$ is taken into account by adding a term to the respective Lagrangians:

$$L^{(1)} = L^{(1)}_{\text{free}} + \frac{e}{c} A_a \frac{dx^a}{d\tau}, \qquad \tau = \frac{l}{c}; \tag{5}$$

$$L^{(2)} = L^{(2)}_{\text{free}} + \frac{e}{c} A_a \frac{dx^a}{dt} = L^{(2)}_{\text{free}} + \frac{e}{c} A_\alpha \frac{dx^\alpha}{dt} + eA_0, \qquad t = \frac{x^0}{c}. \tag{6}$$

Thus the incorporation of the external electromagnetic field is equivalent to the addition of the term $(e/c)A_a(x)$ to each component p_a of the momentum 4-covector of the free particle.

In terms of Hamiltonian formalism (see §33) the Hamiltonian of a free relativistic particle is in view of (4) given by

$$H = E(p) = c\sqrt{p^2 + m^2c^2} = cp_0,$$

where $p^2 = \sum_{\alpha=1}^{3} (p_\alpha)^2$. In the presence of an electromagnetic field $A_a = (A_0, A_1, A_2, A_3) = (A_0, A_\alpha)$, the Hamiltonian is rather

$$H(x, p) = c\sqrt{\sum_\alpha \left(p_\alpha + \frac{e}{c} A_\alpha\right)^2 + m^2c^2} + eA_0(x).$$

Recall from §37.3 that the electromagnetic field tensor is given in terms of the vector-potential by $F_{ba} = \partial A_a/\partial x^b - \partial A_b/\partial x^a$, and that the appropriate action of the field by itself is

$$S\{A\} = -\frac{1}{16\pi c} \int F_{ab} F^{ab}\, d^4x.$$

We know that F_{ab} satisfies the following conditions:

(i) $d(F_{ab}\,dx^a \wedge dx^b) = \left(\frac{\partial F_{ab}}{\partial x^c} - \frac{\partial F_{ac}}{\partial x^b} + \frac{\partial F_{bc}}{\partial x^a}\right) dx^a \wedge dx^b \wedge dx^c = 0$

(constituting the "first pair" of Maxwell's equations—see §25.2(18));

(ii) in the absence of charge, $\partial F^{ab}/\partial x^b = 0$ (which constitutes the "second pair" of Maxwell's equations—see §25.2(21) or §37.3).

If we have N particles in $\mathbb{R}^4_1$ with masses $m_1, \ldots, m_N$, charges $e_1, \ldots, e_N$, and world-lines $\gamma_1, \ldots, \gamma_N$, in the presence of a field $A_a(x)$, then for the total action of particles and field we can use one or the other of the following:

$$S^{(1)} = \sum_{i=1}^{N} \alpha_i \int_{\gamma_i} \langle v_i, v_i \rangle\, d\tau + \sum_{i=1}^{N} \int_{\gamma_i} \frac{e_i}{c} (A_a(x) v_i^a)\, d\tau - \frac{1}{16\pi c} \int F_{ab} F^{ab}\, d^4x; \tag{7}$$

$$S^{(2)} = -\sum_{i=1}^{N} m_i c^2 \int_{\gamma_i} \sqrt{1 - \frac{w_i^2}{c^2}}\, dt + \sum_{i=1}^{N} \int_{\gamma_i} \left[\frac{e_i}{c} A_\alpha(x) w_i^\alpha\, dt + e_i A_0(x)\, dt\right] - \frac{1}{16\pi c} \int F_{ab} F^{ab}\, d^4x. \tag{8}$$

Here w_i and v_i are respectively the ordinary 3-velocity and the 4-velocity of the ith particle. Taking the Lagrangian in (7) in the case of a single particle, i.e.

$$L^{(1)} = \alpha\langle v, v\rangle + \frac{e}{c} A_a v^a,$$

we find that the energy is given by

$$v^a \frac{\partial L^{(1)}}{\partial v^a} - L^{(1)} = 2\alpha\langle v, v\rangle + \frac{e}{c} A_a v^a - L^{(1)} = \alpha\langle v, v\rangle.$$

Hence for a particle in an external electromagnetic field we have that $\langle v, v\rangle$ is constant along each trajectory, whence, as in the case of a free particle, the extremals again have the property that the parameter τ is natural for them.

39.2

The above procedure for including an electromagnetic field in STR turns out not to work for gravitational fields. More precisely, suppose that, on the basis of Poincaré's hypothesis to the effect that gravitational interactions are propagated with the speed of light, we introduce into the Lagrangian a 4-dimensional scalar potential of the form $\varphi(x - ct)$ representing the gravitational field, or a term corresponding to a "gravitational vector-potential" A_a^G, with now the masses m_i in the roles of the charges e_i (the idea being that the mass of a particle is just its "gravitational charge"); experiment shows that in terms of suitable local co-ordinates this vector-potential would have to be of the form $A^G = (\varphi, 0, 0, 0)$, where φ is the usual gravitational potential. Thus the (tentative) procedure would be to consider the action given by the first two terms in the right-hand side of (7) with A replaced by A^G and the e_i by the m_i, and hope that the corresponding extremals turn out to be the correct trajectories. However (as was shown already by Poincaré) the resulting correction to Newton's law of gravitation leads to a value for the amount of precession of the perihelion of the orbit of the planet Mercury, which does not agree with observation (though it has the right order of magnitude). Hence a different approach is needed for incorporating gravitation into the theory of relativity. This problem was solved by Einstein, who proposed the following as the basic hypothesis of his general theory of relativity (GTR): The gravitational field is simply a metric g_{ab} of signature $(+ - - -)$ in four-dimensional space-time $M^4(x^0, x^1, x^2, x^3)$, this metric being such that the curvature is non-zero. (The magnitude of the curvature at each point characterizes the strength of the gravitational field at the point.) Thus a test particle in an external gravitational field is from this point of view simply a free particle in a space with metric g_{ab}, and therefore such a particle will (if its mass m is positive) move along a time-like geodesic $\gamma(\tau) = (x^a(\tau))$, i.e. along an extremal corresponding to the Lagrangian

$$L^{(1)}_{\text{free}} = m\langle v, v\rangle = m\frac{dx^a}{d\tau}\frac{dx^b}{d\tau}g_{ab}. \tag{9}$$

(If $m = 0$ the particle moves along a light geodesic.) As above we have $|v| = \text{const.}$ along each extremal, and we take τ to be the proper time, i.e. $\tau = l/c$.

The inclusion of an electromagnetic field $A_a(x)$ follows the same scheme as before, except that the Minkowski metric is now replaced by the metric g_{ab}:

$$S = \int_{\gamma(\tau)} L^{(1)}\,d\tau = \int_{\gamma(\tau)} L^{(1)}_{\text{free}}\,d\tau + \frac{e}{c}\int_{\gamma(\tau)} A_a\,dx^a \tag{10}$$

The action of the field alone is in this more general context given by

$$S_{\text{field}} = -\frac{1}{16\pi c}\int F_{ab}F^{ab}\sqrt{-g}\,d^4x, \qquad F_{ba} = \frac{\partial A_a}{\partial x^b} - \frac{\partial A_b}{\partial x^a}, \tag{11}$$

where of course $d\sigma = \sqrt{-g}\,d^4x$ is the volume element, $g = \det(g_{ab})$, and $F^{ab} = g^{ac}g^{bd}F_{cd}$. It follows without difficulty that Maxwell's equations $\delta S_{\text{field}}/\delta A_a = 0$ (in the absence of charge), now take the form

$$\begin{aligned} d(F_{ab}\,dx^a \wedge dx^b) &= 0 \quad (\text{"first pair"}); \\ \nabla_b F^{ab} &= 0 \quad (\text{"second pair"}). \end{aligned} \tag{12}$$

Before investigating the equations for the gravitational field g_{ab} itself, we point out a simple consequence of Einstein's hypothesis. (This consequence has in fact already been examined (in Example (b) of §32.2) but bears recapitulation.) Consider a "weak" (in a sense to be made precise) gravitational field-metric g_{ab}, and a "slow" particle of positive mass moving through the field along a geodesic $x(\tau)$; thus

$$\ddot{x}^a + \Gamma^a_{bc}\dot{x}^b\dot{x}^c = 0, \qquad a, b, c = 0, 1, 2, 3, \tag{13}$$

where the Γ^a_{bc} are the Christoffel symbols, i.e. the components of the symmetric connexion compatible with (and so determined by) the metric g_{ab} (see §29.3), and the derivatives are taken with respect to τ, the proper time along the geodesic (i.e. $\tau = l/c$). As in STR we write $t = x^0/c$. By a *weak metric* g_{ab} we shall mean a metric which can be represented as a power series in $1/c$ of the form

$$g_{ab} = g^{(0)}_{ab} + c^{-2}g^{(2)}_{ab} + c^{-3}g^{(3)}_{ab} + \cdots = g^{(0)}_{ab} + O\left(\frac{1}{c^2}\right), \tag{14}$$

where $g^{(1)}_{ab} = 0$, $g^{(0)}_{ab}$ is the Minkowski metric, and for $n \geq 3$ the $g^{(n)}_{ab}(x)$ and their first partial derivatives with respect to $t = x^0/c$, x^1, x^2, x^3 are (for all relevant points x) small in comparison with c. In saying that the particle is *slow* we mean that $dx^a/dt \ll c$. Since τ is assumed to be the proper time along the geodesic, it follows that

$$d\tau = \frac{dl}{c} = \sqrt{\frac{1}{c^2}\frac{dx^a}{dt}\frac{dx^b}{dt}g_{ab}}\,dt \qquad \left(t = \frac{x^0}{c}\right), \tag{15}$$

whence

$$d\tau = \sqrt{1 + O\left(\frac{1}{c^2}\right)}\,dt = \left[1 + O\left(\frac{1}{c^2}\right)\right]dt. \tag{16}$$

By Theorem 29.3.2 the Christoffel symbols are given by

$$\Gamma^a_{bc} = \tfrac{1}{2}g^{ad}\left(\frac{\partial g_{bd}}{\partial x^c} + \frac{\partial g_{cd}}{\partial x^b} - \frac{\partial g_{bc}}{\partial x^d}\right). \tag{17}$$

It follows from (14), (16), (17) and the conditions on the $g^{(n)}_{ab}$ and their first derivatives that in (13) (with $a = 1, 2, 3$) all terms involving a Christoffel symbol are of order at most $O(1/c)$ except for $\Gamma^\alpha_{00}\dot{x}^0\dot{x}^0 = \Gamma^\alpha_{00}c^2 + O(1/c)$, $\alpha = 1, 2, 3$ (the details are given in the proof of Proposition 32.2.2). Now by (14) (and the condition on $\partial g^{(n)}_{ab}/\partial t$, $n \geq 3$) the quantities $\partial g_{ab}/\partial x^0$ have order $O(1/c^3)$ whence

$$\Gamma^\alpha_{00} = \tfrac{1}{2}g^{\alpha\alpha}\left(-\frac{\partial g_{00}}{\partial x^\alpha}\right) + O\left(\frac{1}{c^3}\right),$$

so that

$$\Gamma^\alpha_{00} = \frac{1}{2}\frac{\partial g_{00}}{\partial x^\alpha} + O\left(\frac{1}{c^3}\right), \qquad \alpha = 1, 2, 3. \tag{18}$$

Thus (for $a = 1, 2, 3$) equation (13) takes the form

$$\frac{d^2x^\alpha}{dt^2} = -\frac{1}{2}\frac{\partial g_{00}}{\partial x^\alpha}c^2 + O\left(\frac{1}{c}\right). \tag{19}$$

For a slow particle in a weak field these equations should approximate (to within $O(1/c)$) Newton's equations $d^2x^\alpha/dt^2 = -\partial\varphi/\partial x^\alpha$ for a particle in a classical gravitational field with potential $\varphi(x)$. It follows that we must have

$$-\frac{1}{2}\frac{\partial g_{00}}{\partial x^\alpha}c^2 = -\frac{\partial\varphi}{\partial x^\alpha} + O\left(\frac{1}{c}\right),$$

whence (cf. §32.2(21))

$$g_{00} = 1 + \frac{2\varphi(x)}{c^2} + O\left(\frac{1}{c^3}\right). \tag{20}$$

Thus, turning Proposition 32.2.2 on its head, we have

39.2.1. Proposition. *It is a consequence of Einstein's hypothesis that the component g_{00} of a weak gravitational field g_{ab} must have the form*

$$g_{00} = 1 + \frac{2\varphi(x)}{c^2} + O\left(\frac{1}{c^3}\right),$$

where $\varphi(x)$ is the Newtonian gravitational potential.

From (20) we can obtain in particular a more accurate relationship (than (16)) between the proper time $\tau = l/c$ and the "world" time t of a fixed particle. Thus since x^1, x^2, x^3 are constant we have from (15) that

$$c \cdot d\tau = \sqrt{g_{00}\left(\frac{dx^0}{dt}\right)^2}\, dt = c\sqrt{g_{00}}\, dt,$$

whence by (20)

$$d\tau \approx dt\left(1 + \frac{\varphi(x)}{c^2}\right). \tag{21}$$

Since it is always the case that $\varphi(x) \leq 0$, we conclude that: *In a weak gravitational field the proper time between two events corresponding to a particle fixed in space is less than the world time*:

$$\tau < t \quad \textit{if } \varphi < 0.$$

(The two events may be taken as having co-ordinates respectively $x^0 = x_1^0$, $x^\alpha = 0$, and $x^0 = x_2^0$, $x^\alpha = 0$; the proper time $\tau = l/c$ where l is the length of the world-line joining the two events.)

39.3

Consider now the case of a gravitational field (g_{ab}) (i.e. a metric of signature $(+ - - -)$) in an otherwise empty region of 4-dimensional space M^4, i.e. free of particles and of all other fields. We make the further basic assumption that our theory of gravitation should be "universally covariant", as they say, meaning that the equations for the gravitational field should have the same form in all co-ordinate systems. We shall not go into the details of Einstein's derivation of his equation for the gravitational field (in the form of conditions on the curvature tensor R^a_{bcd} corresponding to the metric g_{ab}) but shall simply state it. (See however §37.4 where we derived Einstein's equation from Hamilton's principle.) The equation(s) in question are:

$$R_{ab} - \tfrac{1}{2}Rg_{ab} = 0, \tag{22}$$

where $R_{bc} = R^a_{bac}$ is the Ricci tensor and $R = g^{ab}R_{ab}$ is the scalar curvature. (Alternatively we may use the equivalent equation $R_{ab} = 0$; this equivalence was shown at the end of §37.4.)

It follows from Bianchi's identities (see Exercises 7 and 8 of §30.5) that the expression $R_{ab} - \frac{1}{2}Rg_{ab}$ ("Einstein's operator"), where g_{ab} is here arbitrary, has the following important property:

$$\nabla_a(R_b^a - \tfrac{1}{2}R\delta_b^a) \equiv 0. \tag{23}$$

Einstein's equation (22) as an equation in the components g_{ab} (in terms of which, of course, the components of the Riemann curvature tensor can be expressed—see §30.1(3)) has the following three properties:

(i) It is a second-order partial differential equation;
(ii) In the case of weak fields (for which, as we saw above, $g_{00} = 1 + 2\varphi(x)/c^2 + O(1/c^3)$) from Einstein's equation one can deduce (we shall show how below) the familiar Poisson equation for the potential φ:

$$\Delta\varphi = \sum_{\alpha=1}^{3} \frac{\partial^2\varphi(x)}{(\partial x^\alpha)^2} = 0. \tag{24}$$

(iii) By Theorem 37.4.1 Einstein's equation (22) is equivalent to the Euler–Lagrange equations for the extremals of some action; in fact (as Hilbert showed) the following action will serve:

$$S = \int R\sqrt{-g}\, d^4x, \qquad g = \det(g_{ab}), \qquad R = R_a^a, \tag{25}$$

where the integral is taken over the region of M^4 defined by $x_1^0 \le x^0 \le x_2^0$. Thus according to Theorem 37.4.1 the variational derivative $\delta S/\delta g^{ab}$ is given by

$$\frac{1}{\sqrt{-g}} \frac{\delta S}{\delta g^{ab}} = R_{ab} - \tfrac{1}{2}Rg_{ab}. \tag{26}$$

Exercise

Show that for an expression of the form $R_{ab} - \gamma R g_{ab}$ to be a variational derivative it is necessary (as well as sufficient) that $\gamma = \frac{1}{2}$. (Use the fact that the divergence of a variational derivative vanishes—cf. (23).)

Remark. The above three conditions are also satisfied by any equation of the form

$$R_{ab} - \tfrac{1}{2}Rg_{ab} = \lambda g_{ab}, \tag{27}$$

where λ is a constant (the so-called "cosmological constant"). However up to the present time no reason has emerged for supposing $\lambda \neq 0$, so that it is generally assumed that in fact $\lambda = 0$; cosmological considerations suggest that $\lambda < 10^{-56}\,\text{cm}^{-2}$.

The simplest non-trivial solution of Poisson's equation $\Delta\varphi = 0$ (which holds for the classical gravitational potential throughout the region of space outside the bodies whose masses are responsible for the gravitational field) is the stationary, spherically symmetric solution

$$\varphi = \frac{\text{const.}}{r}, \qquad r^2 = \sum_{\alpha=1}^{3} (x^\alpha)^2,$$

which corresponds to the situation of a spherically symmetric distribution of mass. If the total mass is M then the constant is $-GM$ where G ($\approx 6.67 \times 10^{-8}$ cm^3/gram sec^2) is Newton's universal constant of gravitation, so that

$$\varphi = -G\frac{M}{r}. \tag{28}$$

What is the analogue in GTR of this Newtonian potential? To answer this question it is clearly appropriate to consider a spherically symmetric metric g_{ab} which is independent of the time $t = x^0/c$ (i.e. is "stationary"), and has the form

$$dl^2 = c^2\,dt^2 g_{00} + g_{11}\,dr^2 - r^2\,d\Omega^2, \tag{29}$$

where $g_{00} > 0$, $g_{11} < 0$, g_{00} and g_{11} depend only on r, and $d\Omega^2 = (d\theta)^2 + \sin^2\theta(d\varphi)^2$ is the element of length on the unit sphere in terms of the spherical co-ordinates θ, φ.

In order to see what explicit form Einstein's equation $R_{ab} = 0$ takes in this case we first need to calculate the components Γ^a_{bc} of the connexion, since R_{ab} is defined in terms of them:

$$R_{ab} = \frac{\partial \Gamma^c_{ab}}{\partial x^c} - \frac{\partial \Gamma^c_{ac}}{\partial x^b} + \Gamma^c_{ab}\Gamma^d_{cd} - \Gamma^c_{ad}\Gamma^d_{bc}. \tag{30}$$

We write for simplicity $g_{00} = \nu$, $g_{11} = -\lambda$. From Christoffel's formula (§29.3(13)) with $x^0 = ct$, $x^1 = r$, $x^2 = \theta$, $x^3 = \varphi$, keeping in mind that $\partial g_{ab}/\partial x^0 = 0$, and denoting differentiation with respect to r by a prime, we obtain by direct calculation that:

$$\Gamma^1_{11} = \frac{\lambda'}{2\lambda}, \qquad \Gamma^0_{10} = \frac{\nu'}{2\nu}, \qquad \Gamma^2_{33} = -\sin\theta\cos\theta,$$

$$\Gamma^1_{22} = -\frac{r}{\lambda}, \qquad \Gamma^1_{00} = \frac{\nu'}{2\lambda}, \qquad \Gamma^2_{12} = \Gamma^3_{13} = \frac{1}{r},$$

$$\Gamma^3_{23} = \cot\theta, \qquad \Gamma^1_{33} = -\frac{r}{\lambda}\sin^2\theta,$$

and that all remaining components are zero. From this and (30) it follows, again by direct calculation, that the only non-zero components of the Ricci tensor are R_{00}, R_{11}, R_{22}, R_{33}, and that these are given by:

$$R_{00} = \frac{v''}{2\lambda} - \frac{(v')^2}{4v^2} - \frac{v'\lambda'}{4\lambda^2} + \frac{v'}{r\lambda},$$

$$R_{11} = -\frac{v''}{2v} + \frac{(v')^2}{4v^2} + \frac{v'\lambda'}{4v\lambda} + \frac{\lambda'}{r\lambda},$$

$$R_{22} = \frac{r\lambda'}{2\lambda^2} - \frac{rv'}{2v\lambda} - \frac{1}{\lambda} + 1,$$

$$R_{33} = R_{22} \sin^2 \theta.$$

It is not difficult to see (for example by putting $v = e^{\mu}$, $\lambda = e^{\varkappa}$) that of the resulting four equations $R_{ii} = 0$, only two are independent, and that their solution is given by

$$v = g_{00} = 1 - \frac{a}{r}, \qquad -\lambda = g_{11} = \frac{-1}{1 - (a/r)};$$

where a is a constant of integration. Hence the metric (known as the *Schwarzschild metric*) outside a spherically symmetric body has the form

$$dl^2 = \left(1 - \frac{a}{r}\right)c^2\, dt^2 - \frac{1}{1 - (a/r)}\, dr^2 - r^2\, d\Omega^2. \tag{31}$$

For large r the metric is weak, i.e.

$$g_{ab} \approx g_{ab}^{(0)} + \frac{1}{c^2} g_{ab}^{(2)} + O\left(\frac{1}{c^3}\right),$$

where from (31) we have $g_{00}^{(2)} = -ac^2/r$. Comparing this with (20) and taking into account the fact that $\varphi = -GM/r$, we obtain $a = 2GM/c^2$. This constant (of the body of mass M) is called the *gravitational Schwarzschild radius*. (For bodies of the mass of the earth $a \approx 0.44$ cm, while for the sun $a \approx 3$ km.) From (31) it is evident that as the radius of a body decreases through the value a some fundamental change occurs. We shall consider this phenomenon in greater detail in §31 of Part II. For the time being we shall rest content with the validity of formula (31) in the region $r > a$ (outside the body).

Since for appropriate (and obvious) co-ordinates x^1, x^2, x^3 we have $dr^2 + r^2\, d\Omega^2 = (dx^1)^2 + (dx^2)^2 + (dx^3)^2$, and since for large enough r

$$\frac{1}{1 - (a/r)} = 1 + \frac{a}{r} + O\left(\frac{a^2}{r^2}\right),$$

it follows from (31) and the fact that $-ac^2/r = 2\varphi$, that for large r the Schwarzschild metric has the following form (in terms of the co-ordinates $x^0 = ct$, and the new x^1, x^2, x^3 just chosen):

$$(dl)^2 = g_{ab}\, dx^a\, dx^b = (dx^0)^2 - \sum_{\alpha=1}^{3} (dx^\alpha)^2 + \frac{2\varphi}{c^2}\left(\sum_{\alpha=0}^{3} (dx^\alpha)^2\right) + O\left(\frac{a^2}{r^2}\right), \quad (32)$$

where $a = 2GM/c^2$.

Remark. It can be shown that this formula, with $O(a^2/r^2)$ replaced by $O(1/c^3)$, holds for weak gravitational metrics generally, and not just for the Schwarzschild metric. (We showed part of this in the preceding subsection, namely that the coefficient of $(dx^0)^2$ has the form $1 + 2\varphi/c^2 + O(1/c^3)$; this was a consequence of the expected agreement to within $O(1/c)$ of the equation for the trajectory of a slow relativistic particle in a weak field, with that of a classical particle in a field of potential φ.)

It is of interest to note that in (32) the correction to the Minkowski metric is (to within $O(1/c^3)$) proportional to the 4-dimensional Euclidean metric (with variable proportionality factor $2\varphi/c^2$). Thus to within $O(1/c^3)$ the metric (32) corresponding to a weak gravitational field is not invariant with respect to the Lorentz group but only with respect to its subgroup $SO(3)$, and is calculated (to within $O(1/c^3)$) in terms of the classical Newtonian potential.

The above-mentioned fact that formula (32) holds generally for the metric of a weak gravitational field, can be used to obtain a correction to the classical trajectories of fast-moving particles, in particular of particles of zero mass. The trajectory $x = x(\eta)$ of a photon satisfies two equations, namely the equation for the geodesics:

$$\frac{d^2x^a}{d\eta^2} + \Gamma^a_{bc}\frac{dx^b}{d\eta}\frac{dx^c}{d\eta} = 0, \qquad a = 0, 1, 2, 3,$$

and the equation $g_{ab}(dx^a/d\eta)(dx^b/d\eta) = 0$, corresponding to the fact that the photon moves with the speed of light. We shall omit the details of the solution of these equations, contenting ourselves with stating the formula for the light paths in a (spherically symmetric) Schwarzschild metric. The formula in question is, in terms of polar co-ordinates r, $\hat{\varphi}$ for the plane in which the particular trajectory lies, as follows (here the hat serves to distinguish the angular co-ordinate $\hat{\varphi}$ from the potential φ):

$$\hat{\varphi} = \int \frac{dr}{r^2\sqrt{\dfrac{1}{\rho^2} - \dfrac{1}{r^2}\left(1 - \dfrac{a}{r}\right)}}, \qquad \rho = \text{const.} \quad (33)$$

Note that in the limit as $a \to 0$ (i.e. as the mass of the spherical body approaches zero) the paths are given by $r \cos \hat{\varphi} = \rho$, i.e. they are straight lines. For small a one can calculate from (33) the correction to the classical straight-line trajectory of a light ray.

39.4

In GTR it is assumed that the interaction between any system of particles or any field (apart from the gravitational field!) and the metric g_{ab}, i.e. the gravitational field, is determined by the so-called energy-momentum tensor T_{ab} in accordance with an equation (the full Einstein equation) of the following form:

$$R_{ab} - \tfrac{1}{2}Rg_{ab} = \text{const.} \times T_{ab}, \tag{34}$$

or

$$R_a^b - \tfrac{1}{2}R\delta_a^b = \text{const.} \times T_a^b, \tag{35}$$

where the constant is assumed to be a universal constant.

The value of this constant can be found by considering the situation of a static dust-cloud of zero pressure in the presence of a weak gravitational metric (as given by (32)). The energy-momentum tensor is in this case given by (see (43) below)

$$T_{ab} = \begin{pmatrix} \rho c^2 & 0 & 0 & 0 \\ 0 & & & \\ 0 & & 0 & \\ 0 & & & \end{pmatrix}, \tag{36}$$

where ρ is the density of the dust-cloud. It follows by direct calculation from Christoffel's formula and the formula (32) (with a^2/r^2 replaced by $1/c^3$) for a weak metric, that among the Christoffel symbols Γ_{bc}^a occurring in the formula for R_{00} (given in §37.4), the only non-zero ones (i.e. non-zero to within $O(1/c^3)$) are given by

$$\Gamma_{00}^\alpha = \frac{1}{c^2}\frac{\partial\varphi}{\partial x^\alpha} + O\left(\frac{1}{c^3}\right), \qquad \alpha = 1, 2, 3,$$

whence

$$R_0^0 = \frac{1}{c^2}\left(\sum_{\alpha=1}^{3} \frac{\partial^2\varphi}{(\partial x^\alpha)^2}\right) + O\left(\frac{1}{c^3}\right) = \frac{1}{c^2}\Delta\varphi + O\left(\frac{1}{c^3}\right). \tag{37}$$

Now from (35) we have $R_a^a - \frac{1}{2}R\delta_a^a = \text{const.} \times T_a^a$, whence it follows, since $R_a^a = R$, that $-R = \text{const.} \times T_a^a$. Hence (35) with $a = b = 0$ becomes

$$R_0^0 = \text{const.} \times (T_0^0 - \tfrac{1}{2}T_a^a),$$

which by (36) and (37) yields in our present context

$$\frac{1}{c^2}\Delta\varphi + O\left(\frac{1}{c^3}\right) = R_0^0 = \text{const.} \times \frac{\rho c^2}{2}. \tag{38}$$

From classical physics we know that the classical gravitational potential φ satisfies Poisson's equation

$$\Delta\varphi = 4\pi G\rho.$$

Comparing this with (38) and bringing in the assumption that the constant we are seeking is universal, we deduce that its value is $8\pi G/c^4$. Thus the full Einstein equation is

$$R_{ab} - \tfrac{1}{2}Rg_{ab} = \frac{8\pi G}{c^4} T_{ab}. \tag{39}$$

Note that in view of (23) the following identity must hold (cf. §37.2(8)):

$$\nabla_b T^b_a = 0. \tag{40}$$

This identity represents the law of conservation in GTR, replacing the analogous conservation laws of classical mechanics and special relativity.

We now give the explicit forms of two particular energy-momentum tensors of the very first importance, namely that of an electromagnetic field (the form of the energy-momentum tensor in this case was derived in the context of special relativity at the end of §37.3), and that of a dense isotropic medium.

(i) The energy-momentum tensor of an electromagnetic field is given by (cf. §37.3(24)):

$$T^{ab} = \frac{1}{4\pi c}(-F^{ac}F^b_c + \tfrac{1}{4}g^{ab}F_{cd}F^{cd}). \tag{41}$$

(ii) The energy-momentum tensor of an isotropic dense medium (or in other words the "hydrodynamic" energy-momentum tensor) is given by

$$T_{ab} = (p + \varepsilon)u_a u_b - pg_{ab}, \tag{42}$$

where p and ε are respectively the pressure and energy density at each point of the medium, in terms of a co-ordinate system "accompanying" or "attached to" the medium, i.e. relative to which the medium does not move (spatially), or, more precisely, in terms of which $u_0 = 1, u_1 = u_2 = u_3 = 0$, where $u = v/c$, v being the 4-covector of velocity of the medium at each point. From (42) it is clear that in terms of such an attached co-ordinate system T_{ab} has the form

$$T_{ab} = \begin{pmatrix} \varepsilon & & & 0 \\ & p & & \\ & & p & \\ 0 & & & p \end{pmatrix}. \tag{43}$$

In order to completely specify the Einstein equation (39) for a given dense isotropic medium, we need to know the "state equation" of the medium, i.e.

the relationship between p and ε. In the case of the dust-nebula considered above, we had $p = 0, \varepsilon = \rho c^2$. The so-called "ultra-relativistic" state equation is $p = \varepsilon/3$, i.e. $T_a^a = 0$.

EXERCISE

Prove that the energy-momentum tensor of an electromagnetic field satisfies (cf. §37.4(36) *et seqq.*)

$$\tfrac{1}{2}T_{ab} = \frac{1}{\sqrt{-g}} \frac{\delta S}{\delta g^{ab}},$$

where

$$S = -\frac{1}{16\pi c} \int F_{ab} F_{cd} g^{ac} g^{bd} \sqrt{-g}\, d^4x, \qquad g = \det(g_{ab}). \tag{44}$$

Remarks. 1. It is usual in expositions of GTR for the symmetric energy-momentum tensor T_{ab} of a material system (which term is understood to subsume (non-gravitational) "field") to be defined by

$$\frac{\delta S_{\text{system}}}{\delta g^{ab}} = \tfrac{1}{2}\sqrt{-g}\, T_{ab}, \tag{45}$$

where the action of the fields comprising the system are supposed given beforehand in the form of a functional depending both on those fields and on the underlying gravitational metric. (For example in the Lagrangians of scalar and vector fields (see (44) above or the examples in §38) the metric enters explicitly in particular in terms involving scalar products of gradients. The situation for "spinor" fields (whose role in GTR will be discussed in §41) is more complicated. Note that the formula (42) for the energy-momentum tensor of a dense isotropic medium also directly involves the metric.) The combined action of a material system and the underlying gravitational field is then

$$S_{\text{total}} = \int R\sqrt{-g}\, d^4x + S_{\text{system}}. \tag{46}$$

Thus if for instance the "material system" is an electromagnetic field F_{ab}, then

$$S_{\text{total}} = \int (R\sqrt{-g} + \text{const.} \times F_{ab}F^{ab}\sqrt{-g})\, d^4x. \tag{47}$$

The full system of equations (Maxwellian and Einsteinian) can in this case be written as

$$\frac{\delta S_{\text{total}}}{\delta A_a} = 0, \qquad \frac{\delta S_{\text{total}}}{\delta g^{ab}} = 0. \tag{48}$$

2. The question has been much debated as to whether a precise meaning can be given to the term "energy-density of a gravitational field" (or indeed to the full concept of a gravitational energy-momentum tensor as a "universally covariant" entity). From the standpoint which views (45) as the general, physically valid definition (for all material systems) of the energy-momentum tensor, the gravitational field plays a special role, distinguished by the fact that the use of (45) as a definition presupposes a metric. (Here we are of course assuming the truth of Einstein's basic postulate that the gravitational field is the metric; this assumption is supported, however, by a whole series of successful experimental verifications.) The upshot of this debate seems to be that there most probably does not exist any universally covariant gravitational energy-momentum tensor (apart from the left-hand side of Einstein's equation). However there is one important exceptional situation (not examined in the present text), namely that of a "localized gravitational packet" in an ambient Minkowski space. Comparison with the Newtonian gravitational theory shows that the amounts by which the components of the metric differ from constants must decrease in a sufficiently regular manner at a rate of the order of r^{-1} (but no faster!) as $r \to \infty$. This allows the definition of the total "mass of the field" in a manner analogous to the classical definition of the mass of a body in terms of the asymptotic behaviour at spatial infinity of the potential of the body's own gravitational field. This quantity (the "mass of the field"), considered as a functional defined on a certain 3-dimensional metric, turns out to be the Hamiltonian of the system, and can therefore be regarded as the physical energy of the gravitational packet. This "gravitational energy" is invariant under arbitrary (internal) coordinate changes, provided that the changes attenuate sufficiently rapidly towards infinity. Thus in this formalism a gravitational field is considered as a new object situated in an underlying Minkowski space, in relation to which (i.e. to the Minkowski metric) its energy is defined.

§40. The Spinor Representations of the Groups $SO(3)$ and $O(3, 1)$. Dirac's Equation and Its Properties

40.1. Automorphisms of Matrix Algebras

Consider the full matrix algebra $M(n, \mathbb{C})$ acting on the n-dimensional complex space $\mathbb{C}^n$. Recall that an *automorphism* of an algebra (or any other algebraic structure) is just an isomorphic self-map, and that an *inner automorphism* of an algebra A is an automorphism h of the form $h(x) = gxg^{-1}$ where g is any fixed unit (i.e. invertible element) in A. Thus if $A = M(n, \mathbb{C})$

then $g \in GL(n, \mathbb{C})$.) The construction of the spinor representations of $SO(3)$ and $O(3, 1)$ rests on the following property of the matrix algebra $M(n, \mathbb{C})$.

40.1.1. Lemma. *Every automorphism of the associative algebra $M(n, \mathbb{C})$ is inner.*

PROOF. Recall that an element P of an algebra $M(n, \mathbb{C})$ is called an *idempotent* (or *projector*) if $P^2 = P$, and that two idempotents P, Q are *orthogonal* if $PQ = QP = 0$. It is easy to see that the images (under their action on $\mathbb{C}^n$) of two orthogonal idempotents intersect in the nullspace of $\mathbb{C}^n$. An idempotent P is said to be *one-dimensional* if the image subspace $P(\mathbb{C}^n)$ is one-dimensional. The matrices $P_1, \ldots, P_n$ defined as follows, form a set of one-dimensional, pairwise orthogonal idempotents:

$$(P_i)_l^k = \begin{cases} 1, & \text{if } k = l = i, \\ 0, & \text{if } k \neq i \text{ or } l \neq i. \end{cases} \tag{1}$$

Thus P_i has entry 1 in its (i, i)th place and zeroes elsewhere. Clearly

$$P_i^2 = P_i, \qquad P_i P_j = 0 \quad \text{if } i \neq j, \qquad P_1 + \cdots + P_n = 1. \tag{2}$$

Consider now an arbitrary automorphism $h: M(n, \mathbb{C}) \to M(n, \mathbb{C})$, and write $h(P_i) = P_i'$. Since h is an isomorphism (so that in particular $h(1) = 1$), the relations (2) must hold also for the P_i':

$$(P_i')^2 = P_i', \qquad P_i' P_j' = 0 \quad \text{if } i \neq j, \qquad P_1' + \cdots + P_n' = 1. \tag{3}$$

Hence the P_1' are also (nonzero) pairwise orthogonal idempotents. Since (by the last relation in (3)) they also satisfy $P_1'(\mathbb{C}^n) + \cdots + P_n'(\mathbb{C}^n) = \mathbb{C}^n$, and since their images have null intersection, they must also be one-dimensional. Write $P_i'(\mathbb{C}^n) = C_i'$.

For each i, j, $1 \leq i \leq n$, $1 \leq j \leq n$, denote by t_{ij} the matrix with entry 1 in the (i, j)th place and zeroes elsewhere (i.e. $(t_{ij})_l^k = 1$ if $i = k, j = l$, otherwise $(t_{ij})_l^k = 0$). (These matrices are sometimes called "transvections".) It is immediate that $t_{ij}(e_j) = e_i$ and $t_{ij}(e_r) = 0$ if $r \neq j$, where $e_1, \ldots, e_n$ are the standard basis vectors for the space $\mathbb{C}^n$. Clearly

$$t_{ii} = P_i, \qquad t_{ij} t_{rs} = 0 \quad \text{if } j \neq r, \qquad t_{ji} t_{ir} = t_{jr}. \tag{4}$$

Writing $h(t_{ij}) = t_{ij}'$, and applying the automorphism h to the relations (4) we obtain

$$t_{ii}' = P_i', \qquad t_{ij}' t_{rs}' = 0 \quad \text{if } j \neq r, \qquad t_{ji}' t_{ir}' = t_{jr}'. \tag{5}$$

Since $P_k' t_{ij}' = 0$ if $k \neq i$, and

$$\mathbb{C}^n = P_1'(\mathbb{C}^n) \oplus \cdots \oplus P_n'(\mathbb{C}^n) = C_1' \oplus \cdots \oplus C_n',$$

it follows that the image space $t_{ij}'(\mathbb{C}_n)$ is one-dimensional, and in fact must co-incide with C_i'. Since

$$t_{ij}'(\mathbb{C}^n) = t_{ij}' P_j'(\mathbb{C}^n) = t_{ij}'(C_j'),$$

it follows that t'_{ij} restricts to an isomorphism from C'_j to C'_i. Let e'_1 be any fixed vector in C'_1, and define $e'_i = t'_{i1}(e'_1)$. The vectors $e'_1, \ldots, e'_n$ are then non-zero vectors in $C'_1, \ldots, C'_n$ respectively, and therefore form a basis for $\mathbb{C}^n$. Define a linear transformation $g: \mathbb{C}^n \to \mathbb{C}^n$ by

$$g(e_i) = e'_i. \tag{6}$$

We shall now verify that for all $x \in M(n, \mathbb{C})$ we have

$$h(x) = gxg^{-1}. \tag{7}$$

To this end note first that

$$t'_{ij}(e'_j) = t'_{ij}t'_{j1}(e'_1) = t'_{i1}(e'_1) = e'_i,$$
$$t'_{ij}(e'_r) = t'_{ij}t'_{r1}(e'_1) = 0 \quad \text{if } r \neq j,$$

whence it is easily verified that $t'_{ij}(e'_r) = gt_{ij}g^{-1}(e'_r)$ for all r. Since the e'_r form a basis for $\mathbb{C}^n$ it follows that

$$h(t_{ij}) = gt_{ij}g^{-1}, \quad \text{for all } i, j.$$

Since every matrix x is a linear combination over $\mathbb{C}$ of the t_{ij}, it follows finally that $h(x) = gxg^{-1}$ for all x, completing the proof. □

In the following two sections we shall be concerned initially with presentations of the matrix algebras $M(2, \mathbb{C})$, $M(4, \mathbb{C})$ in terms of particular matrices, called respectively the Pauli and Dirac matrices.

40.2. The Spinor Representation of the Group $SO(3)$

Consider the following generators of the algebra $M(2, \mathbb{C})$:

$$1, \quad \sigma_x = \sigma_1 = \begin{pmatrix} 0 & 1 \\ 1 & 0 \end{pmatrix}, \quad \sigma_y = \sigma_2 = \begin{pmatrix} 0 & -i \\ i & 0 \end{pmatrix},$$
$$\sigma_z = \sigma_3 = \begin{pmatrix} 1 & 0 \\ 0 & -1 \end{pmatrix}. \tag{8}$$

As was noted in §14.3 the matrices $\sigma_1, \sigma_2, \sigma_3$ are called the *Pauli matrices*. Together with the identity matrix they form an additive basis for $M(2, \mathbb{C})$ as a 4-dimensional vector space. They satisfy the following relations (which yield a presentation of the algebra $M(2, \mathbb{C})$):

(i) $\sigma_q\sigma_l - \sigma_l\sigma_q = 2i\sigma_k$ where (q, l, k) is an even permutation;
(ii) $\sigma_q\sigma_l + \sigma_l\sigma_q = 2\delta_{ql}$.

We rewrite these relations as follows:

(i) $[\sigma_q, \sigma_l] = 2i\sigma_k$, (q, l, k) an even permutation;
(ii) $\{\sigma_q, \sigma_l\} = 2\delta_{ql}$.

It follows from the relations (i) and §24.2(22) that the Lie algebra generated by the matrices $(i/2)\sigma_j$ is isomorphic to the Lie algebra of the group $SO(3)$ (and therefore, by Theorem 24.2.4, also to the Lie algebra of $SU(2)$).

We now derive an important consequence of the relations (ii). For this purpose we first endow the 3-dimensional subspace of $M(2, \mathbb{C})$ spanned by $\sigma_1, \sigma_2, \sigma_3$, with the Euclidean metric with Euclidean co-ordinates x^1, x^2, x^3 where (x^1, x^2, x^3) is the "point" $x^1\sigma_1 + x^2\sigma_2 + x^3\sigma_3$. Let $\Lambda \in O(3)$ be an orthogonal transformation of this 3-dimensional Euclidean space $\mathbb{R}^3(x^1, x^2, x^3)$:

$$\Lambda: \mathbb{R}^3 \to \mathbb{R}^3, \qquad x^{\alpha'} = \lambda^\alpha_\beta x^\beta. \tag{9}$$

Write

$$\sigma'_q = \lambda^\alpha_q \sigma_\alpha; \tag{10}$$

thus the σ'_q are the transforms of the basis vectors (points) σ_q. It follows almost immediately from the orthogonality of Λ that the relations (ii) are preserved:

$$\{\sigma'_q, \sigma'_l\} = 2\delta_{ql}.$$

On the other hand for the relations (i) we have

$$[\sigma'_\alpha, \sigma'_\beta] = [\lambda^\gamma_\alpha \sigma_\gamma, \lambda^\delta_\beta \sigma_\delta] = \lambda^\gamma_\alpha \lambda^\delta_\beta [\sigma_\gamma, \sigma_\delta] = 2i(\lambda^\gamma_\alpha \lambda^\delta_\beta \varepsilon^t_{\gamma\delta} \sigma_t),$$

where as usual

$$\varepsilon^t_{\gamma\delta} = \varepsilon_{\gamma\delta t} = \begin{cases} 1, & \text{if } (\gamma, \delta, t) \text{ is an even permutation,} \\ -1, & \text{if } (\gamma, \delta, t) \text{ is an odd permutation,} \\ 0, & \text{if } \gamma, \delta, t \text{ are not all distinct.} \end{cases}$$

It is not difficult to see (verify it for instance in the case $\alpha = 1, \beta = 2$) that

$$\lambda^\gamma_\alpha \lambda^\delta_\beta \varepsilon^t_{\gamma\delta} \sigma_t = \varepsilon^r_{\alpha\beta} C^t_r \sigma_t,$$

where C^t_r is the (t, r)th cofactor of the matrix $\Lambda = (\lambda^k_l)$. Hence if $\Lambda \in SO(3)$, then $C^t_r = \lambda^t_r$, and

$$[\sigma'_\alpha, \sigma'_\beta] = 2i\varepsilon^r_{\alpha\beta} \sigma'_r.$$

We have thus shown that the relations (i) are also preserved by orthogonal transformations (10) with determinant $+1$.

It is clear that the two sets of relations (i) and (ii) together form a full set of defining relations for the matrix algebra $M(2, \mathbb{C})$, since they allow any product $\sigma_i \sigma_j$ to be expressed as a linear combination of the basis elements $1, \sigma_1, \sigma_2, \sigma_3$. We therefore have the following result.

40.2.1. Theorem. *If the linear transformation of $M(2, \mathbb{C})$ defined by* (10) *has its matrix Λ in $SO(3)$, then it is an algebra automorphism.*

From Lemma 40.1.1 we then obtain the

40.2.2. Corollary. *If the linear transformation h of $M(2, \mathbb{C})$ defined by* (10) *has its matrix Λ in $SO(3)$, then there is an invertible linear transformation $g = g(\Lambda)$ of $\mathbb{C}^2$ such that $h(x) = gxg^{-1}$ for all matrices x in $M(2, \mathbb{C})$, i.e. h is an inner automorphism of $M(2, \mathbb{C})$.*

40.2.3. Definition. The map $\Lambda \mapsto g(\Lambda)$ is called the *spinor representation* of $SO(3)$ in $GL(2, \mathbb{C})$.

Since an inner automorphism determines the conjugating matrix only modulo the centre of $GL(2, \mathbb{C})$, which centre is comprised of the non-zero scalar matrices, it follows that $g(\Lambda)$ is not uniquely defined, in fact is defined only to within a non-zero factor $\lambda \in \mathbb{C}$. The spinor representation is, strictly speaking, the homomorphism from $SO(3)$ to the group $SL(2, \mathbb{C})/\{\pm 1\}$, determined by the many-valued relation $\Lambda \mapsto g(\Lambda)$.

EXERCISES

1. Show that the image of $SO(3)$ under this homomorphism is precisely $SU(2)/\{\pm 1\}$, and deduce that $SO(3) \simeq SU(2)/\{\pm 1\}$. (This was the content of Corollary 13.2.2.)
2. Show that "the" transformation $g(\Lambda)$ corresponding to the rotation Λ through an angle φ about an axis with direction vector $n = (n_x, n_y, n_z)$, $n_x^2 + n_y^2 + n_z^2 = 1$, has the form

$$g(\Lambda) = g(n, \varphi) = \exp\left\{-i\frac{\varphi}{2}(n_x\sigma_x + n_y\sigma_y + n_z\sigma_z)\right\}.$$

40.3. The Spinor Representation of the Lorentz Group

We now transfer our attention to the algebra $M(4, \mathbb{C})$. We seek matrices $\gamma^0, \gamma^1, \gamma^2, \gamma^3$ which generate the algebra $M(4, \mathbb{C})$, and which satisfy the relations

$$\gamma^a\gamma^b + \gamma^b\gamma^a = 2g^{ab} \cdot 1, \tag{11}$$

where g^{ab} is the Minkowski metric. The essence of the next lemma is that the following four matrices fulfil these requirements, and that the relations (11) form a full set of defining relations for $M(4, \mathbb{C})$:

$$\gamma^0 = \begin{pmatrix} 1 & 0 \\ 0 & -1 \end{pmatrix}, \qquad \gamma^2 = \begin{pmatrix} 0 & \sigma_2 \\ -\sigma_2 & 0 \end{pmatrix},$$
$$\gamma^1 = \begin{pmatrix} 0 & \sigma_1 \\ -\sigma_1 & 0 \end{pmatrix}, \qquad \gamma^3 = \begin{pmatrix} 0 & \sigma_3 \\ -\sigma_3 & 0 \end{pmatrix}. \tag{12}$$

(Here what appear as entries are 2×2 blocks, so that the matrices are 4×4, as they should be; the 2×2 matrices $\sigma_1, \sigma_2, \sigma_3$ are as defined in (8) above.) It is straightforward to verify that these matrices satisfy the relations (11).

40.3.1. Lemma. *All of the* 4×4 *matrices* $1, \gamma^a, \gamma^a\gamma^b(a < b), \gamma^a\gamma^b\gamma^c(a < b < c)$, *and* $\gamma^0\gamma^1\gamma^2\gamma^3$, *are linearly independent. It follows that* $\gamma^0, \gamma^1, \gamma^2, \gamma^3$ *generate the algebra* $M(4, \mathbb{C})$, *and further that the relations* (11) *form a full set of defining relations for that algebra.*

PROOF. It is clear that by means of the relations (11) any products of the γ^a of arbitrary length and with repetitions allowed, can be expressed as linear combinations of the 16 particular products figuring in the statement of the lemma. Since $M(4, \mathbb{C})$ has dimension 16, the lemma will follow once we have established the linear independence of those 16 products. We leave to the reader the laborious but straightforward task of computing these 16 matrices and verifying their linear independence. □

Having found appropriate generators of $M(4, \mathbb{C})$ we are now in a position to construct the spinor representation of the group $O(3, 1)$ by exploiting once again Lemma 40.1.1. Consider the Minkowski space $\mathbb{R}^4_1(x^0, x^1, x^2, x^3)$ and an arbitrary matrix $\Lambda \in O(1, 3)$, $\Lambda = (\lambda^a_b)$. Define primed matrices γ'^a by

$$\gamma'^a = \lambda^a_b \gamma^b. \tag{13}$$

From the relations (11) and the fact that by definition the matrix Λ preserves the Minkowski metric, it follows that the "primed" analogues of those relations also hold:

$$\{\gamma'^a, \gamma'^b\} = 2g^{ab} \cdot 1.$$

Hence by Lemma 40.3.1 the map $h = h(\Lambda)$: $M(4, \mathbb{C}) \to M(4, \mathbb{C})$, defined by $1 \mapsto 1$, $\gamma^a \mapsto \gamma'^a$, is an automorphism of the full matrix algebra $M(4, \mathbb{C})$. Therefore by Lemma 40.1.1 there exists $g = g(\Lambda) \in GL(4, \mathbb{C})$ such that $h(x) = gxg^{-1}$ for all $x \in M(4, \mathbb{C})$. The correspondence $\Lambda \mapsto g(\Lambda)$ is called the *spinor representation* of the group $O(1, 3)$ in the group $GL(4, \mathbb{C})$. As in the case of the spinor representation of $SO(3)$, this correspondence is many-valued, since for each $\Lambda \in O(1, 3)$ the matrix $g(\Lambda)$ is defined only to within a non-zero scalar factor. If we require that $g(\Lambda)$ be in $SL(4, \mathbb{C})$ then the correspondence is two-valued:

$$\Lambda \mapsto \pm g(\Lambda) \in SL(4, \mathbb{C}).$$

40.3.2. Definition. The 4-dimensional complex space $\mathbb{C}^4$ with the above spinor representation (i.e. $O(1, 3)$ via $SL(4, \mathbb{C})$) acting on it, is called the *space of* (*4-component*) *spinors*. Thus the elements of this space, written as column vectors, are *spinors*.

Note that the representation of the subgroup $SO(3)$ of $O(1, 3)$ obtained by restricting the above spinor representation, decomposes as the direct sum of two irreducible representations (i.e. without proper invariant subspaces) each equivalent to the spinor representation of $SO(3)$ defined in the preceding subsection. We invite the reader to show this. (Hint: Consider the action on the "semi-spinors" defined below and note that here $SO(3)$ is taken to be the group of transformations $\gamma^1 \mapsto \gamma^1, \gamma^\alpha \mapsto \lambda_k^\alpha \gamma^k, \alpha, k = 1, 2, 3$, where (λ_j^i) is a 3×3 special orthogonal matrix.)

Remark. In terms of the generators $\tilde{\gamma}^0 = \gamma^0, \tilde{\gamma}^\alpha = i\gamma^\alpha, \alpha = 1, 2, 3$, the relations (11) become

$$\{\tilde{\gamma}^a, \tilde{\gamma}^b\} = 2\delta^{ab} \cdot 1.$$

Thus using these new generators one can construct an analogous spinor representation of $SO(4)$.

Exercises

1. If $\Lambda \in SO(3)(\subset O(1, 3))$ is the rotation of $\mathbb{R}^3$ through an angle φ about an axis with unit direction vector $n = (n_1, n_2, n_3)$, then its image $g(\Lambda)$ under the spinor representation of $O(1, 3)$ is (up to a scalar factor) given by

$$g(\Lambda) = g(\varphi, n) = \exp\left\{-i\frac{\varphi}{2}(n_1\Sigma_1 + n_2\Sigma_2 + n_3\Sigma_3)\right\},$$

where $\Sigma_j = \begin{pmatrix} \sigma_j & 0 \\ 0 & \sigma_j \end{pmatrix}$.

2. For the hyperbolic rotation in the plane (x^0, an) through an imaginary angle $i\varphi$, where n is a fixed unit 3-vector (i.e. the elementary Lorentz transformation (corresponding to φ) preserving the quadratic form $(x^0)^2 - a^2((n^1)^2 + (n^2)^2 + (n^3)^2)$ in x^0 and a—see §6.2), the value of the spinor representation is given (up to a scalar factor) by

$$g(\varphi, n) = \exp\left\{-\frac{\varphi}{2}(n_1\alpha_1 + n_2\alpha_2 + n_3\alpha_3)\right\},$$

where $\alpha_j = \begin{pmatrix} 0 & \sigma_j \\ \sigma_j & 0 \end{pmatrix}$.

3. Show that the value of the spinor representation at a spatial reflection $P(x^0, x) = (x^0, -x)$, where x is a (variable) 3-dimensional vector, has (in $SL(4, \mathbb{C})$) the form

$$g(P) = \eta_P\gamma^0, \quad \text{where } \eta_P = \pm i \text{ or } \pm 1.$$

4. Show that under the spinor representation the time-inversion operator $T(x^0, x) = (-x^0, x)$ is represented in $SL(4, \mathbb{C})$ by

$$g(T) = \eta_T\gamma^0\gamma^1\gamma^3, \quad \text{where } |\eta_T| = 1.$$

We now introduce the "semi-spinor" representations of $SO(1, 3)$. Note first that from the way in which the matrices $\gamma^0, \gamma^1, \gamma^2, \gamma^3$ are formed out of the "blocks" $\pm 1, \pm\sigma_1, \pm\sigma_2, \pm\sigma_3$, (see (12)), it follows that in the decomposition of the vector space $\mathbb{C}^4$ in the obvious way as $\mathbb{C}^2 \oplus \mathbb{C}^2$, which we indicate by

$$\psi = \begin{pmatrix} \varphi \\ \chi \end{pmatrix}, \qquad \psi \in \mathbb{C}^4, \qquad \varphi \in \mathbb{C}^2, \qquad \chi \in \mathbb{C}^2, \tag{14}$$

the two summands are each invariant under γ^0 (in fact γ^0 fixes each vector in the first summand, and sends each vector in the second to its negative), and are interchanged by $\gamma^1, \gamma^2, \gamma^3$. Hence if we decompose $\mathbb{C}^4$ instead into the two 2-dimensional subspaces of vectors η, ξ respectively, where

$$\eta = \frac{\varphi + \chi}{\sqrt{2}}, \qquad \xi = \frac{\varphi - \chi}{\sqrt{2}} \tag{15}$$

(by which we mean that we choose the bases

$$\left\{\left(\frac{1}{\sqrt{2}}, 0, \frac{1}{\sqrt{2}}, 0\right), \left(0, \frac{1}{\sqrt{2}}, 0, \frac{1}{\sqrt{2}}\right)\right\}, \quad \left\{\left(\frac{1}{\sqrt{2}}, 0, -\frac{1}{\sqrt{2}}, 0\right), \left(0, \frac{1}{\sqrt{2}}, 0, -\frac{1}{\sqrt{2}}\right)\right\} \tag{16}$$

for the respective summands), then it follows that each of these summands is invariant under $\gamma^1, \gamma^2, \gamma^3$, and that they are interchanged by γ^0, and therefore, in view of Exercise 3 above, also by the matrix representing the spatial reflection:

$$g(P)\colon \eta \mapsto \xi, \ \xi \mapsto \eta,$$

$$P(x^0, x) = (x^0, -x).$$

It follows without difficulty that these two 2-dimensional subspaces of "semi-spinors" are acted on independently by the direct subgroup $SO(1, 3)$ of the Lorentz group $O(1, 3)$ (and are interchanged by the spatial reflection).

40.3.3. Definition. The actions induced by the spinor representation of the group $SO(1, 3)$ on the two subspaces of semi-spinors η and ξ, are called the *semi-spinor representations* of that group. They are denoted by g_+ and g_- respectively. (In §41.3 we shall give an alternative definition of the semi-spinor representations.)

Note that neither of the semi-spinor representations extends to a (2-dimensional) representation of the full Lorentz group $O(1, 3)$.

The spinor representation of the Lorentz group $O(1, 3)$ is not unitary. However it does preserve the indefinite scalar product defined by

$$\langle \psi, \psi \rangle = \psi^* \gamma^0 \psi, \tag{17}$$

where $\psi^* = (\bar{\psi}_1, \bar{\psi}_2, \bar{\psi}_3, \bar{\psi}_4)$ is the complex-conjugate row-vector derived from the column-vector ψ. In terms of the basis comprised of the vectors in (16) (in order), the matrix of γ^0 becomes

$$\gamma^0 = \begin{pmatrix} 0 & 1 \\ 1 & 0 \end{pmatrix}.$$

Hence in terms of the semi-spinors η, ξ, we have that for $\psi = \begin{pmatrix} \eta \\ \xi \end{pmatrix}$, $\langle \psi, \psi \rangle = \xi^*\eta + \eta^*\xi$. (Verify it!) (The invariance of the form (17) under $O(1, 3)$ follows easily from this.)

40.3.4. Definition. The row-spinor $\bar{\psi} = \psi^*\gamma^0$ is called the *Dirac conjugate* of the spinor ψ. (Thus the form $\langle \psi, \psi \rangle$ which as we have just seen is invariant under the spinor representation, can in this notation be rewritten as $\bar{\psi}\psi$.)

EXERCISES (continued)

5. Show that the quantities $\bar{\psi}\gamma^a\psi$ transform (under the action of $O(3, 1)$) like the components of a vector, the quantities $\bar{\psi}\gamma^a\gamma^b\psi$ like the components of a rank-two tensor, the $\bar{\psi}\gamma^a\gamma^b\gamma^c\psi$ like those of a rank-three tensor, and $\bar{\psi}\gamma^4\psi = \bar{\psi}\gamma^0\gamma^1\gamma^2\gamma^3\psi$ like a rank-four tensor (or "pseudo-scalar").
6. Prove that there is no (non-zero) scalar product on $\mathbb{C}^2$, invariant under either of the semi-spinor representations g_+ or g_- of the group $SO(1, 3)$ in $GL(2, \mathbb{C})$. (Hint. This can be inferred from the fact that one of the semi-spinor representations is isomorphic to the standard representation of $SL(2, \mathbb{C})$, and the other to the representation obtained from the standard one by means of complex conjugation.)

Finally we note that the isomorphism between the Lie algebras of the groups $SL(2, \mathbb{C})$ and $SO(1, 3)$ is the particular case $n = 2$ of Proposition 24.6.1 where the Lie algebra of conformal transformations of the sphere S^2 is shown to be isomorphic to the Lie algebra of $SO(1, 3)$. (Cf. Exercise 3(iii) of §24.7.)

40.4. Dirac's Equation

The presentation of $M(4, \mathbb{C})$ in terms of the generators $\gamma^0, \gamma^1, \gamma^2, \gamma^3$ defined by (12), with relations (11), arises naturally in connexion with the following question: Can the Klein–Gordon operator $\Box + m^2$, where

$$\Box = g^{ab} \frac{\partial}{\partial x^a} \frac{\partial}{\partial x^b}$$

(see §38), be expressed as a product of two first-order operators as follows:

$$-(\Box + m^2) = \left(i\gamma^a \frac{\partial}{\partial x^a} + m\right)\left(i\gamma^b \frac{\partial}{\partial x^b} + (-m)\right)? \tag{18}$$

EXERCISE

Show that the existence of this decomposition is equivalent to the relations (11), i.e. is possible if and only if

$$\{\gamma^a, \gamma^b\} = 2g^{ab} \cdot 1. \tag{19}$$

Thus we can achieve the decomposition (18) by taking in particular our γ-matrices (12) as the coefficients.

40.4.1. Definition. The equation (for the 4-spinor ψ)

$$\left(i\gamma^b \frac{\partial}{\partial x^b} - m\right)\psi = 0, \tag{20}$$

is called *Dirac's equation.*

EXERCISE

Show that the Dirac conjugate $\bar{\psi} = \psi^*\gamma^0$ of a solution ψ of (20), satisfies the "conjugate" equation

$$i \frac{\partial \bar{\psi}}{\partial x^a} \gamma^a + m\bar{\psi} = 0. \tag{21}$$

Dirac's equation turns out to be the Euler–Lagrange equation corresponding to the action

$$S = \int \left[\frac{i}{2}\left(\bar{\psi}\gamma^a \frac{\partial \psi}{\partial x^a} - \frac{\partial \bar{\psi}}{\partial x^a} \gamma^a \psi\right) - m\bar{\psi}\psi\right] d^4x, \tag{22}$$

where ψ and $\bar{\psi}$ are regarded as independent (and of course $\bar{\psi}\psi = \psi^*\gamma^0\psi$). It follows easily from its definition (see §37.2(8)), taking into account the obvious fact that Dirac's equation (20) and its conjugate (21) imply the vanishing of the Lagrangian in (22), that the energy-momentum tensor is given by

$$T^{ab} = \frac{i}{2} g^{ac}\left(\bar{\psi}\gamma^b \frac{\partial \psi}{\partial x^c} - \frac{\partial \bar{\psi}}{\partial x^c} \gamma^b \psi\right) = T^{ba}. \tag{23}$$

The current vector has components

$$J^a = \bar{\psi}\gamma^a\psi, \qquad a = 0, 1, 2, 3. \tag{24}$$

The charge density is the first component of the current:

$$J^0 = \bar{\psi}\gamma^0\psi = \psi^*(\gamma^0)^2\psi = \psi^*\psi.$$

The total charge is therefore

$$Q = \int \psi^*\psi \, d^3x = \int J^0 \, d^3x, \tag{25}$$

which is clearly non-negative ($Q \geq 0$), since $\psi^*\psi$ is positive definite. On the other hand the energy density T^{00} need not in general be positive definite; this leads to several difficulties, which we shall not however examine here.

Analogously to §38, for the solutions (for the components of ψ) of Dirac's equation of the plane-wave type, i.e. of the form const. $\times\ e^{i\langle k, x\rangle}$, we must have $\langle k, k\rangle = m^2$ (where here, as throughout, we are setting for simplicity $\hbar = c = 1$). In the case $k^0 > 0$, such a solution is taken as describing a free particle of mass m, whose momentum k lies therefore on the upper half of the mass surface $\langle k, k\rangle = m^2$. The condition $k^0 > 0$ follows from the requirement that the energy of the particle be positive.

If we rewrite Dirac's equation in terms of the semi-spinors η and ξ (see (15)), using the simple actions on these of $\gamma^1, \gamma^2, \gamma^3$ and the interchanging action of γ^0, then we obtain easily the following two equations:

$$\begin{aligned} i\frac{\partial\eta}{\partial t} &= \langle\sigma, p\rangle\eta + m\xi, \\ i\frac{\partial\xi}{\partial t} &= -\langle\sigma, p\rangle\xi + m\eta, \end{aligned} \tag{26}$$

where $p = i(\partial/\partial x)$, and $\langle\sigma, p\rangle = i(\sigma_1(\partial/\partial x^1) + \sigma_2(\partial/\partial x^2) + \sigma_3(\partial/\partial x^3))$. If the mass $m = 0$, then these become two independent equations ("Weyl's equations"). If Weyl's equations are taken as describing physical particles, then these particles must be such that their laws of motion are not invariant under spatial reflections (since the semi-spinor representations do not extend to representations of the full Lorentz group), and they must have zero mass (since by Exercise 6 of the preceding subsection no invariant scalar product of semi-spinors exists; if there did exist an invariant scalar product it could be used to furnish the Lagrangian with a mass-related term).

40.5. Dirac's Equation in an Electromagnetic Field. The Operation of Charge Conjugation

The inclusion of an electromagnetic field given by its vector-potential A, is effected by the standard rule (cf. §38) whereby p_a is replaced by $p_a + eA_a$ in the Lagrangian (here $\hbar$ and c have been put equal to 1), i.e. $\partial/\partial x^a$ is replaced by $\partial/\partial x^a - ieA_a(x)$, where e is the charge. The Euler–Lagrange equation (or "Dirac's equation in an electromagnetic field") is

$$\left[\tilde{\gamma}^a\left(\frac{\partial}{\partial x^a} - ieA_a(x)\right) + m\right]\psi = 0, \tag{27}$$

and the conjugate equation is then

$$\bar{\psi}\left[(\tilde{\gamma}^a)^{\mathrm{T}}\left(\frac{\partial}{\partial x^a} + ieA_a(x)\right) - m\right] = 0, \tag{28}$$

where $\tilde{\gamma}^0 = \gamma^0, \tilde{\gamma}^\alpha = -i\gamma^\alpha$ (see the remark in §40.3), and T denotes as usual the operation of taking the transpose.

Direct calculation using (11) and (12) shows that the matrix $C = \tilde{\gamma}^2\tilde{\gamma}^0$ satisfies

$$C^{-1}\tilde{\gamma}^a C = -(\tilde{\gamma}^a)^{\mathrm{T}}, \qquad a = 0, 1, 2, 3. \tag{29}$$

Substituting in (27) for $\tilde{\gamma}^a$ (from (29)), we obtain

$$\left[C(\tilde{\gamma}^a)^{\mathrm{T}}C^{-1}\left(\frac{\partial}{\partial x^a} - ieA_a(x)\right) - CC^{-1}m\right]\psi = 0,$$

whence

$$\left[(\tilde{\gamma}^a)^{\mathrm{T}}\left(\frac{\partial}{\partial x^a} - ieA_a(x)\right) - m\right]C^{-1}\psi = 0. \tag{30}$$

Now from (12) it follows easily that the conjugate transpose of $\tilde{\gamma}^a$ is $-\tilde{\gamma}^a$ ($a = 0, 1, 2, 3$). On taking the complex conjugate of equation (30) we therefore obtain

$$\left[\tilde{\gamma}^a\left(\frac{\partial}{\partial x^a} + ieA_a(x)\right) + m\right]C^{-1}\bar{\psi} = 0, \tag{31}$$

where here $\bar{\psi} = \psi^*\gamma^0$ is written as a column-vector. Write ψ^C for the column vector $C^{-1}\bar{\psi}$. Comparison of equations (27) and (31) finally yields the following

40.5.1. Theorem. *If the vector field $\psi(x)$ satisfies Dirac's equation* (27) *for a charge e in a field $A(x)$, then the field $\psi^C(x) = C^{-1}\bar{\psi}$ satisfies the same equation with e replaced by $-e$.*

The transformation $\psi \to \psi^C$ is called "charge conjugation", since it entails the reversal of the sign of the charge on the particle described by the field ψ. This theorem has the important consequence that Dirac's equation for the spinor field ψ describes simultaneously two sorts of particle, one with charge e and the other with charge $-e$. A single solution of Dirac's equation (27) yields the wave function $\psi(x)$ of an electron, and the wave function $\psi^C(x)$ of a positron.

Exercise

Verify that the operation of charge conjugation on spinors:

(i) commutes with the action of the direct group $SO(1, 3)$ of Lorentz transformations;
(ii) commutes with the action of the spatial reflection $g(P) = \eta_P\tilde{\gamma}^0$ if $\eta_P = \pm i$, and does not so commute if $\eta_P = \pm 1$ (see Exercise 3 in §40.3);
(iii) commutes with the time-inversion operator provided $\eta_T = \pm 1$ (see Exercise 4 in §40.3).

§41. Covariant Differentiation of Fields with Arbitrary Symmetry

41.1. Gauge Transformations. Gauge-Invariant Lagrangians

We begin with an important special case, exemplified in §38.1. Consider a complex scalar field ψ and a Lagrangian of the form

$$L = L\left(\psi, \bar{\psi}, \frac{\partial \psi}{\partial x^\alpha}, \frac{\partial \bar{\psi}}{\partial x^\alpha}\right), \qquad \alpha = 1, \dots, n, \tag{1}$$

where the bar denotes complex conjugation, and $\psi, \bar{\psi}$ are regarded as formally independent. We suppose that the Lagrangian is invariant under the group of all transformations of the form

$$\psi \to e^{i\varphi}\bar{\psi}, \qquad \bar{\psi} \to e^{-i\varphi}\bar{\psi} \qquad (\varphi = \text{(real) const.}). \tag{2}$$

Thus

$$L\left(e^{i\varphi}\psi, e^{-i\varphi}\bar{\psi}, e^{i\varphi}\frac{\partial \psi}{\partial x^\alpha}, e^{-i\varphi}\frac{\partial \bar{\psi}}{\partial x^\alpha}\right) = L\left(\psi, \bar{\psi}, \frac{\partial \psi}{\partial x^\alpha}, \frac{\partial \bar{\psi}}{\partial x^\alpha}\right). \tag{3}$$

(For example the Lagrangian

$$L = \hbar^2 g^{\alpha\beta}\frac{\partial \bar{\psi}}{\partial x^\alpha}\frac{\partial \psi}{\partial x^\beta} - m^2c^2\bar{\psi}\psi, \tag{4}$$

which we considered in §38.1, is invariant under such transformations.)

In keeping with the general "local principle", we wish now to construct from the given Lagrangian (1) a new Lagrangian invariant under the more general transformations

$$\psi \to e^{ie\varphi(x)}\psi, \qquad \bar{\psi} \to e^{-ie\varphi(x)}\bar{\psi}, \tag{5}$$

where φ is now permitted to vary with x. (Thus the group of transformations (2) will now act independently at each point.) We describe in three steps a procedure for arriving at such a Lagrangian:

Step 1. We introduce n supplementary variables in the form of components of a covector field A_α, and the following n supplementary (to (5)) "gradient" transformation rules for these new variables:

$$A_\alpha \to A_\alpha - \frac{\partial \varphi}{\partial x^\alpha}. \tag{6}$$

(Our context being implicitly quantum-mechanical, normally the constants $\hbar$ and c would appear; we have set them equal to 1 for simplicity.)

Step 2. We define a new Lagrangian $\tilde{L}$ by

$$\tilde{L}\left(\psi, \bar{\psi}, \frac{\partial \psi}{\partial x^\alpha}, \frac{\partial \bar{\psi}}{\partial x^\alpha}, A^\alpha\right) = L(\psi, \bar{\psi}, \nabla_\alpha \psi, \overline{\nabla_\alpha \psi}), \tag{7}$$

where

$$\nabla_\alpha \psi = \frac{\partial \psi}{\partial x^\alpha} + ieA_\alpha \psi. \tag{8}$$

(Note that since $\nabla_\alpha \psi$ is a covector, definition (7) is valid.)

Step 3. The full Lagrangian of our invariant theory then has the form $L(\psi, \bar{\psi}, \nabla\psi, \overline{\nabla\psi}) + L_1(A, \partial A/\partial x)$, where the term L_1 is "gradient-invariant", i.e. invariant under the transformations (6). We shall consider a particular form of this term in §42.

41.1.1. Theorem. *The Lagrangian* (7) *is invariant under the* (*local*) *transformation given by* (5) *and* 6).

Proof. We first see how our new "covariant derivative" $\nabla_\alpha \psi$ transforms under the transformation defined by (5) and (6):

$$\nabla_\alpha \psi \to \frac{\partial}{\partial x^\alpha}(e^{ie\varphi(x)}\psi) + ie\left(A_\alpha - \frac{\partial \varphi}{\partial x^\alpha}\right)e^{ie\varphi(x)}\psi$$

$$= e^{ie\varphi(x)}\left[\frac{\partial \psi}{\partial x^\alpha} + ieA_\alpha \psi\right] = e^{ie\varphi(x)}\nabla_\alpha \psi.$$

In view of the assumed invariance of L given by (3) (where, as usual, ψ, $\bar{\psi}$, $\partial\psi/\partial x^\alpha$, $\partial\bar{\psi}/\partial x^\alpha$ are regarded as formally independent variables), we have

$$L(e^{ie\varphi}\psi, e^{-ie\varphi}\bar{\psi}, e^{ie\varphi}\nabla_\alpha \psi, e^{-ie\varphi}\overline{\nabla_\alpha \psi}) = L(\psi, \bar{\psi}, \nabla_\alpha \psi, \overline{\nabla_\alpha \psi}),$$

so that the new Lagrangian $\tilde{L}$ is invariant as claimed. □

We now turn to the general situation of a vector field $\psi(x) = (\psi^1(x), \dots, \psi^N(x))$ defined on $\mathbb{R}^n$ and taking its values in a (real) N-dimensional vector space (i.e., in $\mathbb{R}^N$). Suppose we are given a Lagrangian $L(\psi, \partial\psi/\partial x^\alpha)$ which is invariant under some (standard) group G of $N \times N$ matrices:

$$L\left(g\psi, g\frac{\partial \psi}{\partial x^a}\right) = L\left(\psi, \frac{\partial \psi}{\partial x^a}\right), \qquad g \in G. \tag{9}$$

Our aim (as in the above special case $N = 1$) is to construct from this L a Lagrangian invariant at each point x under matrices $g(x) \in G$ which depend on x.

Let $A_\alpha(x)$, $\alpha = 1, \ldots, n$, be n (matrix-valued) functions from $\mathbb{R}^n$ to the Lie algebra of the group G, which transform under co-ordinate changes on the underlying space $\mathbb{R}^n$ like the components of a covector:

$$A_\alpha(x) = A_\beta(y)\frac{\partial y^\beta}{\partial x^\alpha}, \qquad y = y(x). \tag{10}$$

We then define the *covariant derivative of the vector field* $\psi\colon \mathbb{R}^n \to \mathbb{R}^N$, *relative to the* A_α by

$$\nabla_\alpha \psi = \frac{\partial \psi}{\partial x^a} + A_\alpha(x)\psi. \tag{11}$$

41.1.2. Theorem. *Under transformations of the form*

$$\psi(x) \to g(x)\psi(x), \tag{12}$$

$$A_\alpha(x) \to g(x)A_\alpha(x)g^{-1}(x) - \frac{\partial g(x)}{\partial x^a}g^{-1}(x), \tag{13}$$

($g(x) \in G$) *the covariant derivative* (11) *transforms according to the rule*

$$\nabla_\alpha \psi \to g(x)\nabla_\alpha \psi. \tag{14}$$

It follows that the Lagrangian $\tilde{L}$ *defined by*

$$\tilde{L}\left(\psi, \frac{\partial \psi}{\partial x^a}, A_\alpha\right) = L(\psi, \nabla_\alpha \psi)$$

is invariant under such transformations.

PROOF. From (11), (12) and (13) we have

$$\nabla_\alpha \psi(x) \to \frac{\partial}{\partial x^a}(g(x)\psi) + \left[gA_\alpha g^{-1} - \frac{\partial g(x)}{\partial x^a}g^{-1}\right]g\psi$$

$$= g(x)\left[\frac{\partial \psi}{\partial x^a} + A_\alpha \psi\right] = g(x)\nabla_\alpha \psi,$$

establishing (14). As to the invariance of $\tilde{L}$, note first that the covariant derivative of a vector-function ψ is a covector:

$$\nabla_\alpha \psi(x) = \nabla_\beta \psi(y)\frac{\partial y^\beta}{\partial x^a}, \qquad y = y(x);$$

this is a consequence of (10), and the fact that the gradient $(\partial\psi/\partial x^\alpha)$ transforms like a covector. Hence by virtue of the invariance of L (see (9)), and the formal independence of ψ and $\partial\psi/\partial x^\alpha$ in $L(\psi, \partial\psi/\partial x^\alpha)$, it follows that

$$L(g\psi, g\nabla_\alpha \psi) = L(\psi, \nabla_\alpha \psi).$$

The desired invariance of $\tilde{L}$ is now immediate from this and the already-established (14). □

The field A_α is called a *gauge* (or *compensating*) *field*, or a *connexion*. Transformations of the form given by (12) and (13) are termed *gauge transformations*, and the group which (as is easily seen) they comprise, a *gauge group*.

Thus the requirement of invariance of the Lagrangian under the group of local transformations (12) leads in the manner described to a Lagrangian $\tilde{L}$ corresponding to the mutual interaction of the field ψ and the gauge field A. The question of the definition of the Lagrangian of the gauge field alone will be considered in §42.

Remark. Consider the special case of a vector-valued field $B(x)$ taking its values in the Lie algebra of a given matrix group G. Given a gauge field $A_\alpha(x)$ whose values lie in the same Lie algebra, we define the *covariant derivative of the field B relative to the connexion* A_α by the formula

$$\nabla_\alpha B = \frac{\partial B}{\partial x^a} + [A_\alpha, B], \tag{15}$$

where $[A_\alpha, B] = A_\alpha B - BA_\alpha$ is the commutator in the Lie algebra.

Exercises

1. Verify that under the "gauge transformations"

$$B(x) \to g(x)B(x)g(x)^{-1},$$

$$A_\alpha(x) \to g(x)A_\alpha(x)g^{-1}(x) - \frac{\partial g(x)}{\partial x^\alpha} g^{-1},$$

the covariant derivative $\nabla_\alpha B$ transforms as follows:

$$\nabla_\alpha B(x) \to g(x)(\nabla_\alpha B(x))g(x)^{-1}.$$

2. Show that infinitesimal gauge transformations (of the form given by (12) and (13) above) can be expressed in the form

$$\psi(x) \to \psi(x) + B(x)\psi(x),$$

$$A_\alpha(x) \to A_\alpha(x) + \nabla_\alpha B(x),$$

where $B(x)$ is a field whose values lie in the Lie algebra of G.

41.2. The Curvature Form

The commutator of a pair of components of a covariant differential operator calculates out as follows:

$$[\nabla_\mu, \nabla_\nu]\psi = \nabla_\mu\nabla_\nu\psi - \nabla_\nu\nabla_\mu\psi = \frac{\partial}{\partial x^\mu}\left(\frac{\partial\psi}{\partial x^\nu} + A_\nu\psi\right)$$

$$+ A_\mu\left(\frac{\partial\psi}{\partial x^\nu} + A_\nu\psi\right) - \frac{\partial}{\partial x^\nu}\left(\frac{\partial\psi}{\partial x^\mu} + A_\mu\psi\right) - A_\nu\left(\frac{\partial\psi}{\partial x^\mu} + A_\mu\psi\right)$$

$$= \left(\frac{\partial A_\nu}{\partial x^\mu} - \frac{\partial A_\mu}{\partial x^\nu} + [A_\mu, A_\nu]\right)\psi.$$

If we denote the resulting operator briefly by $F_{\mu\nu}$:

$$F_{\mu\nu} = \frac{\partial A_\nu}{\partial x^\mu} - \frac{\partial A_\mu}{\partial x^\nu} + [A_\mu, A_\nu], \tag{16}$$

then we can state our conclusion as follows: *The commutator of the operators ∇_μ, ∇_ν is just (left) multiplication by the matrix $F_{\mu\nu}$.*

41.2.1. Theorem. *Under the gauge transformations* (13) *the quantities $F_{\mu\nu}$ transform as follows*:

$$F_{\mu\nu} \to gF_{\mu\nu}g^{-1}. \tag{17}$$

Under co-ordinate changes the $F_{\mu\nu}$ transform according to the rule

$$F_{\mu\nu}(x) = F_{\alpha\beta}(y)\frac{\partial y^\alpha}{\partial x^\mu}\frac{\partial y^\beta}{\partial x^\nu}, \qquad y = y(x).$$

It follows (from this and the definition (16)) *that the $F_{\mu\nu}$ are the components of a skew-symmetric, rank-two tensor (taking its values in the Lie algebra of the group G).*

The proof of (17) consists in directly applying the transformation (13) to the right-hand side of the defining formula (16). The fact that $F_{\mu\nu}$ is a tensor is an easy consequence of (10). The detailed calculations are left to the reader.

The form $\Omega = \sum_{\mu<\nu} F_{\mu\nu}\, dx^\mu \wedge dx^\nu$, with its coefficients in the Lie algebra g of the group G, is called the *curvature form* of the connexion A_α. (Compare (16) with the formula (3) given in §30.1 for the curvature tensor.)

A connexion A_μ is called *trivial* if there exists a function $g_0(x)$ from $\mathbb{R}^n$ to the group G, such that

$$A_\mu(x) = g_0^{-1}(x)\frac{\partial g_0(x)}{\partial x^\mu}\left(= -\frac{\partial g_0^{-1}}{\partial x^\mu}g_0\right). \tag{18}$$

41.2.2. Theorem. *The curvature form of a trivial connexion is zero. Conversely if the curvature form is zero, then the connexion is (locally) trivial.*

We prove only the first part of the theorem, leaving the converse as an exercise for the reader. Applying the gauge transformation (13) with $g = g_0$ to A_μ, we obtain

$$A_\mu \to \tilde{A}_\mu = g_0 A_\mu g_0^{-1} - \frac{\partial g_0}{\partial x^\mu} g_0^{-1},$$

whence, substituting from (18) for A_μ, we deduce that $\tilde{A}_\mu \equiv 0$. Hence by (16) the transformed components $\tilde{F}_{\mu\nu}$ of $F_{\mu\nu}$ also vanish identically. Since by (17), $\tilde{F}_{\mu\nu} = g_0 F_{\mu\nu} g_0^{-1}$, it follows that $F_{\mu\nu} \equiv 0$, as required.

EXERCISE

Define the "covariant differential" of a 2-form Ω by

$$D\Omega = \sum_{\lambda<\mu<\nu} (\nabla_\lambda F_{\mu\nu} + \nabla_\nu F_{\lambda\mu} + \nabla_\mu F_{\nu\lambda})\, dx^\lambda \wedge dx^\mu \wedge dx^\nu. \tag{19}$$

Show that if Ω is the curvature form of some connexion then "Bianchi's identity" holds: $D\Omega \equiv 0$. (Cf. Exercise 7 of §30.5.)

41.3. Basic Examples

(a) *The case* $G = U(1) \simeq SO(2)$. As noted in §38, the interaction of an electromagnetic field A_μ with a complex scalar field ψ is taken into account by replacing in the Lagrangian all derivatives $\partial\psi/\partial x^\alpha$ by the corresponding covariant derivatives $\nabla_\alpha \psi = \partial\psi/\partial x^\alpha + ieA_\alpha\psi$. (In fact we began this section with what was essentially just this example.) Here the group G is the one-dimensional abelian group $G = U(1) = \{e^{ie\varphi}\}$. Its Lie algebra is also one-dimensional, and is commutative (i.e. the Lie commutators of pairs of its elements are all zero); in fact it consists of all the purely imaginary complex numbers. The connexion is $ieA_\mu(x)$, which takes its values in this Lie algebra. From (16) it follows that the curvature form of this connexion is (omitting the constant factor ie) given by $F_{\mu\nu} = \partial A_\nu/\partial x^\mu - \partial A_\mu/\partial x^\nu$, which is the familiar (from §37.3) electromagnetic field tensor. Bianchi's identity (see the exercise immediately preceding this subsection) is in this context equivalent to the closure of the form $\Omega = \sum_{\mu<\nu} F_{\mu\nu}\, dx^\mu \wedge dx^\nu = d(A_\mu\, dx^\mu)$. (Recall that by Theorem 25.2.2, $d^2 = 0$.)

The gauge groups of the succeeding examples are all non-abelian.

(b) *Linear connexions* ($G = GL(n, \mathbb{R})$). Given a region U of $\mathbb{R}^n$ with co-ordinates $x^1, \ldots, x^n$, we may consider the operators $\partial/\partial x^1, \ldots, \partial/\partial x^n$ as forming a basis for the tangent space at each point of U (cf. §7.2). Thus a tangent-vector field ξ defined on U is just a function from U to $\mathbb{R}^n$: $\xi(x) = (\xi^1, \ldots, \xi^n)$.

Under a change from the co-ordinates x^ν on U to new co-ordinates $x^{\nu'} = x^{\nu'}(x)$, the components of the vector $\xi(x)$ transform (locally) according to the usual rule

$$\xi^\nu \to \xi^{\nu'} = \frac{\partial x^{\nu'}}{\partial x^\nu} \xi^\nu = g(x)\xi, \tag{20}$$

where the matrix $g(x) = (\partial x^{\nu'}/\partial x^\nu)$ is invertible (with inverse $g^{-1}(x) = (\partial x^\nu/\partial x^{\nu'}))$), and so lies in $GL(n, \mathbb{R})$. As we saw in §14.1, the Lie algebra (i.e. tangent space at the identity) of the group $GL(n, \mathbb{R})$ comprises all $n \times n$ matrices, i.e. coincides with $M(n, \mathbb{R})$. Hence in this context the components $A_\mu(x)$ of a connexion will be arbitrary $n \times n$ matrices. Given a connexion A_μ we denote its entries by $(A_\mu)^\nu_\lambda = \Gamma^\nu_{\lambda\mu}$. By definition (see (11) above) the covariant derivative of a tangent-vector field ξ with respect to such a connexion is given by

$$\nabla_\mu \xi = \frac{\partial \xi}{\partial x^\mu} + A_\mu \xi, \quad \text{i.e.} \quad (\nabla_\mu \xi)^\nu = \frac{\partial \xi^\nu}{\partial x^\mu} + \Gamma^\nu_{\lambda\mu}\xi^\lambda. \tag{21}$$

The gauge transformation (13) with $g(x) = (\partial x^{\nu'}/\partial x^\nu)$ is given by (note here that $-(\partial g/\partial x^\mu)g^{-1} = g(\partial g^{-1}/\partial x^\mu)$, since $gg^{-1} = 1$):

$$\Gamma^\nu_{\lambda\mu} \to \Gamma^{\nu'}_{\lambda'\mu} = \frac{\partial x^{\nu'}}{\partial x^\nu} \Gamma^\nu_{\lambda\mu} \frac{\partial x^\lambda}{\partial x^{\lambda'}} + \frac{\partial x^{\nu'}}{\partial x^\nu} \frac{\partial}{\partial x^\mu}\left(\frac{\partial x^\nu}{\partial x^{\lambda'}}\right). \tag{22}$$

Since a connexion is by definition a covector we must have $A_{\mu'} = (\partial x^\mu/\partial x^{\mu'})A_\mu$. Hence on multiplying both sides of the equality in (22) by $\partial x^\mu/\partial x^{\mu'}$ (and summing over μ) we obtain

$$\Gamma^{\nu'}_{\lambda'\mu'} = \frac{\partial x^\mu}{\partial x^{\mu'}} \frac{\partial x^{\nu'}}{\partial x^\nu} \Gamma^\nu_{\lambda\mu} \frac{\partial x^\lambda}{\partial x^{\lambda'}} + \frac{\partial x^{\nu'}}{\partial x^\nu} \frac{\partial^2 x^\nu}{\partial x^{\lambda'}\, \partial x^{\mu'}}. \tag{23}$$

We have thus arrived at the transformation law for the Christoffel symbols (see §28.1(22)). It can be verified also that the entries in the matrix values of the curvature form $F_{\mu\nu}$ of the connexion $A_\mu = (\Gamma^\nu_{\lambda\mu})$ are given by

$$(F_{\mu\nu})^\varkappa_\lambda = R^\varkappa_{\lambda,\mu\nu}, \tag{24}$$

where $R^\varkappa_{\lambda\mu\nu}$ is the Riemann curvature tensor corresponding to the connexion $\Gamma^\nu_{\lambda\mu}$ (see §30.1).

Suppose next that U is a region of $\mathbb{R}^n$ endowed with a Riemannian metric $g_{\alpha\beta}(x)$. At each non-singular point of U choose as a basis for the tangent space a set of pairwise orthogonal, unit vectors $\xi_1, \ldots, \xi_n$:

$$\langle \xi_\alpha, \xi_\beta \rangle = \delta_{\alpha\beta},$$

and consider the quantities $\hat{\Gamma}_{\beta\alpha\gamma}$ (determining the connexion compatible with the metric) given by

$$\nabla_{\xi_\alpha} \xi_\beta = \sum_\gamma \hat{\Gamma}_{\beta\alpha\gamma} \xi_\gamma, \tag{25}$$

where $\nabla_{\xi_\alpha}\xi_\beta$ is the usual (directional) covariant derivative (with respect to the symmetric connexion compatible with the metric) defined in §29.3. (Recall that in Theorem 30.1.5 we derived a formula for the coefficients $\hat{\Gamma}_{\beta\alpha\gamma}$ in terms of the vector fields ξ_α.) Here the connexion has the form

$$(A_\alpha)_{\beta\gamma} = \hat{\Gamma}_{\beta\alpha\gamma}.$$

It follows from §30.1(13) that for each α the matrix A_α is skew-symmetric. The group G in this situation consists of the transformations preserving the inner product $\langle \eta_1, \eta_2 \rangle = \delta_{\alpha\beta}\eta_1^\alpha \eta_2^\beta$, where $\eta_i = \eta_i^\alpha \xi_\alpha$, $i = 1, 2$; thus G is the orthogonal group, and the action of the gauge transformations on a tangent vector $\eta(x)$ is in this context given by

$$\eta(x) \to g(x)\eta(x), \qquad \eta = \eta^\alpha \xi_\alpha,$$

where $g(x)$ is for each x an orthogonal $n \times n$ matrix. Since, as already noted, A_α is a skew-symmetric $n \times n$ matrix, it lies in the Lie algebra of G, as required. The values of the curvature form $(F_{\mu\nu})$ are also skew-symmetric (for instance because they also lie in the Lie algebra of G). In the case $n = 2$, the curvature form can be shown to be

$$\Omega = K\sqrt{g}\, dx^1 \wedge dx^2, \tag{26}$$

where K is the Gaussian curvature.

These considerations carry over analogously to the case of a pseudo-Riemannian matric.

(c) *Cartan connexions.* Here G is the affine group (consisting of combinations of linear transformations and translations). Let A_μ be any linear connexion, and ∇_μ the corresponding covariant differential operator (on vector fields $\xi(x)$ tangent to some region U of $\mathbb{R}^n$). The corresponding *Cartan connexion* is defined by the formula

$$\tilde{\nabla}_\mu \xi^\nu = \nabla_\mu \xi^\nu + \delta_\mu^\nu. \tag{27}$$

Exercises

1. Show that the connexion defined by $\tilde{\nabla}_\mu$ is invariant with respect to local affine transformations of the form

$$\xi \to g\xi + y, \tag{28}$$

where $g = (\partial x^{\alpha'}/\partial x^\alpha)$ is the Jacobian matrix of a co-ordinate change $x' = x'(x)$, and $y = (y^1(x), \ldots, y^n(x))$ is an arbitrary vector.

2. Show that the curvature form of the Cartan connexion defined by $\tilde{\nabla}_\mu$ (with its values in the Lie algebra of the affine group—see Exercise 3 of §4.5 and Exercise 11 of §24.7) is given by

$$F_{\mu\nu} = (R^\varkappa_{\lambda,\mu\nu}, T^\lambda_{\mu\nu}), \tag{29}$$

where $R^\varkappa_{\lambda,\mu\nu}$ is the curvature tensor, and $T^\lambda_{\mu\nu}$ the torsion tensor, of the connexion A_μ.

(d) *Covariant differentiation of spinors with respect to a metric.* Let M^4 denote 4-dimensional space-time, endowed with a metric $g_{\alpha\beta}$ of type (1, 3). In terms of the connexion compatible with this metric one can define a covariant derivative of spinors. We shall consider in detail only the semi-spinor representation of $SO(1, 3)$ defined in §40.3 (it was noted there that the two semi-spinor representations are conjugate). That representation can be obtained explicitly as follows. An arbitrary 2×2 Hermitian matrix U (i.e. $\bar{U}^T = U$) can be written in the form

$$U = \begin{pmatrix} u^0 + u^3 & u^1 + iu^2 \\ u^1 - iu^2 & u^0 - u^3 \end{pmatrix} = u^\alpha \sigma_\alpha, \tag{30}$$

where $\sigma_0 = \begin{pmatrix} 1 & 0 \\ 0 & 1 \end{pmatrix}$, and $\sigma_1, \sigma_2, \sigma_3$ are the Pauli matrices (see §40.2(8)). Thus the four matrices $\sigma_0, \sigma_1, \sigma_2, \sigma_3$ form a basis for the real linear space of all Hermitian matrices. The determinant $\det U$ is clearly preserved under transformations of the form $U \mapsto gU\bar{g}^T$, where g lies in $SL(2, \mathbb{C})$. Since

$$\det U = (u^0)^2 - (u^1)^2 - (u^2)^2 - (u^3)^2,$$

such transformations of the linear space of Hermitian matrices may be regarded as isometries of Minkowski space (with co-ordinates u^0, u^1, u^2, u^3), i.e. as elements of $SO(1, 3)$. In this way we obtain a homomorphism (which turns out to be onto) from $SL(2, \mathbb{C})$ to $SO(1, 3)$, under which g and $-g$ have the same image. The inverse map $SO(1, 3) \to SL(2, \mathbb{C})$ is two-valued; it turns out to be (up to conjugacy) the semi-spinor representation.

We shall call the elements of the space $\mathbb{C}^2$ acted on by the group $SL(2, \mathbb{C})$, *(two-component) spinors* (rather than "semi-spinors" as we termed them in §40.3). A field $\xi(x)$ taking its values in this space is then a *spinor field*. The action of $SL(2, \mathbb{C})$ on spinors is the usual one:

$$\xi(x) \to g(x)\xi(x), \qquad g(x) \in SL(2, \mathbb{C}), \tag{31}$$

and corresponding to each $g(x)$ there is the conjugate action $\xi \to \bar{g}\xi$. The covariant derivative of a spinor field $\xi(x)$ with respect to a connexion $A_\alpha(x)$ (whose values of course lie in the Lie algebra of $SL(2, \mathbb{C})$, i.e. are complex zero-trace matrices) is by definition (see (11))

$$\hat{\nabla}_\alpha \xi = \frac{\partial \xi}{\partial x^\alpha} + A_\alpha \xi. \tag{32}$$

Associated with the conjugate action $\xi \to \bar{g}\xi$, there is the corresponding covariant derivative

$$\hat{\nabla}_\alpha \xi = \frac{\partial \xi}{\partial x^\alpha} + \bar{A}_\alpha \xi. \tag{33}$$

Similarly we define the covariant derivative relative to the connexion A_α of a Hermitian matrix $U(x)$ via Leibniz' rule:

$$\hat{\nabla}_\alpha U = \frac{\partial U}{\partial x^\alpha} + A_\alpha U + U\bar{A}_\alpha^{\mathrm{T}}. \tag{34}$$

We can identify each (point-dependent) Hermitian matrix $U(x)$ with the vector (u^α) defined by (30), i.e. by $U = u^\alpha \sigma_\alpha$, where (u^α) is regarded as a tangent vector at x (to the space M^4 with metric $g_{\alpha\beta}$), and the u^α as the components of that vector relative to a "tetrad" of basis vectors which at each point are pairwise orthogonal and satisfy $\xi_0^2 = -\xi_1^2 = -\xi_2^2 = -\xi_3^2 = 1$ (cf. the "application" in §30.1). We impose the requirement that the covariant derivative $\hat{\nabla}_\alpha U$ of the form (34) relative to a spinor connexion A, have the form

$$\hat{\nabla}_\alpha U = \sigma_\beta \nabla_\alpha u^\beta, \qquad U = u^\beta \sigma_\beta, \tag{35}$$

where $\nabla_\alpha u^\beta$ is the usual covariant derivative with respect to the symmetric connexion compatible with the metric $g_{\mu\nu}$.

Exercise

Show that the connexion A_α defined by (35) (in accordance with (34)) is given by

$$A_\alpha = -\tfrac{1}{2}\hat{\Gamma}^\gamma_{\beta\alpha}\sigma_\gamma\sigma_\beta,$$

where both repeated indices γ and β are summed over, but in summing over β the summands corresponding to $\beta = 1, 2, 3$ are taken with negative signs (while that corresponding to $\beta = 0$ is taken with positive sign).

As noted in §40.3 (in the hint to Exercise 6), the 4-dimensional spinor representation of $SO(1, 3)$ (acting on the space $\mathbb{C}^4$ of 4-spinors) decomposes as the direct sum of two 2-dimensional representations, one given by the standard representation of $SL(2, \mathbb{C})$, and the other obtained from this one by complex conjugation. It follows that (32) and (33) together define a covariant differention of 4-spinors.

Exercise

Derive Dirac's equation in the case when a metric is present.

Remark. With each 2-spinor $\xi = (\xi^0, \xi^1)$ one can associate the Hermitian matrix $U = (\xi^\alpha \bar{\xi}^\beta)$. Since U is singular (i.e. $\det U = 0$), the corresponding vector $u^k = \sigma^k_{\alpha\beta}\xi^\alpha\bar{\xi}^\beta$ is isotropic.

EXERCISE

Choose a basis $\xi = (\xi^0, \xi^1)$, $\eta = (\eta^0, \eta^1)$ for the space $\mathbb{C}^2$ of 2-spinors, satisfying $\xi^0\eta^1 - \xi^1\eta^0 = 1$. Show that with such a basis there is associated in a canonical manner a basis ("tetrad") for the tangent space of Minkowski space, in terms of which the metric has the form

$$(g_{ik}) = \begin{pmatrix} 0 & 1 & 0 & 0 \\ 1 & 0 & 0 & 0 \\ 0 & 0 & 1 & 0 \\ 0 & 0 & 0 & 1 \end{pmatrix}.$$

§42. Examples of Gauge-Invariant Functionals. Maxwell's Equations and the Yang–Mills Equation. Functionals with Identically Zero Variational Derivative (Characteristic Classes)

We now consider Lagrangians $L = L(A_\mu)$ corresponding to a gauge field A_μ alone (and to a prescribed matrix group G). Such a Lagrangian must satisfy the following two conditions:

(i) $L(A_\mu)$ is a scalar.
(ii) $L(A_\mu)$ is invariant under gauge transformations.

The simplest functional satisfying these requirements has the form

$$L = -\tfrac{1}{4}g^{\mu\lambda}g^{\nu\varkappa}\langle F_{\mu\nu}, F_{\lambda\varkappa}\rangle, \tag{1}$$

where $F_{\mu\nu}$ is the curvature form of the connexion A_μ (see §41.2(16)), $g_{\mu\nu}$ is an arbitrary metric on the region of interest of the underlying space, and $\langle\ ,\ \rangle$ denotes the Killing form on the Lie algebra of the group G, defined by

$$\langle X, Y\rangle = -\operatorname{tr}(\operatorname{ad} X \operatorname{ad} Y), \tag{2}$$

ad X being the linear transformation of the Lie algebra defined by ad $X(A) = [X, A]$ (see the footnote to §24.4). (We shall assume in what follows that this form is non-degenerate.)

This Lagrangian is a scalar since it is constructed by means of permitted algebraic operations on tensors. To see that it is gauge-invariant, recall first from Theorem 41.2.1 that under gauge transformations the curvature form $F_{\mu\nu}$ is sent to $gF_{\mu\nu}g^{-1}$, whence

$$\langle F_{\mu\nu}, F_{\lambda\varkappa}\rangle \to \langle gF_{\mu\nu}g^{-1}, gF_{\lambda\varkappa}g^{-1}\rangle.$$

Hence the gauge-invariance of the Lagrangian (1) follows from the invariance of the Killing form (which is in turn obvious from its definition (2)). Thus the Lagrangian (1) does indeed have the desired properties (i) and (ii).

Suppose that the metric $g_{\mu\nu}$ is Euclidean or pseudo-Euclidean, i.e. that (after choosing co-ordinates suitably) $g_{\mu\nu} = \varepsilon_\mu \delta_{\mu\nu}$, $\varepsilon_\mu = \pm 1$. We shall now derive the Euler-Lagrange equations corresponding to this case of the Lagrangian (1), i.e. the equations for the extremals of the functional

$$S[A_\mu] = \int -\tfrac{1}{4}\langle F_{\mu\nu}, F_{\mu\nu}\rangle \, d^n x, \tag{3}$$

where here the subscripts μ, ν are summed over, and moreover with the signs $\varepsilon_\nu, \varepsilon_\mu = \pm 1$ taken into account.

42.1. Theorem. *The extremals of the functional* (3) *satisfy the equations*

$$\nabla_\mu F_{\mu\nu} = 0, \tag{4}$$

where $\nabla_\mu F_{\mu\nu} \equiv \partial F_{\mu\nu}/\partial x^\mu + [A_\mu, F_{\mu\nu}]$ (*cf. the remark concluding* §41.1), *and where* μ *is summed over as before.*

Proof. For a small local variation δA_μ we have (as usual to within quantities of the second-order of smallness, i.e. in the conventional formalism of the calculus of variations)

$$\delta S = -\frac{1}{2}\int \langle F_{\mu\nu}, \delta F_{\mu\nu}\rangle \, d^n x,$$

where

$$\delta F_{\mu\nu} = \frac{\partial}{\partial x^\mu}\delta A_\nu - \frac{\partial}{\partial x^\nu}\delta A_\mu + [\delta A_\mu, A_\nu] + [A_\mu, \delta A_\nu]. \tag{5}$$

Using integration by parts and invoking the assumption that δA_μ vanishes on the boundary of the region of integration (or behaves suitably towards ∞), we obtain

$$\int\left\langle F_{\mu\nu}, \frac{\partial}{\partial x^\mu}\delta A_\nu\right\rangle d^n x = -\int\left\langle \frac{\partial F_{\mu\nu}}{\partial x^\mu}, \delta A_\nu\right\rangle d^n x. \tag{6}$$

From the skew-symmetry of the Killing form (cf. §24.4(58)) we also have that

$$\langle F_{\mu\nu}, [\delta A_\mu, A_\nu]\rangle = -\langle [F_{\mu\nu}, A_\nu], \delta A_\mu\rangle. \tag{7}$$

Equations (5), (6) and (7) together imply that

$$\delta S = \frac{1}{2}\int\left\{\left\langle \frac{\partial F_{\mu\nu}}{\partial x^\mu}, \delta A_\nu\right\rangle - \left\langle \frac{\partial F_{\mu\nu}}{\partial x^\nu}, \delta A_\mu\right\rangle + \langle [F_{\mu\nu}, A_\nu], \delta A_\mu\rangle - \langle [F_{\mu\nu}, A_\mu], \delta A_\nu\rangle\right\} d^n x,$$

which after a rearrangement of the indices yields finally

$$\delta S = \int\left\langle \frac{\partial F_{\mu\nu}}{\partial x_\mu} + [A_\mu, F_{\mu\nu}], \delta A_\nu\right\rangle d^n x = \int \langle \nabla_\mu F_{\mu\nu}, \delta A_\nu\rangle \, d^n x.$$

Provided the Killing form is non-degenerate it follows in the usual way (see §37.1) from this and the arbitrariness of the variation δA_ν that the extremals of the functional (3) satisfy the equation $\nabla_\mu F_{\mu\nu} = 0$. This completes the proof. □

Remark. One may of course consider the equations $\nabla_\mu F_{\mu\nu} = 0$ (and Bianchi's identity $\nabla_\lambda F_{\mu\nu} + \nabla_\mu F_{\nu\lambda} + \nabla_\nu F_{\lambda\mu} = 0$) even if the Killing form is degenerate. However they will not in this case arise, as they did above, from a Lagrangian (or at least not from one of such simple form).

EXERCISE (from the recent literature)

For the Cartan connexion (with G the affine group—see Example (c) of §41.3) defined by the connexion compatible with the metric, the equations (4) take the form of Einstein's equations:

$$R_{ab} - \tfrac{1}{2}Rg_{ab} = 0. \tag{8}$$

Examples. (a) In the case of the abelian group $G = U(1)$ (see Example (a) of §41.3) the Lagrangian becomes

$$L = -\frac{1}{4}\left(\frac{\partial A_\nu}{\partial x^\mu} - \frac{\partial A_\mu}{\partial x^\nu}\right)^2 = -\frac{1}{4}F_{\mu\nu}^2, \tag{9}$$

which the reader will recognize from §37.3 as the standard Lagrangian for an electromagnetic field. (Here as usual by $F_{\mu\nu}^2$ we mean the trace $F_{\mu\nu}F^{\mu\nu}$.) The Euler–Lagrange equations (4) take the form (also familiar from §37.3)

$$\frac{\partial F_{\mu\nu}}{\partial x^\mu} = 0. \tag{10}$$

(b) Gauge fields corresponding to the group $SU(2)$ are generally known as *Yang–Mills fields*.

EXERCISE

Derive the Euler–Lagrange equations $\delta S/\delta A = 0$ corresponding to the Lagrangian $L = -\frac{1}{4}g^{\mu\lambda}g^{\nu\varkappa}\langle F_{\mu\nu}, F_{\lambda\varkappa}\rangle$, where the metric $g_{\mu\nu}(x)$ is arbitrary (and fixed, i.e. not subject to variation, representing for instance an external gravitational field).

The action $S[A] = \int F_{\mu\nu}^2\, d^4x$ (in Euclidean $\mathbb{R}^4$ or in $\mathbb{R}^4_{1,3}$) has additional symmetry: it is invariant under the group of all conformal transformations of Euclidean $\mathbb{R}^4$ (or of $\mathbb{R}^4_{1,3}$ as the case may be). (These transformations are described in §15.) To see this, note first that it can be shown (on the basis of §§15, 22.2) that the differential of a conformal transformation involves dilations and isometries only. Since $F_{\mu\nu}^2$ is a trace (relative to the prevailing metric

—Euclidean or Minkowskian), it is invariant under isometries. On the other hand under an arbitrary dilation $x \to \lambda x$, the quantities $F_{\mu\nu}$, being the components of a tensor, transform as follows:

$$F_{\mu\nu} \to \lambda^{-2} F_{\mu\nu}.$$

Since d^4x transforms to $\lambda^4 d^4x$ under this dilation, it follows that $F_{\mu\nu}^2\, d^4x$ is preserved by dilations and isometries. Hence the above action $S[A]$ is invariant under all conformal transformations of $\mathbb{R}^4$ (or $\mathbb{R}^4_{1,3}$), as claimed. Since by the exercise at the end of §15 the group of conformal transformations of $\mathbb{R}^4_{1,3}$ is isomorphic to $O(4, 2)$, we conclude that:

The functional (3) *and Maxwell's equations or the Yang–Mills equations* (4) (*depending on whether the group in question is* $U(1)$ *or* $SU(2)$) *in Minkowski space, are conformally invariant, i.e. have symmetry group isomorphic to* (*at least*) $O(4, 2)$.

Also of importance are scalar-valued gauge-invariant differential forms defined in terms of a given gauge-field A_μ. For example the rank-two form

$$c_1 = \operatorname{tr} \Omega = \sum_{\mu<\nu} \operatorname{tr} F_{\mu\nu}\, dx^\mu \wedge dx^\nu \tag{11}$$

is (clearly) gauge-invariant, i.e. $\operatorname{tr}(g\,\Omega\, g^{-1}) = \operatorname{tr} \Omega$. (Here tr denotes the operation of taking the trace of a matrix.) The form c_1 is (locally) exact (i.e. a differential), since

$$\begin{aligned} \operatorname{tr} \Omega &= \sum_{\mu<\nu} \left(\operatorname{tr}\left(\frac{\partial A_\nu}{\partial x^\mu} - \frac{\partial A_\mu}{\partial x^\nu}\right) + [A_\mu, A_\nu]\right) dx^\mu \wedge dx^\nu \\ &= \sum_{\mu<\nu} \operatorname{tr}\left(\frac{\partial A_\nu}{\partial x^\mu} - \frac{\partial A_\mu}{\partial x^\nu}\right) dx^\mu \wedge dx^\nu = d \operatorname{tr} A, \end{aligned}$$

where we have used the fact that the trace of a commutator of matrices is zero.

Under a local variation $A_\mu \to A_\mu + \delta A_\mu$ of the gauge field, the variation of the form $\operatorname{tr} \Omega$ is given by $\operatorname{tr} \Omega + \operatorname{tr} \delta\Omega$. Since $\operatorname{tr} \delta\Omega$ is a differential ($\operatorname{tr} \delta\Omega = d \operatorname{tr} \delta A$), it follows that, since the variation is local, the variational derivative of the functional $S_1[A_\mu] = \int_{\mathbb{R}^2} \operatorname{tr} \Omega$ (over a 2-dimensional space) is identically zero:

$$\delta S_1[A_\mu] = \int_{\mathbb{R}^2} d(\operatorname{tr} \delta A) = 0. \tag{12}$$

42.2. Definition. Closed gauge-invariant forms ω whose functionals $\int \omega$ have identically zero variational derivatives, are called (*differential-geometric*) *characteristic classes*.

The following are all characteristic classes:

$$c_i = \operatorname{tr}(\Omega \wedge \cdots \wedge \Omega) = \operatorname{tr} \Omega^i, \qquad i \geq 1.$$

(Here the components of the form Ω multiply like matrices.) Thus c_i is a gauge-invariant form of rank $2i$.

EXERCISE

(i) Prove that all of the forms c_i are closed.
(ii) Prove that the variational derivative of the functional $S_i[A_\mu] = \int_{\mathbb{R}^{2i}} c_i$ (over a space of dimension $2i$) vanishes identically: $\delta S_i[A]/\delta A_\mu \equiv 0$. Prove this also for functionals of the forms $\int_{\mathbb{R}^4} (\alpha(c_1 \wedge c_1) + \beta c_2)$, $\int_{\mathbb{R}^6} (\alpha c_1^3 + \beta(c_1 \wedge c_2) + \gamma c_3)$, and so on, for arbitrary (appropriate) polynomials in the c_i.

For the group $SO(2n)$ the odd-indexed c_i are all zero: $c_1 = c_3 = c_5 = \cdots = 0$. Apart from those of the $c_{2i} = p_i$ that are nonzero, there is the further characteristic class

$$\chi_n = \varepsilon^{i_1 \cdots i_{2n}} \Omega_{i_1 i_2} \wedge \Omega_{i_2 i_3} \wedge \cdots \wedge \Omega_{i_{2n-1} i_{2n}}, \tag{13}$$

where $\varepsilon^{i_1 \cdots i_{2n}}$ is the sign of the permutation $\begin{pmatrix} i \cdots 2n \\ i_1 \cdots i_{2n} \end{pmatrix}$, and the Ω_{ij} are the matrix entries in the curvature form $\Omega = \sum_{\mu < \nu} F_{\mu\nu}\, dx^\mu \wedge dx^\nu$.

Examples. (a) In the case $n = 1$ we obtain (cf. §41.3(26))

$$\chi_1 = \varepsilon^{i_1 i_2} \Omega_{i_1 i_2} = 2K\sqrt{g}\, dx^1 \wedge dx^2. \tag{14}$$

By Theorem 37.4.2 the variational derivative with respect to the metric, of the functional

$$S[g_{ij}] = \int_{\mathbb{R}^2} K\sqrt{g}\, dx^1 \wedge dx^2 = \int_{\mathbb{R}^2} \tfrac{1}{2} R\sqrt{g}\, dx^1 \wedge dx^2,$$

vanishes identically.

(b) In the case $n = 2$ we have of course only c_2 (and χ_2) possibly non-zero. For a Riemannian metric these are given by

$$\chi_2 = \varepsilon^{ijkl} R_{ij\mu\nu} R_{kl\rho\sigma}\, dx^\mu \wedge dx^\nu \wedge dx^\rho \wedge dx^\sigma,$$
$$c_2 = R_{ij\mu\nu} R^{ij}_{\rho\sigma}\, dx^\mu \wedge dx^\nu \wedge dx^\rho \wedge dx^\sigma.$$

EXERCISE

Prove that each of the forms χ_n is closed and is a characteristic class.

The general characteristic class for $G = SO(2n)$ has the form of a "polynomial" in $\chi_n, c_2, c_4, \ldots, c_{2n-2}$.

EXERCISE

Show that for $G = SO(4)$ the functionals

$$\int_{\mathbb{R}^4} \chi_2, \quad \int_{\mathbb{R}^4} c_2, \quad \int_{\mathbb{R}^8} \chi_2^2, \quad \int_{\mathbb{R}^8} \chi_2 c_2, \quad \int_{\mathbb{R}^8} c_2^2$$

have identically zero variational derivative, i.e. that the forms $\chi_2, c_2, \chi_2^2, \chi_2 c_2, c_2^2$ are characteristic classes.

Bibliography

(i) Textbooks on Geometry and Topology

*1. Aleksandrov, A. D., 1948. *Intrinsic Geometry of a Convex Surface*. Gostehizdat: Moscow–Leningrad.

2. Bishop, R. L. and Crittenden, R. J., 1964. *Geometry of Manifolds*. Pure and Applied Math., Vol. 25. Academic Press: New York–London.

3. Chern, S. S., 1959. *Complex Manifolds*. Textos de Matemática, No. 5. Instituto de Física e Matemática, Universidade do Recife.

*4. Efimov, N. V., 1971. *Higher Geometry*. Nauka: Moscow.

*5. Finikov, S. P., 1952. *A course in Differential Geometry*. Gostehizdat: Moscow.

6. Gromoll, D., Klingenburg, W., and Meyer, W., 1968. *Riemannsche Geometrie im Grossen*. Lecture Notes in Mathematics, No. 55. Springer-Verlag: Berlin–New York.

7. Helgason, S., 1962. *Differential Geometry and Symmetric Spaces*. Academic Press: New York.

8. Hilbert, D. and Cohn-Vossen, S., 1952. *Geometry and the Imagination*. Chelsea Publishing Co.: New York. (Translated from the German by P. Neményi.)

9. Lefschetz, S., 1942. *Algebraic Topology*. Am. Math. Soc. Colloquium Publications, Vol. 27.

10. Milnor, J. W., 1963. *Morse Theory*. Ann. of Math. Studies, No. 51. Princeton Univ. Press: Princeton, N.J.

11. Milnor, J. W., 1968. *Singular Points of Complex Hypersurfaces*. Ann. of Math. Studies, No. 61. Princeton Univ. Press: Princeton, N.J.

12. Nomizu, K. 1956. *Lie Groups and Differential Geometry*. Math. Soc. Japan.

*13. Norden, A. P. 1956. *The Theory of Surfaces*. Gostehizdat: Moscow.

*14. Novikov, S. P., Miščenko, A. S., Solov'ev, Ju. P., and Fomenko, A. T., 1978. *Problems in Geometry*. Moscow State University Press: Moscow.

15. Pogorelov, A. V., 1967. *Differential Geometry*. P. Noordhoff: Groningen. (Translated from the first Russian edition by L. F. Boron.)

16. Pogorelov, A. V., 1973. *Extrinsic Geometry of Convex Surfaces*. Translations of Math. Monographs, Vol. 35. A.M.S.: Providence, R.I.

* In Russian.

17. Pontrjagin, L. S., 1959. *Smooth Manifolds and Their Applications to Homotopy Theory*. Am. Math. Soc. Translations, Series 2, Vol. 11, pp. 1–114. A.M.S., Providence, R.I.
18. Pontrjagin, L. S., 1966. *Topological Groups*. Gordon & Breach: New York–London–Paris. (Translation of the second Russian edition by Arlen Brown.)
*19. Raševskiĭ, P. K., 1956. *A Course in Differential Geometry*. Nauka: Moscow.
*20. Raševskiĭ, P. K., 1967. *Riemannian Geometry and Tensor Analysis*. Nauka: Moscow.
*21. Rohlin, V. A. and Fuks, D. B., 1977. *A Beginning Course in Topology. Chapters in Geometry*. Nauka: Moscow.
*22. Rozendorn, E. R., 1971. *Problems in Differential Geometry*. Nauka: Moscow.
23. Seifert, H. and Threlfall, W., 1980. *A Textbook of Topology*. Academic Press: New York. (Translated from the German by M. A. Goldman.)
24. Seifert, H. and Threlfall, W., 1932. *Variationsrechnung im Grossen*. Hamburger Math. Einzelschr., No. 24. Teubner: Leipzig. (Reprinted by Chelsea: New York, 1951.)
25. Serre, J.-P., 1964. *Lie Algebras and Lie Groups*. Lectures given at Harvard Univ. W. A. Benjamin: New York.
26. Springer, G., 1957. *Introduction to Riemann Surfaces*. Addison-Wesley: Reading, Mass.
27. Steenrod, N. E., 1951. *The Topology of Fibre Bundles*. Princeton Math. Series, Vol. 14. Princeton Univ. Press: Princeton, N.J.
28. Struik, D. J., 1950. *Lectures on Classical Differential Geometry*. Addison-Wesley: Reading, Mass.

(ii) Texts on Differential Equations and Classical Mechanics

29. Arnol'd, V. I., 1978. *Mathematical Methods of Classical Mechanics*. Graduate Texts in Math. Springer-Verlag: New York–Heidelberg–Berlin. (Translated from the Russian by K. Vogtmann and A. Weinstein.)
*30. Arnol'd, V. I., 1978. *Supplementary Chapters to the Theory of Ordinary Differential Equations*. Nauka: Moscow.
*31. Golubev, V. V., 1953. *Lectures on the Integration of the Equations of Motion of a Heavy Rigid Body About a Fixed Point*. Gostehizdat: Moscow.
32. Landau, L. D. and Lifšic, E. M., 1960. *Mechanics*. Course of Theoretical Physics, Vol. 1. Pergamon Press: Oxford–London–New York–Paris; Addison-Wesley: Reading, Mass. (Translated from the Russian by J. B. Sykes and J. S. Bell.)
*33. Pontrjagin, L. S., 1970. *Ordinary Differential Equations*. Nauka: Moscow.
34. Coddington, E. A. and Levinson, N., 1955. *Theory of Ordinary Differential Equations*. McGraw-Hill: New York–Toronto–London.

(iii) Supplementary Texts

35. Ahiezer, A. I. and Beresteckiĭ, V. B., 1965. *Quantum Electrodynamics*. Interscience Monographs and Texts in Physics and Astronomy, Vol. 11. Interscience Publishers (John Wiley and Sons): New York–London–Sydney. (Translated from the second Russian edition by G. M. Volkhoff.)
*36. Bogojavlenskiĭ, O. I., 1980. *Methods of the Qualitative Theory of Dynamical Systems in Astrophysics and the Dynamics of Gases*. Nauka: Moscow.
37. Bogoljubov, N. N. and Širkov, D. B., 1959. *Introduction to the Theory of Quantum Fields*. Interscience: New York. (Translation.)

38. Coxeter, H. S. M., and Moser, W. O. J., 1972. *Generators and Relations for Discrete Groups.* Ergebnisse der Mathematik und ihrer Grenzgebiete, Bd. 14. Springer-Verlag: New York–Heidelberg–Berlin.
*39. Delone, B. N., Aleksandrov, A. D., and Padurov, N. N., 1934. *Mathematical Foundations of the Lattice Analysis of Crystals.* ONTI: Leningrad-Moscow.
40. Feynman, R. P., Leighton, R. B., and Sands, M., 1963. *The Feynman Lectures on Physics.* Addison-Wesley: Reading, Mass.
41. Landau, L. D. and Lifšic, E. M., 1971. *The Classical Theory of Fields.* Third revised English edition. Course of Theoretical Physics, Vol. 2. Addison-Wesley: Reading, Mass.; Pergamon Press: London. (Translated from the Russian by Morton Hamermesh.)
42. Landau, L. D. and Lifšic, E. M., 1960. *Electrodynamics of Continuous Media.* Course of Theoretical Physics, Vol. 8. Pergamon Press: Oxford–London–New York–Paris; Addison-Wesley: Reading, Mass. (Translated from the Russian by J. B. Sykes and J. S. Bell.)
43. Misner, C. W., Thorne, K. S., and Wheeler, J. A., 1973. *Gravitation.* W. H. Freeman: San Francisco.
44. Peierls, R. E., 1960. *Quantum Theory of Solids. Theoretical Physics in the Twentieth Century* (Pauli Memorial Volume), pp. 140–160, Interscience: New York.
*45. Sedov, L. I., 1976. *Mechanics of a Continuous Medium.* Nauka: Moscow.
*46. Slavnov, A. A. and Faddeev, L. D., 1978. *Introduction to the Quantum Theory of Gauge Fields.* Nauka: Moscow.
47. Wintner, A., 1941. *Analytical Foundations of Celestial Mechanics.* Princeton Univ. Press: Princeton, N.J.
*48. Zaharov, V. E., Manakov, S. V., Novikov, S. P., and Pitaevskiĭ, L. P., 1980. *The Theory of Solitons* (under the general editorship of S. P. Novikov). Nauka: Moscow.
*49. Zel'dovič, Ja. B. and Novikov, I. D., 1977. *Relativistic Astrophysics.* Nauka: Moscow.

Index